Hormones and the Endocrine System

Bernhard Kleine • Winfried G. Rossmanith

Hormones and the Endocrine System

Textbook of Endocrinology

Springer

Bernhard Kleine
Lenzkirch, Germany

Winfried G. Rossmanith
Ettlingen, Germany

ISBN 978-3-319-79212-5 ISBN 978-3-319-15060-4 (eBook)
DOI 10.1007/978-3-319-15060-4

Springer Cham Heidelberg New York Dordrecht London

Softcover re-print of the Hardcover 1st edition 2016

Printed on acid-free paper

Springer International Publishing AG Switzerland is part of Springer Science+Business Media (www.springer.com)

Contents

1 Introduction ... 1
- 1.1 Whom To Thank ... 2

2 History of Hormones and Endocrinology ... 3
- 2.1 Ancient Times ... 3
- 2.2 New Age ... 6
- 2.3 Modern Times ... 7
- 2.4 Problems Remaining ... 7
- 2.5 Historical Summary ... 8

Part I Hormone Biochemistry

3 Hormones: Definitions ... 11
- 3.1 The Nature of the Endocrine System ... 11
- 3.2 What Is a Hormone? ... 12
  - 3.2.1 Hormonal Effect ... 12
  - 3.2.2 Neurotransmitters ... 13
  - 3.2.3 Cytokines/Lymphokines ... 13
  - 3.2.4 Prostaglandins/Thromboxanes ... 13
  - 3.2.5 Pheromones ... 13
- 3.3 How Do We Define the Endocrine System? ... 14
- 3.4 Endocrine Hierarchy ... 15
- 3.5 Hormone Classes ... 16
  - 3.5.1 Protein/Peptide Hormones ... 16
  - 3.5.2 Terpenes: Juvenile Hormones and Steroids Hormones ... 17
  - 3.5.3 Amino Acid Derivatives ... 17

4 Vertebrate Protein and Peptide Hormones ... 19
- 4.1 Translation ... 21
  - 4.1.1 Reading Genetic Information: Transcription ... 21
  - 4.1.2 Coding and Other Sequences ... 21
  - 4.1.3 Splicing ... 21
  - 4.1.4 RNA Cap ... 22
  - 4.1.5 Nuclear Export of Messenger RNA ... 22

4.1.6 Docking to Ribosomes 22
4.1.7 Translational Termination 22
4.1.8 Membrane and Secretory Proteins 22
4.2 Posttranslational Modification: Hormone Maturation 23
4.2.1 Removal of the Signal Peptide 23
4.2.2 Folding and Disulfide Bridges 23
4.2.3 Protein Complexes 24
4.2.4 Glycosylation 24
4.2.5 Prohormone Convertases 25
4.2.6 Monobasic and Dibasic Sequence Motifs in Invertebrates and Vertebrates 27
4.2.7 Chopping the C-Terminus 28
4.2.8 Oxidation of the Terminal Glycine 28
4.2.9 Cyclization of the N-Terminal Glutamine 29
4.2.10 Esterification of Ghrelin 29
4.3 Peptide Hormones of the Hypothalamus and the Brain 30
4.3.1 Hypothalamic-Releasing Hormones 30
4.3.2 Gonadotropin-Inhibiting Hormone 42
4.3.3 Neuropeptide Y 45
4.3.4 Agouti-Related Protein 45
4.3.5 Somatostatin 46
4.3.6 Substance P 49
4.3.7 Proopiomelanocortin 49
4.3.8 Ghrelin 49
4.3.9 Kisspeptin 50
4.3.10 Galanin 51
4.3.11 Melanin-Concentrating Hormone 54
4.3.12 Orexins 56
4.4 Anterior Pituitary Hormones 59
4.4.1 POMC 59
4.4.2 TSH 67
4.4.3 LH, FSH, CG 69
4.4.4 Growth Hormone 76
4.4.5 Prolactin 79
4.5 Hormones of the Pars Intermedia 81
4.5.1 Somatolactin 81
4.5.2 Prolactin-Releasing Peptide 83
4.6 Hormones of the Posterior Pituitary: Oxytocin and Vasopressin 84
4.6.1 Introduction 84
4.6.2 Structure and Genes 86
4.6.3 Physiology of Oxytocin 87
4.6.4 Physiology of Vasopressin (AVP, Adiuretin) 87
4.6.5 Phylogeny 88

4.7 Regulators of Sugar and Energy Metabolism ... 89
4.7.1 Insulin ... 89
4.7.2 Glucagon ... 93
4.8 Regulators of Food Intake ... 95
4.8.1 Leptin ... 95
4.8.2 Ghrelin ... 96
4.9 Nonsteroidal Regulators of Bone ... 99
4.9.1 Introduction ... 99
4.9.2 Structure and Genes ... 99
4.9.3 Physiology ... 100
4.9.4 Phylogeny ... 101
4.10 Digestive Tract Hormones ... 101
4.10.1 Gastrin ... 101
4.10.2 Cholecystokinin ... 103
4.10.3 Secretin ... 104
4.10.4 VIP ... 106
4.10.5 GIP ... 107
4.10.6 PNP, NPY, PYY ... 108
4.11 Neuropeptides in the Enteric Nervous System ... 112
4.11.1 Endorphins and Enkephalins ... 112
4.11.2 Tachykinins: Substance P, Neurokinin, and Endokinin ... 113
4.11.3 Gastrin-Releasing Peptide/Bombesin ... 115
4.12 Nonsteroidal Regulators of Reproduction ... 116
4.12.1 Activin/Inhibin ... 116
4.12.2 Follistatin ... 119
4.12.3 Antimüllerian Hormone ... 120
4.13 Angiotensins and Renin ... 122
4.13.1 Introduction ... 122
4.13.2 Structure and Genes ... 123
4.13.3 Physiology ... 123
4.13.4 Phylogeny ... 123
4.14 Atrial Natriuretic Peptides ... 124
4.14.1 Introduction ... 124
4.14.2 Structure and Genes ... 124
4.14.3 Physiology ... 126
4.14.4 Phylogeny ... 126
4.15 Additional Neuropeptides ... 127
4.15.1 Parathormone ... 127
4.15.2 Stanniocalcin ... 128
4.15.3 Erythropoietin ... 130

5 Invertebrate Protein and Peptide Hormones 133
5.1 Metabolically Active Peptide Hormones 135
5.1.1 Crustacean Hyperglycemic Hormone 135
5.1.2 Bombyxin and Insulin-Like Peptides (ILP) 137
5.1.3 AKH, RPCH, and HrTH 139
5.2 Regulation of Heart Frequency and Pressure by Neuropeptides 141
5.2.1 Cardioacceleratory Peptides: CAP 141
5.2.2 Cardioexcitatory Peptide, NDWF-Amide 143
5.2.3 Enterins 144
5.2.4 *Mytilus* Inhibitory Peptides (MIP; AMrP) 145
5.2.5 Diuretic Hormones 147
5.3 Kinins 148
5.3.1 Pyrokinin/Myotropin and Pheromone Biosynthesis Activating Neuropeptide (PBAN) 149
5.3.2 Orcokinins 151
5.3.3 Leucokinins/Lymnokinins 153
5.3.4 Tachykinin-Related Peptides (TRP) 155
5.3.5 Sulfakinins 157
5.4 Neuropeptides of Reproduction 158
5.4.1 PTTH 158
5.4.2 PTSH; MIP 160
5.4.3 Pheromonostatic Peptide, Sex Peptides 161
5.4.4 GIH, VIH 163
5.4.5 TMOF 164
5.4.6 Nebcolloostatin 165
5.5 Peptide Hormone of Metamorphosis and Molting 166
5.5.1 Molt-Inhibiting Hormone (MIH) 166
5.5.2 Corazonin 168
5.5.3 Ecdysis-Triggering Hormone (ETH) 169
5.5.4 Eclosion Hormone (EH) 170
5.5.5 Bursicon 172
5.6 Regulators of Food Intake: RFamide and FMRFamide 174
5.6.1 Introduction 174
5.6.2 Biochemistry and Structure 175
5.6.3 Physiology 177
5.6.4 Phylogeny 178
5.7 Neuropeptide Regulators of Juvenile Hormone Metabolism 178
5.7.1 Allatotropins 178
5.7.2 Allatostatins 179
5.8 Peptide Hormones of Skin: Pigment-Dispersing Hormone 181
5.8.1 Introduction 181
5.8.2 Biochemistry and Structure 182
5.8.3 Physiology 182
5.8.4 Phylogeny 183

5.9 Other Neuropeptides ... 184
5.9.1 Neuropeptide-F—Two Peptide Genes ... 184
5.9.2 Proctolin ... 187
5.10 Summary and Overview ... 188

6 Hormones from Mevalonate: JH and Steroids ... 191
6.1 Introduction ... 192
6.2 Pathways to Juvenile Hormone and Steroids ... 192
6.2.1 JH Synthesis ... 192
6.2.2 From Farnesyl Pyrophosphate via Squalene to Cholesterol ... 194
6.3 Juvenile Hormones ... 196
6.3.1 Introduction ... 196
6.3.2 Biochemistry and Structure ... 196
6.3.3 Physiology ... 199
6.4 Steroids ... 201
6.5 Steroidogenic Enzymes ... 203
6.5.1 CYP11A1 ... 203
6.5.2 HSD3B ... 204
6.5.3 CYP17 ... 205
6.5.4 HSD17B ... 206
6.5.5 5α Reductase ... 211
6.5.6 CYP21 ... 213
6.5.7 CYP11B1 ... 214
6.5.8 CYP11B2 ... 215
6.5.9 CYP19 ... 216
6.5.10 Concluding Remark ... 220
6.6 Progesterone ... 221
6.7 Sex Hormones ... 223
6.7.1 Gonadal Development ... 224
6.7.2 Gonadal Hormone Production ... 225
6.7.3 Androgen Production in the Adrenal Glands ... 225
6.8 Corticoids ... 226
6.8.1 Cortisol ... 226
6.8.2 Aldosterone ... 228
6.9 Brain Steroids ... 229
6.10 Additional Steroid Hormones ... 230
6.10.1 Insect and Crustacean Steroid Hormones ... 230
6.10.2 Steroids in Plants and Fungi ... 230
6.11 Dihydroxyvitamin D_3 (Calcitriol) ... 232
6.11.1 Vitamin D_3 Synthesis ... 233
6.11.2 Vitamin D Uptake from Food ... 235
6.11.3 Synthesis of 25-Hydroxyvitamin D_3 ... 235
6.11.4 Synthesis of Calcitriol ... 235
6.11.5 Calcitriol Degradation ... 236

7 Hormones Derived by Amino Acid Conversion 237
7.1 Catecholamines 237
7.1.1 Dopamine 239
7.1.2 Noradrenaline and Adrenaline 240
7.2 Thyroxine: The Thyroid Hormone 240
7.3 Melatonin 243

8 Receptors 247
8.1 Nuclear Receptors 248
8.2 Heptahelical Transmembrane Receptors 249
8.2.1 G Proteins 252
8.2.2 Receptor–G Protein Interactions 253
8.2.3 Target of G Proteins 254
8.2.4 Variability by Differentially Expressed Receptor Subtypes: Somatostatin Receptors 254
8.3 Receptors with Tyrosine Kinase Activity 255
8.4 Membrane Receptors with Serine/Threonine Kinase Activity 257
8.5 Membrane Receptors Without Kinase Activity 257
8.6 Membrane Steroid Receptors: Still Unknown? 258

Part II Endocrine Physiology

9 Synthesis, Release, and Action 263
9.1 The Three Steps of Hormone Regulation 263
9.2 Effective Hormone Concentrations of the Hypothalamic–Pituitary–Gonadal Axis 264

10 Endocrine Active Organs 269
10.1 Overview 270
10.2 The Endocrine System of the Brain 270
10.2.1 The Hypothalamus 270
10.2.2 The Pituitary Stalk 272
10.2.3 The Pituitary 273
10.2.4 The Pineal Gland 277
10.3 The Thyroid and Parathyroid Glands 278
10.3.1 The Thyroid Gland 278
10.3.2 The Parathyroid Glands 279
10.4 The Adrenal Glands 280
10.4.1 The Adrenal Cortex 281
10.4.2 The Adrenal Medulla 282
10.5 Endocrine Cells of the Gastrointestinal Tract 283
10.5.1 The Stomach 284
10.5.2 Duodenum and Ileum 285
10.6 Langerhans Islets of the Pancreas 285
10.7 Endocrine Cells of the Kidney 287

10.8 Gonads 288
10.8.1 Gonad Development 288
10.8.2 Male Gonads: Testes 288
10.8.3 Female Gonads: Ovaries 290
10.9 Selected Endocrine Active Organs in Invertebrates 292
10.9.1 Neurosecretion in Mollusks 292
10.9.2 Endocrine Glands in Crustaceans 294
10.9.3 Neurosecretion and Endocrine Organs in Insects 294

11 Regulatory Patterns 299
11.1 Feedback Control 300
11.2 Regulator Circuits 301
11.2.1 Pressure and Stress 301
11.2.2 Calcium Metabolism 304
11.3 Regulation of Reproduction 305
11.3.1 Regulation of GnRH 305
11.3.2 Regulation of the Menstrual Cycle 307
11.3.3 Puberty 309
11.3.4 Menopause 310
11.4 Glucose Metabolism 311
11.4.1 The Origin of Blood Glucose 311
11.4.2 Regulators and Control Variables 311
11.4.3 Glucose-Dependent Gene Expression in the Liver 311
11.4.4 Glucose-Dependent Insulin Secretion in the Pancreas 312
11.4.5 Insulin-Dependent Procedures 313
11.4.6 Glucagon and Blood Glucose Level Increase 313
11.5 Appetite and Hunger 313
11.5.1 Central Nervous System 314
11.5.2 Parasympathetic Fibers of the Vagus Nerve 315
11.5.3 Sympathetic Fibers of the Splanchnic Nerve 315
11.5.4 Enteric Nervous System 315
11.5.5 Endocrine Mediators and Neuropeptides 315
11.5.6 Hormone Receptors on Nerve Cells 317
11.5.7 Mechanoreceptors 317
11.5.8 Feeding Circuits in the Brain 318
11.5.9 Hunger and Food Intake in *Drosophila melanogaster* 319
11.6 Growth 321
11.6.1 Epiphyseal Cartilages 321
11.6.2 Zonal Organization of the Epiphysis 322
11.6.3 Regulation by Hormones 322
11.7 Growth and Molt in Ecdysozoans 324
11.7.1 Regulation of Growth in Insects 326
11.7.2 Hormones and Postembryonic Development 326
11.7.3 Linkage of Growth and Metamorphosis 328

11.7.4 Regulation of Ecdysis 330
11.7.5 Postembryonic Development in Holometabolous Species 332
11.8 Regulation of Blood Pressure, Osmolarity and Blood Volume 334
11.8.1 Integration of Several Control Circuits 334
11.8.2 Osmoreceptors at the Blood–Brain Barrier 335
11.8.3 Angiotensin II Receptors at the Blood–Brain Barrier 336
11.8.4 AVP Release in the Posterior Pituitary 337
11.8.5 The Role of Oxytocin 337
11.8.6 Thirst and the Endocrine System of the Brain 337
11.8.7 Biochemistry of Water and Sodium Resorption 338

12 Endocrine Rock 'n Roll: Rhythms and Secretion 339
12.1 A Universal Pacemaker 340
12.2 Circadian Rhythms (Pulse Frequency 24 ± 2 h) 342
12.3 Ultradian Rhythms (Pulse Frequencies Below 22 h) 343
12.4 Seasonal (Annual) Rhythms 345

13 Evolution of the Endocrine System 347
13.1 Division of Labor 349
13.2 Evolution of Neuropeptide Hormones 349
13.3 Evolution of Glycoprotein Hormones 350
13.4 Insulin and Insulin-Like Proteins 351
13.5 Evolution of CYP Enzymes and Steroid Hormones 351
13.6 Catecholamine Evolution 353
13.7 Conclusion 353

Part III Hormones and Medicine

14 Diseases of the Endocrine System 357
14.1 Defects in the CNS/Hypothalamus 359
14.1.1 Kallmann Syndrome 359
14.1.2 Craniopharyngioma 359
14.2 Pituitary Defects 359
14.2.1 Genetic Deficits 359
14.2.2 Pituitary Tumors 360
14.2.3 Disturbances of Water Homeostasis 362
14.3 Thyroid Diseases 363
14.3.1 Lack of Thyroxine Synthesis: Hypothyroidism 363
14.3.2 Thyroxine Excess: Thyrotoxicosis 363
14.4 Dysfunctions of the Endocrine Pancreas 364
14.4.1 Pancreatic Tumors 364
14.4.2 Diabetes Mellitus 364
14.5 Adrenal Dysfunctions 365
14.5.1 Congenital Adrenal Hyperplasia 365
14.5.2 Hypercorticism 366

14.5.3 Catecholamine-Releasing Tumors ... 367
14.5.4 Autoimmune Adrenalitis (Addison Disease) ... 367
14.5.5 Aldosterone Deficiencies ... 368
14.6 Multiple Endocrine Neoplasia ... 368
14.7 Fertility Dysfunctions and Malfunctions of Reproductive Organs ... 369
14.7.1 Genetic Defects Affecting the Formation of Sex Organs ... 371
14.7.2 Gene Defects Influencing the Hypothalamus and Pituitary ... 372
14.7.3 Gene Defects of GnRHR, Gonadotropin Synthesis, and Recognition ... 373
14.7.4 Gene Defects Affecting Steroidogenesis ... 374

15 Boosting Performance: Legally or Not ... 377
15.1 Anabolic Steroids ... 378
15.2 β_2-Agonists ... 379
15.3 Peptide Hormones ... 379
15.3.1 Gonadotropins ... 379
15.3.2 Corticotropin ... 380
15.3.3 GH and Insulin-Like Growth Factor ... 380
15.3.4 Hypothalamic Releasing Hormones ... 380
15.4 Beta Blockers ... 381
15.5 Erythropoietin ... 381

16 Appendix ... 383
16.1 List of Species ... 383
16.2 Glossary ... 385
16.2.1 Cell Components ... 385
16.2.2 Cell Structure ... 386
16.2.3 Intracellular Vesicles and Their Transport ... 387
16.2.4 Additional Definitions ... 388
16.2.5 Amino Acids ... 390
16.3 PyMOL Scripts ... 392
16.3.1 Glycoprotein Hormone α Chain ... 392
16.3.2 Growth Hormone/Prolactin Using PyMOL ... 393
16.3.3 CYP51 ... 394
16.3.4 CYP19 ... 395

Erratum to: Hormones and the Endocrine System ... E1

References ... 397

Index ... 419

Abbreviations

(names of species are at the end of the list)

3D	Three-dimensional
aa	Amino acid
AANAT	Arylalkylamine *N*-acetyltransferase
ACE	Angiotensin-converting enzyme
ACTH	Adrenocorticotropic hormone
AgRP	Agouti-related peptide
AMH	Antimüllerian hormone
ANP	Atrial natriuretic peptide
AP	Anterior pituitary
APUD	Amine precursor uptake and decarboxylation
Aq2	Aquaporin 2
AR	Androgen receptor
ARNT	Aryl hydrocarbon receptor nuclear translocator
ATP	Adenosine triphosphate
AVP	Arginine vasopressin
AV3V	Anterior ventricular part of the third ventricle
BCP	Bag cell peptide
bHLH	Basic helix–loop–helix
BMAL	Brain and muscle aryl hydrocarbon receptor nuclear translocator like protein 1
BMP	Bone morphogenetic protein
bp	Base pair
CAH	Congenital adrenal hyperplasia
cAMP	Cyclic adenosine monophosphate
CAP	Cardioacceleratory peptide
CART	Cocaine- and amphetamine-regulated transcript
CBG	Cortisol-binding globulin
CBP	CREB-binding protein
CC	Chief cell
CCK	Cholecystokinin
CDC	Caudodorsal cell
CFTR	Cystic fibrosis transmembrane conductance regulator

CG	Choriogonadotropin
cGMP	Cyclic guanosine monophosphate
CGrP	Calcitonin-gene-related peptide
CK1ϵ	Casein kinase 1 epsilon
CLOCK	Circadian locomotor output cycles kaput
CNS	Central nervous system
CPHD	Combined pituitary hormone deficiency
CR	Cortisol receptor
CRE	cAMP-reactive element
CREB	CRE-binding protein
CRH	Corticotropin-releasing hormone
CRY	Cryptochrome
CST	Cortistatin
CT	Calcitonin
C-terminal	Carboxy terminal
CYP	Cytochrome P450
DBP	Vitamin D_3 binding protein
DDT	Dichlorodiphenyltrichloroethane
DH	Diapause hormone
DHEA	Dehydroepiandrosterone
DHT	Dihydrotestosterone
DI	Deiodinase
DiaH	Diapause hormone
Di(OH)vitD$_3$	1,25-Dihydroxyvitamin D_3
DiuH	Diuretic hormone
DM	Diabetes mellitus
DNA	Deoxyribonucleic acid
DOPA	3,4-Dihydroxyphenylalanine
EH	Eclosion hormone
ELH	Egg-laying hormone
eNAC	Endothelial sodium channel
ENS	Enteric nervous system
EPO	Erythropoietin
ER	Endoplasmic reticulum
ETH	Ecdysis-triggering hormone
FC	Follicular cell
FGF	Fibroblast growth factor
FSH	Follicle-stimulating hormone
GABA	γ-Aminobutyric acid
GC	Granulosa cell
GCG	Glucagon
G-CSF	Granulocyte colony stimulating factor
GDP	Guanosine diphosphate
GH	Growth hormone

GHRH	Growth-hormone-releasing hormone
GHS-R	Growth hormone secretagogue receptor
GI	Gastrointestinal
GIH	Gonad-inhibiting hormone
GIP	Gastric inhibitory peptide
GK	Glucokinase
GLC	Granulosa lutein cells
Glc	Glucose
Glc6P	Glucose 6-phosphate
Glc6Pase	Glucose 6-phosphatase
GLP	Glucagon-like peptide
GluT	Glucose transporter
GnRH	Gonadotropin-releasing hormone
GnRHR	Gonadotropin-releasing hormone receptor
GPCR	G-protein-coupled receptor
G protein	GTP/GDP-binding protein
GR	Glucocorticoid receptor
GRP	Gastrin-releasing peptide
GRPP	Glicentin-related peptide
GTP	Guanosine triphosphate
GZ	Growth zone
hCG	Human choriogonadotropin
HDL	High-density lipoprotein
HIOMT	Hydroxyindole *O*-methyltransferase
HLA	Human lymphocyte antigen
HSD	Hydroxysteroid dehydrogenase
5-HT	5-Hydroxytryptamine
HZ	Hypertrophic zone
ICER	Inducible cyclic AMP early repressor
IGF	Insulin-like growth factor
IGFBP	Insulin-like growth factor binding protein
Ihh	Indian hedgehog
IL-1	Interleukin-1
ILP	Insulin-like peptide
Ins-R	Insulin receptor
IP_3	Inositol trisphosphate
JH	Juvenile hormone
Kal	Kallman syndrome protein
kb	Kilobase
kDa	Kilodalton
LC	Locus coeruleus
LDL	Low-density lipoprotein
LH	Luteinizing hormone
LHA	Lateral hypothalamic area

LPH	Lipocortin
MCH	Melanin-concentrating hormone
MC4-R	Melanocortin 4 receptor
MC-R	Melanocortin receptor
MEN	Multiple endocrine neoplasia
MIH	Molt-inhibiting hormone
MIP	Myoinhibitory peptide
MR	Mineralocorticoid receptor
mRNA	Messenger RNA
MSH	Melanocyte-stimulating hormone
MTC	Mdeullary thyroid carcinoma
NA	Noradrenaline
NAD	Nicotine amide dinucleotide
NADP	Nicotine amide dinucleotide phosphate
N-CAM	Neuronal cell adhesion molecule
NGF	Nerve growth factor
NH	Neurohypophysis
NPF	Neuropeptide F
NPY	Neuropeptide Y
N-terminal	Amino terminal
OC	Oxyphilic cell
OMIM	Online Mendelian Inheritance in Man
OXT	Oxytocin
PACAP	Pituitary adenylate cyclase activating peptide
PAM	Peptidylglycine alpha-amidating monooxygenase
PAS	PER–ARNT–SIM
PBAN	Pheromone-biosynthesis-activating neuropeptide
PC	Prohormone convertase
PDB	Protein Data Bank
PDF	Pigment-dispersing factor
PDGF	Platelet-derived growth factor
PDH	Pigment-dispersing hormone
PEPCK	Phosphoenolpyruvate carboxykinase
PER	Period
PETH	Pre-ecdysis-triggering hormone
PHM	Peptidylglycine alpha-hydroxylating monooxygenase
PKA	Protein kinase A
PM	Plasma membrane
PNP	Pancreatic polypeptide
POMC	Proopiomelanocortin
PoP	Posterior pituitary
PNP	Pancreatic polypeptide
PRL	Prolactin
PSP	Pheromonostatic peptide
PTH	Parathyroid hormone

PTHrP	Parathyroid-hormone-related peptide
PTSH	Prothoracicostatic hormone
PTTH	Prothoracicotropic hormone
PYY	Peptide tyrosine tyrosine
rER	Rough endoplasmic reticulum
RET	Receptor protein tyrosine kinase
RNA	Ribonucleic acid
RPCH	Red-pigment-concentrating hormone
RX2	Runx2
RZ	Rest zone
SCC	Side chain cleavage monooxygenase
sER	Smooth endoplasmic reticulum
SF1	Steroidogenic factor 1
SG	Stratum granulosum
SGLT	Sodium–glucose cotransporter
SH2	Src homology domain 2
SHBG	Sex-hormone-binding globulin
SIM	Single-minded homolog
SNARE	Soluble *N*-ethylmaleimide sensitive factor attachment receptor
SP	Substance P
SRC	Rous sarcoma virus oncogen
SRDa	5-α-Reductase
SRIF	Somatostatin
SRY	Sex-determining region Y
SST	Somatostatin
StAR	Steroidogenic acute regulatory protein
STAT	Signal transducer and activator of transcription
STH	Somatotropin
T_3	Triiodothyronine
T_4	Thyroxine
TBG	Thyroxine-binding globulin
TGF	Transforming growth factor
TIDA	Tuberoinfundibular dopaminergic
TLC	Theca lutein cells
TMOF	Trypsin-modulating oostatic factor
TPO	Thyroid peroxidase
TR	Triiodothyronine/thyroxine receptor
TRH	Thyrotropin-releasing hormone
tRNA	Transfer RNA
TSH	Thyroid-stimulating hormone
VIH	Vitellogenesis-inhibiting peptide
VIP	Vasoactive intestinal peptide
VLDL	Very low density lipoprotein

A. aegyptii	*Aedes aegyptii* (yellow fever mosquito)
A. gambiae	*Anopheles gambiae*
A. mellifera	*Apis mellifera* (honeybee)
A. californica	*Aplysia californica* (California sea hare)
A. kurodai	*Aplysia kurodai* (another sea hare)
B. mori	*Bombyx mori* (silkworm)
C. maenas	*Carcinus maenas* (European shore crab)
C. elegans	*Caenorhabditis elegans*
C. intestinalis	*Ciona intestinalis* (vase tunicate)
D. maximus	*Dipetalogaster maximus*
D. melanogaster	*Drosophila melanogaster* (common fruit fly)
D. rerio	*Danio rerio* (zebrafish)
H. americanus	*Homarus americanus* (American lobster)
H. pomatia	*Helix pomatia* (Burgundy snail)
L. maderae	*Leucophaea madera* (Madeira cockroach)
L. stagnalis	*Lymnaea stagnalis* (great pond snail)
L. migratoria	*Locusta migratoria* (migratory locust)
M. sexta	*Manduca sexta* (tobacco hornworm)
O. immunis	*Orconectes immunis* (calico crayfish)
O. limosus	*Orconectes limosus* (spinycheek crayfish)
P. bernardus	*Pagurus bernhardus* (hermit crab)
P. clarkii	*Procambarus clarkii* (red swamp crayfish)
R. prolixus	*Rhodnius prolixus*
S. purpurata	*Strongylocentrotus purpuratus* (California purple sea urchin)
T. castaneum	*Tribolium castaneum* (red flour beetle)

Introduction 1

Zum Geleit
Lass die Moleküle rasen,
was sie auch zusammenhobeln
Lass das Tüfteln, lass das Knobeln!
Heilig halte die Ekstasen!
(Christian Morgenstern, *Galgenlieder*)[1]

Content

1.1 Whom To Thank 2

Why? Endocrinology is the study of the mechanisms that regulate essential body functions such as reproduction, metabolism, water balance, feeding, and growth. Research in these fields is fascinating and has widened our view of human physiology. New drugs for the control of feeding and food uptake or sexual activity have reached the medical practice but have also resulted in a gray market for substances of doubtful efficiency. E-mail folders choke on pseudomedical spam. Neuropeptide Y, a nerve cell product which is involved in the generation of hunger, has been the focus of some 2,000 articles since the beginning of the millennium.

Since the introduction of molecular biological techniques into biomedical research, many old questions have been answered, and new questions have arisen.

[1]Preface
Let the molecules race free,
however much they dash and spring!
Stop your meddling, stop your modeling
save this sacred ecstasy!
(translation by K. Moorwood)

B. Kleine, W.G. Rossmanith, *Hormones and the Endocrine System*,
DOI 10.1007/978-3-319-15060-4_1

Today, the regulation of a cell is discussed on the level of gene activity and signal transduction from the surface to the nucleus. Endocrinologists have engaged in this new method and have thus acquired many new and sophisticated tools. Without these advances in molecular biological tools, could we have imagined, for example, that variants of the leptin receptor missing the part for signal transduction might still serve as a vehicle for leptin uptake through the blood–brain barrier? In this book, we aim to integrate such new results into an enhanced understanding of endocrinology.

For Whom? If you like science, if you do not regard biology and chemistry as of satanic provenience, if you do not painfully avoid modern genetics, then you are the one for whom we wrote this book. If you are willing to delve into cell biology, the better. Basic knowledge of DNA, RNA, and proteins is very helpful. Any student of biology, medicine, or related areas is a potential reader. Anyone working in this field might consult this book for reference.

About What? After briefly looking into the history of hormones, we will dive into the basics of endocrinology: hormones and their receptors, not only from *Homo sapiens* and its close relatives, but also from invertebrate species such as flies and crayfish. Later we will deal with regulatory circuits fundamental for controlling blood pressure, water balance, digestion, or reproduction, all controlled by hormones. In the final chapters, we will present the endocrine system by its failures and look at the origins of endocrine-related diseases. A brief look at the use of hormones and their derivatives to boost performance in sport by means of doping completes the book.

1.1 Whom To Thank

We thank all the readers of our book in Germany who have helped by their critique to enhance the content and the design of our book. To all colleagues who have freely provided articles and manuscripts to expand individual chapters, thank you very much. Several friends read large parts of the manuscript and provided valuable comments. Thanks to all of you. A special thank you goes to Radivoj Krstić, who let us use the anatomical plates in his book *Human Microscopic Anatomy* for the anatomy of the endocrine active organs.

History of Hormones and Endocrinology 2

Contents

2.1 Ancient Times 3
2.2 New Age 6
2.3 Modern Times 7
2.4 Problems Remaining 7
2.5 Historical Summary 8

Medical history is built on traditions from the Middle East (Egyptian, Jewish, Arabian), the Far East (mostly Chinese), and Europe, with Greek, Roman, medieval, and modern elements. Because of geographical distances, language or religious communication barriers, or rejection of older traditions, the European medical community had to rediscover facts which were known in ancient Egypt, China, or Arabia a long time ago. Whether an endocrinological profession was known in ancient times may be doubted.

Below, we will separate ancient times, the new age, and modern times without any assessment in mind. The search for disease causes by scientific means—hypothesizing, performing experiments, refuting or corroborating the hypothesis—was established by the Greeks.

Table 2.1 does not cover the richness of modern discoveries for lack of space. Therefore, we conclude the historical overview with the reevaluation of the hypothalamic–pituitary portal system.

2.1 Ancient Times

As endocrinology carved in stone, a bust of pharaoh Amenhotep IV (Fig. 2.1) bears all the signs of acromegaly, which is mostly induced by tumor-secreted growth hormone. Thyroid gland malfunctions were known long ago, and as early as the second pre-Christian millennium were treated with (iodine-containing) seaweed or

B. Kleine, W.G. Rossmanith, *Hormones and the Endocrine System*,
DOI 10.1007/978-3-319-15060-4_2

Table 2.1 Historical summary

Year	Organ, action, hormone	Discoverer
?	Role of testes in animal husbandry	Humans
4000–3000 B.C.	Human ovariectomy	Ancient Egypt
3000 B.C.	Amenhotep IV and acromegaly (?)	Ancient Egypt
1600 B.C.	Seaweed and heated sponges for treatment of goiter	Chinese, mentioned by Plinius, Juvenal, and Galen (50 B.C. to A.D. 300)
1550 B.C.	Diabetic polyuria and treatment	Egyptian papyrus
460–400 B.C.	Relationship between mumps and infertility	Hippocrates
	Relationship between galactorrhea and amenorrhea	Hippocrates
A.D. 81–138	Description of diabetes mellitus	Aretaeus
7th century	Sweet taste of urine in diabetes mellitus	Chen Chuan; rediscovered by T. Willis (1621)
1135	Sheep testis extract against male infertility	Hsu Shu-Wei
1561	Description of ovaries, oviduct, corpus luteum	Vesalius
1563	Description of adrenal glands	B. Eustachio
17th century	Estrogens from urine	Chinese iatrochemists
1664	Pancreatic juice	R. de Graaf
1772	Report of acromegaly	N. Saucerotte
1802, 1835, 1840	Report of thyrotoxicosis	G. Flajani, R.J. Graves, and C. von Basedow
1855	Diseases of the adrenal cortex	T. Addison
1864	Report of acromegaly and pituitary tumor	A. Verga
1865	Description of the hypothalamus	J.B. Luys
1869	Pancreatic islet cells	P. Langerhans
1893	Langerhans cells producing hormones	G.-E. Laguesse
1894	Blood-pressure-enhancing compounds the adrenal extract	G. Oliver and E.E. Schaefer
1895	Vasopression from pituitary extracts	G. Oliver and E.E. Schaefer
1898–1904	Structural determination and synthesis of adrenaline	J.J. Abel, A.C. Crawford, J. Takamine, T.B. Aldrich, and F. Stolz
1902	Discovery of secretin	W. Bayliss and E.H. Starling
1903–1906	Discovery of gastrin	J.S. Edkins
1905	First use of the term "hormone"	E.H. Starling, proposed by W.B. Hardy
1906	Transphenoidal surgery of pituitary tumors	H. Schloffer
1906	Oxytocin action from pituitary extracts	H. Dale
1912	Relationship between the posterior pituitary and diabetes insipidus	A.E. Frank

(continued)

Table 2.1 (continued)

Year	Organ, action, hormone	Discoverer
1914	Thyroxine crystals	E.C. Kendall
1926	Insulin crystals	J.J. Abel
1921	Report on the discovery of insulin	N.C. Paulesco (pancreatin) and F.G. Banting and C.H. Best (insulin)
1926	Isolation of oxytocin and vasopressin	O. Kamm
1928	Isolation of LH and FSH	B. Zondek and S. Aschheim
1929	Discovery of prolactin	P. Stricker and F. Grueter
1929	Report on the action of TSH	M. Aron, L. Loeb, and R.B. Basset
1932	Relationship between polyglandular syndrome and pituitary–adrenal hyperactivity	H. Cushing
1934	Description of the crustacean pigment-concentrating hormone in *Pandalus borealis*	L.H. Kleinholz
1935	Testosterone isolated from testes	E. Laqueur
1936	Description of stress	H. Selye
1937–1952	Isolation of adrenal steroids	E.C. Kendall
1939	Renin synthesis from the juxtaglomerular apparatus	A. Goormaghtigh
1940–1949	Isolation of LH, ACTH, GH, and FSH	Choh Hao Li and Evans
1943	Use of radioactive iodine in thyroid diseases	S. Hertz, A. Roberts, and C.P. Leblond
1944	Comparison of insects and vertebrates leads to the term "neurosecretion"	B. and E. Scharrer
1945	Discovery of noradrenaline	P. Holtz, K. Credner, and G. Kronenberg
1948	Neuronal control of the pituitary	G.W. Harris
1951	Oxytocin and vasopressin released by the pituitary	W. Bargmann and E. Scharrer
1956	Autoantibodies in Hashimoto thyroiditis	I.M. Roitt and D. Doniach
1958	Isolation and structure of melatonin	A.B. Lerner
1966	Isolation and structure of gastrin	R. Gregory
1966–1971	GH structure and synthesis	Choh Hao Li and colleagues
1968	TRH	R. Guillemin, A.V. Schally and colleagues
1971	Elucidation of the structure of TSH	J.G. Pierce and colleagues
1971	GnRH	A.V. Schally, A. Arimura, and colleagues
1973	Somatostatin	I. Brazeau, R. Guillemin, and colleagues
1975	Reevaluation of the hypothalamic–pituitary portal system	P.M. Daniel and M. Pritchard

ACTH adrenocorticotropic hormone, *FSH* follicle-stimulating hormone, *GH* growth hormone, *GnRH* gonadotropin-releasing hormone, *LH* luteinizing hormone, *TRH* thyrotropin-releasing hormone, *TSH* thyroid-stimulating hormone

Fig. 2.1 Pharaoh Akhenaton (Amenhotep IV): possibly the first known acromegaly patient (From the Egyptian Museum Berlin)

dried sponges. Since the beginning of animal husbandry, castration of male cattle has been known to make them infertile and to enhance their size and fat content. That bullocks are easier to handle than bulls must also have been apparent. From ancient times, documents have been passed on with descriptions of diseases and therapies therefor which fall into endocrine categories. Whether endocrine secretion or interconnection of the pituitary and the gonads or the pituitary and the adrenal glands was known 2,000 years ago is not known. Any such knowledge was not passed to an endocrinological middle age and had to be rediscovered.

2.2 New Age

Unfortunately, we cannot go back to medical sources from the Arabian world, for which reason an endocrinological middle age has to be excluded. The endocrinological new age apparently started in the sixteenth century and with the description of female reproductive organs by Vesalius. The invention of the microscope in about 1590 offered detailed insight surpassing the macroscopic facts. After an extended period where the various endocrine active organs and their pathological states were analyzed, with the help of some experiments with organ extracts, at the end of the nineteenth century, reports of endocrine actions by gland extracts (adrenal, pituitary) and the very first chemical synthesis of a hormone (adrenaline, also known as epinephrine) were published.

2.3 Modern Times

The modern era of endocrinology is characterized by the use of isolated and chemically defined or synthesized hormones. Within the last 20 years, additional hormones have been identified—for example, galanin in the pig gut—but modern times started with crystals of adrenaline, thyroxine, and insulin. By modern analytical chemistry, the structures of protein hormones such as luteinizing hormone, follicle-stimulating hormone, human choriogonadotropin, and thyroid-stimulating hormone were observed. Later, it was the turn of hypothalamic neuropeptides: thyrotropin-releasing hormone, gonadotropin-releasing hormone,[1] growth-hormone-releasing hormone, and corticotropin-releasing hormone.

The discovery of hormone receptors, be they nuclear steroid and thyroxine receptors or be they heptahelical G-protein-coupled membrane receptors, is an achievement of the last third of the twentieth century.

2.4 Problems Remaining

The sequences of cytochrome P450 enzymes, which are particularly responsible for steroid hormone biosynthesis, have been evaluated, and the crystal structures of individual family members have been published recently. To complete the picture, more time and more effort will be necessary.

Also incomplete is the understanding of numerous endocrine interactions, biosynthetic regulations, and hormone release with respect to influences by other hormones or intracellular regulators. This is the subject of very ambitious research in animal models and with tumor cells.

Many elements of the endocrine system have been discovered and are synthetically available for experimental purposes. Animal models provide new findings. Whether these models can be applied in humans has to be investigated for each new substance or method. The basics of the endocrine system have been conserved in humans and other mammals; however, there are relevant differences: for example, rodents do not use cortisol, but use corticosterone; and after the evolutionary separation of primates and humans, a characteristic reorganization of the Y chromosome occurred. The development of male gonads during fetal development, therefore, cannot be studied in closely related nonhuman primates. In humans, however, given legal and ethical restrictions because of individual human rights, it is with very few exceptions possible to envisage the normal situation by analyzing only malfunctions or diseases.

[1] Also known as luteinizing-hormone-releasing hormone or follicle-stimulating-hormone-releasing hormone

2.5 Historical Summary

Since recent developments in endocrinology demolish any summary, Table 2.1 ends at 1975. Recent summarizing information is freely available in the PubMed library of the National Institutes of Health (Bethesda, MD, USA).

Part I

Biochemistry of Hormones: The Basics

Hormones: Some Definitions

3

Contents

3.1 The Nature of the Endocrine System 11
3.2 What Is a Hormone? 12
3.2.1 Hormonal Effect 12
3.2.2 Neurotransmitters 13
3.2.3 Cytokines/Lymphokines 13
3.2.4 Prostaglandins/Thromboxanes 13
3.2.5 Pheromones 13
3.3 How Do We Define the Endocrine System? 14
3.4 Endocrine Hierarchy 15
3.5 Hormone Classes 16
3.5.1 Protein/Peptide Hormones 16
3.5.2 Terpenes: Juvenile Hormones and Steroids Hormones 17
3.5.3 Amino Acid Derivatives 17

3.1 The Nature of the Endocrine System

Protozoans sense signals from the environment on their cell surface and adapt to these messages. They do not require the kinds of messages which help cells in multicellular organisms to communicate information to one another about their conditions, their requirements, and how they can serve other cells. Once multicellular organisms with different organs and their different functions had developed from protozoans, it became necessary to develop message transport within the organism. Two systems evolved: the nervous system and the endocrine system. There is a third system, the immune system, which also processes information.

B. Kleine, W.G. Rossmanith, *Hormones and the Endocrine System*,
DOI 10.1007/978-3-319-15060-4_3

3.2 What Is a Hormone?

Hormones are signals. They announce to the organism the state of the secreting cell as well as the state of the hormone-producing organ; they are like semaphore flags. There is a set of cells restricted to the target organ to recognize the semaphore flag. Thus, the effect of a hormone is solely within the target organ. If cells other than those intended start to react, pathological disorders might develop—for example, endometriosis: only the epithelium of the uterus should react to estrogens by growing in order to allow nidation of the trophoblast. When cells outside the uterus, such as in the peritoneum and in other organs of a woman, start to react to the periodic estrogen signals by growing and ablation, the woman feels great pain and experiences internal bleeding.

Hormones are transferred from the organ of release to the target organ by means of the blood. Some hormones need a specific vehicle for this transport; however, most hormones move without a vehicle through the circulation. During transport, hormones are prone to enzymatic degradation. To reach the target organ in sufficient concentrations, many hormones are released by the coordinated action of many cells. The release of these hormones is achieved in pulses.

3.2.1 Hormonal Effect

The process by which a substance is released in one organ and transferred via the blood to its target organ is called an *endocrine* process: at one place synthesis and release occur, then undirected transport in the blood occurs, and finally reactivity occurs at a distant cell or organ. The science of hormones (of endocrine-acting drugs, of internal secretion) was thus named endocrinology.

Exocrine release, as opposed to endocrine secretion into the blood, characterizes gland production—for example, release into the oral cavity, into the gut lumen, or across the skin. Furthermore, odorant release, territory marking, and pheromone secretion are exocrine processes as well.

A mediator which affects neighboring cells—for example, within a gland or in the intestinal wall, in a lymph node, or within the placenta—acts in a paracrine manner. Most regulators with very short biological half-lives show this paracrine effect.

Hormones, however, are not very stable in blood. Hypothalamic releasing (RH) hormones such as corticotropin RH, thyrotropin RH, and gonadotropin RH have half-lives in blood below 10 min; owing to the short distance to the target organ—the pituitary gland (only 2–3 cm away)—the life span of these hormones is sufficiently long for them to reach the target organ in pulses. The simultaneous arrival of hormones in concentrations above the level needed for activation results in the endocrine functions.

When an effector substance expresses its effects on the releasing cell—for example, in tumor cells, in activated lymphocytes, or again, in the placenta at the interface of maternal and fetal blood—these actions are called autocrine actions.

3.2.2 Neurotransmitters

Hormones which act via the blood circulation must be distinguished from neurotransmitters such as GABA, acetylcholine, and opioids, whose effects are concentrated on the synaptic cleft between nerve cell and target cell. There are, however, some substances that act as hormones distantly from where they have been released and function as neurotransmitters as well: dopamine and noradrenaline (norepinephrine). Other hormones such as vasopressin (antidiuretic hormone) and neuropeptide Y are also active as neurotransmitters.

3.2.3 Cytokines/Lymphokines

Cytokines and lymphokines are messengers and regulators within the immune system. Cells of the immune system are widespread over the entire organism and are often mobile, in contrast to cells of the endocrine system, which are fixed in defined organ structures. With respect to their distant action, cytokines and lymphokines resemble hormones, and since the detection of T-cell growth factors more than 30 years ago, the similarity between cytokines and interleukins, and hormones has been established. Hormones, in addition, act on cells of the immune system, and the latter express hormone receptors. We would like to maintain the distinction of the very complex immune system and its mediators from the endocrine system in order to distinguish the fundamental mechanisms from less obvious mechanisms.

3.2.4 Prostaglandins/Thromboxanes

Prostaglandins and thromboxanes derived from arachidonic acid should also be distinguished from hormones. By the action of cyclooxygenases and other enzymes, these short-lived messengers are generated, but their rapid degradation does not allow distant action and they are, therefore, not as effective as hormones. These substances are mainly made by cells of the immune system, and they are also excluded from the endocrine system.

3.2.5 Pheromones

Pheromones are very special messengers which do not act within an organism. They relay information from one individual of a species to others: a female butterfly, for example, lets males know of her existence and her readiness to mate. The extreme sensitivity of its pheromone receptors triggers the male butterfly to fly toward the faint pheromone gradient in order to copulate. Pheromone-like actions have also be reported from fungi: male and female fungal steroids are secreted from respective sprouts and trigger the formation of fruit bodies (see Fig. 6.24 and Sect. 6.10.2). A so-called sex peptide in insects transferred with the sperm from a male to a female blocks further pheromone release by female insects as long as the fertilized eggs have not been deposited and also acts in a pheromone-like manner. Some pheromone receptors are part of the superfamily of heptahelical G-protein-coupled membrane receptors. However, the known structures of almost all pheromones differ from those of hormones, although some fungal steroids act as pheromones.

3.3 How Do We Define the Endocrine System?

When comparing the endocrine system with the nervous system or the immune system, we should recognize that the endocrine system is distinguished by the following two characteristics:

1. *Single organs at defined places*. Very few single organs are found in the endocrine system, rather some of them are doubled and are present on both sides of the body. In contrast, in the immune system, there are some single organs such as the spleen and thymus, many lymph nodes with similar functions are distributed across the whole organism, and Peyer's patches, for example, are present in the gut not once, but manifold. In the endocrine system, soluble hormones drift through the organism, whereas in the immune system, cells wander throughout the circulation and the lymphatics as well. A similar migration of endocrine cells occurs only in metastatic tumors.
2. *Soluble mediators with distant effects*. This contrasts with the nervous system, where information is transferred across the very narrow synaptic cleft. Nerve cells themselves with their axons connect the distant organs and body regions: that is, far-reaching cells and short ways for the mediators (neurotransmitters) in the nervous system, narrowly defined organs, and far-reaching mediators (hormones) in the endocrine system.

All three systems have in common that the activity of all cells is controlled by either synaptic contacts or soluble mediators. Hormones act on nerve cells, immune cells, and endocrine cells, and nerves reach nerve cells, endocrine cells, and immune cells; the mediators of the immune system, on the other hand, have receptors at nerve cells and endocrine cells. The three systems are intimately linked and mutually influence each other.

The release of hormones by neurosecretory cells demonstrates these intimate connections. Neurosecretion in humans is observed in the hypothalamus. The axons of these cells reach into the median eminence and via the pituitary stalk (infundibulum) to the posterior pituitary. Stimulated by other nerve cells and the neurotransmitters released from these cells, these neurosecretory cells release the preformed hormones, which then enter the blood in fenestrated areas. An organ with fenestrated capillaries and without the blood–brain barrier is called a neurohemal organ. Such neurohemal organs are not only developed by mammals, but were also present in the common precursors of invertebrates such as insects and vertebrates such as mammals.

The endocrine system can be characterized unambiguously. Owing to the intimate functional relationship between the endocrine and nervous systems, it cannot be regarded separately. Therefore, we will name the relevant parts of the nervous system in this book, but we will not describe them in detail. The reader is referred to specialized books on these matters.

3.4 Endocrine Hierarchy

Hormonal regulation in endocrine organs is controlled by the brain: within neurosecretory cells of the hypothalamus, signals are integrated and processed, resulting in release of hormones into the capillaries of the median eminence. From there, they reach the pituitary, where other tropic hormones are released, which then induce the hormonal (synthesis and) release in the thyroid gland, the adrenal glands, or the gonads. These hierarchies have been named the hypothalamic–pituitary–thyroid, hypothalamic–pituitary–adrenal, and hypothalamic–pituitary–gonadal axes (Fig. 3.1).

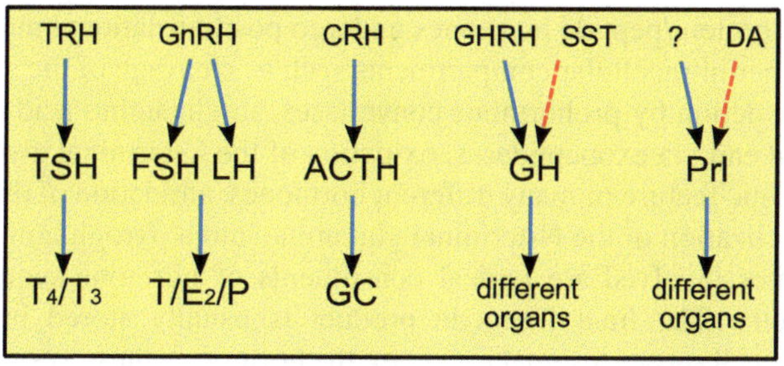

Fig. 3.1 Endocrine regulation via hypothalamic–pituitary axes. The *upper row* shows the hypothalamic releasing hormones which induce or inhibit tropic hormone release in the pituitary—thyrotropin, gonadotropins, adrenocorticotropic hormone (*ACTH*), somatotropin (*SST*), and prolactin (*Prl*). These in turn regulate the release of thyroid hormones, steroids, glucocorticoids (*GC*), and insulin-like growth factor, and other factors. The *question mark* indicates a so far unidentified prolactin-releasing hormone. *CRH* corticotropin-releasing hormone, *DA* dopamine, E_2 estradiol, *FSH* follicle-stimulating hormone, *GH* growth hormone, *GHRH* growth-hormone-releasing hormone, *GnRH* gonadotropin-releasing hormone, *LH* luteinizing hormone, *P* progesterone, *T* testosterone, T_3 triiodothyronine, T_4 thyroxine, *TRH* thyrotropin-releasing hormone, *TSH* thyroid-stimulating hormone

3.5 Hormone Classes

3.5.1 Protein/Peptide Hormones

This first category comprises hormones regulating reproduction (gonadotropins), energy balance (insulin), or blood pressure (angiotensins).[1] Regulators of hormone-synthesizing cells such as gonadotropin-releasing hormone and growth-hormone-releasing hormone, and hormones controlling feeding—for example, leptin and neuropeptide Y—fall into this class. Most vertebrates have hormones of similar structure. Some of these, such as insulin, however, are not restricted to vertebrates, but are widespread in the animal kingdom—for example, as insulin-like peptides and bombyxin in invertebrates. In invertebrates, several other hormones with structures similar to those of hormones in vertebrates have been observed. In the insect *Drosophila melanogaster*, 40 protein/peptide hormones have been counted, the majority without homology to vertebrate hormones.

These hormones are produced like any other protein: RNA is translated from the hormone gene. This RNA is then processed in the cell nucleus and exported from there as messenger RNA. In the cytosol, this messenger RNA associates with ribosomes. After the initiation of peptide synthesis, the RNA–ribosome complex binds to the endoplasmic reticulum (ER), the signal peptide reaches the inner ER via a pore, and the remainder of the protein is translated into the ER.

Almost all protein/peptide hormones undergo posttranslational modifications in the ER and other intracellular compartments such as cleavage of the signal peptide, endopeptidase action by prohormone convertases, single amino acid removal from the C-terminal end via exopeptidases, oxidation of the C-terminal glycine resulting in a characteristic feature of many different hormones, amidation of the C-terminus, and finally cyclization of the N-terminal glutamine into a pyroglutamate residue.

The enzymes involved are typical constituents of hormone-synthesizing (i.e., endocrine) cells. The final hormone product is usually stored in intracellular vesicles. If the cell senses a signal to release the hormones, these vesicles fuse to the cell membrane and thus release their content, the hormones.

Some hormones exhibit endocrine effects, and act as neurotransmitters. A distinction appears difficult since it is not clear whether, for example, noradrenaline is released by the autonomic nervous system or by the adrenal medulla. This difficulty can also be found in insects and crustaceans. Sometimes a neuropeptide is released into the hemolymph, whereas in the same organism the target organ has axons from the hormone-producing cells. This might mirror the intimate connection between the endocrine system and the nervous system. Many different neuropeptides, however, are released within neurohemal organs and act in an endocrine fashion as in vertebrates. The common origin of these structures confirms that neurohemal organs are an important common feature of metazoans.

---

[1] Peptides and proteins are distinguished by their chain length: below 100 amino acids, polypeptides are called peptides; polypeptides with more than 100 amino acids are called proteins.

3.5.2 Terpenes: Juvenile Hormones and Steroids Hormones

In these two hormone classes we find estradiol and testosterone (female and male sex hormones), corticosteroids and mineralocorticosteroids such as cortisol and aldosterone, and gestagens—for example, progesterone. Insects have their own steroids—ecdysone (the hormone that controls molting; ecdysis) and its precursor. All these steroids are derived from cholesterol. Vitamin D_3 is produced similarly.

In complex biochemical reactions, the precursor squalene is converted into cholesterol. This cholesterol is present in the plasma membrane. A characteristic feature of steroid-forming cells is the expression of steroidogenic acute regulatory protein, which transfers cholesterol from the plasma membrane into mitochondria, where the first step of vertebrate steroid synthesis occurs: conversion of cholesterol into pregnenolone. Starting from pregnenolone, all the other human steroids are synthesized in a series of conversions. These steps require several enzymes. The presence or absence of these enzymes is the critical determinant which decides whether in a given cell cortisol or estradiol is made.

Sometimes the final steroid product is stored in the cell. More often these steroids are released immediately after synthesis by diffusion into the cellular environment. Thus, the release of a steroid hormone is regulated not by a signal for release of preformed molecules, but by activating the gene for the rate-limiting or final enzyme in the synthesis cascade.

Juvenile hormones are derived like steroids from the farnesyl pyrophosphate precursor of cholesterol. These characteristic sesquiterpenes play an essential role in the development of invertebrates and their molts, but they are also required for gonadal development or the cast determination in social insects. In crustaceans, farnesoate methyl ester—an intermediate in juvenile hormone biosynthesis in insects—has been described as a hormone that slows molting. In crustaceans and insects, juvenile hormones are distributed via the hemolymph.

3.5.3 Amino Acid Derivatives

The fourth class of hormones is derived from amino acids. Triiodothyronine and thyroxine are derivatives of tyrosine, as are the catecholamines dopamine, noradrenaline, and adrenaline. Indolamines such as melatonin and related molecules such as serotonin are made from tryptophan in successive steps.

4 Vertebrate Protein and Peptide Hormones

Contents

4.1 Translation 21
  4.1.1 Reading Genetic Information: Transcription 21
  4.1.2 Coding and Other Sequences 21
  4.1.3 Splicing 21
  4.1.4 RNA Cap 22
  4.1.5 Nuclear Export of Messenger RNA 22
  4.1.6 Docking to Ribosomes 22
  4.1.7 Translational Termination 22
  4.1.8 Membrane and Secretory Proteins 22
4.2 Posttranslational Modification: Hormone Maturation 23
  4.2.1 Removal of the Signal Peptide 23
  4.2.2 Folding and Disulfide Bridges 23
  4.2.3 Protein Complexes 24
  4.2.4 Glycosylation 24
  4.2.5 Prohormone Convertases 25
  4.2.6 Monobasic and Dibasic Sequence Motifs in Invertebrates and Vertebrates 27
  4.2.7 Chopping the C-Terminus 28
  4.2.8 Oxidation of the Terminal Glycine 28
  4.2.9 Cyclization of the N-Terminal Glutamine 29
  4.2.10 Esterification of Ghrelin 29
4.3 Peptide Hormones of the Hypothalamus and the Brain 30
  4.3.1 Hypothalamic-Releasing Hormones 30
  4.3.2 Gonadotropin-Inhibiting Hormone 42
  4.3.3 Neuropeptide Y 45
  4.3.4 Agouti-Related Protein 45
  4.3.5 Somatostatin 46
  4.3.6 Substance P 49
  4.3.7 Proopiomelanocortin 49
  4.3.8 Ghrelin 49

The original version of this chapter was revised.
An erratum to this chapter can be found at DOI 10.1007/978-3-319-15060-4_17

B. Kleine, W.G. Rossmanith, *Hormones and the Endocrine System*,
DOI 10.1007/978-3-319-15060-4_4

4.3.9 Kisspeptin.... 50
4.3.10 Galanin.... 51
4.3.11 Melanin-Concentrating Hormone.... 54
4.3.12 Orexins.... 56
4.4 Anterior Pituitary Hormones.... 59
4.4.1 POMC.... 59
4.4.2 TSH.... 67
4.4.3 LH, FSH, CG.... 69
4.4.4 Growth Hormone.... 76
4.4.5 Prolactin.... 79
4.5 Hormones of the Pars Intermedia.... 81
4.5.1 Somatolactin.... 81
4.5.2 Prolactin-Releasing Peptide.... 83
4.6 Hormones of the Posterior Pituitary: Oxytocin and Vasopressin.... 84
4.6.1 Introduction.... 84
4.6.2 Structure and Genes.... 86
4.6.3 Physiology of Oxytocin.... 87
4.6.4 Physiology of Vasopressin (AVP, Adiuretin).... 87
4.6.5 Phylogeny.... 88
4.7 Regulators of Sugar and Energy Metabolism.... 89
4.7.1 Insulin.... 89
4.7.2 Glucagon.... 93
4.8 Regulators of Food Intake.... 95
4.8.1 Leptin.... 95
4.8.2 Ghrelin.... 96
4.9 Nonsteroidal Regulators of Bone.... 99
4.9.1 Introduction.... 99
4.9.2 Structure and Genes.... 99
4.9.3 Physiology.... 100
4.9.4 Phylogeny.... 101
4.10 Digestive Tract Hormones.... 101
4.10.1 Gastrin.... 101
4.10.2 Cholecystokinin.... 103
4.10.3 Secretin.... 104
4.10.4 VIP.... 106
4.10.5 GIP.... 107
4.10.6 PNP, NPY, PYY.... 108
4.11 Neuropeptides in the Enteric Nervous System.... 112
4.11.1 Endorphins and Enkephalins.... 112
4.11.2 Tachykinins: Substance P, Neurokinin, and Endokinin.... 113
4.11.3 Gastrin-Releasing Peptide/Bombesin.... 115
4.12 Nonsteroidal Regulators of Reproduction.... 116
4.12.1 Activin/Inhibin.... 116
4.12.2 Follistatin.... 119
4.12.3 Antimüllerian Hormone.... 120
4.13 Angiotensins and Renin.... 122
4.13.1 Introduction.... 122
4.13.2 Structure and Genes.... 123
4.13.3 Physiology.... 123
4.13.4 Phylogeny.... 123
4.14 Atrial Natriuretic Peptides.... 124
4.14.1 Introduction.... 124
4.14.2 Structure and Genes.... 124

4.14.3 Physiology 126
4.14.4 Phylogeny 126
4.15 Additional Neuropeptides 127
4.15.1 Parathormone 127
4.15.2 Stanniocalcin 128
4.15.3 Erythropoietin 130

Readers familiar with how proteins are made may skip the following section. For those who are not familiar, we provide a short introduction to this process from which all life has arisen. The mechanism of forming structures from the genetic blueprint is obviously as old as life itself because it is common to all forms of life on Earth.

4.1 Translation

4.1.1 Reading Genetic Information: Transcription

Genetic information if encoded in the chromosome by means of the sequence of four bases—adenine, cytosine, guanine, and thymine—in the double strand of deoxyribonucleic acid genetic information is coded for in the chromosomes. This information is transcribed into a single-stranded ribonucleic acid when a gene is activated. In the case of bacterial, viral, or many yeast genes, the RNA is directly coupled to ribosomes with whose help single amino acids are added to a protein sequence according to the code in the DNA.

4.1.2 Coding and Other Sequences

In eukaryotic cells the coding information on the DNA double strand is interspersed with noncoding chromosomal regions, which will never be used for protein synthesis. The coding sequences are called exons, those without coding information, introns.

4.1.3 Splicing

The primary RNA transcript still contains exons and introns. By a process called splicing the introns are removed. Splicing is performed using enzymatically active RNAs and proteins. These proteins are called splicing factors.

Many RNAs can be spliced to different products, alternative splicing; for obtaining differentially spliced RNA the just-mentioned splicing factors are responsible that are found in a cell-type-specific manner. Thirty different splicing factors have been found; their regulation is not yet well understood.

4.1.4 RNA Cap

Eukaryotic RNA has on its 5′ end an additional structure, the so-called RNA cap that reduces the RNA degradation in the cytoplasm.

4.1.5 Nuclear Export of Messenger RNA

RNA when spliced and capped is called messenger RNA (mRNA). This mRNA is exported through the nuclear membrane with the help of transfer proteins and thus reaches the cytosol.

4.1.6 Docking to Ribosomes

In the cytosol two ribosomal subunits aggregate with the mRNA. Transfer RNA will load the amino acids into the ribosomes that will then be added to the protein sequence according to the genetic code. This process is called translation.

4.1.7 Translational Termination

A termination signal within the RNA sequence lets the ribosomal subunits fall off the mRNA. mRNA and ribosomes can be reused.

4.1.8 Membrane and Secretory Proteins

In the case of secreted proteins or membrane proteins this general translation pathway is extended. During translation membrane proteins are integrated into membranes although secretory proteins are not translated into the cytosol, but into special, membrane-sealed, cellular compartments, from where the secreted proteins, for example, hormones, are finally secreted.

These compartment are vesicles of the ER where synthesis of membrane proteins and secretory cell products takes place. These vesicles themselves are enclosed with a double membrane like the cell membrane. Other cellular compartments with separate double membranes are the eukaryotic nucleus, prokaryotes (i.e., bacteria or blue algae do not possess a nucleus), mitochondria, where energy is gained from sugars, and the Golgi apparatus, where protein maturation occurs. Secretory granules that contain the mature hormones ready for secretion are also separated from the cytosol by a double membrane.

Later, we demonstrate that some proteins of the steroid synthetic pathway are localized to mitochondria, some are found in the ER, and others stay in the cytosol. There exists a topological separation of different enzymatic functions of steroid-forming cascades.

4.2 Posttranslational Modification: Hormone Maturation

Precursors of protein/peptide hormones are formed at the membrane of the ER and they are translocated through this membrane into the ER vesicles. Therein and in other matured vesicles hormone maturation will occur.

4.2.1 Removal of the Signal Peptide

The first 22–30 amino acids of a precursor protein that is formed at the ER membrane are called signal peptides. Once the growing polypeptide chain has reached the interior of the ER the enzyme signal peptidase cleaves off this signal peptide, a process that is performed for membrane and secretory proteins.

4.2.2 Folding and Disulfide Bridges

The growing polypeptide chain moves through the ER pore as a linear strand. Within the ER this strand is folded into the three-dimensional structure characteristic for any protein. Folding is achieved with the help of chaperones, for example, heat shock proteins.

The three-dimensional protein structure resulting from folding contains mainly helices and β sheets. Other areas exist in an unordered form. Hydrogen bonds are essential structural elements for the maintenance of a given three-dimensional structure, as well as ionic and nonionic interactions between the amino acids of an individual protein. The use of supercomputers has not yet made it possible to create a general algorithm for protein folding: folding prediction has only been possible with varying success.

Coupled with folding is the generation of intramolecular disulfide bonds thereby covalently linking two cysteine residues. These disulfide bonds together with the interactions just mentioned determine the three-dimensional protein structure. The glycoprotein hormones such as thyroid-stimulating hormone (TSH), luteinizing hormone (LH), follicle-stimulating hormone (FSH), choriogonadotropin (CG) or nerve growth factor, as well as the insect hormone bursicon, form a special cysteine knot (Fig. 4.1); two pairs of disulfide bonds with short amino acid sequences between adjacent cysteines form a belt. The third disulfide bond is directed through this belt (Fig. 4.1). Modifications of this knot structure render the protein nonfunctional. The proper formation of the cysteine knot is indispensable for these hormones. It appears almost self-evident that this structure had been conserved during evolution whereas other amino acids were exchanged. The distances between two cysteine residues and thus the chain length in between remained constant during vertebrate evolution, which gives a clue for the conservation of the functional properties, too.

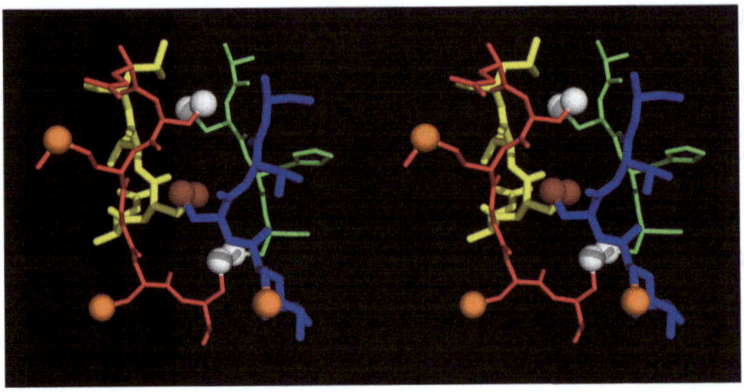

Fig. 4.1 Stereo view of the cysteine knot of the gonadotropin α-chain: two disulfide bonds (*white*: sulfur atoms) between Cys28 (*red* chain, amino acids 27–32) and cys82 (*green* chain, amino acids 81–84), and between Cys32 (*red*) to Cys84 (*green*) form a ring through which reaches the third disulfide bond between Cys10 (*yellow* chain, amino acids 9–12) and Cys60 (*blue* chain, amino acids 58–62) (*brown*: sulfur atoms) (Source: GenBank 1HRP and PyMOL)

4.2.3 Protein Complexes

The next step during hormone formation is the aggregation of identical or different polypeptides to larger complexes. This is a general feature not only of hormones, but also of many other proteins.

Within hormones the glycoprotein hormones are complexes of two different polypeptides. The first chain, the α chain is the common chain of four different glycoprotein hormones, and the β chain is characteristic for the four hormones: LH, FSH, TSH, and CG. Singular α or β chains are nonfunctional. The complex of both chains is necessary to give rise to the proper structure that triggers the hormone receptor on the target cell.

Oxytocin has equally been found in a complex with other peptides, the neurophysins. Whereas α-glycoprotein and β-glycoprotein hormone chains are transcribed from different genes, the oxytocin and the neurophysins are coded for in the same gene and transcribed into a single protein which is then processed during hormone maturation, however, the separated peptides stay together in a complex. At the final stage in the secretory granule, the mature oxytocin is no longer complexed to neurophysins.

4.2.4 Glycosylation

This step again primarily concerns the glycoprotein hormones. Several asparagine residues are substituted with oligosaccharides. In the Golgi apparatus these contain mannose-rich oligosaccharides. These mannoses form a sorting signal that leads the way to the secretory granules. In later vesicles the mannoses are partially replaced with other sugars and acquire fucoses and terminal *N*-acetylneuraminic acids, the latter the characteristics of mature glycoproteins. Addition of sugars and replacement of mannoses are common processes of glycoprotein synthesis and not restricted to hormones.

4.2.5 Prohormone Convertases

4.2.5.1 Introduction

We now discuss the special pathways of hormone maturation. As shown above, the first processing of the newly formed polypeptide chain is performed by the signal peptidase which removes the signal peptide. Whenever a signal peptide reaches the interior of the ER it is quickly and reliably removed but the chain itself is not yet finished.

By cleaving off the signal peptide, one end of many protein and peptide hormones is exposed. This end is called the amino terminus or N-terminal end. Here we find the name giving α-amino group at carbon atom 1 of the terminal amino acid. Because all the other α-amino groups are involved in the peptide bonds (boxes in Fig. 4.2), there is only this single α-amino residue in any polypeptide chain. The opposite end of the polypeptide chain is called the carboxy-terminal or C-terminus due to the free carboxy group there, a characteristic feature of organic acids. There is only one free C-terminal carboxy group in any polypeptide inasmuch as all the others are also part of the peptide bonds.

The C-terminus of almost any vertebrate protein/peptide hormone is exposed by enzymes that recognize dipeptide motifs formed by lysine (**K**) and arginine (**R**) and cleave the polypeptide chain behind these dipeptides. These enzymes were labeled prohormone convertases (PC) because they convert the precursor chains into functional hormones (at least sometimes).

$$NH_2\text{–}CHR_1\text{–}COOH + NH_2\text{–}CHR_2\text{–}COOH \rightarrow NH_2\text{–}CHR_1\text{–}\boxed{CO\text{–}NH}\text{–}CHR_2\text{–}COOH$$

$$NH_2\text{–}CHR_1\text{–}\boxed{CO\text{–}NH}\text{–}CHR_2\text{–}COOH + NH_2\text{–}CHR_3\text{–}COOH \rightarrow NH_2\text{–}CHR_1\text{–}\boxed{CO\text{–}NH}\text{–}CHR_2\text{–}\boxed{CO\text{–}NH}\text{–}CHR_3\text{–}COOH$$

Fig. 4.2 Forming peptide bonds. R_1, R_2, and R_3 represent different amino acid side-chains (Table 16.2)—for example, R_1 is CH_3; this amino acid is called alanine; two alanines (R_1 and R_2 are CH_3) give rise to alanylalanine; and by adding a third alanine (R_3 is CH_3), we obtain alanylalanylalanine (Ala–Ala–Ala or **AAA**)

4.2.5.2 Sequences and Genes

PC1

The human *PC1* gene (other names: neuroendocrine convertase 1 (NEC1); prohormone convertase 3 (PC3)) is found on chromosome 5 at locus 5q15–21 and holds 14 exons. Its promoter is preferentially stimulated by cAMP, and also by, for example, CRH. This suggests coordinated activation of the hormone precursor proopiomelanocortin (POMC) and of its processing enzyme PC1.

The protein PC1 is a serine protease of the subtilisin/kexin type.[1]

PC2

Twelve exons of the human *PC2* gene are distributed on chromosome 20 (20p11.2). The protein precursor is formed in an inactive form and requires for its activation the coexpression of the protein 7B2 (SGNE1). Defects in either of these two genes result in hypoglycemia, hyperinsulinemia, and hypoglucagonemia, indicating that these enzymes participate in insulin precursor processing. The pathological effects are more pronounced when 7B2 is defective compared with PC2 defects.

4.2.5.3 Properties and Physiology

Prohormone convertases cleave an inner peptide bond of polypeptides. Thus, they belong to the large group of endopeptidases. Some endopeptidases cleave the bond between any two amino acids, for example, proteinase K. Others such as trypsin or chymotrypsin recognize single amino acids and cleave the polypeptide chain after these monoamino acid motifs. Prohormone convertases 1 and 2, however, recognize diamino acid motifs with lysine (**K**) and arginine (**K**).[2] The dibasic amino acid motifs are **KK**, **KR**, **RK** and **RR**.

While PC1 preferentially cleaves behind the motif **KR**(= lysyl-arginyl)[3] all four motifs are recognized and cleaved by PC2. During processing of the POMC precursor this is most important. POMC gives rise to different peptides depending on the PC active in a cell. In addition to adrenocorticotropic hormone (ACTH) β-lipotropin (β-LPH), γ-LPH, β-endorphin, and three distinct melanocyte-stimulating hormones (MSH) are formed by alternative splicing. A cell with only PC1 derives only ACTH and β-LPH from POMC-like corticotropic cells of the pituitary. Other cells in the brain express PC2. These cells form γ-LPH, β-endorphin and MSHs (Figs. 4.17 and 4.19).

The POMC example shows that PC1 or PC2 may cut a precursor chain severalfold. The neuropeptide TRH, for example, from the hypothalamus which induces TSH release in the pituitary exists in six copies in the TRH precursor. Each copy of the peptide sequence **QHPG** is preceded by a **KR** motif and followed by a **KR** or **RR**

---

[1]Subtilisin is a protease of *Bacillus subtilis*, kexin the prohormone convertase from *Saccharomyces*.

[2]Amino acids names using the single-letter code are always printed in bold letters of equal spacing.

[3]Amino acids within a polypeptide chain are called, for example, lysyl or arginyl.

mature TRH: **pEHP-NH_2**

pre-pro-TRH: (human TRH precursor)

```
MPGPWLLLAL ALTLNLTGVP GGRAQPEAAQ QEAVTAAEHP GLDDFLRQVE
RLLFLRENIQ RLQGDQGEHS ASQIFQSDWL S
KR:QHPGKR: EEEEEEGVEEEEEEEGGAVGPH
KR:QHPGRR: EDEASWSVDVTQH
KR:QHPGRR: SPWLAYAVP
KR:QHPGRR: LADPKAQRSWEEEEEEEEEREEDLMPE
KR:QHPGKR: ALGGPCGPQGAYGQAGLLLGLLDDLSRSQGAEE
KR:QHPGRR: AAWVREPLEE
```

Fig. 4.3 Thyrotropin-releasing–hormone (TRH). The precursor sequence starts at **MPG** in the *upper line*, and ends in *line 8*. The motifs **KR** and **RR** are emphasized by *colons* which also indicates cleavage sites where Prohormone convertase 1 or Prohormone convertase 2 process the precursor. Amino acids are depicted by letters (See Table 16.2)

motif (Fig. 4.3). PC1 and PC2 produce six oligopeptides from the TRH precursor. TRH is essential for metabolic regulation. Multiplication of its sequence ensures that a single point mutation induces only a gradual loss thus protecting against a dominant TRH defect. Production of multiple copies of a peptide from the same precursor is economical and reduces the energy required for formation of ribosomal complexes and translational start because they are only to be complexed once.

4.2.5.4 Phylogeny

The mechanisms of hormone formation have not changed much from the very first days of primordial neuropeptides. Thus prohormone convertases are among the primordial enzymes of hormone formation, already found in invertebrates.

4.2.6 Monobasic and Dibasic Sequence Motifs in Invertebrates and Vertebrates

Viewing the many neuropeptide precursors of vertebrate and invertebrate species the common **KR** sequences are striking. These constitute by far the most frequent peptide motif recognized by prohormone convertase. **KR** is the PC1 motif. Much less frequent are the other three dibasic motifs **KK**, **RK**, or **RR** recognized by PC2. Sometimes (more in invertebrates, less in vertebrates) we find motifs for furin-like peptidases **RxxxxR** with two to four variable (= **x**) amino acids. Very rarely there are monobasic **K** or **R** cleavage sites where in mammals trypsin or chymotrypsin would cleave the chain.

Veenstra (2000) and Southey et al. (2008, 2006) have reviewed that **KR** sites are always used whereas**RR**, **KK**, or **RK** sites are less frequently used. The utilization of monobasic recognition sites is not yet understood because the identity and even more the specificity of enzymes in different taxonomical orders are far from being fully understood. Sometimes furin-like and dibasic sites in the same precursor are used: for example, the short neuropeptide F (sNPF) precursor from the mosquito Anopheles gambiae (Fig. 5.31) is cleaved into five, from the same extract chemically identified oligopeptides. Three of these are cleaved after a dibasic, however, two of them in a furin-like motif.

4.2.7 Chopping the C-Terminus

By cleaving the TRH precursor chain at the **KR** motif, the maturation process is not yet finished. In many cases this peptide is still nonfunctional. Comparing the different TRH-cleavage products with mature TRH, we observe that they are still extended at the C-terminus.

Other peptidases than those described thus far will now chop off all amino acids from the C-terminus until they encounter a glycine residue. Glycine cannot be removed by these enzymes. Thus **QHPGKR** will be left from **QHPG**, however, **QHPG** will also remain from **QHPGRR:EDEASWSVDVTQH** because there is no glycine but that in position 4; all the other amino acids will be sequentially removed from the C-terminus. After similar processing of the other four oligopeptides six **QHPG** peptides are present.

4.2.8 Oxidation of the Terminal Glycine

The peptidyl glycine-α-amidating monooxygenase (PAM) oxidizes the terminal glycine into an amide residue (Fig. 4.4). At first the α-C atom of glycine will be oxidized. This reaction is only possible in glycine with its two hydrogen atoms at the Cα atom. The second step involves removal of glyoxal and leaves the NH_2 function. Because this is coupled to a carbonyl double bond the structural name is amide. Amides are less prone to chemical attack than amino groups. Amidation of the C-terminus increases the overall survival of a peptide in the body where many enzymes are ready to digest a lonely peptide.

Remember that precursor sequences such as peptide-**GxxKR** peptide-**GxxRR**, peptide-**GxxRK**, or peptide-**GxxKK** will result in a peptide-amide at the C-terminal end of the hormone (xx indicates small or larger peptide sequences and may also be missing).

$$\text{Polypeptide: } \boxed{\text{CO–NH}}\text{–}CH_2\text{–COOH} \rightarrow$$

$$\text{polypeptide: } \boxed{\text{CO–NH}}\text{–CHO–COOH} \rightarrow$$

$$\text{polypeptide: CO–}NH_2 \quad + \quad \text{OHC–CHO}$$

Fig. 4.4 Amidation of the C-terminal glycine

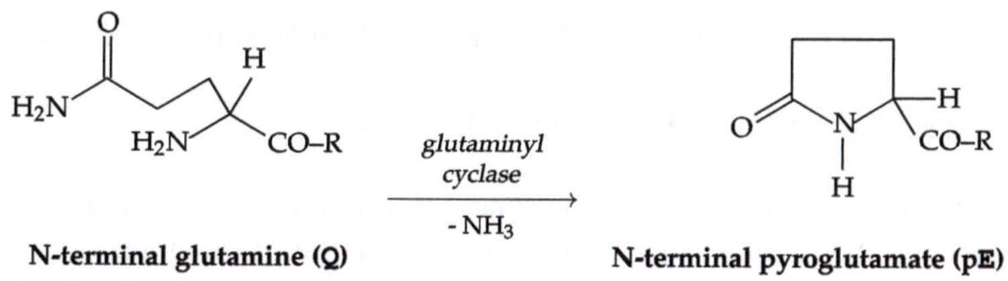

Fig. 4.5 Cyclization of the N-terminal glutamine

4.2.9 Cyclization of the N-Terminal Glutamine

The perseverance of a peptide hormone in the circulation will be further enhanced by one additional step of hormone maturation: the N-terminal glutamine (**Q**) will undergo intramolecular cyclization giving rise to a N-terminal pyroglutamic acid group (p**E**; Fig. 4.5).

The neuropeptide gonadotropin-releasing hormone thus loses its last free amino group. Such a peptide, especially if no lysyl residue is present that has an additional ϵ-amino group, is better armed against enzymatic degradation. Given that GnRH need only survive for a little more than 2 cm of bloodstream its half-life of 5 min in blood is sufficiently long enough to ensure that the receptors on the pituitary cells get triggered.

4.2.10 Esterification of Ghrelin

Without precedent among secreted peptides is esterification of ghrelin by octanoic acid. *O*-Acyltransferase—that is, the enzyme that transfers octanoic acid to the hydroxyl group of serine at position 3 of ghrelin—has been identified (Yang et al. 2008; Gutierrez et al. 2008).

Other peptides with long-chain fatty acid substitutions have these at their N-terminus or at the free lysyl amino groups. The reversible transfer of palmitic acid to the cysteine of guanosine nucleotide-binding proteins (G proteins) while forming a thioester bond suggests a special function for this modification. The acylated G-protein complexes associate with membranes and may thus facilitate hormone receptor interactions (see also Sect. 8.2.1). Gene activation by acetylation of histones also belongs to these mechanisms. β-Endorphins and α-melanocortins are also N-terminally acetylated.

Apart from octanoic acid other fatty acids including decanoic acid and its unsaturated decenoic acid have also been found as substituents of ghrelin. We would assume that further chain elongation might result in strong unspecific interactions of ghrelin with any membranes. A hormone with such long-chain fatty acids will never reach its receptor because it previously got stuck somewhere.

4.3 Peptide Hormones of the Hypothalamus and the Brain

4.3.1 Hypothalamic-Releasing Hormones

GnRH, TRH, CRH, GHRH (Table 4.1): These four neuropeptides stimulate release of hormones in the pituitary: GnRH induces the release of the gonadotropins, LH and FSH, TRH of thyrotropin (thyroid-stimulating hormones, TSH), CRH boosts corticotropin (ACTH; adrenocorticotropic hormone) release, and GHRH stimulates growth hormone (GH; older term somatotropin) secretion. After being formed in neurosecretory cells of the hypothalamus (see Sect. 10.2.1), the four neuropeptides are transferred via axonal transport into the median eminence where they will be released by appropriate stimuli. The blood capillaries will be reached by diffusion through little windows in the capillary wall. By direct transport through a portal system the four releasing hormones reach the anterior pituitary and their cellular targets leaving the capillaries again in fenestrated areas.

Such fenestrated passages between brain cells and blood vessels are called neurohemal organs. Usually the vessels in the brain are covered with a thickened layer of cells, the blood–brain barrier (BBB); in neurohemal organs the BBB is missing and a direct transport of hormones into and from the blood is permitted.

Released in the median eminence the four neuropeptides reach the pituitary straight via a portal system. The distance is not much larger than 2 cm. During this short passage the peptides are stable. In the pituitary there are again fenestrated capillaries allowing the hormones to reach the receptors on the target cells.

4.3.1.1 TRH

Introduction

TRH was the first hypothalamic neuropeptide whose structure could be determined in 1969 (Boler et al. 1969; Burgus et al. 1969). About 500 tons of sheep brain were used to extract the peptide and identify the structure pyrGlu-His-Pro-NH_2. Compared to usual peptides TRH shows three distinct characteristics:

1. It is very short, only a tripeptide.
2. The C-terminus is amidated.
3. The N-terminus is a pyroglutamic acid.

Table 4.1 The hypothalamic-releasing hormones (RH)

Name	Abbreviation	Sequence
Corticotropin RH	**CRH**	**SEEPPISLDL TFHLLREVLE MARAEQLAQQ AHSNRKLMEI I**-NH_2
Thyrotropin RH	**TRH**	**pEHP**-NH_2
Gonadotropin RH	**GnRH**	**pEHWSY GLRPG**-NH_2
Growth-hormone RH	**GHRH**	**YADAIFTNSY RKVLGQLSAR KLLQDIMSRQ QGESNQERGA RARL**-NH_2

Fact sheet 4.1: Thyrotropin-Releasing Hormone

Gene:	Chromosome 3; locus 3q13.3-q21; three exons.
Sequence:	**pEHP-NH$_2$**.
Synthesis and target:	TRH is preferentially synthesized in the paraventricular nucleus and acts via the median eminence on thyrotropic and lactotropic cells of the pituitary. TRH is also active as a neurotransmitter in many neurons.
Function:	Releasing hormone for thyrotropin and prolactin; major stimulator of metabolism; controls thyroid gland functions; equally active as neurotransmitter.
Receptor:	Heptahelical GPC membrane receptor.

Inasmuch as TRH was the first neuropeptide whose structure was determined, these features were very new; TRH appears to be the proverbial needle in the haystack to be looked for. The problems the protagonists in the race for the first neuropeptide structure, Schally and Guillemin, encountered can be studied in the book by Crapo (1985)

Biochemistry and Genes

On chromosome 3 (3q13.3-a21) the singular gene for the TRH precursor was found to contain three exons. After splicing and translation, the precursor contains several copies of the **QHPG** sequence; the **KR** prohormone convertase 1 recognition site is present several-fold, too. By PC1 the precursor is cleaved and the several precursor peptides then undergo maturation to the final TRH: p**EHP**-NH$_2$ (see Sect. 4.2). There are six copies of the **QHPG** sequence in the TRH precursor in humans, five in rats, and seven in frogs.

Physiology

TRH is the major regulator of the thyroid hormone and thus of energy homeostasis. In “lower” vertebrates TRH functions as a neurotransmitter because these animals do not synthesize thyrotropin. This neurotransmitter function is also retained in mammals, independently of the hypothalamic–pituitary–thyroidal axis.

Apart from the hypothalamus pro-TRH is synthesized in many brain regions: in the reticular nucleus of the thalamus, in the cerebral cortex, in pyramidal cells of the hippocampus, in external “plexiformal” layers of the olfactory bulb, in the sexually dimorphic nucleus, in the preoptic area, in the supraoptic nucleus, and in the substantia nigra, as well as in the pineal gland and the spinal cord.

Nonneural tissues where TRH is expressed are the mammalian pancreas and normal thyroid tissue. Frogs express TRH in their skin.

Human TRH regulates a circadian TSH rhythm with maximal release at midnight and minimal concentrations in the late afternoon. There are additional ultradian TSH peaks in 2- to 4-h intervals (see also Chap. 12). These rhythms are controlled by the suprachiasmatic nucleus and other cerebral pacemakers (Chap. 12). The limbic system, the pineal gland, and CNS regions involved in stress responses (Sect. 11.2.1) co-influence the pulsatile TRH/TSH release.

Catecholamines are further important regulators of hypothalamic TRH neurons: α_1-adrenergic neurons from the brainstem activate hypothalamic TRH neurons. Noradrenaline induces in vitro TRH secretion and dopamine inhibits TSH release. Application of the tyrosine hydroxylase inhibitor α-methyl-*p*-tyrosine diminishes the TSH release triggered by chilling (compare catecholamine biosynthesis, Fig. 7.1).

Endogenous opioids as well as somatostatin block TRH release; the latter blocks TSH release as well.

By glucocorticoids TRH mRNA transcription is directly blocked and by stimulating somotastatin indirectly. Dexamethason, a synthetic glucocorticoid, however, stimulates TRH transcription. In vivo such directly stimulating effects are counteracted by inhibitory neural influences from, for example, the hippocampus.

In its role as neurotransmitter TRH is involved in thermoregulation and in the amplification of noradrenergic and dopaminergic effects. By stimulating the preoptic area a direct influence on the regulation of body temperature is exerted. While activating the thyroid gland and thus metabolic activity, TRH indirectly enhances the body temperature and the activity of *sympathetic neurons* in the brainstem and the spinal cord.

Phylogeny

TRH is a characteristic vertebrate hormone. In agnathans (hagfish and lampreys) TRH positivity has been observed by immunocytology. Related peptides have been observed in lancelets and echinoderms: p**ESP**-amide in lancelets and p**EWP**-amide and p**EYP** together in a common precursor protein. Multiple copies of the sequences are found as in the human TRH precursor. Teleosts, frogs, birds, and mammals all express one homologous gene in the brain with the translated sequence **QHPG**. Maturation to the active p**EHP**-NH_2 is found in all these vertebrates. In *Xenopus laevis*, a second gene was identified that also shows seven peptide copies. This gene, however, has a different promoter and is predominantly expressed in the frog's skin. At least one of the TRH receptors is expressed in the *Xenopus laevis* skin, which suggests that color adaption to the environment might be regulated by TRH.

4.3.1.2 CRH

Introduction

An adequate response to stress in mammals depends on a functional hypothalamic–pituitary–adrenal axis (HPA). CRH, its receptors on corticotropic cells in the pituitary, ACTH released by these cells and its receptors, together with cortisol synthesis and release in the adrenal constitute this HPA. The indispensable role of CRH was demonstrated in the analysis of children suffering from congenital isolated adrenocorticotropic hormone deficiency where an abnormal CRH gene structure or expression was observed.

Fact sheet 4.2: Corticotropin-releasing hormone

Gene:	Chromosome 8; locus 8q13; two exons
Sequence:	**SEEPPISLDL TFHLLREVLE MARAEQLAQQ AHSNRKLMEI I**-NH_2.
Synthesis and target:	CRH is predominantly synthesized in the paraventricular nucleus, released in the median eminence and targets corticotropic cells of the pituitary; CRH is released from the placenta.
Function:	Releasing hormone for ACTH; central regulator of neuroendocrine reactions and behavior in response to stress; during gestation a potential indicator of preterm delivery.
Receptor:	Two heptahelical GPC membrane receptors, CRHR1 and CRHR2, with alternatively spliced products.

Biochemistry and Genes

CRH is derived from a preprohormone in a classical way processed as shown in the earlier chapters. The amidated C-terminus is a prerequisite for CRH receptor binding whereas the N-terminus is not required. Thus N-terminally shortened CRH peptides such as the CRH-9–41 peptide are fully biologically active. Oxidation of the methionine residue in position 38 destroys any biological activity which is a way as to inactivate CRH. The human CRH gene is found on chromosome 8 (locus 8q13) (Kellogg et al. 1989).

Physiology

CRH and vasopressin are the primary hormonal regulators of the human stress response. The observation of CRH and its receptors in the brain region apart from the hypothalamus, for example, as in the limbic system, in the central, stimulating sympathetic system of the brainstem and the spinal cord suggest this role. Intracerebral injection of CRH in animals leads to a coordinated sequence of physiological and behavioral reactions. These comprise:

- Activation of the hypothalamic–pituitary–adrenal axis
- Activation of the system of the nervus sympathicus
- Enhanced alertness
- Suppression of feeding and sexual activity
- Hypothalamic hypogonadism
- Changes in locomotor activity

These items characterize the usual behavior when stressed.

There are additional allies of this response which function as important regulators of corticotropic cells. A mutual positive interaction exists between CRH and vasopressin (AVP) release in the hypothalamopituitary unit: AVP induces CRH release and CRH stimulates AVP release. Without stress the pulses of these two hormones are more than 80 % overlapping. During stress the amplitude enlarges and if magnocellular AVP neurons are involved a continuous increase of the AVP level in plasma is observed.

CRH as well as AVP are released after stimulation by catecholamines (dopamine, noradrenaline, and adrenaline). AVP/CRH neurons on the one hand and the locus coeruleus plus noradrenergic neurons of the central stress response system are intimately mutually innervated and are regulated by the same factors in parallel. There are some ultrashort feedback loops by CRH on CRH neurons and by noradrenalin on noradrenergic neurons (Strakis and Chrousos 1997).

CRH and noradrenergic neurons as well are triggered by serotonin (5-HT) and acetylcholine and inhibited by corticosteroids and by the neurotransmitter γ-aminobutyric acid (GABA). The peptides derived from POMC and released after CRH stimulation in the pituitary such as ACTH, α-MSH, β-endorphin and further opioids such as dynorphin, exert a feedback inhibition on CRH and on noradrenergic neurons. Intracerebral injection of noradrenaline upregulates CRH, AVP, and ACTH release in the CNS, but not the ACTH secretion in the pituitary. Catecholamines, therefore, influence brain regions that are upstream of the pituitary functions and thus enhance AVP and CRH release.

AVP and CRH neurons additionally release products of the dynorphin gene together with AVP or CRH. These products including β-endorphin are potent endogenous opioids and suppress AVP and CRH effects on target cells.

CRH-Binding Protein

In addition to the CRH receptor there is a plasma CRH-binding protein (CRH-BP) with high affinity for CRH. Binding to this binding protein blocks activation of the CRH receptor by CRH. The binding protein is not related to the receptor. It has been found in the CNS, in placenta, in the amnion liquid, and in human plasma. In mice and cows, however, the binding protein has not been identified in plasma. Expression of the CRH-BP in the brain modulates reactions to stress. Deletion of the CRH-BP gene in mice increases anxiety in these mice but not in control animals. As mice do not have CRH-BP in their plasma or in the adrenal glands, there is no effect of the deletion in modified mice (Karolyi et al. 1999).

Phylogeny

Vertebrates possess three further CRH-related genes involved in the response to stress: CRH/CRF, urocortin/urotensin I, stresscopin (SCP)/urotensin III, and stresscopin-related peptide (SRP)/urotensin II. Between mammals and teleosts sequence homology is above 96 % for CRH/CRF, and above 55 % for SCP. There are similar precursor proteins and derived peptides in insects. This suggests that fight-or-flight responses and the handling of stress might have been present early in chordate evolution and that the two CRH receptors have mediated these responses.

Human urocortin (40 amino acids long and coded for on chromosome 2) preferentially binds to CRH receptor 2. Its effect is diminishing appetite and not to mediate stress. Stresscopin (40 amino acids) and stresscopin-related peptide (coded for on chromosome 3 (3q21.3; 43 amino acids)) display similar reactions. This might indicate that CRH mediates immediate reaction to stress by inducing cortisol synthesis and release and managing the metabolic changes depends on CRH, but equally on urocortin, stresscopin, or SCP.

The homology between vertebrates and insects extends to the CRH binding protein (Huising and Flik 2005). This extends the discussion of whether CRH might have already been present in common ancestors of insects and vertebrates.

4.3.1.3 GnRH

Fact sheet 4.3: Gonadotropin-Releasing Hormone

Genes:	GnRH-I: Chromosome 8; locus 8p21-p11.2; four exons GnRH-II: Chromosome 20 ; locus: 20p13; four exons
Sequences:	GnRH I: p**EHWSY GLRPG**-NH_2 GnRH II: p**EHWSH GWYPG**-NH_2.
Synthesis and target:	GnRH-I is expressed in several nuclei of the hypothalamus, released in the median eminence and triggers pituitary gonadotropic cells. Additionally, GnRH is expressed in trophoblastic cells of the placenta, to a lesser degree by T-lymphocytes. GnRH-II is preferentially expressed in kidney, bone marrow, and prostate; in the brain in the caudate nucleus, the hippocampus, and the amygdala.
Function:	GnRH-I: releasing hormone of LH and FSH; central regulator of reproduction; during gestation stimulator of choriogonadotropin . GnRH-II: found to enhance ovarian cancer invasiveness.
Receptor:	One GPCR: GnRHR1 (the human *GnRHR2* gene is a pseudogene).

Introduction

GnRH-I is the major regulator of vertebrate reproduction. Its sequence is identical for almost all mammalian species (with the known exception of guinea pigs). Other vertebrates possess this mammalian GnRH-I such as certain teleost fish or frogs. Another peptide, GnRH-II, very similar to GnRH I, is labeled according the species where it was identified: chicken-II. In fish the first GnRH is seabream GnRH (sbGnRH), the second GnRH-II, also known as chII-GnRH, and a third GnRH-III, also known as salmon-GnRH (smGnRH). Although GnRH-II is predominantly expressed in the forebrain of fish, the other two GnRH are found in midbrain. sbGnRH is the hypothalamic hormone.

Biochemistry and Genes

The human genes on chromosomes 8 and 20 are similarly organized, the introns being larger in the GnRH-I gene on chromosome 8. Alternative splicing of the GnRH-II mRNA enlarges the polypeptide by several amino acids; this alternative splicing appears tissue specific (White et al. 1998).

Using the GnRH precursor the decapeptide GnRH is formed by the sequential action of the following enzymes:

1. the signal peptidase
2. the prohormone convertase PC1

3. the exopeptidase E
4. the peptidylglycyl α-amidating monooxygenase (PAM)
5. the glutaminyl cyclase

A large collection of synthetic structural analogues was instrumental in identifying structure–function relationships:

- N- and C-terminus are required for receptor binding.
- Amino acids (AA) 1–4 are necessary to release LH or FSH.
- The side-chains of His^2–yr^5–Arg^8 are essential for full biological activity.
- Replacement of Arg^8 decreases LH and FSH secretion.
- Changing Gly^6 for Leu^6 influences the capacity for LH secretion in a more profound way than the activity to release FSH.[4]
- The secondary structure of all GnRH peptides is conserved, because the β-turn, formed by amino acids 5–8, induces a hairpin loop required for receptor binding.

Physiology

Mammalian reproduction depends on secretion of hypothalamic GnRH in all species analyzed thus far. Its regulation is controlled by multiple set points, hormones, neurotransmitters, and regulator circuits (see also Sect. 11.3).

Most critical for functional GnRH activity is pulsatile release of GnRH. Without this periodic and (only in adults) fully adjusted release pattern, any LH and FSH secretion does not take place; on the contrary continuously elevated GnRH levels in the blood lead to LH and FSH suppression. This constitutes one mechanism of contraception (see also Sect. 11.3).

During fetal development, GnRH-positive neurons migrate from the olfactory bulb to the hypothalamus. Any disturbance of this migration results due to missing hypothalamic GnRH neurons in infertility frequently combined with olfactory defects (Kallmann syndrome).

The physiology of GnRH-II[5] is far from being understood. Due to its conserved structure for 500 million years a critical role is suggested. On the other hand, in cow and sheep, the gene is present as well as the gene for the GnRH2 receptor. However, there is a mutation in the cow sequence that prohibits receptor binding and the sheep gene harbors a premature stop codon that abolishes any GnRH-II synthesis. The human GnRH2 receptor is afunctional owing to a frameshift mutation. GnRH-II binds the GnRH1 receptor, however, with a signal induction different from GnRH-I induction. An important functional role is thus not evident at all.

The teleost GnRH-III is reported to be expressed in the forebrain whereas GnRH-I and GnRH-II have been found in the diencephalon. The GnRH-III neurosecretory cells are located close to the nervus terminalis, not far from the olfactory bulb. Their axons reach the retina. It has been suggested that GnRH-III may control

[4] Genazzani et al. (1996).

[5] GnRH-I is either mGnRH, chGnRH-I, or sbGnRG; GnRH-II is always chGnRH; GnRH-III is smGnRH.

pattern recognition in animals ready to mate. It is worth noting that GnRH-I neurons originally were observed in the very same brain region. These latter, however, migrate to the hypothalamus whereas the GnRH-III neurons stay in the forebrain. In vertebrates other than teleosts, a GnRH-III gene has not been found. There are further GnRH-like genes in agnathans; these are, however, not related to the GnRH-II gene.

Phylogeny

For a long time, it was assumed that GnRH peptides were characteristic vertebrate hormones. This assumption has been discarded. There are three GnRH peptides in lampetra, presumably evolutionarily older than chondrychthyes and osteichthyes or newer vertebrates; molluscs such as *Aplysia californica* and *Octopus vulgaris* were shown to form a GnRH-like peptide of 12 amino acids, with the insertion at the same place. *Ciona intestinalis* expresses nine GnRH peptides and a 16 amino acid long GnRH-like peptide that differs from all other peptides by an elongation at the C-terminus (Table 4.2). Reports in corals about GnRH activity that could release LH from teleost cells have not been corroborated by peptide sequencing or cDNA cloning (Twan et al. 2006).

Table 4.2 Sequences of GnRH variants with mammalian GnRH as a reference

Species	Abbreviation	Sequence	Reference
Mammalian GnRH	(mGnRH)	p**EHWSYGLRPG**-NH_2	Matsuo et al. (1971)
Chicken I	(chGnRH-I)	p-------**Q**---NH_2	Miyamoto et al. (1983)
Frog	(frGnRH)	p-------**W**---NH_2	Yoo et al. (2000)
Seabream (Sparidae)	(sbGnRH)	p-------**S**---NH_2	Powell et al. (1994)
Salmon	(smGnRH)	p------**WL**---NH_2	Sherwood et al. (1983)
White fish (*Coregonus*)	(whGnRH)	p------**MN**---NH_2	Adams et al. (2002)
Guinea pig	(gpGnRH)	p-**Y**----**V**----NH_2	Jimenez-Liñan et al. (1997)
Medaka (Japanese killifish)	(mdGnRH)	p----**F**--**S**---NH_2	Okubo et al. (2000)
Chicken II	(chGnRH-II)	p----**H**-**WY**---NH_2	Miyamoto et al. (1984)
Seawolf (Anarhichadidae)	(cfGnRH)	p----**H**--**N**---NH_2	Ngamvongchon et al. (1992)
Herring	(hgGnRH)	p----**H**--**S**---NH_2	Carolsfeld et al. (2000)
Dogfish	(dfGnRH)	p----**H**-**WL**---NH_2	Lovejoy et al. (1992)
Lamprey I	(lGnRH-I)	p--**Y**-**LEWK**---NH_2	Sherwood et al. (1986)
Lamprey II	(lGnRH-II)	p----**H**-**WF**---NH_2	GenBank ABE66462
Lamprey III	(lGnRH-III)	p----**HDWK**---NH_2	Sower et al. (1993)
Tunicate I	(tGnRH-I)	p----**DYFK**---NH_2	Powell et al. (1996)
Tunicate II		p----**LCHA**---NH_2	Powell et al. (1996)
Tunicate III		p-----**EFM**---NH_2	Adams et al. (2003)

(continued)

Table 4.2 (continued)

Species	Abbreviation	Sequence	Reference
Mammalian GnRH	(mGnRH)	p**EHWSYGLRPG**-NH_2	Matsuo et al. (1971)
Tunicate IV		p----**NQ-T**---NH_2	Adams et al. (2003)
Tunicate V		p-----**EYM**---NH_2	Adams et al. (2003)
Tunicate VI		p----**K--Y**---NH_2	Adams et al. (2003)
Tunicate VII		p-----**A-S**---NH_2	Adams et al. (2003)
Tunicate VIII		p----**LA-S**---NH_2	Adams et al. (2003)
Tunicate IX		p----**NK-A**---NH_2	Adams et al. (2003)
Tunicate X		p----**NWWI--APGYNG**-NH_2	GenBank AB219239.1: Ikemoto and Park (2006)
Octopus		p**ENY-F-N-WH**---NH_2	Iwakoshi et al. (2002)
Aplysia		p**ENY-F-N-WYA**--NH_2	Zhang et al. (2008a)

Tunicate GnRHs are from *Ciona productum* (I/II), *C. intestinalis*, or *C. savignyi* III–IX

The figure from Guilgur et al. (2006) was used as a template to generate a phylogenetic tree that includes nonvertebrate GnRH sequences.

Figure 4.6 shows the characteristic three GnRH types of fish. Some species have two chII-GnRH precursors, goldfish and carp probably indicating the duplication of the entire genome.

Figure 4.6 Phylogenetic tree of GnRH variants (Sect. 4.6). Precursor sequences from GenBank (query: "gonadotropin releasing hormone" minus "receptor") were imported into ClustalW clustered with the neighbor joining method and 1,000 bootstrappings. The resulting tree was sketched using `treedyn` and illustrated using `inkscape`.
Colorsg: *red lines*: Mammals; *orange lines*: reptiles and amphibians; *blue lines*: fishes; *green lines*: cyclostomata; *dotted black lines*: invertebrates

Abbreviations: ac: Aplysia GnRH; chI, chII: chicken GnRH I/II; ci: Ciona GnRH; cf: catfish GnRH; cp: guinea pig GnRH; fr: frog GnRH; hr: herring GnRH; lp: lamprey GnRH; m: mammalina GnRH; md: medaka GnRH; ov: octopus GnRH; sm: salmon GnRH; sb: sea bream GnRH; wf: white fish GnRH.

A. anser: greylag goose; *A. burtoni*: *Astatotilapia burtoni* (cichlidae); *A. californica*: California sea hare; *A. japonica*: Japanese eel; *A. sapidissima*: American shad; *A. schlegelii*: Japanese black sea bream; *C. auratus*: goldfish; *C. carpio*: carp; *C. clupeaformis*: lake whitefish; *C. gariepinus*: African sharptooth catfish; *C. intestinalis*: vase tunicate; *C. nebulosus*: *Cynoscion nebulosus*; *C. porcellus*: guinea pig; *D. labrax*: European seabass; *D. rerio*: zebrafish; *E. macularis*: leopard gecko; *G. australis*: pouched lamprey; *G. gallus*: chicken; *H. huso*: Beluga sturgeon; *H. sapiens*: Homo sapiens sapiens; *I. fossor*: northern brook lamprey; *I. unicuspis*: silver lamprey; *L. richardsoni*: western brook lamprey; *L. tridentatus*: pacific lamprey; *M. albus*: Asian swamp eel; *M. appendix*: American brook lamprey; *M. auratus*: Golden hamster; *M. cephalus*: flathead mullet; *M. gallopavo*: turkey; *M. mordax*: Australian lamprey; *M. mulatta*: Rhesus macaque; *M. musculus*: house mouse; *M. saxatilis*: Atlantic striped bass; *M. undulatus*: Atlantic croaker; *O. aries*: sheep; *O. bonariensis*: *Odonthesthes bonariensis*; *O. latipes*: medaka, Japanese killifish; *O. masou*: cherry salmon; *O. mossambicus*: Mozambique tilapia; *O. mykiss*: rainbow trout; *O. nerka*: sockeye salmon; *O. niloticus*: Nile tilapia; *O. tshawytscha*: Chinook salmon; *O. vulgaris*: common octopus; *P. major*: Japanese madai; *P. marinus*: sea lamprey; *P. notatus*: bluntnose minnow; *R. catesbeiana*: American bullfrog; *R. dybowskii*: Dybowski's frog; *R. norvegicus*: rat; *R. rutilus*: common roach; *R. sarba*: tarwhine; *S. aurata*: gilt-head (sea) bream; *S. fontinalis*: brook trout ; *S. jardinii*: gulf saratoga; *S. ocellatus*: red drum *S. murinus*: Asian musk shrew; *S. salar*: Atlantic salmon; *S. scrofa*: wild boar; *S. trutta*: brown trout; *T. natans*: rubber eel; *T. belangeri*: northern treeshrew; *T. rubripes*: Fugu; *T. nigroviridis*: green spotted puffer; *T. vulpecula*: common brushtail possum; *T. thynnus*: northern bluefin tuna; *V. moseri*: barfin flounder; *X. laevis*: African clawed frog

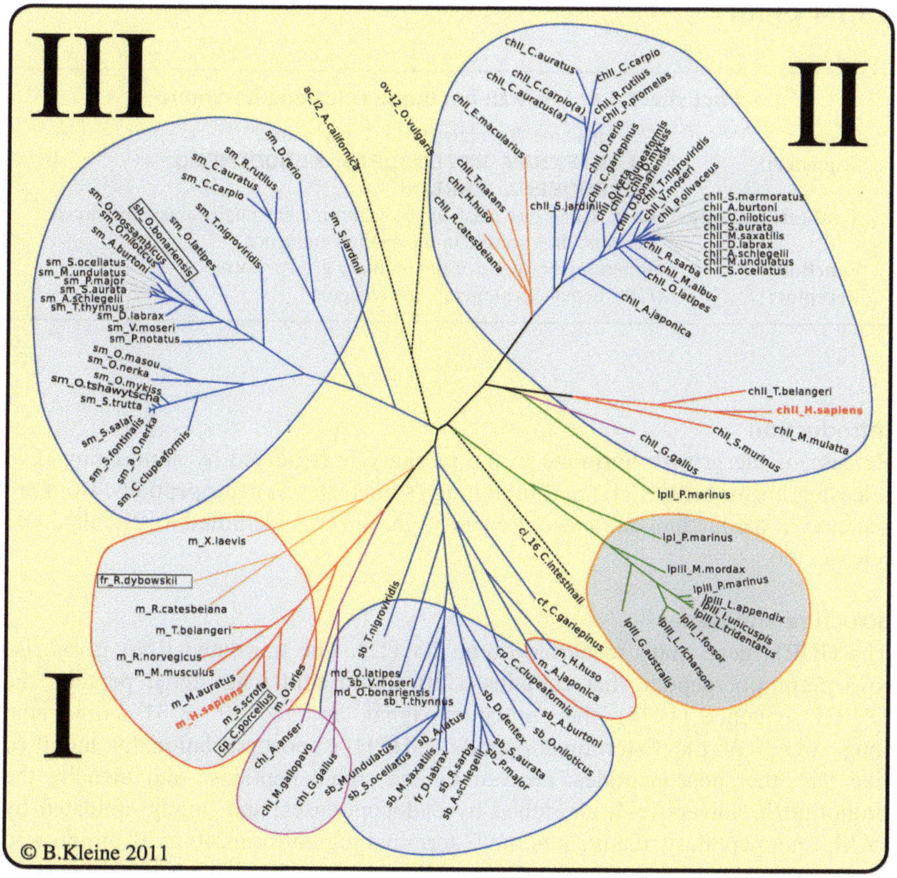

Fig. 4.6 Phylogenetic tree of GnRH variants

The ClustalW algorithm sorts the sequences first of all due to the sequence of the mature peptides (whereas the input comprised the whole precursor proteins). Further differentiation occurs due to differences in the remaining sequences, signal peptide, or associated peptide. There are additional differences separating chII-GnRH of mammals and birds. Although any variations of chII-GnRH have not been observed, smGnRH-II and GnRH (m-, chI- or sb-GnRH) show single amino acid exchanges (boxed in Fig. 4.6).

The pattern gains further complexity if we include the GnRH receptors. Up to five distinct GnRH receptor genes have been found (e.g., in the *Takifugu rupripes* genome (http://genome.jgi-psf.org/Takru4/Takru4.home.html) and in seabream (Moncaut et al. 2005). Some degree of tissue-specific differential expression of different receptor genes does not allow a definite association of receptor type and GnRH variant with any function. For this reason, the situation in mammals with the hypothalamic GnRH secretion and the derived pituitary gonadotropin release appears functionally clear. We discuss it again in Sect. 11.3.

4.3.1.4 GHRH

Fact sheet 4.4: Growth hormone releasing hormone

Gene:	Chromosome 20; locus 20q11.2; five exons.
Sequence:	**YADAIFTNSY RKVLGQLSAR KLLQDIMSRQ QGESNQERGA RARL**-NH_2.
Synthesis and target	GHRH neurons exist in the ventromedial nucleus and the arcuate nucleus; they secrete in the median eminence.
Function:	Releasing hormone of the pituitary growth hormone.
Receptor:	GPC heptahelical membrane receptor.

Introduction

Release of the growth hormone in the pituitary is regulated by stimulating (GH releasing hormone; GHRH) and inhibiting (somatostatin) neuropeptides. Both are secreted in the median eminence. Recently, ghrelin was identified to stimulate GH release, too.

Structure and Genes

The GHRH gene maps to chromosome 20q11.2. The translated RNA gives rise to a prepropolypeptide that contains a 30 amino acid long signal peptide, the GHRH sequence (1–44), the amidation signal, and the 30 or 31 amino acid long stretch of the C-terminal peptide. GHRH is posttranslationally modified like the other neuropeptides: cleaved by the signal peptidase and then by the prohormone convertase-1, shortened by endopeptidases, and finally amidated by PAM. Endopeptidase treatment at the C-terminal region forms 40 or 37 amino acid long, biologically active peptides. Further digestion to a 29 amino acid long peptide abolishes any biological activity.

Physiology

GHRH release is controlled by product feedback; it is growth hormone regulated. In the majority of brain regions the GH receptor and GHRH RNA were co-localized: in the hypothalamus, thalamus, septal region, hippocampus, dentate gyrus, or amygdala.

GHRH expression is higher in the hypothalami of male rats than in female hypothalami. This sexually dimorphic behavior is controlled by steroid hormones: Dihydrotestosterone (DHT) injection into ovariectomized rats masculinized their GH secretion. Injection of estradiol, however, diminished the GHRH secretion in male rats. In addition, the GH feedback control of GHRH release appears gender specific.

Those neurosecretory cells that release GHRH in the median eminence are found in the ventromedial nucleus and in the arcuate nucleus of the hypothalamus. They are interconnected with different CNS areas: signals from the "sleep center(s)" are stimulating and coupled to the sleep rhythm. Signals from the amygdala and from ascending noradrenergic neurons of the brain stem are related to the activation of the stress reaction. These mediate stress-induced GH release. The ventromedial nucleus processes the secretion of hormones involved in blood glucose regulation and thus influences the GHRH release in reaction to hypoglycemia (see also Sect. 11.4).

GH release is regulated by GRHR stimulation and somatostatin (SST) inhibition. Functional and anatomically reciprocal interactions exist between the ventromedial nucleus, arcuate nucleus, and paraventricular nucleus: endogenous SST inhibits GHRH secretion from median eminence, whereas intracerebral SST injection stimulates GHRH release. GHRH neurons of the arcuate nucleus express high-affinity SST receptors. In addition to SST regulation, circadian GHRH pulses are controlled by zeitgeber of the suprachiasmatic nucleus. This circadian GHRH rhythm is synchronized to the sleep rhythm: elevated GH secretion during sleep, reduced GH release while awake.

GHRH neurons are further influenced by other neurons and their neurotransmitters: sleep-induced GH release is modulated by serotonergic and cholinergic neurons. Circadian GH pulses mediated by GHRH may be inhibited by α-antagonists *i.e.* inhibitors of catecholamine α-receptors, or substances directly blocking catecholamine biosynthesis. β_2-Agonists, stimuli of the β_2-catecholamine receptor, induce GH release presumably by stopping SST secretion. Anticholinergic drugs inhibit all GH stimulating effects but hypoglycemia. L-DOPA[6] as well as dopamine increase GH release most probably due to their local conversion into noradrenaline.

Apart from SST other CNS neuropeptides interact with GHRH neurons and contribute to GH release:

- Endogenous endorphins, in particular β-endorphin, increases GHRH and GH release.
- TRH, injected in to rat brain enhances GH release by a Ca^{2+}-dependent, cAMP-independent mechanism. In humans, TRH injection increases GH levels only in patients with acromegaly.
- Galanin, motilin and NPY[7] stimulated GH secretion from isolated rat pituitary cells. A subgroup of GHRH neurons themselves expresses neuropeptide Y which *in vitro* appears to upregulate GH release. Applied into a cerebral ventricle NPY quenches GH secretion suggesting additional regulators of GHRH and SST neurons by the inhibiting ascending noradrenergic neurons from the brainstem usually stimulating GH secretion via GHRH.

Phylogeny

Phylogenetically GHRH is closely related to another neuropeptide, PACAP.[8] By gene duplication at the beginning of mammalian evolution two different genes for

[6] 3,4-Dihydroxyphenylalanine.

[7] Neuropeptide Y.

[8] Pituitary adenylate cyclase activating peptide.

GHRH and PACAP were formed. Nonmammalian vertebrates form GHRH and PACAP by alternatively splicing the same RNA precursor. It is worth noting that the PACAP sequence is more strongly conserved in mammals than is the GHRH sequence (Montero et al. 2000).

After the event of gene duplication the GHRH exon in the PACAP gene gave rise to the PACAP-related peptide (PRP) whereas the PACAP exon of the GHRH gene mutated to the C-peptide of GHRH. PACAP is, like GHRH, amidated.

PACAP has been reported as an additional inducer of noradrenaline secretion in the adrenal glands and as a mediator of the metabolic response to elevated blood glucose levels. PACAP knockout mice and flies with PACAP defects show behavioral disorders, in PACAP$^{-/-}$ mice the metabolite 5-hydroxyindoleacetate was reduced.

PACAP and GHRH belong to the family of secretin-like peptides.

4.3.2 Gonadotropin-Inhibiting Hormone

Fact sheet 4.5: Neuropeptide VF; RF-related peptide; aka gonadotropin-inhibiting hormone (GnIH)

Gene: Chromosome 7; locus 7p15; three exons.

Structure: See Fig. 4.7; three peptides with a similar C-terminus, LPLRFamide, LPQRFamide, and LPLRSamide.

Synthesis and target: Neuropeptide VF is expressed in dorsomedial neurons and in gonads and acts as a neurotransmitter preferentially on GnRH neurons in the median eminence.

Function: Inhibitor of GnRH release and of gonadal activity.

Receptor: Neuropeptide FF receptor 1; GPCR147 from the rhodopsin receptor family.

4.3.2.1 Introduction

Any preproproteins of human peptide hormones cleaved to release peptide hormones possess basic dipeptide motifs where either PC1 or PC2 may act. Later in this book we show that such motifs do exist in arthropods and thus most presumably already in common ancestors of vertebrates and invertebrates some 600 million years ago. There are, however, in flies, snails, or shellfish additional neuropeptides endocrine-active where monobasic recognition sites are used, however, close inspection showed that most of the peptides can be cleaved at recognition sites composed of the two basic amino acids **R/K** spaced 0, 2, 4, or 6 amino acids apart: $\mathtt{R/Kx}_N\mathtt{R/K}$ ($n = 0, 2, 4, 6$). The lack of usual or unusual cleavage sites directly raises the question of whether peptides where an N-terminal cleavage site is missing are active as neuropeptides if at all.

In Chap. 12 we deal with circannual rhythms. In all but a few species reproduction is coupled to the annual seasons. Any activity has its time, copulation, breeding, upbringing, hibernation, or bird migration. Whereas in humans female reproductive activity cycles with a period of 28 days and humans can be sexually active on any day, in birds, for example, such behavior inhibits successful raising of the litter and would commit to much energy required, for example, for migration or survival in the cold season. In most animals reproductive activity is related to the annual cycle of seasons in males and females, at least in nondomesticated ones. A gonadotropin-inhibiting hormone in men appears dispensable, but it might definitely have a role in wild animals.

GnRH release has been previously shown to be inhibited by some neurotransmitters, however, a negative regulator had not been shown in vertebrates. Tsutsui et al. (2000) decided to look for hypothalamic peptides blocking GnRH release and identified in quail a dodecapeptide doing just that. They called this peptide gonadotropin inhibitory hormone (GnIH). Because this peptide has never been shown released into the circulation we would call the naming premature. GnIH belongs to the RFamide family of peptides present in vertebrates and invertebrates. The human GnIH homologue is cleaved from the neuropeptide-VF precursor protein.

4.3.2.2 Structure and Gene

The gene for neuropeptide VF is located on the short arm of chromosome 7. GnIH and their mammalian homologues RF-related peptides (RFRP) are characterized by the C-terminal **LPxRF** (x = **L** or **Q**). The human neuropeptide VF precursor contains in contrast to the chicken one only two GnIH homologues, the RFRP-2 (on *blue* background) has a C-terminal serine (**S**) instead of phenylalanine (**F**) (Fig. 4.7).

RFRP have unusual prohormone convertase motifs in contrast to other vertebrate hormones in this book: C-termini possess a rarely used **Rx**$_N$**R** (n = 0, 2, 4, or 6) motif where the PC1 can act. Some singular arginines (**R**) might be recognized by other endopeptidases. However, inspecting the precursor sequences in this

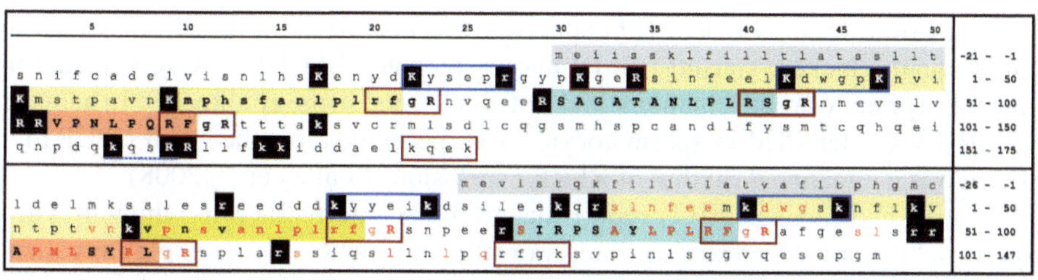

Fig. 4.7 Chicken gonatotropin inhibiting hormone (GnIH) and the human homologue, RFamide-related peptide (RFRP). The *upper* sequence displays the human preproprotein, the *lower* that of chicken. Signal peptides are on *gray* background, RFRP-1 on *yellow*, RFRP-2 on *blue*, and RFRP-3 on *orange* background. Monobasic and dibasic peptid motifs. are *inverted*. *Red* and *blue* frames indicated unusual PC1 motifs. Conserved amino acids are shown *red* in the chicken sequence. RFRP-1 is labeled by Tsutsui as a presumable **LPLRF**amide peptide due to the unknown cleavage mechanisms for the N-terminus (Tsutsui 2009) (Source: Swiss-Prot Q9HCQ7 (human) and GenBank BAE17049 (chicken))

book, hormonal activity does not rely on enzymes cutting after singular arginines. Whenever these are present in a precursor, there are multiple copies of peptide sequences and there are other peptides cleaved at dibasic sites of the **R/Kx**$_N$**K/R** with n = 0 the most obvious. Whether RFRP-1 is an active neurotransmitter is not clear and is obviously questioned by the most active group which called it only a presumable neuropeptide (Tsutsui 2009).

The GnIH/RFRP receptor is called in mammals the neuropeptide FF receptor (OT7T0222 or GPCR147) which is triggered by other RFamides, too.

In a very recent review, Tsutsui and Ubuka (2014) report an overview of GnIH in birds and mammals. They demonstrate that GnIH receptors on GnRH neurons influence the release of this hormone. What is actually blocked is the interaction of GnRHR with the $Gs_{\alpha\ s}$ subunit, which is responsible for activating adenylate cyclase and cAMP.

4.3.2.3 Physiology

The original observation in quail that GnIH directly inhibits pituitary LH/FSH release (Tsutsui et al. 2000) could not be confirmed in mammals. Isolated quail pituitaries where incubated with GnIH and showed reduction of LH/FSH secretion. Whereas quail GnIH is expressed in the paraventricular nucleus (Tsutsui et al. 2000) RFRP e.g. in rats it is expressed in the dorsomedial hypothalamus (Rizwan et al. 2009. RFRP axons originating in dorsomedial hypothalamus cell bodies reach in rats to the median eminence, however, not to its external border from where neuropeptides would be released into the hypothalamic–pituitary portal system (Rizwan et al. 2009).

Inhibition of gonadotropin secretion in mammals is achieved by inhibiting not the gonadotropin releasing cells in the pituitary but by blocking the secretion of GnRH in the pituitary. RFRP influences the membrane firing of GnRH neurons, mostly inhibitory: 40 % of the investigated neurons reduced ion channel openings when treated with RFRP-3, in 10 % of these firing was enhanced and about half of neurons remained unchanged (Ducret et al. 2009).

Apart from brain GnIH/RFRP expression has also been observed in gonads. In birds (quail, chicken, starling) GnIH and its receptor GPCR147 were detected in theca and granulosa cells, in interstitial testes cells, and in the epididymis. In hamster RFRP-3 was identified in spermatocytes and spermatids together with GPCR147. There was a circannual rhythm of RFRP expression (Bentley et al. 2008).

Tsutsui and Ubuka (2014) report that by RNA interference in white-crowned sparrow, "Birds reduced resting time, spontaneous production of complex vocalizations, and stimulated brief agonistic vocalizations. GnIH RNAi further enhanced song production of short duration in male birds when they were challenged by playbacks of novel male songs. These behaviors resembled those of breeding birds during territorial defense. The overall results suggested that GnIH gene silencing induces arousal." They have other facts about blocking influence on arousal.

GnIH and its mammalian analogues might thus fulfill a role in the circannual regulation of reproductive activity which, however, has not been sufficiently analyzed in order to give a general picture. Such regulation as mentioned above might be necessary for the survival of wildlife species. The fact that GnIH expression is induced by melatonin would fit (Ubuka et al. 2005) into the scheme, melatonin being the hormone used to estimate in molecular terms the duration of nighttime in mammals.

4.3.2.4 Phylogenesis

Thus far LPxRF amides have been observed only in vertebrates (mammals, amphibians, birds, and fish) and in hagfish, but not in lamprey, where a similar precursor was found, but with slightly modified peptides (Tsutsui and Osugi 2009; Tsutsui and Ubuka 2014).

4.3.3 Neuropeptide Y

Neuropeptide Y (NPY) neurons are broadly distributed a protein. The structure and functions are discussed in Sects. 4.10 and 11.5. In the hypothalamus NPY neurons localize predominantly to the arcuate nucleus. NPY release from there controls feeding and CRH release. In the periphery NPY is often formed in noradrenergic neurons.

4.3.4 Agouti-Related Protein

Fact sheet 4.6: Agouti-related

Gene:	Chromosome 16 ; locus 16q22; four exons
Sequence:	See Fig. 4.8
Synthesis and target:	AgRP is predominantly formed in the arcuate nucleus and controls feeding by binding to the melanocortin receptor 4 (MC-R4). This receptor is found on cells of the paraventricular nucleus, the dorsal motor nucleus of the vagus, and in the raphe nucleus, areas involved in energy homeostasis.
Function:	AgRP inhibits activation through MC4-R and thus permits enhanced feeding.
Receptor:	GPC heptahelical receptor: melanocortin 4 receptor.

4.3.4.1 Introduction

By discovering the *Agouti* gene regulation of skin pigmentation could be better understood. In humans the agouti protein is—in contrast to rodents—not restricted to the skin, but equally expressed in adipose tissue, in the testes and ovaries, in heart, and in kidney and liver (Dinulescu and Cone 2000). The agouti protein functions as an MSH antagonist at the melanocortin receptor I (MC-R1) in rodent melanocytes.

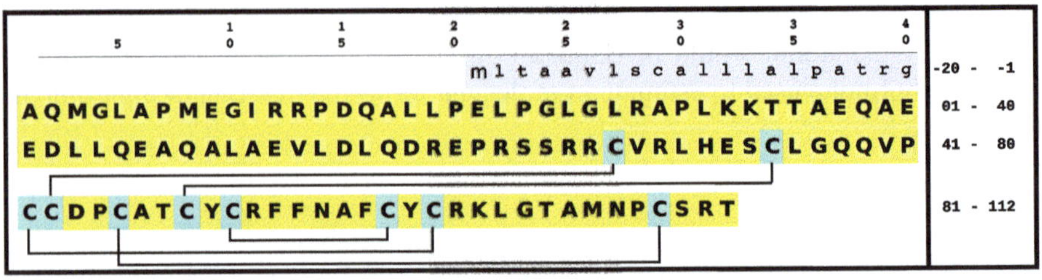

Fig. 4.8 Sequence and disulfide bridges of agouti-related protein. The signal peptide is shown in *lowercase*. The intramolecular disulfide bonds are depicted by *lines* between cysteine residues (Source: NP_001129; disulfide bonds were identified by Bures et al. 1998)

Agouti-related protein (AgRP; Fig. 4.8) has a similar antagonistic role as agouti, but at the hypothalamic MC-R4. It is involved in the regulation of feeding.

4.3.4.2 Structure and Gene

The AgRP gene was mapped to chromosome 16 (16q22). AgRP is 131 amino acids long, and its C-terminal region (82–131) is antagonistically active as is AgRP. The intramolecular cysteine bonds are functionally indispensable.

4.3.4.3 Physiology

Neurons in the arcuate nucleus produce agouti-related protein (AgRP). This protein is a specific antagonist of the melanocortin 4 receptor MC4-R. By inhibiting MSH association with the MC4-R and thus the suppression of feeding AgRP stimulates feeding. Mice with defects of MC4-R develop gluttony and adiposity.

4.3.4.4 Phylogeny

Until now, AgRP sequences are limited to vertebrates (Klovins et al. 2004). The proteins are characterized by 10 cysteine residues. It has not been possible thus far to delineate the divergence of agouti and AgRP.

4.3.5 Somatostatin

4.3.5.1 Introduction

The releasing hormones thus far mentioned mediate hormone secretion in the pituitary. In contrast to these hypothalamic somatostatin (SST or SRIF as *somatotropin release inhibitory* factor) blocks the pituitary release of the growth hormone. GH secretion is thus controlled by the balance of activating GHRH and inhibiting somatostatin. Other hypothalamic inhibitory peptides for any of the other hormones in the pituitary have not (yet?) been found, however, prolactin release is tonically suppressed by the catecholamine dopamine.

Fact sheet 4.7: Somatostatin

Gene: Chromosome 3; locus 3q28; two exons

Sequence: SS14: **AGCKNFFWKTFTSC**
SS28: **SANSNPAMAPRERKAGCKNFFWKTFTSC**
(see Fig. 4.9); an intramolecular disulfide bond forms a cyclic peptide; amino acids **FWK** are essential for receptor binding.

Synthesis and target: Hypothalamic somatostatin is formed mainly in the paraventricular nucleus, to a lesser degree in the arcuate nucleus or ventromedial nucleus. Its targets are somatotropic cells in the pituitary and GHRH neurons in the arcuate nucleus and ventromedial nucleus. Somatostatin synthesis in the gastrointestinal tract serves regulation of endocrine cells therein, often in a paracrine way.

Function: Somatostatin is an inhibitor of multiple endocrine functions.

Receptor: Five human somatostatin receptors have been identified. Their cell type specific expression may explain the divergent somatostatin effects on target cells.

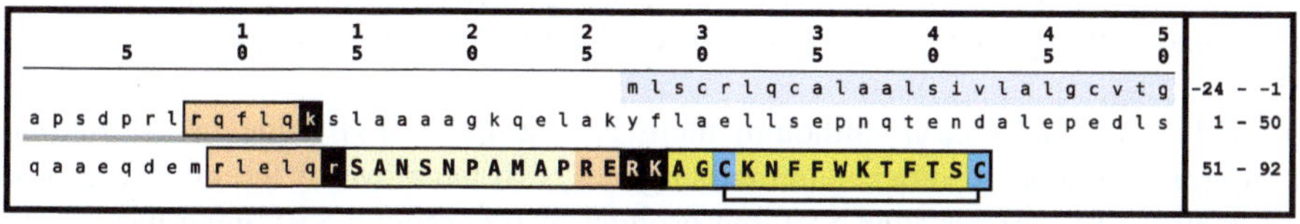

Fig. 4.9 The somatostatin precursor and its derived peptides. From the precursor prosomatostatin (PSS) furin cleaves off somatostain-28 (SST-28) and SST-14 apart from a short N-terminal peptide PSS(1–10) whereas by Prohormone convertase1 only SST-14 can be released. The two cysteines by generating an intramolecular disulfide bond form the ring structure of SST (Source: GenBank NP_001039)

In the search for GnRH, Burgus, Ling, Butcher, and Guillemin (1973) isolated from about 500,000 ovine hypothalami a cyclic tetradecapeptide inhibiting GH release from the pituitary. At the same time they were able to report the isolation of human somatostatin.

4.3.5.2 Structure and Gene

Somatostatin is derived from a precursor by proteolytic cleavages by either PC1 or furin (Fig. 4.9). PC1 can only cleave off the short SST-14 variant, whereas furin liberates the longer SST-28, SST-14 and an additional N-terminal peptide.

Expression of the somatostatin gene on chromosome 3 is controlled by stimulating signals increasing intracellular cAMP and by repressive influences of thus far unknown character.

4.3.5.3 Physiology

Somatostatin generation is not restricted to the hypothalamus. SST is an inhibiting agent of different endocrine and neuronal processes. In the GI tract SST attenuates multiple hormones (see Sect. 4.10); in the mammary glands milk ejection is suppressed. Apart from the hypothalamus SST neurons are located to other brain areas. Different SST functions do not originate from SST variants themselves (no functional differences found between SST-28 and SST-14). These different functions arise by differentially expressed somatostatin receptors that mediate type-specific signal transduction pathways and are specifically expressed by cell type on various target cells (Sect. 8.2.4).

The independence of the multiple SST functions is due to the short SST half-life in blood (below 3 min) and due its rapid inactivation. For therapeutic reason an SST agonist was developed with similar SST receptor binding but an enhanced life span in blood: octreotide (Fig. 14.1).

Pituitary GH release is controlled twice by SST. SST secretion into the median eminence will inhibit GH release by SST receptor-mediated suppression of GH release in somatotropic cells. The second inhibitory signal is through direct SST action on GHRH secreting neurons still in the hypothalamus (see Müller et al. 1999).[9]

4.3.5.4 Phylogeny

SST is present in a variety of invertebrates. The paracrine gastrointestinal regulation from and within pancreatic islets is, however, a vertebrate achievement.

- *SST gene duplication:* although there is a singular SST gene in the human chromosome as in other mammalian genes, fish have two different SST genes; compared to mammalian SST the product of this second SST gene has one to four amino acid exchanges (Sheridan et al. 2000).
- *Cortistatin:* In 1996 de Lecea et al. reported another peptide with strong homology to somatostatin: cortistatin (CST), which appears to play an important role for sleep regulation. The peptide homology is 10 of 14 amino acids; all residues involved in receptor coupling are conserved (Fig. 4.10). The cysteine forming the intracellular disulfide bond and thus the cyclic peptide are conserved as well. The cortistatin gene maps to chromosome 1 (1p36.22) and comprises two exons. Three different CST peptides have been isolated: CST-14, CST-17, and CST-29.

  CST is expressed in several tissues: in the cerebral cortex and in the hippocampus, furthermore in pancreas, gut, kidneys, testis, and leukocytes. The final proof, however, is lacking for some of these inasmuch as sometimes only the presence of RNA, but not of the CST peptide has been confirmed.

  Unlike SST CST binds not only to SST receptors but the growth hormone secretagogue receptor, too (GHS-R): this receptor was first observed more

```
DRMPCRNFFWKTFSSCK
    | ||||||| ||
   AGCKNFFWKTFTSC
```

Fig. 4.10 Cortistatin-17 and somatostatin-14: sequence comparison

[9]SST secretion is inhibited by the pancreatic polypeptide (PP) (section 4.10.6; Kim W, et al., FEBS Letters, 588:3233–3239)

than 20 years ago and has recently been identified as a ghrelin receptor. The MrgX2 (mas-related gene), first shown to mediate pain and nociception, binds CST as well (Robas et al. 2003). Proadrenomedullar peptides after binding to MrgX2 generate elevated blood pressure by inhibiting catecholamine release from sympathetic neurons or chromaffine medullary adrenal cells. In contrast to SST CST is expressed by several types of immune cells (Gonzalez-Rey et al. 2006) inhibiting endotoxin-induced cytokine release and thus protecting against lethal outcome of endotoxic shock.

In spite of these differences the endocrine functions of SST and CST are very similar with respect to central GH regulation, prolactin control, and GI-driven insulin release. The two peptides appear to be mutually restorable.

- *Somatostatin in invertebrates:* By immunological means somatostatin or somatostatin-like peptides (already with the disulfide bond) have been found in neurons of protostomes; in deuterostomes SST was equally found in neurons, and in the gut mucosa, too: in singular neuroendocrine cells in invertebrates, however, in vertebrates in the known Langerhans islets combined with insulin (and glucagon and PNP[10]; Conlon et al. (1988); Falkmer et al. (1985)).
- *SST family in vertebrates* Tostivint et al. have recently demonstrated that the genes for SST, CST, and urotensin II/urotensin-related peptide (UII/URP) arose by two gene duplications. The original precursor gave rise to a tandem of SST/CST or UII/URP genes. Such a tandem exists in very early vertebrates suggesting duplication early in development. The tandem may then be duplicated with the entire genome, an event which is timed to early fish evolution (Tostivint et al. 2006).

4.3.6 Substance P

Substance P (SP) is used in the brain to adapt to stress (Sect. 4.11.2).

4.3.7 Proopiomelanocortin

POMC neurons synthesize the β-endorphine contributing to the reaction to stress as well as α-MSH which is involved in control of food uptake. Alternative processing of POMC is described in Sect. 4.4.1.

4.3.8 Ghrelin

Ghrelin is also a mediator of food intake. It is discussed in Sect. 4.8.2.

[10]Pancreatic polypeptide.

4.3.9 Kisspeptin

Fact sheet 4.8: Kisspeptin

Gene:	Chromosome 1 (1q32); three exons
Sequence:	See Fig. 4.11
Synthesis and target:	Kisspeptin is formed by neurons in the arcuate nucleus and the paraventricular nucleus; it is released via synapses in the preoptic region and controls GnRH secretion.
Receptor	heptahelical GPC receptor: GPR54

4.3.9.1 Introduction

While studying tumor metastasis Lee et al. (1996) identified a protein fully blocking metastasis without inhibiting melanoma cell proliferation. GPR54 was identified as the receptor for this kisspeptin protein (Kotani et al. 2001). GPR54 knockout mice were viable but did not demonstrate sexual maturation which in turn led to kisspeptin's role in GnRH secretion.

4.3.9.2 Structure and Genes

The *KISS1* gene mapping to chromosome 1 has three exons, the first one noncoding.

Kisspeptins (Fig. 4.11) are derived from the primary KISS-1 protein by post-translational modifications. Aside from kisspeptin 54 the literature reports smaller kisspeptins with chain length of 10 to 14 amino acids. The peptidase for tissue-specific processing is not yet identified.

4.3.9.3 Physiology

Kisspeptins supposedly have two roles: they block metastases of tumor as well as of placenta cells and, centrally, they control GnRH release in the median eminence:

1. *Metastasis inhibiting function:* From the first description on, the number of reports on suppression of invasive tumor migration has been ever increasing. In

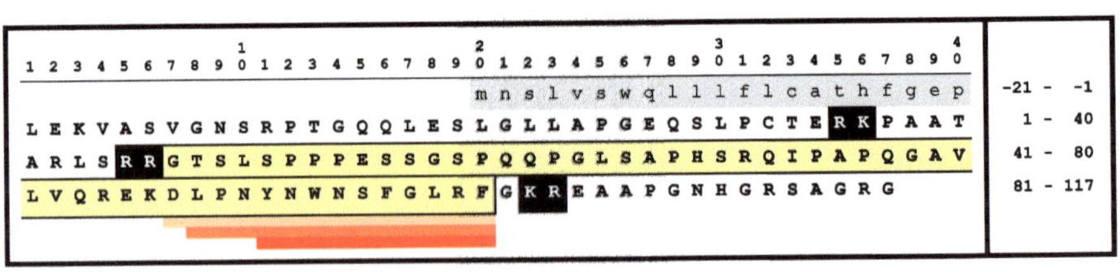

Fig. 4.11 Primary sequences of the KISS-1 protein and of derived kisspeptins. By Prohormone convertases the Kiss-1 gene product (*upper case*) is processed giving rise to the kisspeptin-54 *yellow*; after further C-terminal amidation, smaller kisspeptins 14, 13, and 10 (*orange to red*) are cleaved off by proteolytic digestion (Source: GenBank: NP_002247)

some tumors suppression of NF-κB translocation into the nucleus by kisspeptin was shown. In addition different signal pathways via protein kinase A or protein kinase C were shown. Blocking metastasis may be related to CXCR4 signal transduction; CXCR4 is seen as an important player in metastasis and in the interactions of cells with the environment (Navenot et al. 2005). In mice Bilban et al. (2004) have shown that kisspeptin and its receptor regulating trophoblast invasion into the maternal endometrium are predominantly expressed in early gestation: at term only the measurable Kiss RNA was 30 times less than in the third month of gestation.

2. *Regulation of GnRH release:* Kisspeptin is expressed in neurons of the arcuate nucleus and released in the paraventricular nucleus and the preoptic nuclei. Estradiol is known to control kisspeptin release; whether GABA or further neurotransmitters modulate kisspeptin secretion is thus far unknown. The control of the pulsatile secretion is further unknown. Neurons in the paraventricular nucleus are at least involved in the preovulatory LH peak; the regulation of this is not known either in women nor in any animal.

4.3.9.4 Phylogeny

Thus far kisspeptins and the kisspeptin receptor are found in vertebrates: primates, rodents, fish. **RF**amides, however, are among the primordial neuropeptides and found in those species with the earliest existence of neurons and neurosecretion.

Similarly kisspeptin during maturation of gonads was observed in fish as in man: male *D. rerio* express kisspeptin maximally during the first formation of sperm.

4.3.10 Galanin

Fact sheet 4.9: Galanin

Gene:	Chromosome 11; locus 11q13.2; six exons
Sequence:	see Fig. 4.12
Synthesis and target:	Galanin is released from CNS neurons, from neurons of the enteric nerve system, as well as by cytotrophoblastic cells of the placenta, and acts in endocrine and paracrine fashions on cells regulating feeding, insulin release, memory, and reproduction.
Function:	Galanin is a regulator of different aspects of homeostasis and of reproduction. Its presence is essential for maturation of mammary glands and of milk synthesis.
Receptor:	The three human galanin receptors are heptahelical GPC membrane receptors.

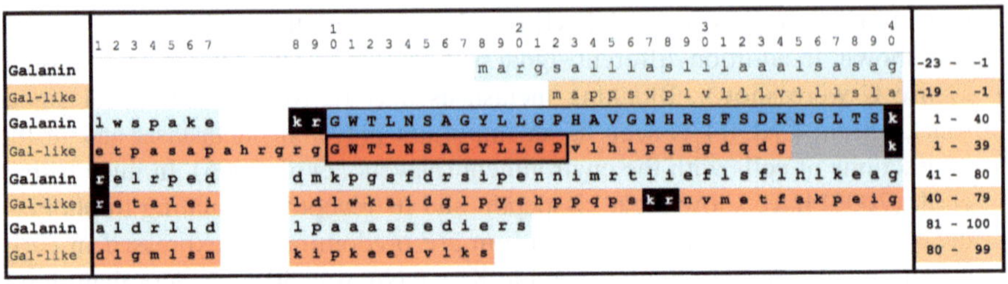

Fig. 4.12 The galanin prepproproteins: after removal of the signal peptide(*upper line*) the galanin peptide (*in uppercase* and *blue*) is liberated from the proprotein by Prohormone convertase1. Dibasic peptide motifs are shown *inversely*. An alternative galanin-related peptide lacks the first Prohormone convertase1 motif; additional spaces are added to superimpose the homologous partial galanin sequence and the second **KR** motif (Source: GenBank CAA01907 and NP_149097)

4.3.10.1 Introduction

Galanin was first identified as a gut peptide (Tatemoto et al. 1983). Later on it was shown to be expressed in multiple types of neurons. Its expression in the human placenta could also be shown (Kleine et al. 2001). Recent literature demonstrates divergent roles of galanin.[11] Experimentally proven is a direct relationship to uptake of a fatty enriched diet. The indispensable role of galanin for mammary gland maturation and function has already been shown before.

4.3.10.2 Structure and Genes

The preprotein is transcribed from the gene on chromosome 11 (close to a metalloproteinase gene). After removal of the signal peptide the PC1 cuts galanin from the proprotein. In animals the terminal glycine is amidated. Due to an amino acid change from glycine to serine, nonamidated human galanin remains and its sequence is prolonged by one amino acid.

First in swine and then in other mammals an alternative galanin-related peptide (GalrP) was found. The human GalrP lacks the first PC1 motif. The sequence identical to galanin is thus N-terminally elongated (Fig. 4.12).

Three galanin receptors belong to GPC heptahelical membrane receptor families. Gal-R1 and Gal-R3 induce adenylate cyclase and thus cAMP elevation, and Gal-R2 signals via phospholipase C.

4.3.10.3 Physiology

Experiments in mice have demonstrated galanin's functions in at least two pathways: galanin-defective (*Gal*$^{-/-}$) mice were viable and fertile which precludes any essential role in reproduction. However, once these *Gal*$^{-/-}$ females had given birth they were unable to feed their offspring because the mammary glands were afunctional. *Gal*$^{-/-}$ mice utilized a fat-enriched diet to a lesser extent than

---

[11]Issue 12 of Cellular and Molecular Life Sciences in the year 2008 (Vol. 65) was dedicated to modern galanin research and comprises a couple of reviews on the subject.

comparable wildtype mice. Whether the phenomena are important in humans is not yet proven.

In addition, *Gal*$^{-/-}$ mice do not show estradiol-induced prolactin stimulation. Thus, galanin appears to be an estrogen-dependent stimulator of pituitary prolactin synthesis and release. Galanin receptor mutations may be factors of prolactinoma development.

Neurons exclusively forming galanin may not exist: galanin has been co-localized to neurons and cells expressing multiple other hormones or neurotransmitters: GnRH, GHRH, prolactin, vasopressin, CRH, oxytocin, substance P, CGRP, noradrenaline, or acetylcholine. Using patch clamp techniques, galanin was shown to inhibit neurons by elevating potassium currents (triggered by Gal-R1 and Gal-R3) and by downregulating calcium currents. Galanin also blocks synaptic plasticity, for example, when memory develops (*long-term potentiation*). Especially in the arcuate nucleus galanin presynaptically reduces GABA release as well as postsynaptically via galanin receptors. Singular effects by galanin stimulation have been observed in the dorsal–vagal complex where calcium currents were decreased.

Apart from the CNS is galanin expressed in the anterior pituitary, in the adrenal medulla, in the pancreas, the urogenital tract, and in skin (Wynick and Bacon 2002; Tortorella et al. 2007; Bauer et al. 2008). Neurons innervating the heart, kidneys, or gut have also been found to be galanin positive by immunocytochemistry.

When axons are cut, galanin and its receptors are upregulated. Galanin also seems relevant for nociception and neuronal development in spinal ganglia in mice (Hobson et al. 2008). Animal models suggest effects of galanin on learning and memory (Miller 1998; Rustay et al. 2005).

Galanin might also play a role during reproduction: it is hypothalamically co-expressed together with GnRH; the count of galanin plus GnRH double positive neurons was fivefold elevated in female rats compared to male. This difference may be due to a testis-dependent epigenetic regulation (Merchenthaler 1998).

Most recent articles suggest that galanin characterizes metastatic breast cell tumors. Further such markers are vascular endothelial growth factor (VEGF) and related drugs induced during tissue hypoxia by hypoxia-induced factor 1 (HIF1). Given that galanin is required for mammary gland maturation such a role is feasible (Bertucci and Birnbaum 2009).

4.3.10.4 Phylogeny

Galanin has been identified thus far in vertebrates.[12] However, proteins with homology have been found in placozoans, that is, before protostomes and deuterostomes developed separately. Allatostatin receptors in insects share up to 50 % homology with vertebrate galanin receptors.

---

[12]In the protein library there are some sequences labeled "potentially galanin" from bacteria; these sequences, according to our own ClustalW analysis, do not show any homology with vertebrate galanins.

4.3.11 Melanin-Concentrating Hormone

Fact sheet 4.10: Melanin concentrating hormone (MCH)

Gene:	Chromosome: 12 ; locus: 12q23.2 ; three exons
Sequence:	**DFDMLRCMLGRVYRPCWQV** (Fig. 4.13).
Synthesis and target:	MCH is released from the lateral hypothalamic area and acts on MCH receptors in brain and in the periphery.
Function:	Mammalian MCH is a neuropeptide regulator of feeding behavior and of energy expenditure; in fish and amphibia it is active as a melanophore-concentrating hormone and triggers lightening of the skin.
Receptor:	Two GPCR MCH-R1 (aka orphan receptor SLC-1) and MCH-R2 with different functional and topological characteristics.

4.3.11.1 Introduction

The adaption of the fish skin to the environment expressed as darkening or lightening has been suggested to be controlled by mutually antagonistic peptides since the year 1931 (Hogben and Slome 1931). The hormone for darkening turned out to be the MSH (from POMC), however, a melanin-concentrating hormone could only be identified in 1983 by Kawauchi et al. (1983) in salmon. A homologous hormone was later isolated and cloned from different mammals including humans. Several mammalian MCH share the same structure; in mice there are two amino acid exchanges and in sheep the sequence appears truncated (Pissios et al. 2006).

4.3.11.2 Structure and Gene

The original MCH from salmon is a 17 amino acid long cyclic peptide with an intramolecular disulfide bridge; human (and common mammalian) MCH has two additional N-terminal amino acids and few exchanges compared to the salmon sequence. The peptide is released from a precursor which in addition to the signal peptide and the MCH bears two other peptides called neuropeptide GE and neuropeptide EI (due to the termini), the latter being amidated (Fig. 4.13). The human *MCH* gene on chromosome 12 has three exons; the next neighbors are a nucleoporin and IGF1.

The MCH receptor was identified to be the orphan GPCR SLC-1 which was relabeled to MCH-R1. The receptor couples to different G- proteins and activates different signaling pathways: increase in intracellular free calcium, suppression of forskolin stimulated cAMP, stimulation of phophoinositol pathways, and triggering of extracellular-signal-regulated kinases (ERK). The receptor is preferentially expressed in the brain, highly in the piriform cortex and the olfactory tubercle, and with lower density in the nucleus accumbens and the amygdala. Further expression has been found in the arcuate nucleus and the ventromedial nucleus.

5 10 15 20 25 30 35 40 45 50	
m a k m n l s s y i l i l t f s l f s q g	-21 - -1
i l l s a s k s i r n l d d d m v f n t f r l g k g f q k e d t a e k s v i a p s l e q y k n d e s	1 - 50
s f m n e e e n k v s k n t g s k h n f l n h g l p l n l a i k p y l a l k G S V A F P A E N G V Q	51 - 100
N T E S T Q E k r E I G D E E N S A K F P I g r r D F D M L R C M L G R V Y R P C W Q V	101 - 144

Fig. 4.13 Primary sequence of human melanin concentrating hormone (MCH). The signal peptide is highlighted *light gray*, the MCH peptide *yellow* and two other peptides *dark gray*. Endopeptidase motifs are *inverted*. The disulfide bridge of MCH is indicated by *blue lines* (Source: GenBank NM_002665.2)

By Northern blot and in situ hybridization analysis of human and monkey tissue, Sailer et al. (2001) showed that expression of MCH-R2 mRNA is restricted to several regions of the brain, including the arcuate nucleus and the ventral medial hypothalamus, areas implicated in regulation of body weight.

4.3.11.3 Physiology

The original function in fish (lightening of skin) has been lost during further vertebrate development. Today human MCH is considered an important regulator of food intake and energy expenditure. This has been shown by different authors, either in model animals or in humans (for reviews, see Nahon 2006; Flier 2004; Pissios et al. 2006). The first hint for this event stems from direct intracerebral injections of MCH into rat brains (Qu et al. 1996) which induced food intake. Careful analysis by a couple of laboratories identified the lateral hypothalamic area (LHA) as the place where most MCH perikarya were located with projections into many other brain areas. This is conclusive for a neurotransmitter role of MCH in induction of feeding and energy expenditure. The LHA, in addition, comprises another type of cell directly related to appetite and feeding behavior: orexin neurons and orexins likewise active as feeding control neurotransmitters.

Whereas POMC and other neurons almost always show a membrane potential, MCH neurons are mostly quiet and become active upon stimulation, either by synaptic contacts or, as shown recently, by elevated glucose levels. Upon stimulation MCH neurons secrete MCH over synapses and, to some lower degree, into the circulation. The role of MCH in periphery, however, has not attracted attention as do the neurotransmitter actions, and it thus largely speculative. Skin cells and some other cell types have been shown to express MCH-R1.

Most instructive have been MCH knockout mice: These mice were hypophagic (reduced food intake) and lean compared to wildtype littermates. This leanness was attributed to an enhanced energy consumption in these MCH-deficient mice.

Long-term exposure to intracerebroventricular MCH led to an increased body weight in mice, which increased even further with a parallel 33 % fat diet (one fifth compared with 1/20 (with normal diet) within 14 days of MCH exposure). Rats kept in the cold appear to depend on MHC to adapt to the low temperature by activating brown adipose tissue's fat consumption: blocking MHC expression by antisense RNA resulted in a dramatic weight loss compared to controls. Using an MCH receptor antagonist it could be shown in obese mice that this treatment reduced food uptake, stopped body weight gain, instead reduced body weight, lowered the overall fat content, and reduced hypercholesterolemia, hyperinsulinemia, hyperglycemia, and hyperleptinemia associated with obesity in these animals.

These facts demonstrate clearly a neuropeptide control of MCH on feeding, energy mobilization, and accumulation of energy stores.

The interaction of MCH neurons with other central centers of feeding are dealt with in the chapter about feeding (Sect. 11.5.8).

4.3.11.4 Phylogeny

Any MCH-like protein has only been described in vertebrates.

4.3.12 Orexins

Fact sheet 4.11: Orexins/Hypocretin (HCRT)

Gene:	Chromosome 17; locus 17q21; two exons.
Synthesis and target:	Two orexins, A and B, cleaved from the same precursor; orexins are formed in neurons (neurosecretory cells?) of the lateral hypothalamus and the enteric nervous system (ENS) and almost exclusively released as neurotransmitters.
Structure:	The peptide structures have been determined by NMR analysis (Kim et al. 2004a).
Function:	Orexins are involved in regulation of feeding and the maintenance of a regular sleep cycle.
Receptor:	Two heptahelical GPCR, HCRT-R1 and HCRT-R2; HCRT-R1 is recognized by orexin-A, whereas HCRT-R2 is bound by both orexins.

4.3.12.1 Introduction

Orexins, also known as hypocretins (i.e., incretins of the hypothalamus, belonging to the secretin protein family), were identified in 1998 as peptides contributing to the regulation of feeding (de Lecea et al. 1998). In the meantime it has been additionally found that they play a role in the regulation of sleep because a defect in the orexin gen triggers familiar narcolepsy, and acquired narcolepsy can be induced by destruction of orexin neurons in the lateral hypothalamus. The original assumption of an exclusive presence of orexin neurons in the brain has been discarded when orexin formation could also be observed in the enteric nerve system (ENS) (Kirchgessner 2002).

4.3.12.2 Structure and Gene

The gene for the orexin precursor has been found on the long arm of chromosome 17. After removal of the 33 amino acids of the signal peptide orexin A is released

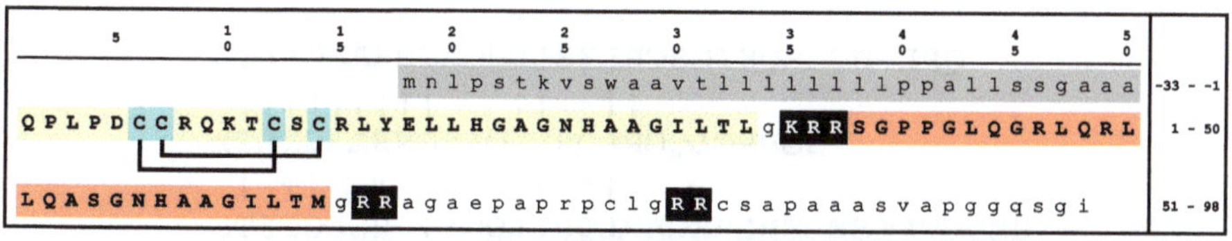

Fig. 4.14 Orexin precursor and its primary sequence. Following the 33 amino acids of the signal peptide (highlighted *gray*) there are two orexin peptides: orexin A (on a *yellow background*) harboring two disulfide bridges (*lines* connecting *blue* cysteines) and the cysteine lacking orexin B (on *orange*) (Source Swiss-Prot: O43612.1.; PyMOL)

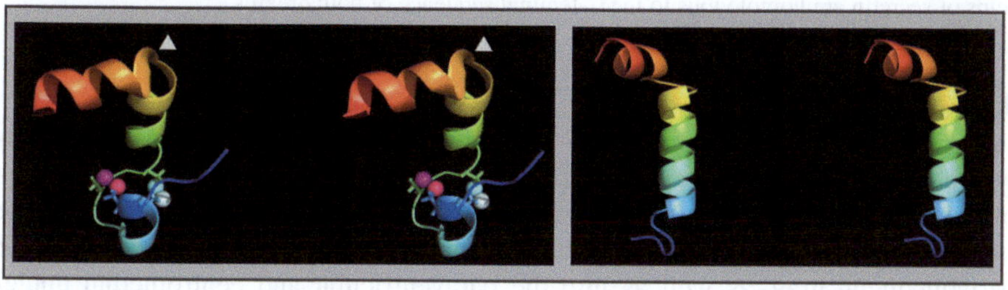

Fig. 4.15 Structures of the human orexins. The stereo models of orexin A (*left*) and orexin B (*right*) show a *blue* colored N-terminus (in the lower part of the image) and a *red* colored C-terminus. The *arrowheads* in the orexin A image indicate a β-turn

by PC1 and orexin 2 by PC2. Therefore the release of orexin B is coupled to the expression of the PC2. The extent of expression of this enzyme in the lateral hypothalamus and its regulation in the brain region has not yet been conclusively analyzed. Other prohormone convertase—for example, furin—do not play any role due to the lack of peptide motifs in the orexin preprotein.

Orexin-A contains two N-terminal disulfide bridges missing in orexin-B. This difference is already present in fish *e.g.* ABF29871.1 from codfish. The two peptides are C-terminally amidated; orexin-A has an N-terminal pyroglutamate (p**E**; Fig. 4.15). These peptides have been analyzed by NMR (Fig. 4.16). The C-terminal helices share structural homology; the sequence homology is enhanced in the C-terminal region compared to the N-terminus.

Two heptahelical GPCR have been shown to be orexin receptors. Their sequence homology is 64 %. The HCRT-R1 activates upon ligand binding hypoxia-inducible factor (HIF-1a). This HIF-1a activation leads to increased glucose uptake and enhanced glycolysis. In contrast to the hypoxia-induced activation ATP is not formed by anaerobic glycolysis but by stimulating the citrate pathway and oxidative phosphorylation (Sikder and Kodadek 2007). HRCT-2 is linked in dogs to inherited narcolepsy (Lin et al. 1999): an insertion of 226 base pairs in intron 3 of the *HCRT-2* gene causes aberrant splicing and a shortened and afunctional receptor protein.

```
                ========   =======
   QPLPDCCRQKTCSCRLYELLHGAGNHAAGILTL
                 ||  ||   |||||||||
      SGPPGLQGRLQRLLQASGNHAAGILTM
                 ||**||*    |+ |+ |
HSDGTFTSELSRLREGARLQRLLQGLV HSDGTFTS
```

Fig. 4.16 Homology of orexins and secretin. The homology of both orexins is highest in those helices Kim et al. (2004a) determined. Also the homology to the eponymous secretin is obvious (compare the fact sheet 4.36). Orexins are indeed the secretine of the hypothalamus. This homology is mostly due to the helices *Doppelmarkierung* above the orexin-A sequence. The first eight amino acids of secretin are homologous to the C-teminal sequence of both orexins

4.3.12.3 Physiology

In the brain orexin neurons are almost exclusively found in the lateral hypothalamus which is among others a center for the regulation of feeding. The orexin neurons of rat project their axon in different brain regions (Hagan et al. 1999): the lateral hypothalamus itself, the perifornical nucleus, the dorsal raphe nucleus, the periaqueductal gray, as well as into the paraventricular and centromedial nuclei of the thalamus. Most intensively orexin neuron axon were found in the *locus coeruleus*. Here mostly probably noradrenergic neurons are synaptically connected to orexin-A. The neurons in the LC[13] are especially involved in the regulation of stress (see Sect. 11.2.1) and in the control of alertness and sleep. Orexin intracerebroventricularly injected prolonged states of alertness, enhanced locomotor activity, and reduced sleep periods. There are related findings that lack of orexin or a defective orexin receptor are at the origin of narcolepsy (Lin et al. 1999; Chemelli et al. 1999; Ebrahim et al. 2003, 2005). Orexin is thus applied when alertness should be enhanced.

Orexin positive cells are additionally found in the enteric nervous system in the mucosa of humans and animals: in the *plexus myentericus* and the *plexus submucosus* about one of four cells was orexin and leptin positive (Kirchgessner and Liu 1999). The authors suggest, therefore, that orexins have an important role in energy homeostasis.

4.3.12.4 Phylogeny

Thus far orexin sequences have been found in vertebrates and lampreys. In the latter case (XP_002598524) an orexin is not formed when a PC2 is active, only fragments thereof. In skate (*Leucoraja ocellata*) two orexins are coded for, disulfated bridges are present, and intramolecular PC2 motifs are absent. In the codfish Xu and Volkoff (2007) there are two peptides, 50 and 29 amino acids long, released by PC1. In humans, only the orexin-A is released by the PC1, not orexin-B, which raises doubts about its role as a neuropeptide, and both peptides can be targeted by the PC2 (see Fig. 4.14).

[13]Locus coeruleus.

4.4 Anterior Pituitary Hormones

4.4.1 POMC

Fact sheet 4.12: Proopiomelanocortin (POMC)

Gene:	Chromosome 2 ; locus: 2p23; three or four exons.
Sequence:	See Fig. 4.17.
Synthesis and target:	The prohormone is formed in corticotropic cells of the anterior pituitary, within melanotropic cells of the pars intermedia, in the hypothalamus and in skin. Due to the expression of PC1 and/or PC2 ACTH, β-endorphins and/or MSH peptides are synthesized.
Receptor:	Heptahelical G protein-coupled receptors: melanocortin receptors, opioid receptors.

Fig. 4.17 (Page 59). Primary sequences of POMC from vertebrates.
Colours code: peptide motifs *white, bold and uppercase*: on *red* **KR**, on *blue* **RR**, on *orange* **KK**, on *dark gray* **RK**; signal peptide: *gray* background; peptides: MSHs: *bold red*; ACTH: α-MSH plus *green*; lipocortin: *pink*; β-endorphin: on *yellow*; amino acids on *black*: changes compared to the human sequence. The POMC analogue from *P. marinus* (AAC59724.1) is not shown due to missing space. This peptide contains a much elongated α MSH-like peptide, a β endorphin-like peptide, but no ACTH. The remaining part of that peptide cannot be aligned, the cysteines in the N-terminal peptide are out of place and additional pro-hormone convertase motives are missing .

Introduction

Proopiomelanocortin (POMC) constitutes the precursor of seven different peptide hormones. It is expressed predominantly in corticotropic cells of the adenohypophysis as well as in melanotropic cells of the pars intermedia which in adult life is regressed. After processing POMC in the anterior pituitary adrenocorticotropic hormone (ACTH), β-lipotropin (LPH) as well as β-endorphin are stored in vesicles. In contrast, in the intermediate lobe melanocortins (MSH) and acetyl-β-endorphin are formed. POMC is further expressed in the hypothalamus, in the testis, in the ovary, in the adrenal medulla, in placenta, in the lungs, in skin, and especially in circulating monocytes and in tissue macrophages.

Structure and Genes

The *POMC* gene on chromosome 2 gives rise to two variants differing by an additional exon of 50 bp in the 5′ untranslated region. The POMC mRNA in pituitary and hypothalamus is about 1.1 kilobases (kB) long, whereas extracranial RNA has only 800–900 bp. In tumors there is a further RNA of about 1.4 kB size.

From the different POMC RNAs the same precursor protein is translated: POMC with 267 amino acids in humans. From this precursor, different fragments are cleaved by PC1/PC2. Three different types of peptides are known: melanocortins, adrenocorticotropic hormone (corticotropin, ACTH), and endorphins (see Fig. 4.17). These peptide hormones have different action. Which neuroendocrine cells which of these hormones releases is determined by the expression of the prohormone convertases. In addition, from bovine POMC it is known that

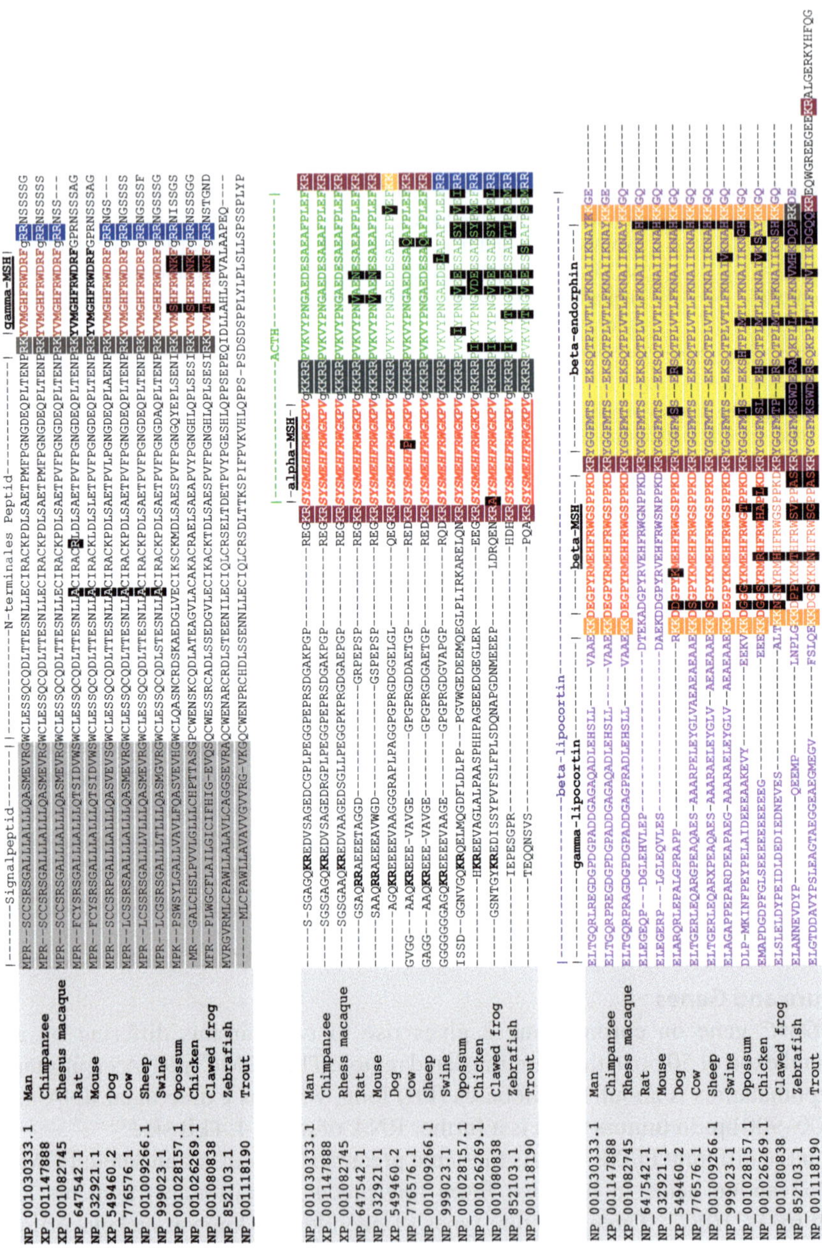

Fig. 4.17 Primary sequences of POMC from vertebrates

different glycosylations influence the use of the peptide motifs by the prohormone convertases (Birch et al. 1991).

The *POMC* gene is predominantly expressed in corticotropic cells of the pituitary, in neurons of the arcuate nucleus and the nucleus of the solitary tract, and in keratinocytes and melanocytes of the skin.

The dog sequence shown in Fig. 4.17 has been retracted from GenBank due to low quality. Another sequence (XP_849463) still shows the **KK** motif (at the end of the *middle block*) which indicates that ACTH may not be a product from the anterior pituitary in dog, swine, opossum, chicken, and frog because the PC1 motif is lacking. In fish, however, there are two *POMC* genes and only the one shown of zebrafish has mutation translated into the **KR** to **KK** change. The gene product of the second gene can be processed to give rise to ACTH (Alsop and Vijayan 2009).

The corticotropic cells of the pituitary express predominantly PC1 forming thus ACTH, β-Lipotropin and β-Endorphin from POMC. In contrast, melanocytes of the intermediate lobe express additional PC2 which in turn forces α-MSH synthesis. POMC neurons of the ventral hypothalamus equally synthesize α-MSH.

Phylogeny

POMC and the peptides derived thereof have thus far been found but in vertebrates and agnathans. The genome of the urochordate *S. purpurata*[14] contains neither a POMC precursor nor any melanocortin receptor (Burke et al. 2006). There are, however, reports that parasitic *Schistosoma mansoni* expresses POMC as a defense against vertebrate immune attack and releases MSH, ACTH, and β-endorphin peptides. Since these reports in 1992 (Duvaux-Miret et al. 1992), a POMC gene from *S. mansoni* has not been published. Furthermore, an isolation of γ-MSH from leech has been reported. In the blue mussel POMC derived peptides were reported as well. Again any genomic sequence has yet to be found.

In most vertebrates a single POMC exists; a second POMC gene exists in fish after an additional genome duplication (de Souza et al. 2005). Fish POMCα is expressed in the hypothalamic nucleus lateralis tuberis, in the anterior pituitary, and in the intermediate lobe, whereas POMCβ is found in the preoptic region and faintly in the pituitary stalk. The endorphin sequence is retained only in POMCα.

Comparing selected mammalian and other vertebrate POMC sequences (Fig. 4.17), you may notice that:

- The α-MSH sequence is conserved from teleosts on; there are two amino acid exchanges in clawed frog and in cows. The dibasic sequence motifs are conserved, too. Apart from *X. laevis* the glycines for C-terminal amidation and the N-terminal serine that is acetylated are always present.
- In rats and mice β-MSH is lacking because the **KK**-cleavage site is gone.
- Teleosts do not have a γ-MSH.

[14] Strongylocentrotus purpuratus.

- Within mammals ACTH has two characteristic exchanges in rodents. Nonmammalian ACTH have some constant differences compared to mammals. The lack of a PC1 motif in dog, swine, opossum, or chicken raises the question whether and how these animals make ACTH. There is no report in the literature on the problem. In *D. rerio* a second POMC rescues this defect, however, in tetrapods and birds only a single gene has ever been reported.
- γ-Lipocortin is the most variant part of the different POMC regions.
- Most interesting is the high conservation of the N-terminal peptide: the four cysteines are at similar positions and equally spaced. With the exception of opossum there are but two positions with alternative amino acids although any function of this sequence has not yet been described.

The melanocortin receptors reflect the different genome duplications (Fig. 4.18). In lamprey i.e. before the second genome duplication two MCR were found. After the next genome duplication vertebrates have five MCR genes which in humans are on chromosomes 16 and 18 with synteny tom, for example, fugu (Klovins et al. 2004). After the third genome duplication restricted to fish in the thus far analyzed species several genes got lost including both the MCR3 genes in *Takifugu rubripes* as well as the γ-MSH (as shown in fugu, zebrafish, and trout). MCR3 is the specific receptor for γ-MSH in humans and other species analyzed.

The topological distribution of the hormone targets merits attendance.

In humans melanocortin receptor 1 (MCR1) is expressed in skin: in melanocytes, keratinocytes, fibroblasts, endothelial cells, and antigen presenting cells. This receptor binds α-MSH and ACTH with equal specificity. In leukocytes MCR1 mediates the inflammation inhibiting effects of α-MSH. MCR2 is the receptor in the adrenal cortex mediating the steroidogenic activity of ACTH. This receptor is not activated by MSHs. MCR3 is expressed in CNS, gastrointestinal tract, and in placenta. γ-MSH exhibits the highest activity for this receptor. MCR4 is specially expressed in CNS. α-MSH and ACTH activate this receptor more strongly than β-MSH or γ-MSH. Finally, MCR5 is the α-MSH receptor of sebaceous glands but found in other tissues, too. Obviously the receptor has its role in the regulation of exocrine glands.

Compared with humans, the distribution and functional specificity in *T. rubripes* is different: truMcR1 is faintly expressed but in CNS; truMcR2 in brain and in the adrenal, (truMcR3 is lacking); truMcR4 and truMcR5 are found in brain, in the adrenal, and in gut (truMcR4), respectively; and in the eye (truMcR5). As in humans in *T. rubipes* truMcR2 is stimulated exclusively by ACTH, not by any MSH. truMcR1 and truMcR4 show much stronger activation by ACTH than by α-MSH (Klovins et al. 2004): this is different from humans and interpreted by the authors as a hint to an ancestral function of ACTH.

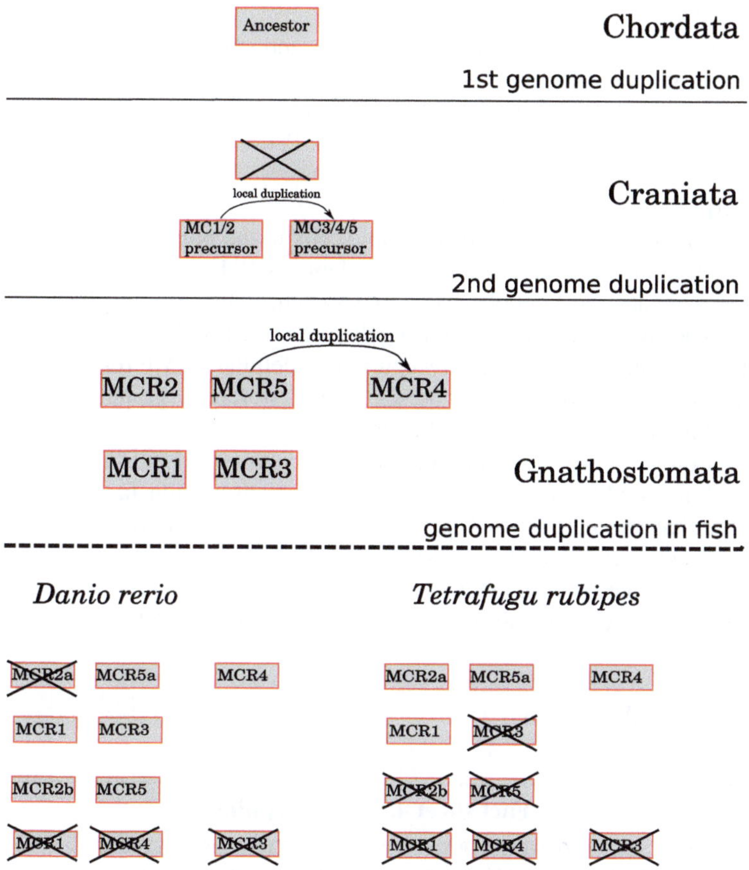

Fig. 4.18 Hypothesis of melanocortin receptor gene (MCR) development. After an initial genome duplication of an unknown ancestor gene one gene is lost and the other locally duplicated again. A second genome duplication was followed by doubling of one of the resulting genes. The pattern of five MCR is found in the majority of vertebrates. In fish a third genome duplication took place. Not all of the genes were maintained which gave rise to characteristic patterns like those of *D. rerio* or *T. rubripes* (Source: redrawn due to Klovins et al. 2004)

4.4.1.1 ACTH

Fact sheet 4.13: Adrenocorticotropic hormone (ACTH)

Sequence:	SYSMEHFRWG KPVGKKRRPV KVYPNGAEDE SAEAFPLEI
Synthesis and target:	In POMC forming cells by activity of prohormone convertase 1
Function:	Stimulation of steroidogenesis in the adrenal
Receptor:	Heptahelical G-protein-coupled receptor: melanocortin 2 receptor

Introduction

ACTH is the effector hormone of the HPA axis and stimulates hormone release in the adrenals: glucocorticoids as well as mineralocorticoids are released after ACTH stimulation, as well as adrenaline whose synthesis is stimulated by ACTH which induces the last enzyme of adrenaline synthesis: PNMT (see Fig. 7.1).

Physiology

ACTH is released from adrenocorticotropic cells in the pituitary by hourly pulses. This secretion thus results from hypothalamic CRH release into the median eminence and into the hypophyseal portal system. During night, the quantity of ACTH is twice to thrice as high as during daytime (see Fig. 12.2). ACTH stimulates by its effect on the adrenal MCR2 receptors intracellular cAMP resulting in steroid hormone synthesis and (because these are not stored in granules) release. Of paramount importance is the cortisol formation.

Regulation of POMC→ACTH release in the pituitary results from stimulation of intracellular cAMP by CRH and binding of transcription factor Nur to Nur-responsive elements in the POMC promoter. Feedback inhibition by glucocorticoids is due to direct interaction of ligand-bound glucocorticoid receptors with Nur, thus blocking its interaction with DNA (Murakami et al. 2007).

ACTH's half-life in human blood was determined to be 19 min (Keenan et al. 2004).

4.4.1.2 Endorphins

Fact sheet 4.14: Endorphins

Sequence:	YGGFMTSEKS QTPLVTLIKN AIIKNAYKKG
Synthesis and target:	In POMC expressing cells by the action of PC1
Function:	Endogenous opioid
Receptor:	Heptahelical G-protein-coupled receptor: μ-opioid receptor (MOR)

Introduction

Endorphins are endogenous opiates . These are released, for example, as a reaction to pain. Inasmuch as β-endorphin release is coupled to ACTH release, reactions to stress are metabolic changes, circulatory reactions mediated by glucocorticoids or adrenaline, and on the other hand analgesic effects by endorphins. Triggering the feeling of happiness by endurance sports appears to be an adaption to stress.

Physiology

As shown in Fig. 4.19 β-endorphin and ACTH are formed in equal amounts by the POMC *processing* by PC1. Endorphin synthesis takes place in all cells where MSH is formed. Endorphins thus can be released from all those cells/tissues where ACTH is formed and from those MSH forming cells in hypothalamus, pituitary, skin, additional tissues, and leukocytes.

β-Endorphin has a role in analgesia: β-endorphin is one endogenous ligand of μ-opioid and δ-opioid receptors (MOR and DOR). Both receptors are found in neurons that trigger pain, so-called nocireceptors communicating injury, chemical attack, heat, or coldness to the nocicenters of the brain. MOR neurons are densely found in the periaqueductal gray of the midbrain. As do other analgesics β-endorphin blocks the activity of nocireceptors.

It is not yet clear whether endorphins act directly or indirectly via other analgesic mediators. Endorphins 1 and 2 have, for example, a greater affinity to MOR than endorphins. However, the proteins where they should be cleaved from have not yet been found. Apart from endorphins and endomorphins, enkephalins and dynorphins are analgesically active as well.

Information about a phylogeny of endomorphins and nocireception has not yet been published.

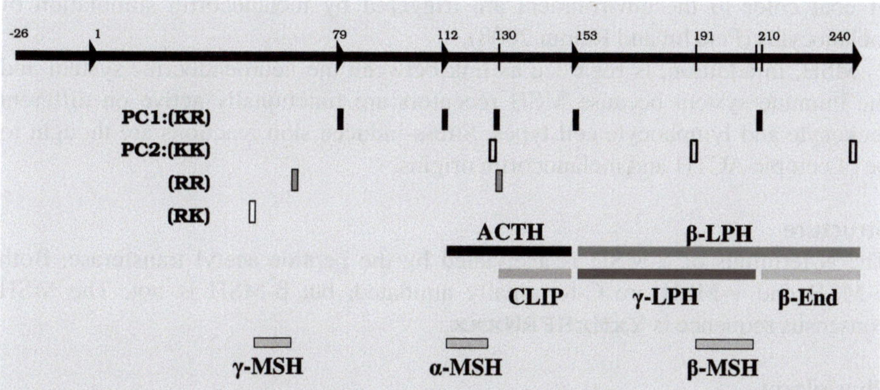

Fig. 4.19 POMC: prohormone convertases and alternative peptides. By the prohormone convertases PC1 and/or PC2, proopiomelanocortin (POMC) peptides are cleaved from the precursor: adrenocorticotropin (ACTH), lipocortin (LPH), melanocortin (MSH), endorphin (End), and corticotropin-like intermediate peptide (CLIP). PC1 cleaves but after Lys-Arg (**KR**), whereas PC2 cleaves after Lys-Lys (**KK**), Arg-Arg (**RR**), or Arg-Lys (**RK**). Which peptides are formed depends on the type of endocrine cell and of its convertase expression

4.4.1.3 Melanocortins

Fact sheet 4.15: α-, β-, and γ-Melanocyte stimulating hormone (MSH)

Sequence:	α-MSH Acetyl-**SYSMEHFRWGKPV**-NH_2 β-MSH **DEGPYRMEHFRWGSPPKD** γ-MSH **YVMGHFRWDRF**-NH_2
Synthesis and target:	MSH are made within POMC neurons in the CNS, in skin, in the cells of the pars intermedia, and in other peripheral tissues by the action of PC1 and PC2, PAM, and other enzymes. They act on melanocytes and centrally on centers of satiation.
Function:	Centrally MSH acts as antagonist of NPY and agouti-like peptides and induces satiety. In melanophorous cells it triggers formation and aggregation of melanin. MSH further inhibits leukocyte activation.
Receptor:	Heptahelical G-protein-coupled receptors: melanocortin receptors (MCR) 1,2,3,4,5.

Introduction

Not long ago, it was observed that in humans melanocortin and its receptors are important regulators of the balance of appetite and satiety. In animals adaptations of coat color to the environment are triggered by melanocortin stimulation of melanocytes (Penzlin and Ramm 2008).

MSH, in addition, is regarded as link between the neuroendocrine system and the immune system because MSH receptors are functionally active on different leukocyte and lymphocyte cell types. Stress-induced skin reactions are thought to be of ectopic ACTH and melanocortin origins.

Structure

The N-terminus of α-MSH is acetylated by the peptide acetyl transferase. Both α-MSH and γ-MSH are C-terminally amidated, but β-MSH is not. The MSH consensus sequence is **YxMxHFRWxxx**.

Physiology

Functions of MSH peptides depend on the target organ: regulation of food uptake occurs in the CNS, induction of melanin synthesis in the skin is dependent on sunlight, and inhibition of leukocyte activation in the circulation.

The latter effect is due to protein kinase A activation by enhanced cAMP after MSH binding to the MCR. This PKA blocks phosphorylation of IκB which in the cytosol is complexed by NFκB. Without this IκB phosphorylation dissociation of the complex after receptor-triggered leukocyte stimulation is blocked and thus the signal transduction by NFκB transport into the nuclease does not take place, blocking several activation patterns (Catania 2007).

The central regulation of food uptake is normally blocked by MSH released in the hypothalamus and by its binding to MCR on NPY neurons in the arcuate nucleus which is called a tonical suppression. By the release of AgRP—triggered by ghrelin, for example—this suppression is removed and NPY neurons become active, triggering appetite.

During control of melanin synthesis in human skin, UV light activates POMC expression in melanocytes and keratinocytes. Due to PC2 expression in these cells MSH is formed which, in turn, stimulates pigmentation.

4.4.2 TSH

Fact sheet 4.16: Thyroid stimulating hormone (TSH)

Gene α chain:	Chromosome 6 (6q12-q21); four exons.
Gene β chain:	Chromosome 1 (1p13); three exons.
Sequence:	Figs. 4.20 and 4.21.
Synthesis and target:	TSH is formed in thyrotropic cells of the pituitary and released by TRH trigger. TSH acts on TSH receptors in the thyroid.
Receptor:	Heptahelical G-protein-coupled receptor: The TSH-R stimulates adenylate cyclase.

4.4.2.1 Introduction

Thyroid-stimulating hormone and the group of gonadotropins described thereafter constitute the group of glycoprotein hormones all built from heterodimers. The larger α-chain has an identical amino acid sequence in the different glycoprotein hormones (LH, FSH, CG, TSH) and is coded for by the TSH-α gene (Fig. 4.20). The glycoprotein hormones differ by their β-chains (Fig. 4.21). In addition they differ in glycosylation and furthermore in sulfation, the latter also rendering α-chains different.

Investigating potential ligands for some leucine-rich G protein-coupled receptors (GPCR), which are evolutionarily ancient among the TSH-R and other glycoprotein hormone receptors, an additional α2 and β5 chain were found in humans whose

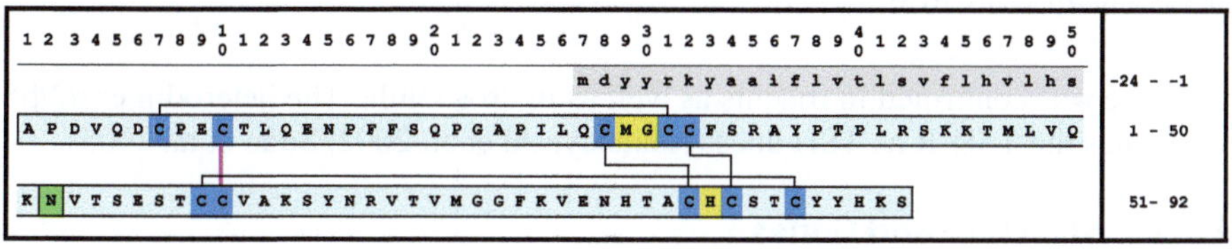

Fig. 4.20 α-chain of thyroid-stimulating hormone (TSH). The signal peptide with 24 amino acids (*gray background*) is followed by 92 amino acids of the mature TSH-α chain; disulfide bridges are indicated by black lines between the different cysteine residues **C**. The boxed asparagine (**N**) is glycosylated (Source: GenBank P01215)

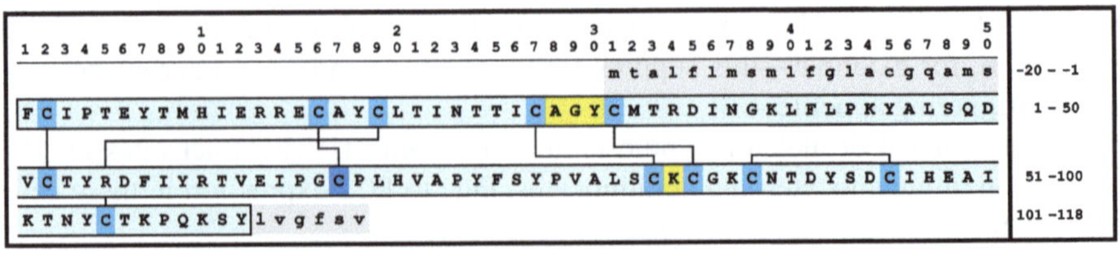

Fig. 4.21 thyroid stimulating hormone (TSH) β-chain. In front of the 112 amino acids of the mature TSH-β-chain there is on the precursor protein a signal peptide of 20 amino acids and at the end an additional six amino acid long peptide (on *gray background*). Disulfide bridges are drawn as lines between *blue* cysteine residues (**C**). The boxed asparagine (**N**) is glycosylated (Source: GenBank P01222)

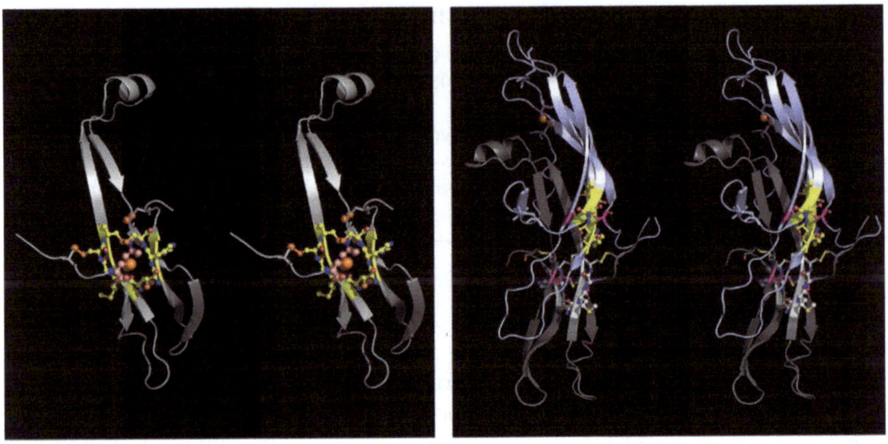

Fig. 4.22 Structure of choriongonadotropin (hCG). ***Left:*** the α chain backbone contains three loops partially twisted held together by beta sheets and five disulfide bridges. Beta sheets are indicated by *arrows*. Cysteines are shown as *spheres/wire frame*, sulfur atoms as *orange spheres*. The cysteine residues marked by *yellow carbon atoms* construct a plain which is pierced by a third disulfide bridge (*ping carbon atoms, large orange sulfur*). The other two disulfide bridges keep the N-terminus and the C-terminus close to this cysteine knot. ***Right:*** in the entire hCG both chains are complexed. The β chain itself has another cysteine knot (*arrow*) formed by three disulfide bridges. The other β chain disulfide bridges maintain a structure into which the α-chain can glide. ***Stereo:*** in order to get the three-dimensional view focus behind the images; both will eventually superimpose and be fused into a single image (Source: 1HRP, und PyMoL. Scripts for the construction are included in Appendix 16.3)

presence was confirmed in rodents as well as in drosophila. The heterodimer α2/β5 stimulates the TSH-R as TSH does (Nakabayashi et al. 2002; Sudo et al. 2005).

4.4.2.2 Structure and Genes

The TSH heterodimer contains the 92 amino acid long α-chain (Fig. 4.20) and the 112 amino acid long β-chain (Fig. 4.21). Structurally characteristic and indispensable is the cysteine knot. This element characterizes the family of nerve growth factors (NGF), too.

The human TSH genes are located on chromosome 6 (α) and chromosome 1 (β) with four, respectively, three exons. The three-dimensional glycoprotein hormone structure was first determined from FSH (Fig. 4.22).

Intact cysteine knots of both chains are required for synthesis and posttranslational sorting into the secretory granules; for receptor binding, not all disulfide bridges are necessary.

4.4.2.3 Physiology

TSH synthesis is stimulated in thyrotropic pituitary cells by the action of thyrotropin-releasing hormone (TRH) and (by TRH triggering again) is released into the portal system. TRH stimulates phosphorylation of cAMP reactive elements binding proteins CREB which recruits CREB-binding protein (CBP) which then, in cooperation with P-LIM,activates the TSH-α promoter. The β-TSH promoter is activated by CBP plus the pituitary transcription factor Pit1.

4.4.2.4 Phylogeny

(is discussed together with all other glycoprotein hormones (Sect. 4.4.3.4).

4.4.3 LH, FSH, CG

Two of the human three gonadotropins are made and released from the pituitary, luteinizing hormone (LH) and follicle stimulating hormone (FSH); the third, choriogonadotropin (CG) is of placental origin and *the* sign of pregnancy. When the possibility of pregnancy can be excluded the presence of CG in blood is indicative of a chorion carcinoma.

4.4.3.1 LH/CG

Fact sheet 4.17: Luteinizing hormone (LH)

Gene α chain:	Chromosome 6 (6q12-q21); four exons.
Gene β chain LH:	Chromosome 19 (19p13.32); three exons.
Sequence:	Fig. 4.23.
Synthesis and target:	LH is synthesized in gonadotropic cells of the pituitary; its release into the portal system is triggered by GnRH. Target of LH/CG is the common LH-receptor on cells in the gonads: for men the Leydig cells and for women theca cells.
Receptor:	Heptahelical GPCR: the LH/CG receptor activates the adenylate cyclase.

Introduction

Inasmuch as the structure of LH and CG is very similar and because they bind to the same receptor, they are treated together here.

Fact sheet 4.18: Choriogonadotropin (CG)

Gene α chain: Chromosome 6 (6q12-q21); four exons.
Gene β chains CG: Chromosome 19 (19p13.32); *CGB*; *CGB1*; *CGB2*; *CGB5 CGB7*; *CGB8*.
Sequence: Fig. 4.23.
Synthesis and target: CG is synthesized in placental cytotrophoblastic cells and in the syncytium (in the presence of GnRH). Targets of CG are CG receptors in the corpus luteum and the endometrium.
Receptor: Heptahelical GPCR: The LH/CG receptor activates the adenylate cyclase.

Structure and Genes

With the crystal structure of human CG determined (Lapthorn et al. 1994; other CG have only been found in primates, the CG of horses is a LH) it became evident that gonadotropins as well as NGF,[15] PDGF,[16] TGF-β display a so-called cysteine knot (see Fig. 4.1). Two disulfide bridges together with some beta sheets form a ring; across the ring the third disulfide bridge is formed (Fig. 4.1). Cysteine residues of gonadotropins are conserved in vertebrates, not only the β-chains, but the α-chains as well. Equally conserved is the asparagine residue at position 30 of LH/CG which is N-glycosylated; other N-glycosylation sites are not fully conserved: compare the sequence of LH (Fig. 4.23) and of TSH (Fig. 4.21).

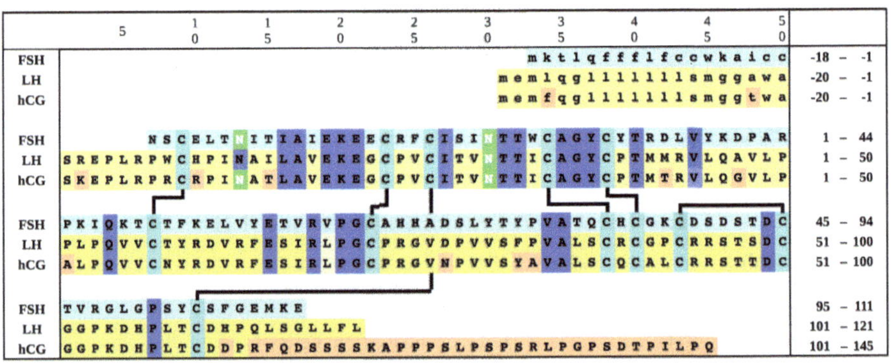

Fig. 4.23 Sequence comparison of three human glycoprotein hormone beta chains: follicle-stimulating hormone (FSH), luteinizing hormone (LH), and choriogonadotropin (hCG). With an N-terminal FSH shift by six Amino acid all cysteine residues are placed at identical positions. Cysteines are on *blue background*. The cysteines in the mature peptide are conserved. Conserved amino acids or groups of amino acids are on *violett background*. The 25 Amino acid that distinguish CG and LH are on *light orange background*. N-glycosylation sites are *white on green*. Disulfide bridges are according to the Swiss-Prot annotations (Source: FSH: NM_001018090, LH: P01229, hCG: P01233)

[15] Nerve growth factor.

[16] Platelet-derived growth factor.

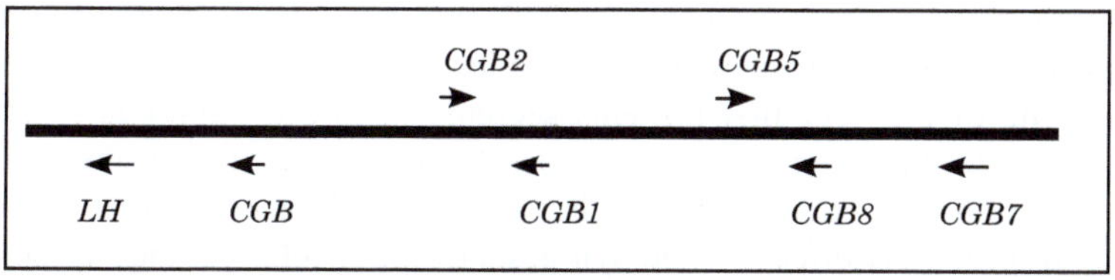

Fig. 4.24 Organization and orientation of the human luteinizing hormone (LH) and choriongonadotropin (CG) genes on chromosome 19 (Source: http://www.ncbi.nlm.nih.gov/projects/mapview/maps.cgi?TAXID=9606&CHR=19&MAPS=genes-r&QSTR=chorionic+gonadotropin&QUERY=uid%28-2038742931%29&BEG=48%2C940K&END=49%2C160K&oview=default)

Those N-glycosylated asparagines are conserved, too. (Primary structures from http://www.chem.gla.ac.uk).

Sequences of CG and LH are more than 90 % identical, whereas FSH is only about 30 % homologues. Within these 30 % are the fully conserved cysteines and the distances in between which helps to retain the conserved structure (Fig. 4.23).

The LH and CG genes are located on chromosome 19, separated only by the pseudogene of neutrophin. The order of the genes is shown in Fig. 4.24.

Physiology of LH

LH release is triggered (like FSH) by pulsatile GnRH. In addition, LH synthesis is regulated by aktivins and inhibins. Increased concentrations of progesterone which are made in the corpus luteum block LH formation.

LH acts on male and female gonads. In the testes testosterone synthesis and release are stimulated in Leydig cells. (The release is immediate by diffusion: steroids are not (except in rare instances) collected in granules and not released from these latter by receptor-mediated vesicle–membrane fusion; they diffuse from the cells and are (often) bound by serum transport proteins.) In the menstrual cycle LH from pituitary stimulates testosterone formation in follicular theca cells. This testosterone diffusing to the granulosa cells is aromatized into estradiol. Due to sufficient elevation of circulating estradiol on cycle day 14 a sharp increase of LH release (LH surge) leads to ovulation. In mice kisspeptin neurons of the anteroventral periventricular nucleus increase, upon elevated estradiol, their GnRH release and thus the LH surge whereas GnRH release from neurons of the arcuate nucleus seems to be inhibited (Han et al. 2005). The increase of serum progesterone once the corpus luteum menstrualis is formed reduces LH and FSH release.

Physiology of CG

When ovulation is not followed by pregnancy, the corpus luteum is spent after 14 days and degenerates. The progesterone release finishes and inhibition of LH and FSH release is abrogated. In the case of pregnancy, however, the trophoblast developing from the fertilized egg releases CG which again stimulates progesterone synthesis from the corpus luteum. This CG-supported progesterone synthesis is accomplished in later pregnancy by the fetus itself.

Being one of the very first proteins secreted from the trophoblast raised the question about CG's role during nidation. During these studies the effects of progesterone and estradiol were confirmed that facilitate nidation. There was, however, a direct CG effect on the synthesis of glycodelin (Cameo et al. 2004). Simultaneous activation of Notch-1 by CG and progesterone should have an antiapoptotic effect on the endometrium. A special role of CG on the endometrium was observed because the LHCGR in endometrial cells does not activate adenylate cyclase and thus increase cAMP but induces via a rapid phosphorylation of ERK 1/2 (*extracellular-signal-regulated kinase*) cyclooxygenase synthesis and thus prostaglandin E_2-synthesis. It could be possible that this signal transduction is due to a splice variant of LHCGR (Cameo et al. 2004).

Phylogeny

This is presented in Sect. 4.4.3.4.

4.4.3.2 FSH

Fact sheet 4.19: Follicle stimulating hormone (FSH)

Gene α chain:	Chromosome 6 (6q12-q21); four exons.
Gene β chain:	Chromosome 11 (11p13); three exons.
Sequence:	Fig. 4.23.
Synthesis and target:	FSH is formed and released by gonadotropic cells of the pituitary upon GnRH stimulation. Target of FSH are FSH receptors (FSHR) in the gonads: in men on Sertoli cells and in women on follicular granulosa cells.
Receptor:	Heptahelical GPCR: FSHR activates the adenylate cyclase.

Introduction

In the ovary FSH (aka follitropin) induces follicle maturation. In testis FSH stimulates sperm-forming Sertoli cells.

Structure and Genes

The FSH β chain is coded for on chromosome 11 in three exons. Its gene product associates, as do other glycoprotein hormones, with the α chain. Dias (2001) and others have shown that the mode of N-glycosylation of the arginines N^7_β and N^{31}_β influences the potency of FSH: different patterns of glycosylation are differentially active. Some patterns in fact are endogenous antagonists and block FSHR activity.

Physiology

In the pituitary LH and FSH are synthesized by the same gonadotropic cells. No hint exists that these two hormones are sorted into selected vesicles which then might secrete either LH or FSH. On the contrary, tracer experiments suggest that all intracellular vesicles contain both LH and FSH. With reference to Sect. 11.3 there are, however, different LH and FSH levels at different stages of the menstrual cycle. Elevated estradiol suppresses serum FSH concentration although LH remains high. In sheep and rats an estrogen-responsive element (ERE) in the FSHR gene promoter is used to block FSH transcription. Such an ERE is lacking in the LH receptor gene. In contrast to FSH regulation is the LH synthesis in pituitary gonadotropic cells not inhibited but stimulated. A blocking effect of estradiol on LH synthesis thus must take place in the hypothalamus where GnRH formation is inhibited by estradiol.

FSHR are present on granulosa cells in the wall of ovarian follicles and on Sertoli cells in testis.

4.4.3.3 GPHA2B5

Fact sheet 4.20: Thyrostimulin (GPHA2B5)

Gen α-chain-2:	Chromosome 11 (11q13); three exons.
Gen β-chain-5:	Chromosome 14; two exons.
Sequence:	α-chain MPMASPQTLV LYLLVLAVTE AWGQEAVIPG CHLHPFNVTV RSDRQGTCQG SHVAQACVGH CESSAFPSRY SVLVASGYRH NITSVSQCCT ISGLKKVKVQ LQCVGSRREE LEIFTARACQ CDMCRLSRY β-chain MKLAFLFLGP MALLLLAGYG CVLGASSGNL RTFVGCAVRE FTFLAKKPGC RGFGSPRMPA GVAVRPGRNP FWNPPILKPI IESVPTTRPN R
Synthesis and target:	GPHA2B5 is expressed in some corticotropic pituitary cells, in the retina, in skin, and in testis. It activates the TSHR.
Receptor:	Heptahelical GPCR: TSHR activates the adenylate cyclase.

Introduction

In 2002 Hsu et al. described an additional glycoprotein hormone that they had identified at first by sequence analysis of the published human genome and known mRNA sequences and whose expression they investigated later on. Most surprisingly, this GPHA2B5 was also found and expressed in invertebrates.

Structure and Genes

The second human GPH α-chain gene was localized to chromosome 11 and contains three exons, and the fifth GPH β-chain to chromosome 14 with two exons. The GPHA2 protein is up to 35 % homologous to the common α-chain of LH/FSH/CG/TSH, however, 9 of 10 cysteine residues are conserved with respect to position and spacing. In GPHB5 10 of 12 cysteines are conserved including those of cysteine knot. Nakabayashi et al. (2002) could show that the two chains associate. The presence of a functional dimer was, however, challenged by Alvarez et al. (2009) showing that the structural requirements for dimerization are most probably not met and if at all only at very high protein concentrations. To date any functional dimer has not been isolated from any species. In invertebrates, the genes for GPHA2 and GPHB5 are linked, although this linkage is lost in vertebrates.

Physiology

GPHA2B5 was shown to activate the TSH receptor (Nakabayashi et al. 2002) and induce thyroid hormone. This led to its name, thyrostimulin.[17] Compared to the expression of GPHA2 in the anterior pituitary the GPHB5 expression in 2000-fold reduced (Nagasaki et al. 2006). Double positive for both chains were only a few corticotropic pituitary cells (Okada et al. 2006). A simultaneous expression of both chains was also found in humans and rat in the eye, skin, ovaries, and in testis with unknown impact. In contrast to TSH release, thyrostimulin release is not stimulated by TRH.

Regulation of both genes seems to be differentially regulated, *e.g.* in lancelets, where the distribution of both genes is shown to differ (Dos Santos et al. 2009) The expression of GPHA2 in human pancreas is regulated by the organ-specific transcription factor isl-1 (Suzuki et al. 2007) and not dependent on triiodothyronine or TRβ1. GPHB5 regulation might probably be modified by inflammatory cytokines that induce NKκB to translocate to the nucleus and bind to a responsive element on the GPHB5 promoter (Suzuki et al. 2009).

Whether homodimers of GPHA2 are by themselves active is an open, but testable, question. In lancelets *i.e.* at the beginning of vertebrate evolution, both GPHA2 and GPHB5 are expressed, GPHA2, however, in a much more restrictive way.

Exploring the role of these novel gonadotropin chains, van Zeijl et al. (2011) recently generated GPHB5 knockout mice. These knockout mice displayed a remarkable reduction of serum thyroxine. The authors concluded that the lack of GPHB5 influences the hypothalamic pituitary thyroid axis and that GPHB5 might have a role during development (Trudeau 2009, see also). Tando and Kubokawa (2009) analyzed expression of GPHA2B5 in the prevertebrate lancelets where the genes were expressed in an analogue of vertebrate pituitary. The finding gives rise to the hypothesis that GPHA2B5 is the precursor of all gonadotropins.

In *D. melanogaster*[18] fly gpha2b5 activated the orphan receptor DLGR1, but not the related DLGR2. Fly gpha2b5 did not activate the human TSHR, but a chimeric fly gpha2/human GPHB5 did which is interpreted as the conservation of the functional structure over millions of years (Sudo et al. 2005).

Sellami et al. (2011) reported expression of Gpha2 and Gphb5 in the same few abdominal neurons in drosophila, in larvae as well as in adults. They did not isolate the protein.

[17]The name thyrostimulin, however, has been used since 1950 for an at that time unknown compound with thyroid-activating properties (Roche et al. 1950).

[18]Drosophila melanogaster.

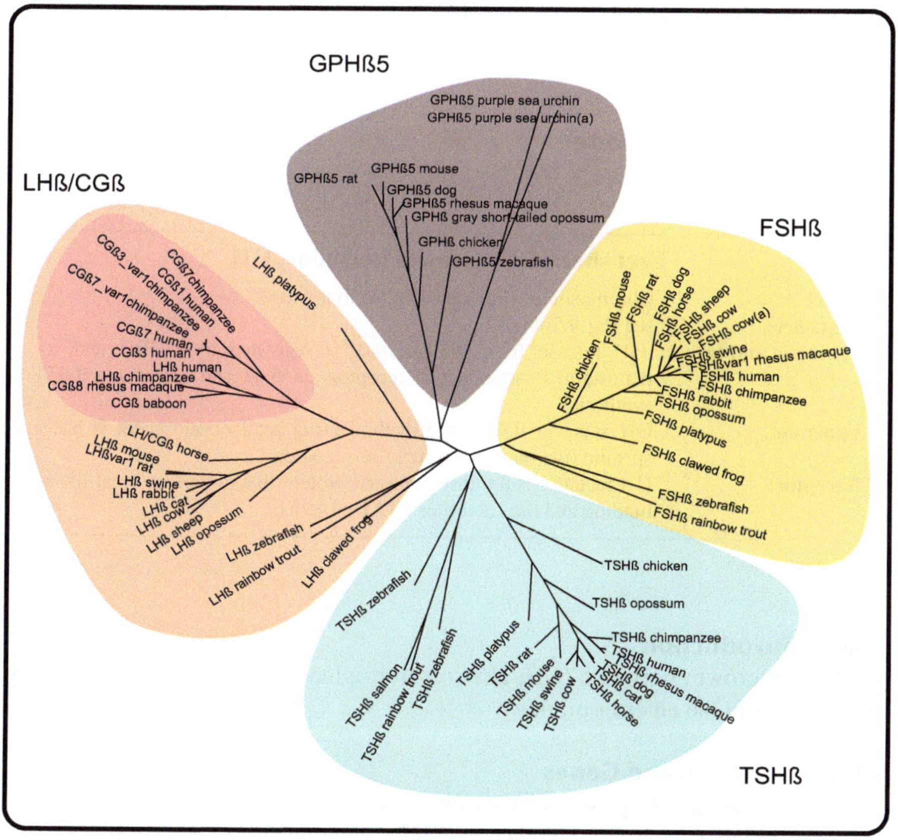

Fig. 4.25 Evolutionary tree of glycoprotein hormone β chains. Using 65 reference sequences from GenBank and ClustalW (neighbor joining method/1,000 bootstrapping steps) phylogenetic dependencies were identified and drawn using the program TreeDyn. Four trees matching the four β chains of follicle-stimulating hormone (FSH), choriogonadotropin/luteinizing hormone (CG/LSH), thyroid stimulating hormone (TSH), and glycoprotein hormone B5 (GPHB5) can be separated. The primate CG subtree is part of the LH tree. Please note that horse CG is a LH. CG from *Oryctolagus cuniculus* (AF362079; rabbit) is obviously a database error. Of closely related variants, only one variant is shown. Some CG-β chains were omitted for clarity reasons

4.4.3.4 Phylogeny of Glycoprotein Hormones

Figure 4.25 demonstrated that separation of LH/FSH/TSH predates vertebrate development or is at its very earliest stage. Dos Santos et al. (2009) have shown that separation of GPHA2 and the common GPHA1 genes coincides with vertebrate origin. They have also shown that the linkage to one chromosome got lost during these stages. The more ancient GPHA2 and GPHB5 genes and proteins are of bilaterial origin lacking in cnidarians and can be found in most of the species analyzed with a few exceptions such as the honeybee although other insects have these genes and GPHA2 is not found in chicken. The uncertain role of thyrostimulin for deiodothyronine/thyroxine release has been discussed in the last chapter.

Note that horse choriogonadotropin does not belong to the CG tree, but belongs to the LH tree. Different CG/LH genes are only present in primates.

4.4.4 Growth Hormone

Fact sheet 4.21: Growth hormone, GH

Gene:	Chromosome: 17; locus: 17q24; four exons.
Sequence:	see Fig. 4.26.
Synthesis and target:	GH release from the pituitary is regulated by GHRH and somatostain. It acts on GH receptors in liver and stimulates IGF-1 synthesis.
Function:	GH acts via IGF1 on growth. During fetal development it has a specific role on nerve system development.
Receptor:	GH receptor is a transmembrane protein that dimerizes upon ligand binding and thus activates STAT proteins.

4.4.4.1 Introduction

The human growth hormone (GH, aka somatotropin, Fig. 4.27) is made by somatotropic cells in the anterior pituitary.

4.4.4.2 Structure and Genes

Within the human *GH* gene locus on chromosome 17 (17q22-q24) for pituitary GH there are also genes for the placental GH variant and three other prolactin-like hormones. (chorionic somatomammotropins (CSH); Chen et al. 1989[19]). Variants arise by alternative splicing, the regulation of which is not yet known. Pituitary GH has 191 amino acids; with alternative splicing a 20 amino acid variant arises instead of the 22 kDa[20] proteins.

In Fig. 4.26 exon-intron boundaries are shown by vertical bars. Variants of GH (lines 2–3) arise by a splicing that removes exon three and/or four (items 01 and 02). The lack of exon three and four modifies the tetrahelical structure (see Fig. 4.27). There some cases of growth hormone deficiency where such an afunctional variant blocks the GH receptor inhibiting the function of GH. GH_var2 (item 03) results from the use of alternative splice acceptor site in exon3.

The GH crystal structure has been determined. In Fig. 4.27 you may notice the high degree of helices characteristic for this hormone. The different helices are

[19] http://www.ncbi.nlm.nih.gov/projects/mapview/maps.cgi?TAXID=9606&CHR=17&MAPS=genec%2Cgenes-r&QSTR=somatotropin&QUERY=uid%28-2146581366%29&BEG=63%2C800K&END=63%2C950K&oview=default

[20] Kilodalton.

[21] Protein Data Bank.

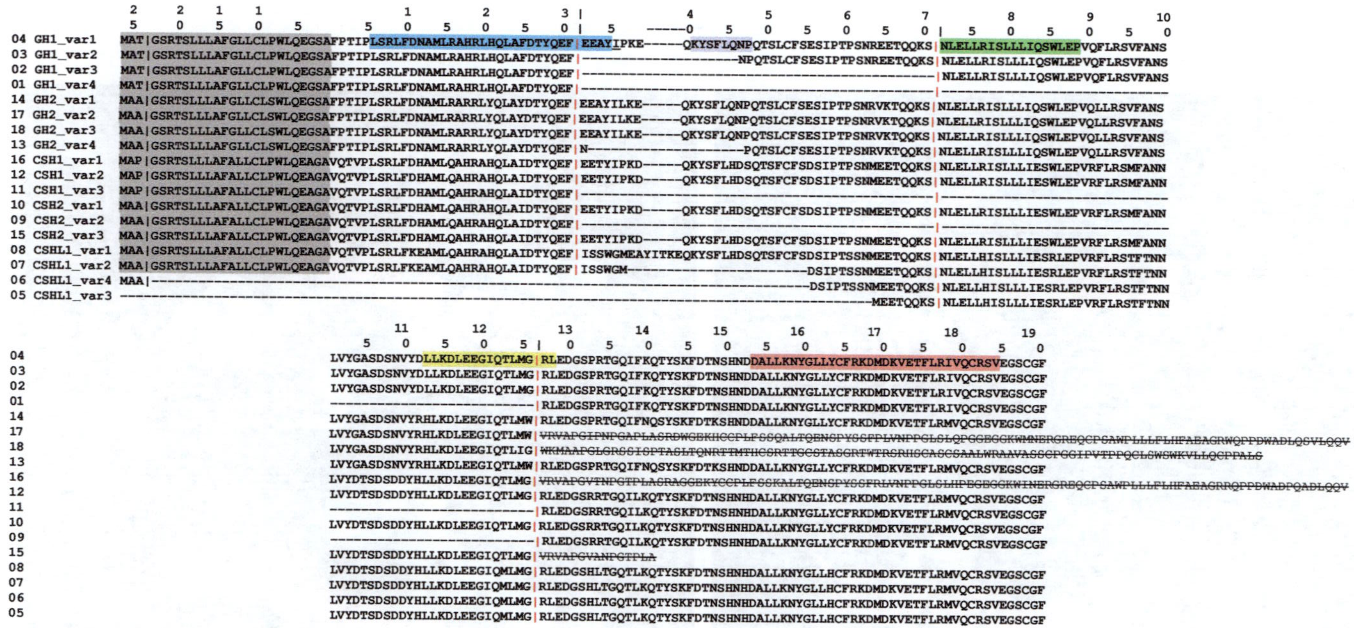

Fig. 4.26 Comparison of human somatotropin splice variants. Exon boundaries are marked by *vertical lines*, lacking amino acids by ---. With the exception of GH1_var2 gaps coincide with exons. Numbering is according to GH_var1. In variants 15, 16, 17, and 18, intron 4 is not removed which results in new unrelated sequences (*crossed out*). The helices of GH!_var1 are *color coded*. (sources: GH1_var1 NP_000506.2 (04); GH1_var2 NP_072053.1 (03); GH1_var3 NP_072054.1 (02); GH1_var4 NP_072055.1 (01); GH2_var1 NP_002050.1 (14); GH2_var2 NP_072051.1 (17); GH2_var3 NP_072052.1 (18); GH2_var4 NP_072050.1 (13); CSH1_var1 NP_001308.1 (16); CSH1_var2 NP_072166.1 (12); CSH1_var3 NP_072167.1 (11); CSH2_var1 NP_066271.1 (10); CSH2_var2 NP_072170.1 (09); CSH2_var3 NP_072171.1 (15); CSHL1_var1 NP_072101.1 (08); CSHL1_var2 NP_072103.1 (07); CSHL1_var3 NP_001309.3 (05); CSHL1_var4 NP_072102.1 (06))

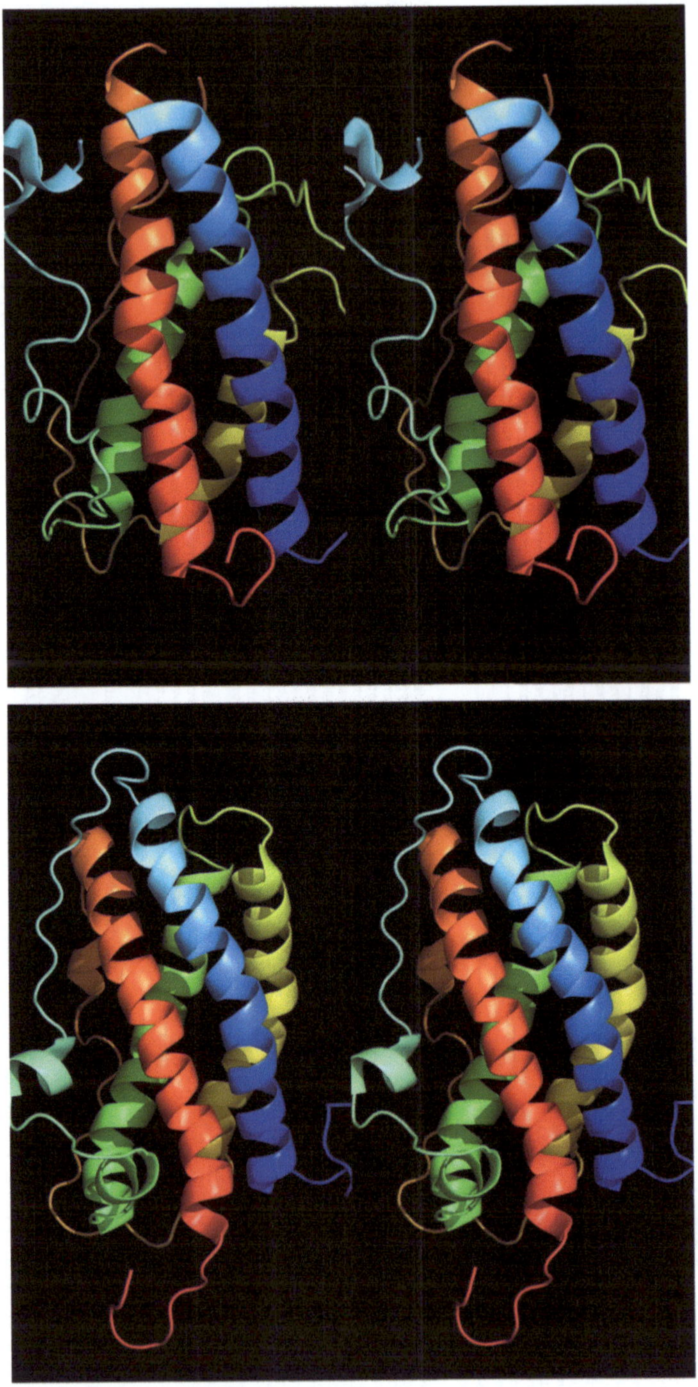

Fig. 4.27 Stereo views of the growth hormone (*above*) and the prolactin molecule (*below*) (Source: Chantalat et al. 1995; PDB: 1HGU (GH); Keeler et al. 2003; PDB: 1N9D, Model 1 (prolactin))

underlined in Fig. 4.26. Activation of the receptor leads to dimerization upon ligand binding: one GH molecule binds to one receptor. The dimer recruits the second receptor molecule.

4.4.4.3 Physiology

GH acts directly using the GH receptor or mediated by insulin-like growth factor 1 (IGF1). IGF1 is formed in hepatocytes stimulated by GH. GH release is pulsatile and reaches (in children) maximal rates about 1 h after the onset of deep sleep. Adult men have six to eight pulses in 24 h whereas women have a less regular pulse rate; in addition women release, per pulse, higher GH rates than men, most probably due to a stronger estrogen influence.

Pertinently elevated GH concentrations *e.g.* due to pituitary tumors may lead to acromegaly, whereas a GH deficit is at the origin of dwarfism. This condition may also be due to nonfunctional GH receptors.

4.4.4.4 Phylogeny

As are prolactin and leptin, GH is a member of the so-called class I cytokines such as interleukin-6, IL11, IL12, erythropoietin, and G-CSG (Huising et al. 2006). The phylogenic context of these proteins suggests a prevertebrate origin (Huising et al. 2006). An IL-6 receptor homologue has been found in flies, but in the literature and in GenBank no class-I cytokine from an invertebrate species has been described although a couple of invertebrate species genomes have been completely sequenced.

The phylogenetic tree of GH has some peculiarities according to Huising et al. (2006): GH genes from sturgeon seem to be more closely related to mammalian GH than to those from amphibians. In order to explain this discrepancy the authors discuss different mutational velocities.

In fish there is another hormone, somatolactin that does not have a homologue in other vertebrate families. This is presented in Sect. 4.5.1.

4.4.5 Prolactin

4.4.5.1 Introduction

Prolactin is produced by lactotropic cells of the pituitary. These develop from common GH- and prolactin-producing cells.

4.4.5.2 Structure and Genes

In the GH gene cluster on chromosome 17 several somatomammotropic, placental, and prolactin-related proteins are coded for (see the previous section), the prolactin gene, however, is found on chromosome 6. Homology between GH and prolactin with respect to amino acids is mainly due to the conservation of disulfide bridges. The crystal structure of prolactin, however, is similar to that of GH: both have the same four long helices (Keeler et al. 2003) (Fig. 4.27). Complexed with zinc GH binds to the prolactin receptor, confirming strong structural similarity between prolactin and GH. Whether prolactin binds to the GH receptor is not described.

Fact sheet 4.22: prolactin (PRL)

Gene: Chromosome: 6; locus: 6p22.2-p21.3; five exons.

Sequence: see Fig. 4.28.

Synthesis and target: Prolactin is formed by lactotropic pituitary cells and acts on prolactin receptors: these are located in the woman's breast, in the ovary, in testis, sexually independent in the CNS, and in adipose tissue and several other organs.

Function: Prolactin regulates milk production and supports the functions of the corpus luteum. Prolactin plays a role in the establishment of mother/child relations.

Receptor: Prolactin receptor is a transmembrane receptor which on ligand-induced dimerization stimulates the JAK/STAT signal transduction pathway.

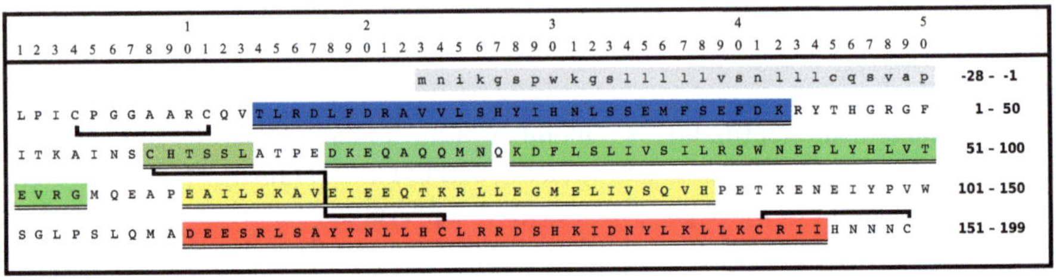

Fig. 4.28 Primary structure of prolactin: disulfide bridges are indicated by *black lines*, helices are *colored* according to Fig. 4.27, arginine 31 (N; *green on black*) is potentially glycosylated (Source: GenBank NP_000939, Swiss-Prot P01236)

Prolactin can be differentially glycosylated (Lewis et al. 1989); N-glycosylated molecules appear to be more active.

4.4.5.3 Physiology

Despite many efforts, a hypothalamic-releasing peptide for prolactin has not been found to date. Prolactin secretion is negatively controlled by the catecholamine dopamine (see Chap. 7). Prolactin is released, like GH, in pulses during sleep. TRH (Sect. 4.3) is one stimulus of prolactin release. Recently, a potent prolactin-releasing peptide (PRLRP) from the pars intermedia has been cloned and functionally tested in humans and rodents (see Sect. 4.5.2). The PRLRP receptor could also be identified. In contrast to the other hypothalamic-releasing hormones, no PRLRP axons could be found in the median eminence, furthermore PRLRP is made in the pars intermedia. Whether PRLRP is thus a true hypothalamic prolactin-releasing hormone is doubtful. In addition, the pars intermedia is degenerated in adults. Whether PRLRP has a different role in nonprimates remains to be answered.

During pregnancy maternal prolactin promotes—in concerted action with estrogens, progesterone, and placental lactogens—functional maturation of mammary glands. After birth prolactin promotes milk production.

The tumor of prolactin-producing lactotropic cells, the prolactinoma, is the most frequent tumor in the pituitary. In a subgroup of cases prolactin is secreted together with GH. Because normal lactotropic cells are controlled by dopamine, in many cases the prolactinoma can be controlled by the dopamine agonist bromocriptin.

Inasmuch as strongly elevated prolactin serum levels caused by prolactinoma cells block pulsatile GnRH release from the hypothalamus and thus pituitary gonadotropin release, prolactin has been regarded as a contraceptive drug. The prolactin levels during lactation, however, are not sufficient to inhibit follicle maturation and ovulation.

In male animals prolactin controls the yearly rhythms of testis function and thus sexual activity: prolactin stimulates testis growth. In men prolactin has been found elevated after orgasm. Prolactin receptors are present on Leydig cells, spermatids, spermatocytes, and widely distributed in testis epithelium (Bartke 2004). For these reasons prolactin is regarded as important for reproductive activity. In addition, prolactin controls maternal functions such as nest building and brood care (Freeman et al. 2000).

4.4.5.4 Phylogeny

Like GH, prolactin is a class-I cytokine which has been found only in vertebrates.

4.5 Hormones of the Pars Intermedia

4.5.1 Somatolactin

Fact sheet 4.23: Somatolactin

Sequence:	Fig. 4.29.
Synthesis and target:	Somatolactin is made in the fish's pars intermedia.
Receptor:	A cloned somatolactin receptor from salmon is a class-1 cytokine receptor.

4.5.1.1 Introduction

As mentioned earlier, in fish an additional pituitary homologue of GH and prolactin has been found, somatolactin. Somatolactin is produced in the intermediate lobe. There are two histologically distinct cell types, one expressing POMC and secreting MSH, whereas the other cells were found in 1990 to produce somatolactin (Ono et al. 1990). As in lactotropic and thyrotropic cells, SL expression is dependent on the pituitary transcription factor Pit1.

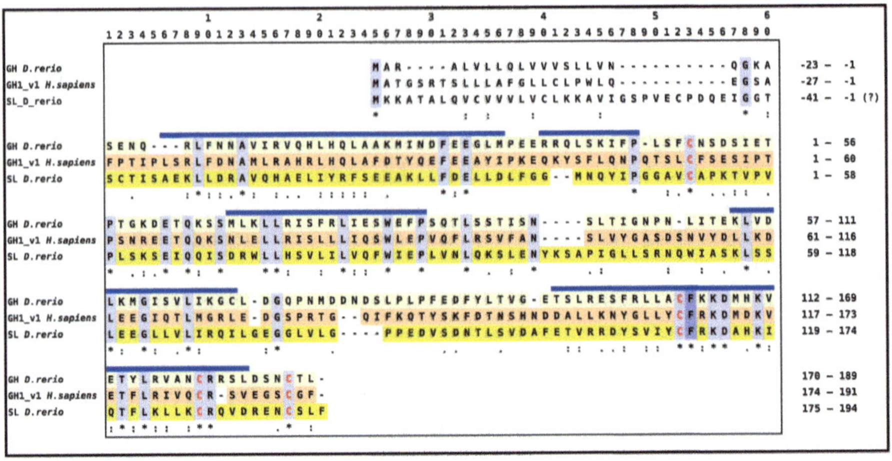

Fig. 4.29 Primary structure of somatolactin (*yellow*) from zebrafish compared to growth hormones (GH) from zebrafish (*light yellow*) and humans (*light orange*). Four cysteine residues (*red on light blue*) are conserved. As does prolactin, somatolactin has an additional N-terminal disulfide bridge. Identical amino acids are in *light blue*. The program ClustalW indicates highly conserved (:) and partially conserved (.) amino acids. Overall homology between GH and SL from zebrafish is 24 % in relation to somatolactin and between human GH and zebrafish somatolactin 22 % respectively. The helices of human GH are indicated by *blue* bars (Sources: NP_001032763, NP_001018328, NP_000506; ClustalW version 2.0.8)

4.5.1.2 Biochemistry and Genes

Somatolactin is a class-I cytokine such as GH and prolactin. The sequence homologies are only marginal; the tetra-helical structure, however, as well as the position and spacing of the cysteine residue are conserved. The somatolactin receptor appears to be related to GH and prolactin receptors.

4.5.1.3 Physiology

The functional role of somatolactin is far from being well understood. It has been observed that it may influence HCO_3^- exchange (Kakizawa et al. 1997); during sexual maturation somatolactin was found to increase (Taniyama et al. 1999). The *ci* mutant from medaka (*Oryzias latipes*) bears a somatolactin deletion that results in inappropriate adaptation of pigment cells to environmental changes. Other changes compared to wildtypes have not been detected. This led the authors to suggest that somatolactin preferentially inhibits proliferation and morphogenesis of leucophores.

Very recently Wan and Chan reported ectopic expression of somatolactin genes in zebrafish that lead to an increased transcription of IGFs, insulin, leptin, sterol regulatory element binding protein 1, and fatty acid synthase, and enhanced expression of vitellogenin and proopiomelanocortin.

The above-mentioned *ci* mutant with reduced orange pigments made the mutant males unattractive to females. This observation prompted experiments where somatolactin was overexpressed (Fukamachi et al. 2009). It could be shown that somatolactin overexpression enhanced orange pigmentation and simultaneously male attractiveness. These findings support the role of somatolactin in pigmentation and reproduction. It should also be mentioned that the somatolactin receptor was not found to be modulated in tilapia when transferred from seawater to freshwater although GH receptors were increasingly expressed and prolactin receptors reduced (Breves et al. 2010).

4.5.1.4 Phylogeny

Somatolactin is fish specific. In some species, two different somatolactin genes are present that are most probably due to the third karyotype duplication which happened only in fish.

4.5.2 Prolactin-Releasing Peptide

Fact sheet 4.24: Prolactin-releasing peptide (PRLRP)

Gene:	Chromosome 2; locus 2q37; two exons.
Sequence:	Fig. 4.30.
Synthesis and target	PRLRP is formed in the brain and the hypophyseal pars media, and perhaps in other tissues, too. PRLRP receptors have been identified in the anterior pituitary as well as in other tissues. How PRLRP can target the anterior pituitary is unknown.
Function:	Release of prolactin.
Receptor:	The GPCR hGR3 specifically binds PRLRP.

4.5.2.1 Introduction

Prolactin-releasing peptide (PRLRP) is the most promising candidate for the long-term evasive hormone that triggers prolactin release from lactotropic pituitary cells (Hnasko et al. 1997). PRLRP has been described as formed in the hypothalamus of different species but it is (if at all) an unusual releasing hormone: liberation of the mature peptide from the precursors requires PC2 instead of PC1 because an **KR** motif is lacking. In the rat brain, remarkably, there were no PRLRP neurons in the median eminence which strongly argues against a normal hypothalamic pituitary transport.

4.5.2.2 Structure and Genes

The RNA transcribed from the gene on chromosome 2 gives rise to an 87 amino acid precursor protein. After the removal of the 21 amino acid long signal peptide the precursor is further processed by PC2 cleaving at the **RR** motif. Thus the 32 amino acid long PRLRP is made. Alternatively the N-terminal **SRTHRHSMEIR** can be removed by a thus far unknown endopeptidase that results

1 2 3 4 5 6 7 8 9 0 1 2 3 4 5 6 7 8 9 0 1 2 3 4 5 6 7 8 9 0 1 2 3 4 5 6 7 8 9 0	
m k v l r a w l l c l l m l g l a l r g a a	-21 – -1
S R T H R H S M E I R T P D I N P A W Y A S R G I R P V G R F g **R R** r a t l g d	1 – 40
v p k p g l r p r l t c f p l e g g a m s s q d g	40 – 65

Fig. 4.30 Primary structure of prolactin-releasing peptides (PRLRP). The precursor protein has 87 amino acids including the signal peptide with 21 amino acids (on *gray background*). Prohormone convertase-2 cleaves after the **RR** motif; the C-terminus of the free peptide *bold, uppercase on blue backround* is amidated. An alternatively shortened PRLRP has only 20 amino acids (on *lighter blue background*) (Source: Genbank P81277)

in the **TPDINPAWYASRGIRPVGRFGRR**. Both peptides are C-terminally amidated and are stored in granules. PRLRP was found to be expressed preferentially in the medulla oblongata, then in the hypothalamus and the pituitary pars media.

4.5.2.3 Physiology

PRLRP was identified as the product of cell lines that originate from the intermediate lobe of the pituitary pars media, Hnasko et al. (1997)). Very soon the role of PRLRP as a PRL-releasing hormone was questioned. PRLRP is expressed in the hypothalamus. But one structural requirement for a releasing hormone acting in the pituitary is not met: axons of PRLRP neurons do not reach the median eminence, and the peptide thus cannot get into the portal system and reach the pituitary (Maruyama et al. 1999). In addition the PRL-releasing activity of PRLRP is much less than that of TRH. It remains doubtful whether PRLRP is a veritable PRL-releasing hormone. The doubts are supported by findings in tilapia where PRLRP did not stimulate PRL secretion (Watanabe and Kaneko 2010).

The most remarkable phenotype of PRLRP knockout mice is obesity; this strongly suggests that PRLRP might be of broader or other activity than previous thought (Mochiduki et al. 2010). Confirming these results in mice, in goldfish PRLRP was found to reduce food uptake after intracerebral or intraperitoneal injection (Kelly and Peter 2006).

4.5.2.4 Phylogeny

PRLRP has been observed only in vertebrates.

4.6 Hormones of the Posterior Pituitary: Oxytocin and Vasopressin

4.6.1 Introduction

Oxytocin (OXT) and vasopressin (AVP) are the two hormones of the neurohypophysis (posterior pituitary (PoP); *lobus nervosus*). Unlike the anterior pituitary hormones they are not transcribed in the pituitary, but in magnocellular neurons

Fact sheet 4.25: Oxytocin (OXT)

Gene: Chromosome 20 (20p13).

Sequence: Table 4.3.

Synthesis and target: Magnocellular neurons in the supraoptic nucleus and in the paraventricular nucleus generate oxytocin which after axonal transport is released in the posterior pituitary. OXT is made in the **CL!**[a] as well. OXT acts on the OXT receptor on smooth muscle cells.

Physiology: Oxytocin is required for relaxation of the pelvis muscles during birth. In addition it acts on the heart. Oxytocin induces brood care in mammalians.

Receptor: Oxytocin receptor is a heptahelical GPCR.

[a]**CL!**.

of the supraoptic nucleus or of the paraventricular nucleus. Axons of these neurons reach the PoP via the infundibulum (Fig. 10.2). Prohormones are axonally transported into the PoP. During the transport the precursors are processed to the mature peptides. Vasopressin's role as regulator of blood osmolarity is known from invertebrates; an oxytoxic-related peptide has been found in echinoderms, that is, invertebrates; the OXT role during birth, however, is restricted to placental mammals. Receptors for both hormones are of early evolutionary origin.

Table 4.3 Sequences of oxytocin/vasopressin analogues and their occurrences

Name	Sequence	Taxon
Arginine vasopressin	CY F QNCP R G-NH_2	Most mammals
Lysine vasopressin	CY F QNCP K G-NH_2	Pigs, marsupials
Oxytocin	CY F QNCP L G-NH_2	Mammals
Mesotocin	CY I QNCP I G-NH_2	Birds, marsupials, reptiles
Arginine vasotocin	CY I QNCP R G-NH_2	All nonmammalian vertebrates
Isotocin	CY I SNCP I G-NH_2	Fish
Glumitocin	CY I SNCP Q G-NH_2	Fish
Valitocin	CY I QNCP V G-NH_2	Fish
Aspartocin	CY I NNCP L G-NH_2	Fish
Inotocin	CL I TNCP R G-NH_2	Insects
Arginine conopressin	CI I RNCP R G-NH_2	Snails
Lysine conopressin	CF I RNCP K G-NH_2	*Hirudo*, snails
Anetocin	CF V RNCP T G-NH_2	Insects
Cephalotocin	CL I RNCP I G-NH_2	Octopus
Octopressin	CF W TSCP I G-NH_2	Octopus

4.6.2 Structure and Genes

OXT and AVP are cyclic peptides with an intramolecular disulfide bridge.

The human genes of OXT and AVP are closely linked on chromosome 20 (p13)[22] with reversed orientations. In humans the distance is 12 kb, and in mice on the syntenic region of chromosome 2 3 kb.[23]

Fact sheet 4.26: Vasopressin/adiuretin (AVP)

Gene:	Chromosome 20 (20p13).
Sequence:	Table 4.3.
Synthesis and target:	Magnocellular neurons in the supraoptic nucleus and in the paraventricular nucleus generate vasopressin that, after axonal transport, is released in the posterior pituitary. AVP acts on AVP receptors in the adrenal cortex.
Physiology:	Vasopressin is the regulator of water homeostasis.
Receptor:	AVP receptor is a heptahelical GPCR.

Despite distinct promoters transcriptional activation of OXT and AVP genes is similar with respect to osmotic stimuli. OXT and AVP bind to the OXT-R as well as to AVP-R. OXT thus might be substituted for by AVP which has been formally proven in OXT knockout mice discussed below.

Apart from OXT and AVP the precursor proteins contain additional associated polypeptides called neurophysins. During maturation and axonal transport in acidified vesicles OXT and AVP are complexed to these neurophysins. On release in a more basic environment, hormones and neurophysins are separated. The AVP prohormone bears a third glycoprotein of unknown function.

Structural analyses have found that OXT/AVP association onto neurophysins depends on the pH of the secretory granula. OXT and bovine neurophysins crystallize in a complex of two neurophysins plus one OXT.

The importance of the AVP neurophysin interaction for proper axonal transport and regulated AVP release is evident from "Brattleboro"-rat which harbors a mutation of neurophysin II with the consequence of disturbed osmolarity regulation and water uptake. Such mutations have been found in families with diabetes insipidus patients.

[22]http://www.ncbi.nlm.nih.gov/projects/mapview/maps.cgi?TAXID=9606&QSTR=5020[gene_id]&chr=20&maps=ugHs-r,genes-r&beg=2950000&end=3130000&links=off&verbose=on&compress=off&width=350&size=30

[23]Syntenic chromosomal region exhibits similar sequences of genes and their orientations.

4.6.3 Physiology of Oxytocin

OXT is the hormone of birth induction. Shortly before parturition uterus OXT receptors are massively increased. OXT is released in large amounts directly prior to birth. It is the arrival of the amnion in the cervix that triggers this sudden burst of preformed oxytocin (and vasopressin). This is called the Ferguson reflex. By intramuscular OXT injection by an obstetrician or a midwife birth is initiated. However, during pregnancy OXT antagonists block the receptor and maintain tocolysis. The prepartal OXT burst overcomes tocolysis.

The uterus muscles are additionally protected from preterm labor by the enzyme oxytocinase. Only the strong prepartal burst of OXT can overcome the enzymatic degradation and let sufficient molecules reach the OXT-R on the myometrium.

In the mammary gland this OXT burst triggers contraction of myoepithelial cells around milk-producing alveolar cells and thus presses the milk into the milk ducts (milk ejection reflex).

The complexity of these actions is further confirmed by the analysis of OXT knockout mice: these mice do not have any difficulties in giving birth. Their pregnancy period of 18.5 ± 1 days does not differ from wildtype mice. Obviously OXT triggers can be induced by other mediators.

OXT knockouts were fully fertile and gave birth to viable offspring, however, it turned out that the knockout foster mice could not provide milk to their offspring. The defect could be overcome by injection of external OXT demonstrating that milk ejection is strongly dependent on OXT.

Apart from the direct effects on the parturition OXT induces in animals social behavior patterns that secure care of the newborn: nest building, carrying the litter together, licking the brood. Such patterns could be induced by OXT in rats and in wild mice which would usually kill foreign offspring. OXT antagonists could inhibit brood care even in suckling animals. However, had brood care been triggered once it could no longer be inhibited.

OXT is not restricted to females but made by males as well. It could be shown (at least in rats) that OXT injection led to erection and ejaculation. OXT has been found together with ANP[24] to regulate natriuresis and kaliuresis in the heart .

Owing to its role as an antistress hormone, OXT facilitates pairing. The full image of OXT effects, in particular its regulation by steroids, merits many further efforts.

4.6.4 Physiology of Vasopressin (AVP, Adiuretin)

Osmotic stimuli—changes in the salt concentration of blood—are the most important regulators of vasopressin release. The magnocellular hypothalamic neurons of the supraoptic nucleus themselves are osmoreceptors. In response to changes

[24] Atrial natriuretic peptide.

in osmolarity they change the openings of their potassium channels. AVP release can also be triggered by blood pressure and blood volume changes. AVP neurons, however, are more sensitive to changes of osmolarity than to those of pressure or volume.

The prominent role of AVP is regulation of the water balance by diuresis (secretion in the kidneys) reflected by the alternative term adiuretin. AVP on the one hand and the kidney, both together with thirst, are *the* regulators of this homeostasis. If increased osmolarity leads to vasopressin release, then kidney channels retain water. The kidney may excrete between 0.5 and 20 l of water per day without influencing homeostasis. If the kidney retains the maximal amount of water and water loss still occurs, for example, by transpiration, then the brain induces thirst to input more water for maintaining the osmolar optimum.

In parallel to osmoregulation vasopressin acts as a neurotransmitter. Magnocellular neurons from the supraoptic nucleus and the paraventricular nucleus not only project into the PoP[25] but into the anterior pituitary, too, and into other parts of the brain. In the anterior pituitary AVP facilitates ACTH release.

4.6.5 Phylogeny

The phylogeny of OXT and AVP reaches back to invertebrates. AVP analogues have been observed in locusts and earthworms (Proux et al. 1987; Oumi et al. 1994). In echinoderms (sea stars), early deuterostomes, octopressin and cephalotocin proteins are already associated with neurophysins having 14 cysteines as in vertebrates.[26] The locupressin (inotocin) from *L. migratoria*[27] exists (with identical amino acid sequence and so far unknown neurophysin to date) in two peptide variants: one is the cyclic peptide as in deuterostomes, the other a peptide dimer where two intermolecular disulfide bridges couple two peptide monomers. The bridges are between cysteine residue on position 1 of the one monomer to the cysteine on position 6 of the other (Proux et al. 1987).

Fish isotocin and vasotocin genes are already on the same chromosome, but not yet in the immediate neighborhood. Agnathans only have a single vasotocin gene. In coelacanths the GnRH2 gene is already linked to the two neuropeptide genes. This conformation is maintained in frogs and chicken. In the mammalian genome there is the change to vasopressin. In marsupials a selective duplication of vasopressin and mesotocin has taken place with one VP bearing a [Lys8] variant, which can be seen in pigs as well (Gwee et al. 2008). The change from mesotocin to oxytocin and the inversion of the OXT gene is characteristic of the clade Eutheria.

[25] Posterior pituitary.

[26] GenBank: BAD93372 und BAD93373.

[27] Locusta migratoria.

Vasotocin and vasopressin have been found in all analyzed species to be functionally active in osmolarity regulation.[28] The roles of mesotocin, oxytocin, and isotocin, in contrast, are more divergent.

In nonmammalian species especially in birds and marsupials oxytocins are important for reproduction: In chicken arginine vasotocin is required for egg laying; marsupials need mesotocin during birth, for example, in *Macropus eugenii* (tammar wallaby) (Siebel et al. 2005). In *Trichosurus vulpecula* (common brushtail possum) mesotocin regulates the seasonal size changes of the prostate (Fink et al. 2005). Analogues are listed in Table 4.3.

4.7 Regulators of Sugar and Energy Metabolism

4.7.1 Insulin

Fact sheet 4.27: Insulin (INS)

Gene:	Chromosome: 11; locus: 11q5 ; three exons.
Sequence:	Fig. 4.32.
Synthesis and target:	Human insulin is formed inβ cells of Langerhans islets and is used by insulin receptors on almost any cell for glucose transport therein.
Function:	Insulin regulates the usage of glucose in liver and other cells. In the liver it stimulates sugar storage and amino acid synthesis. In muscle and adipose tissue the incorporated glucose is used for ATP production and fatty acid synthesis.
Receptor:	Insulin receptor is a transmembrane receptor with tyrosine kinase activity. Because the insulin receptor does not have SH domains, an INS receptor substrate called a molecule is required for the initiation of a signal cascade.

4.7.1.1 Introduction

Insulin, the hormone of β cells in Langerhans islets of the pancreas, is indispensable for sugar metabolism.

4.7.1.2 Biochemistry and Genes

The mature hormone consists of two peptide chains called A and B that are linked via two intermolecular disulfide bridges. Proinsulin bears a C-peptide linking the A and B chains. The name of the chains had already been established before the gene was cloned and it was found that the B chain lies before the A chain (Fig. 4.31). By

[28] Vasotocin regulates oocytes maturation and ovulation in fish. (Joy KP, Chaube R (2015) Vasotocin – A new player in the control of oocyte maturation and ovulation in fish. Gen Comp Endo 221:54–63)

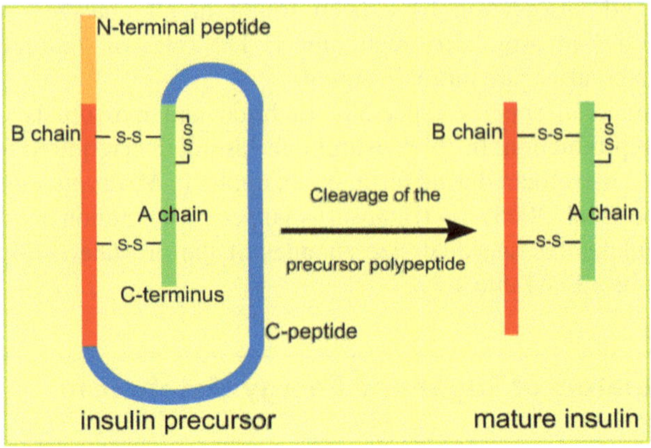

Fig. 4.31 Insulin structure

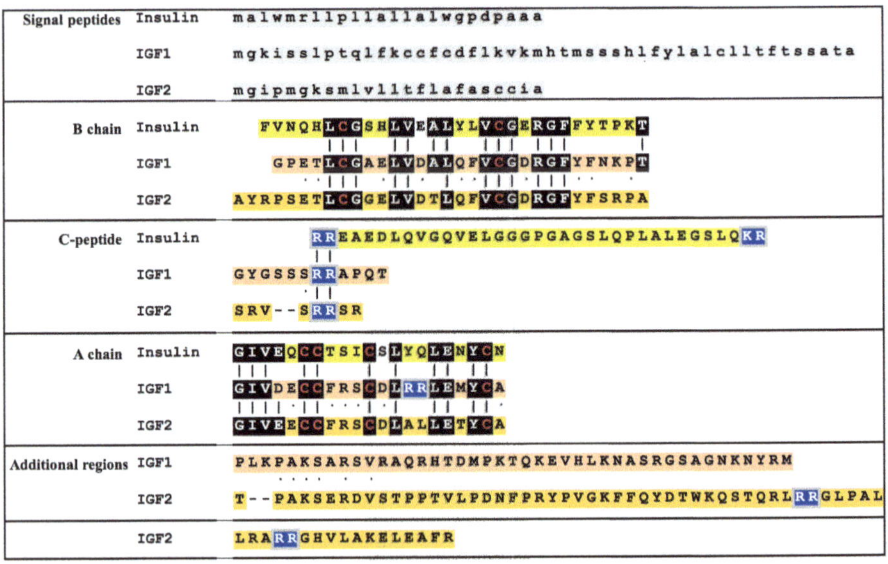

Fig. 4.32 Sequence comparison of insulin and insulin-like growth factors (IGF) (Source: GenBank: NP_000198, NP_000609.1, NP_000603.1)

peptidases (e.g., carbopeptidase-H or PC2) the C-peptide is cleaved, whereas the A and B chains remain linked by the disulfide bridges. These enzymes furthermore remove the N-terminus of proinsulin..

The insulin gene on chromosome 11 is closely linked to tyrosine hydroxylase and IGF2. These genes are oriented in the same direction. Apart from primary insulin and IGF2 transcripts a mixed insulin/IGF2 transcript has been found expressed in limbs and eyes.[29]

Coded for by the same locus there is an antisense IGF2 transcript potently suppressing placental expression of IGF2. On the gene map[30] it is evident that several inherited diseases have been linked to this region. For some time it has been known the 500-kb section on chromosome 11 between the IGF2 gene and CDKN1C (cyclin-dependent kinase inhibitor 1C) is regulated by imprinting: IGF2 and insulin, but not tyrosine hydroxylase, are only transcribed from the paternal allele, not from the maternal one. CDKN1C and genes in its neighborhood are transcribed only from the maternal, but not from the paternal allele. In some diseases, for example, in the case of Wilms tumor, imprinting does not work correctly and genes are transcribed from both alleles. Some insight into the development of imprinting exemplified using the IGF2-CDKN1C chromosomal section can be found from Ager et al. (2008a,b).

4.7.1.3 Physiology

Insulin influences several key reactions of carbohydrate, protein, and lipid metabolism (see Sect. 11.4):

Glucose uptake: In adipose and muscle cells binding of insulin to insulin receptor triggers the transport of preformed glucose transporters into the membrane. This leads directly to an increased glucose uptake into the cells.

Storage of glucose into glycogen: Triggered by insulin, glucose is stored as glycogen.

Glucose oxidation: By the oxidation of glucose to carbon dioxide energy in the form of adenosine triphosphate (ATP) is gained. This ATP is a general energy carrier in cells and forces movements, intracellular transport along the cytoskeleton, ion transports across membranes and many other reactions. ATP is the fuel on which the cell depends.

Anabolism of lipids: In adipose cells insulin triggers mobilization of the glucose transporter GlcT; glucose is utilized for fatty acid synthesis which gives rise to lipids. The lipid formation again is triggered by insulin.

Amino acid and protein synthesis: Protein synthesis from amino acids is enhanced by insulin. Furthermore insulin stimulates amino acid synthesis from liver cells to reach, for example, muscle cells and their usage in translation into proteins.

In general we can note that due to insulin triggering glucose is used for fatty acid synthesis, for storage as glycogen, for protein synthesis, and for ATP formation.

---

[29]http://www.ncbi.nlm.nih.gov/projects/sviewer/?id=NC_000011.8&v=2105322..2140619

[30]http://www.ncbi.nlm.nih.gov/projects/mapview/maps.cgi?TAXID=9606&CHR=11&MAPS=ugHs-r%2Cgenes-r&QSTR=5020[gene_id]&BEG=2M&END=2%2C300K&oview=default

Fact sheet 4.28: IGF1

Gene:	Chromosome 12; locus 12q22; five exons.
Sequence:	Fig. 4.32, two variants due to alternative splicing.
Synthesis and target	IGF1 synthesis preferentially in the liver; IGF1 receptors are ubiquitous on all cells.
Function:	IGF1 is the major mediator of insulin after birth and a regulator of growth, too.
Receptor:	IGF1-R triggers anabolic metabolism of carbohydrates and proteins; it enhances cellular proliferation.

4.7.1.4 Phylogeny

Insulin is member of a family of related proteins. Among these are *androgenic gland factor* of crustaceans, insect bombyxins (silkworm, mosquito, hawk moth, and others), insulin-like proteins from *D. melanogaster*, and relaxins from vertebrates, also further IGF. Insulin has already been isolated from sponges (Porifera). Insulin, IGF1, and IGF2 differ by the fact that the C peptide is cleaved from the insulin precursor whereas the IGF retains the entire sequence. Insect bombyxin, however, as well as the sponge insulin are built of two chains. In humans (as in other mammals) there are additional insulin-like (two chain) proteins: insulin-like 3 (INSL3; chromosome locus 19p13), insulin-like 4 (chromosome locus 9p24), insulin-like 5 (chromosome locus 1p31), insulin-like 6 (chromosome locus 9p24), as well as IGF-like (one chain) proteins: IGF-like 1, 2, 3, and 4 (all on chromosome 19). Trying to prepare a tree showing evolutionary dependencies we failed due to very low bootstrap values which indicate low significance. We therefore have resigned showing such a tree of the insulin/relaxin/IGF family. Using only the sequences of the A and B peptide, Perillo and Arnone (2014) have shown a tree that included echinoderms, hemichordates, cartilaginous and bony fish, birds, and mammals. No protostome sequence was included.

The receptors form insulin and IGF of nonvertebrates and vertebrates are, as far as analyzed, tyrosine kinase receptors.

4.7.1.5 Insulin-Like Hormone: IGF1 und IGF2

Insulin-like growth factors (IFG) differ from insulin by retaining the C peptide due to the lack of the **KR** cleavage site for PC1 or PC2 (Fig. 4.32). Cysteine residues are conserved; in the B peptide are 14 of 30 amino acids between insulin and IGF1 and 12 of 30 amino acids between insulin and IGF2 conserved; in the A peptide 11 of 21 for IGF1 and 12 of 21 for IGF are identical.

The similarity of the molecules lets insulin and IGF bind to the same receptors. Preferentially insulin binds to an insulin receptor and IGF binds to IGF-R. With reduced avidity insulin binds to IGF-R, like the binding of IGF to an insulin receptor. Combination of one extracellular chain of insulin receptor with one chain of IGF-R lets insulin and IGF bind with similar avidity to this recombinant receptor (Jones and Clemmons 1995).

IGF influences several functions in different cells. Some of these are described in Sect. 11.6. For IGF there are binding proteins IGFBP that act as transport proteins

Fact sheet 4.29: IGF2

Gene: Chromosome 11; locus 11q5; eight exons.
Sequence: Fig. 4.32.
Synthesis and target: IGF2 is expressed in many different cells and binds to insulin receptor or IGF1 receptor.
Function: IGF2 is a growth stimulus. Its regulation by imprinting and antisense transcription inhibits overshooting growth and transformation. IGF2 preferentially regulates prenatal growth.
Receptor: IGF2 binds to IGF1-R as well as to the IGF2-R, also known as mannose 6-phosphate receptor. Growth-promoting signals are mediated by IGF1-R; the role of IGF2-R appears to be IGF2 regulating.

and help to keep IGF in the bloodstream. With the help of the binding proteins, binding to the extracellular matrix is possible. Receptor interaction is very often possible once the binding proteins are digested. There are special IGFBP proteases for this digestion. These different proteins establish a regulatory system with many ways of fine tuning (Baxter 2000).

4.7.2 Glucagon

Fact sheet 4.30: Glucagon

Gene: Chromosome 2; locus 2q36; six exons.
Sequence: Fig. 4.33.
Synthesis and target: Glucagon is made in the α cells of pancreatic islets and released into the circulation. It acts preferentially on glucagon receptors on liver cells.
Function: Glucagon stimulates glucose release from glycogen stores in the liver cells.
Receptor: Glucagon receptor is a GPCR.

4.7.2.1 Introduction

The preproglucagon precursor contains apart from glucagon three additional peptides: glicentin-related peptide (GRPP) and the glucagon-like peptide 1 and 2 ((GLP1 and GLP2).

4.7.2.2 Structure and Genes

In the section on prohormone convertases (Sect. 4.2), we discussed preferential cleavage after the amino acids **KR**, **RR**, **RK** and **KK** with PC1 preferentially cleaving after **KR**. In the glucagon precursor there are several such arginine–lysine combinations: 51/52 **KR**, 69/70 **RR**, 82/83 **KR**, 89/90 **KR**, 128/ 129 **RR**, 144/145

12345678901234567890123456789012345678901234567890	
mksiyfvaglfvmlvqgswq	-20 - -1
RSLQDTEEKSRSFSASQADPLSDPDQMNEDKRHSQGTFTSDYSKYLDSRR	1 - 50
AQDFVQWLMNTKRnrnniaKRHDEFERHAEGTFTSDVSSYLEGQAAKEFI	51 - 100
AWLVKGRgRRdfpeevaiveelgRRHADGSFSDEMNTILDNLAARDFINW	101 - 150
LIQTKITDRK	151 - 160

Fig. 4.33 Primary sequence and organization of the glucagon precursor. In α-cells of the pancreas the preproglucagon is cleaved into the glizentin-related peptide (GrPP, amino acids 1–30), glucagon (amino acids 33–61) and the two glucagon-like peptides GLP1 (amino acids 78–108) and GLP2 (amino acids 126–158); in the brain and in the intestinal tract, however, glizentin (amino acids 1–69) and the glizentin cleavage products GrPP and oxyntomodulin (amino acids 33–69) as well GLP1 and GLP2 are formed. The prohormone convertase motifs are *inverted*. Those unused are shown in *light gray* (Source NP_002045)

RR, and 179/180 **RK**. By PC1 only glucagon 53–81, is released. Other prohormone convertases generate the other three peptides. GLP-1 is C-terminally amidated.

4.7.2.3 Physiology

After binding to its GPCR glucagon acts via G-proteins on the adenylate cyclase and stimulates protein kinase A. Thus phosporylase kinase and in turn phosphorylase are stimulated, the latter triggering glycogen degradation whereby glucose is made available. Additionally, gluconeogenesis is induced by activating phosphoenolpyruvate carboxykinase and Glc6Pase. The glucagon actions are exclusively directed to liver cells (see Sect. 11.4.6).

GLP-1 equally excised from the glucagon precursor, is postprandially released in endocrine cells of the intestine. It stimulates insulin release and insulin transcription, and suppresses glucagon synthesis and release.

GLP-2 is most probably a paracrine stimulator of intestinal crypts (Sherwood et al. 2000).

4.7.2.4 Phylogeny

Glucagon belongs to the superfamily of PACAP/secretin/glucagon peptides. Earliest genes for PACAP have been discovered in tunicates, glucagon is described in the earliest vertebrates. In humans apart from PACAP, secretin, glucagon, GLP-1 and GLP-2, GIP, GRF, PHM, and VIP belong to this family. The glucagon sequence is identical in humans, monkey, swine, sheep, and cow; two glucagon genes have been found in some fish species (Sherwood et al. 2000).

4.8 Regulators of Food Intake

4.8.1 Leptin

Fact sheet 4.31: Leptin

Gene:	Chromosome 7; locus 7q13.3; three exons.
Sequence:	Fig. 4.34.
Synthesis and target:	Leptin is made in adipose cells. It binds to brain and peripheral leptin receptors.
Function:	Leptin inhibits food intake and stimulates energy consumption.
Receptor:	Leptin receptor belongs to the cytokine-like receptor family.

4.8.1.1 Introduction

Leptin (Fig. 4.34; from the Greek *leptos*: thin) is the hormone of adipose cells. Using leptin these cells signal their fat content. The more fat is stored, the larger the leptin amounts released. The leptin level in blood is a measure ofthe state of nutrition.

4.8.1.2 Structure and Genes

The protein leptin resembles the growth hormone in its structure. Both belong to the family of tetrahelical cytokines such as the granulocyte-colony-stimulating factor (G-CSF).

4.8.1.3 Physiology

Leptin exhibits two major functions. First, it antagonizes hunger. Hunger is centrally mediated, for example, by neuropeptide-Y (NPY). With NPY applied or intrinsically released there is a demand for food intake. This NPY-mediated demand is blocked by leptin. Second, enhanced leptin stimulates oxygen consumption for glucose metabolism, enhancement of body temperature, and depletion of fat depots in adipose cells.

Leptin is additionally secreted by myofibroblast in order to induce surfactant production in type 2 alveolar cells. There leptin acts in a paracrine manner (Torday and Rehan 2007).

4.8.1.4 Phylogeny

Leptin has only been identified in vertebrates (Huising et al. 2006). The evolutionary tree reflects the majority view of vertebrate evolution. Some teleosts possess two leptin genes.

```
                  1                   2                   3                   4                   5
1 2 3 4 5 6 7 8 9 0 1 2 3 4 5 6 7 8 9 0 1 2 3 4 5 6 7 8 9 0 1 2 3 4 5 6 7 8 9 0 1 2 3 4 5 6 7 8 9 0

                                                            m h w g t l c g f l w l w p y l f y v q a   -21 -  -1
V P I Q K V Q D D T K T L I K T I V T R I N D I S H T Q S V S S K Q K V T G L D F I P G L H P I L T     0 -  50
L S K M D Q T L A V Y Q Q I L T S M P S R N V I Q I S N D L E N L R D L L H V L A F S K S C H L P W    51 - 100
A S G L E T L D S L G G V L E A S G Y S T E V V A L S R L Q G S L Q D M L W Q L D L S P G C           151 - 146
```

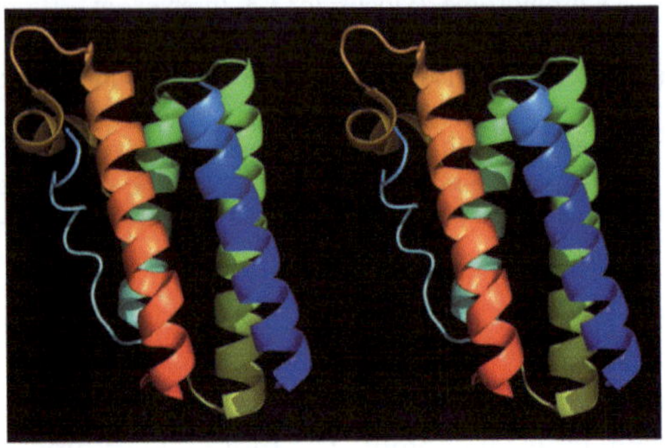

Fig. 4.34 Sequence and three-dimensional structure of leptin. *Upper:* the signal peptide is highlighted *gray*, the mature peptide *yellow*; helices are *red underlined* (Source: Protein Data Bank entry 1AX8 with PyMOL)

4.8.2 Ghrelin

Fact sheet 4.32: Ghrelin	
Gene:	Chromosome 3; locus 3p26; four exons.
Sequence:	Fig. 4.35.
Synthesis and target	Ghrelin is released in the stomach and in hypothalamic neurons. It is involved in control of GH secretion in pituitary somatotropic cells.
Function:	Ghrelin stimulates the transcription of the pit-1 transcription factor required for GH transcription.
Receptor:	Ghrelin receptor (growth hormone secretagogue receptor) is a hepta-helical GPCR.

```
1 2 3 4 5 6 7 8 9 10 1 2 3 4 5 6 7 8 9 20 1 2 3 4 5 6 7 8 9 30 1 2 3 4 5 6 7 8 9 40 1 2 3 4 5 6 7 8 9 50

                                                    mpspgtvcsllllgmlwldlama   -23 - -1
GSSFLSPDHQRVQQRKESKKPPAKLQPRalagwlrpedggqaegaedeme                          1 - 50
vrFNAPFDVGIKLSGVQYQQHSQALgkflqdilweeakeapadk                               51 - 94
```

Fig. 4.35 The primary sequence of ghrelin. The signal peptide (on *gray background*) is followed by the ghrelin sequence (*bold* on *yellow background*). On the same precursor there is an additional peptide with antagonistic actions at the ghrelin receptor: obestatin (on *light blue background*) (Zhang et al. 2005)

4.8.2.1 Introduction

Small synthetic peptides and other nonpeptide molecules had been reported for quite some time to stimulate GH release from the pituitary. The receptor that transfers the signals is called the growth hormone secretagogue receptor GHS-R1. In 1999, the endogenous ligand for this receptor was discovered by a Japanese team: ghrelin (Kojima et al. 1999).

4.8.2.2 Structure and Genes

Ghrelin is a peptide of 28 amino acids (Fig. 4.35) with an octanoic acid ester group on serine in position 3 for maximal functional activity. The esterification by octanoic acid is without precedent in the literature. The analogue des-gln14-ghrelin, by alternative splicing lacking a glutamine residue at position 14, once esterified by octanoic acid is as fully active as ghrelin. The first five amino acids including the esterified serine in position 3 appear sufficient for ghrelin activity.

The O-acyl transferase adding an octanoyl group to ghrelin has recently been described (Yang et al. 2008; Gutierrez et al. 2008).

The majority of ghrelin molecules, however, are not esterified. According to a recent publication, such a desacyl ghrelin is an ghrelin antagonist in mice (Asakawa et al. 2005).

There is a ghrelin-associated peptide cleaved from the common precursor and which may give rise to another peptide with potential C-terminal amidation. Whether this peptide called obestatin is an ghrelin antagonist, as well, is debated.

4.8.2.3 Physiology

Octanoyl-ghrelin is synthesized in endocrine cells of the GI tract, preferentially in the X/A-like stomach cells. This ghrelin release precedes food intake. Postprandially, ghrelin is suppressed.

Ghrelin's role as regulator of food intake is amplified by the fact that certain neurons in between hypothalamic nuclei also release ghrelin, which then triggers release of NPY and/or AgRP from neurons of the arcuate nucleus. These NPY and AgRP are those mediators eliciting hunger.

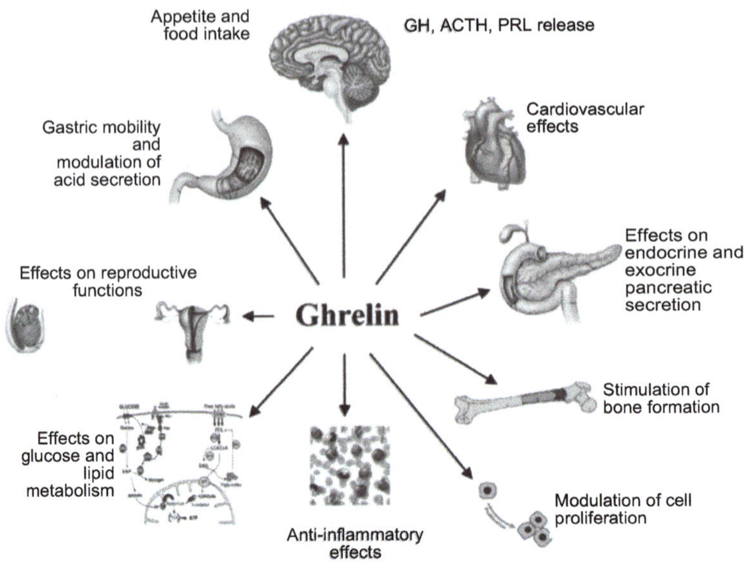

Fig. 4.36 Known effects of ghrelin (From De Vriese and Delporte (2008); Ghrelin: a new peptide regulating growth hormone release and food intake. Int J Biochem Cell Biol 40:1420–1424; with permission of Elsevier)

De Vriese and Delporte review in 2008 different actions of ghrelin: enhancement of GH, ACTH and PRL release, effect on endocrine and exocrine pancreas functions, increase in bone formation, influence on cellular division, anti-inflammatory effects, action of glucose and fat metabolism, stimulus of reproductive activity, triggering gut mobility and acid secretion, appetite stimulus, and boosting of food intake. Because ghrelin was only discovered 12 years ago, further actions might be discovered. Some the actions of ghrelin are sketched in Fig. 4.36.

Totally surprising was the observation that ghrelin knockout mice neither exhibit strong defects, nor are these mice especially obese (Kim et al. 2015).

4.8.2.4 Phylogeny

Until now, ghrelin is only found in vertebrates. In teleosts, too, ghrelin is a stomach hormone triggering pituitary GH and PRL release. In eels ghrelin is esterified by octanoic as well as by decanoic acid; in the cichlid *Tilapia oreochromis* it is exclusively esterified by decanoic acid (Kaiya et al. 2003b,a). A ghrelin-like peptide from shark is likewise acylated; whereas all other vertebrate ghrelin analyzed possess the N-terminal sequence **GSSF**, shark have **GVSF**. Such a peptide was able to enhance intracellular calcium in hamster cells expressing GHS-R. With respect to invertebrates, it has been found that the GHS-R is related to pheromone-biosynthesis–activating neuropeptide (PBAN); PBAN, however, does not show any homology to ghrelin (Choi et al. 2003).

The obestatin peptide from the ghrelin precursor has also been found in teleosts.

4.9 Nonsteroidal Regulators of Bone Formation: Calcitonin/Calcitonin-Gene–Related Peptide

Fact sheet 4.33: Calcitonin (CT)	
Gene:	Chromosome 11; locus 11p15.2; six exons.
Sequence:	Fig. 4.37.
Synthesis and target:	Calcitonin from the thyroid gland acts on bone cells.
Function:	Calcitonin controls the estradiol-dependent bone storage of calcium and blocks bone catabolism.
Receptor:	Calcitonin receptor is a heptahelical GPCR.

4.9.1 Introduction

Calcitonin (CT) is a product of thyroid C cells. It is a major regulator of calcium metabolism. CGrP (calcitonin-gene–related peptide) transcribed from the same gene but derived by alternate splicing, however, is a neuropeptide. Similar peptides such as CGrP are the amylin from pancreatic β cells and adrenomedullin.

4.9.2 Structure and Genes

The primary calcitonin transcript of the *CALCA* gene on chromosome 11 encodes a second peptide, the "calcitonin-gene–related peptide" (CGrP). Owing to alternative splicing the primary transcript is modified either to a calcitonin or CGrP mRNA (Fig. 4.37). The precursor sequences coded for by the exons 2, 3, and 4 are identical; those from exon 5 are calcitonin specific, and those from exon 6 are CGrP specific (Fig. 4.38). The processing of the calcitonin gene is taken as a model for alternative splicing (Lou and Gagel 1998).

Almost in the direct neighborhood of the *CALCA* gene on chromosome 11 the *CALCB* which gives rise only to an additional CGrP due to the lack of the CT exon.[31] There is a *CALCP* pseudogene closely linked on chromosome 11. The gene for amylin (aka islet amyloid polypeptide; IAPP) with its three exons is located on chromosome locus 12p12.1 on the antisense strand and the intron of a anion transport gene; the gene for adrenomedullin with four exons is on chromosome locus 11p15.4 (5 megabases distant from *CALCA*/*CACLB*).

Calcitonin is cleaved from the precursor by PC1, shortened by endopeptidases, and amidated by PAM. CGrP is only cleaved from its precursor by PC2 because the

[31]NCBI mapview: http://www.ncbi.nlm.nih.gov/mapview/maps.cgi?TAXID=9606&CHR=11&MAPS=genes-r,genec[868.93%3A878.71]-r&QSTR=114130[MIM]&QUERY=uid(14176198,12718486)&GOTO=874.75human%3A11%3AISCN&rsize=9.78000000000086

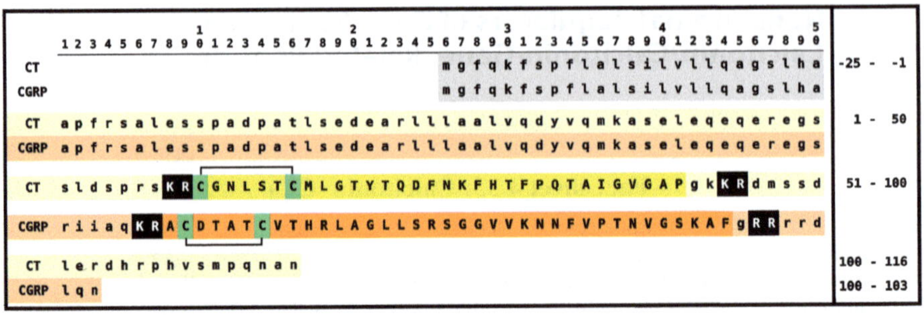

Fig. 4.37 Sequences of calcitonin and calcitonin-gene–related peptide (CGrP) (According to Broad et al. (1989); source: AAA58403)

Fig. 4.38 Organization of the calcitonin/calcitonin-gene–related peptide (CGrP) gene (According to HSCALCAC, GenBank X15943)

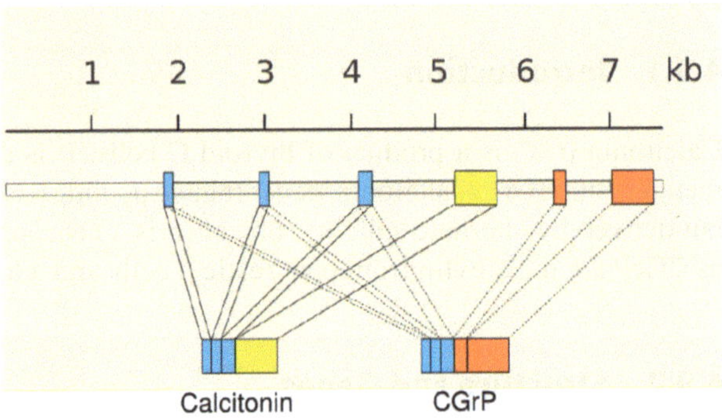

C-terminal **KR** motif for PC1 is lacking. Both peptides have an N-terminal disulfide bridge in common.

CTs and CGrPs action are mediated by specific receptors. The CT receptor is CALC-R, a heptahelical GPCR, like the CGrP receptor. The receptor activity modulating protein (RAMP-1) associates with CALC-R and can bind to CGrP-R. Amylin binds to the complex of CALC-R and RAMP-1 or RAMP-3. Association with RAMP-1 is necessary for an efficient membrane expression of CGrP. Without RAMP-1 the CGrP-R is not functionally active.

Recently it had been found that the serum concentrations of the CT precursor ProCT (amino acids 3–116) are characteristically enhanced during bacterial infections (Muller et al. 2001). ProCT might possibly be a more specific marker for bacterial infections than, for example, C-reactive protein (CRP) (Schuetz et al. 2011).

4.9.3 Physiology

Calcitonin biosynthesis is controlled by estrogens. For the treatment of osteoporosis as a consequence of postmenopausal estradiol deficiency calcitonin is also used.

CT inhibits further bone resorption. CT is further required for the maintenance of adequate calcium levels during stress, pregnancy, or lactation (Wimalawansa 1996).

CGrP is the neuropeptide most active for dilatation. An important role of CGrP is in the heart where CGrP secreted by neurons enhances heartbeat frequency, atrial contraction, and in summary the coronary circulation. The renal blood flow and that of other organs could equally be stimulated by CGrP.

Calca-knockout mice were viable, fertile, and showed enhanced bone mass and bone formation more than wildtype mice. Although $Calca^{+/+}$ mice reduced bone mass after estradiol depletion, the bone mass remained constant in *Calca*-knockout mice (Hoff et al. 2002). The fact that CT knockouts do not exhibit deficiencies led to the conclusion of CT redundancy in humans and rodents. The activity of human CT is considerably reduced compared with, for example, fish CT. Due to the development of parathyroid glands and of the parathormone the action of CT seemed to be confined to bone protection during stress (Hirsch and Baruch 2003).

4.9.4 Phylogeny

Calcitonin and CGrP are vertebrate peptides. In invertebrates only homologues to the calcitonin receptor have been discovered in tunicates, echinodermates, insects, and nematodes.

4.10 Digestive Tract Hormones

In the GI tract there are more than 30 polypeptides, some of them controlling digestion as hormones, others as neurotransmitters. The majority of these hormones are released by specialized GI tract cells. In the stomach, gastrin and somatostatin are released, and in the duodenum somatostatin, cholecystokinin, secretin, and GIP. In the pancreas there are cells that secrete insulin, glucagon, somatostatin, or the pancreatic polypeptide. In the ileum, there is somatostatin, enteroglucagon, and neurotensin release.

4.10.1 Gastrin

4.10.1.1 Introduction

Gastrin is a peptide from so-called G cells in the pyloric antrum, close to the duodenum. The hormone is secreted into the blood in an endocrine way and not exocrine into the gut lumen like digestion enzymes, acid, or carbonate. Gastrin, like cholecystokinin, was discovered almost 100 years ago in animals. Recently its role in triggering gastric ulcer under infection by *Helicobacter pylori* has attracted attention.

Fact sheet 4.34: Gastrin

Gene:		Chromosome 17; locus 17q21; three exons.
Sequence:	Gastrin-71	SWKPRSQQPD APLGTGANRD LELPWLEQQG PASHHRRQLG PQGPPHLVAD PSKKQGPWLE EEEEAYGWMD F-NH_2
	Gastrin-34	QLG PQGPPHLVAD PSKKQGPWLE EEEEAYGWMD F-NH_2
	Gastrin-17	QGPWLE EEEEAYGWMD F-NH_2
	Gastrin-6	YGWMD F-NH_2
		Gastrin-17, the major active molecule, may be sulfated on the tyrosine residue (**Y**). The N-terminal glutamine of gastrin-17 could be cycled to pyroglutamate.
Synthesis and target:		Gastrin is released from G-cells of the stomach antrum and acts on gastrin receptors on parietal cells of the stomach epithelium and on the central nervous system.
Function:		Gastrin-34 and gastrin-17 stimulate gastric acid release from stomach parietal cells. It is further required for cellular differentiation of the gastric epithelium.
Receptor:		Gastrin receptor (cholecystokinin receptor-2) is a heptahelical GPCR. It is expressed in many brain cells.

4.10.1.2 Structure and Genes

The active gastrin-34 and gastrin-17 are cleaved from a precursor. For the formation of Gastrin-17 PC2 is required which is expressed in G cells and can effectively cleave at the **KK** motif. Sulfate substitution of the tyrosine residue might enhance PC2 recognition. Sulfated and nonsulfated gastrins are equally active. Gastrin-34 has a prolonged half-life in the circulation compared with gastrin-17 (Dockray 1999).

4.10.1.3 Physiology

Gastrin is formed in G cells of the stomach and duodenum. It enhances gastric acid release from parietal stomach epithelium cells. Gastrin-gene–defective mice lack maturation of the stomach mucosa. Although acid synthesis and release can be maintained with gastrin synthesis blocked by histamine or acetylcholine, epithelial cells do not mature at all without gastrin.

The gastrin-releasing peptide (bombesin, Sect. 4.11.3) stimulates enhancement of gastrin levels in serum and of acid release in the stomach. Whether bombesin is functional active is not known inasmuch as bombesin is a brain neuropeptide and any secretion into the blood has not been found in humans.

4.10.1.4 Phylogeny

See cholecystokinin (next section).

4.10.2 Cholecystokinin

Fact sheet 4.35: Cholecystokinin (CCK)

Gene:		Chromosome 3; locus 3p22.1; three exons.
Sequence:	CCK-83	QPVPPADPAG SGLQRAEEAP
		RRQLRVSQRT DGESRAHLGA LLARYIQQAR
		KAPSGRMSIV KNLQNLDPSH RISDRDYMGW MDF-NH_2
	CCK-58	VSQRT DGESRAHLGA LLARYIQQAR
		KAPSGRMSIV KNLQNLDPSH RISDRDYMGW MDF-NH_2
	CCK-39	YIQQAR
		KAPSGRMSIV KNLQNLDPSH RISDRDYMGW MDF-NH_2
	CCK-33	KAPSGRMSIV KNLQNLDPSH RISDRDYMGW MDF-NH_2
	CCK-22	NLQNLDPSH RISDRDYMGW MDF-NH_2
	CCK-08	DYMGW MDF-NH_2
Synthesis and target:		The majority of CCK is released from I cells in the duodenum and jejunum and acts on gallbladder muscles, pancreatic acinar cells, and parietal cells of the stomach. CCK is widely found in the CNS.
Function:		By CCK bile release and pancreatic enzyme production are increased, gastric acid synthesis reduced, and (centrally) appetite reduced.
Receptor:		The CCK receptor CCK-R1 is a heptahelical GPCR; CCK-R2 is the gastrin receptor.

4.10.2.1 Introduction

Cholecystokinin (CCK) is released in the duodenum and the jejunun. Due to the common C-terminal **WMDF**amide sequence a close phylogenetic relation to gastrin had been postulated for some time. CCK is probably the original hormone.

4.10.2.2 Structure and Genes

CCK is cleaved from a precursor shortened by PC2 and amidated by PAM. After the release of CCK-33 an endopeptidase might degrade CCK to CCK-8 (Moran and Kinzig 2004). In serum CCK-33 is most abundant (Rehfeld et al. 2001). The tyrosin of CCK-08 is often sulfated. The CCK gene is located on the short arm of chromosome 3.

The preferred CCK receptor is CCK-R1 present on pancreatic acinar cells, on gallbladder muscles, on smooth muscle cells, on D cells of the gut mucosa and on nerve cells of the central and peripheral nervous system.

4.10.2.3 Physiology

CCK-8 was observed as the hormone triggering gallbladder contractions. The action on acinar cells might facilitate the development of pancreatic tumors. Applied into animals CCK reduces food intake, whereas a receptor antagonist increases food intake. CCK is therefore regarded as a satiation signal (Moran and Kinzig 2004).

Table 4.4 Peptides of the gastrin/CCK family

Name	Sequence
Cholecystokinin-8	**DYMGWMDF**-NH_2
Gastrin-17	**QGPWLEE EEEAYGWMDF**-NH_2
Caerulein[a]	p**EQDYTGWMDF**-NH_2
Phyllocaerulein[b]	p**EQYTGWMDF**-NH_2
Cionin[c]	**NYYGWMDF**-NH_2

Tyrosines (**Y**) are frequently sulfated
[a]From the Australian green tree frog (*Litoria caerulea*), Anastasi et al. (1967, 1970)
[b]From the waxy monkey frog (*Phyllomedusa sauvagii*), Anastasi et al. (1969)
[c]From *Ciona intestinalis*, Johnsen and Rehfeld (1990)

4.10.2.4 Phylogeny of the Gastrin Peptide Gene Family

Anti-CCK have been found to bind to nonvertebrate tissues, however, the only peptide identified is cionin from *Ciona intestinalis* which shares the C-terminal sequence with gastrin and CCK. In frog caeruleins with the same sequence have been discovered (see Table 4.4). CCK and gastrin have been found in all vertebrates analyzed: fish, amphibians, reptiles, birds, and mammals (Johnsen 1998).

4.10.3 Secretin

Fact sheet 4.36: Secretin

Gene: Chromosome 11; locus 11p15; five exons.

Sequence: **HSDGTFT SELSRLREGA RLQRLLQGLV**-NH_2.

Synthesis and target: Duodenal S cells stimulated by gastric acid release secretin which acts on the secretin receptor on pancreatic acinar cells and on stomach cells.

Function: Secretin controls gastric acid neutralization by stimulating bicarbonate, water, and salt release from pancreas and gut. It inhibits gastric acid release from parietal stomach epithelium cells.

Receptor: Secretin receptor is a heptahelical membrane GPCR.

4.10.3.1 Introduction

Secretin, discovered on January 16, 1902,[32] passes as the prototype of peptide hormones.

4.10.3.2 Structure and Genes

Human secretin was cloned and sequenced only in 2000 (Whitmore et al. 2000). It is released classically from a precursor protein; most probably PC2 is the active enzyme recognizing the **RR** motif in front and the **KR** motif after the secretin sequence. The C-terminal glycine is amidated.

The human gene for secretin is located on the telomeric end of the short arm of chromosome 11 (11p15) close to dopamine receptor-D4.

4.10.3.3 Physiology

Secretin is released from cells in the duodenum and in the jujenum into the blood. It facilitates water, salt, and enzyme release from the pancreas, stomach, and gut cells into the lumen thus neutralizing gastric acid; secretin inhibits endocrine gastrin and somatostatin release; it advances insulin release. Contractions of stomach and gut are inhibited.

Secretin is preferentially expressed in the jejunum. The expression in other cells, the brain included, is reduced compared to the jejunum. In cells of the gallbladder it triggers membrane fusion of vesicles with preformed aquaporin thus enhancing water pumping into the bile.

4.10.3.4 Phylogeny

Bacterial secretins responsible for export of proteins only share the name with the vertebrate hormone secretin. Secretins have been isolated and cloned but from mammals and birds, not from fish. The secretin receptor family (GPCR-B), however, has been observed in arthropods, nematodes, and tunicates. The protein has not been found in the fully sequenced genome of *Ciona intestinalis* (tunicates) (Burke et al. 2006).

[32]I happened to be present at the discovery. In an anesthetized dog, a loop of jejunum was tied at both ends and the nerves supplying it dissected out and divided so that it was connected with the rest of the body only by its blood vessels. On the introduction of some weak HCl into the duodenum, secretion from the pancreas occurred and continued for some minutes. After this had subsided a few cubic centimeters of acid were introduced into the enervated loop of jejunum. To our surprise a similarly marked secretion was produced. I remember Starling saying: "Then it must be a chemical reflex." Rapidly cutting off a further piece of jejunum he rubbed its mucous membrane with sand in weak HCl, filtered it, and injected it into the jugular vein of the animal. After a few moments the pancreas responded with a much greater secretion than had occurred before. It was a great afternoon (Martin 1927).

4.10.4 VIP

Fact sheet 4.37: Vasoactive intestinal peptide (VIP)

Gene:	Chromosome 6; locus 6q25.2; seven exons.
Sequence:	**HSDAVFTD NYTRLRKQMA VKKYLNSILN**-NH_2.
Synthesis and target:	VIP is made by neurons and mostly acts as a neurotransmitter.
Function	In the GI tract VIP acts to relax smooth muscles. VIP blocks gastric acid synthesis and stimulates water transport into bile, as well as bicarbonate, enzyme, and chloride secretion in the gut.
Receptor:	VCAP receptors 1 and 2 are heptahelical membrane GPCR. They bind VIP and PACAP.

4.10.4.1 Introduction

As are the neuromedins or secretin vasoactive intestinal peptide (VIP) is an member of the evolutionary old glucagon/PACAP protein family already found in tunicates (Sherwood et al. 2000). VIP actions are not restricted to the GI tract, they include brain activity, neuroendocrine functions, heart muscle activity, breathing, and sexual activity. In addition to its expression in the duodenum VIP is made in many brain cells as well as in other organs.

4.10.4.2 Structure and Genes

VIP is released from a precursor by PC1 (or PC2) and C-terminally amidated by PAM. There is an additional peptide on the same precursor, called peptide–histidine–methionine in humans and peptide–histidine–isoleucine in other mammals.

4.10.4.3 Physiology

VIP is a neurotransmitter and neuromodulator of the enteric nervous system (ENS; see Fig. 11.7). VIP is synthesized in neurons, and also in mast cells and granulocytes. VIP makes the smooth muscle cells of the gut relax; it blocks gastric acid formation and stimulates water transport in the bile, as well as release of bicarbonate and pancreatic enzymes and chloride from the gut (Boushey and Drucker 2003).

Because VIP passively transfuses through the blood–brain barrier, VIP from the periphery might act centrally. Recent results suggest that VIP is decisive for light-dark adaption of the biological clock. Mice with a defect of the VCAP2 receptor could not maintain a 24-h rhythm and got arrhythmic (in the dark; Colwell et al. 2003). Furthermore, triggered by VIP, glia cells secrete neuroprotective factors: interleukin-1 (IL-1), IL-6, neurotrophin-3, nexin-1, RANTES and MIP chemokines, activity-dependent neurotrophic factor (ADNF) and activity-dependent neuroprotective protein (ADNP) (Dejda et al. 2005)

In addition, VIP is an important regulator of the immune system and of dendritic cell maturation. VIP exhibits potent anti-inflammatory actions and is efficient against rheumatoid arthritis (Delgado et al. 2004).

4.10.4.4 Phylogeny

VIP has exclusively been found in vertebrates: sharks, bony fishes, and in the other vertebrate classes. The VCAP receptor, too, belongs to an ancient receptor family, but is not older than vertebrates. in sharks it has been observed that VIP stimulates chloride channels.

4.10.5 GIP

Fact sheet 4.38: Glucose-dependent insulinotropic polypeptide, gastroinhibitory polypeptide (GIP)

Gene:	Chromosome 17; locus 17q21.32; six exons.
Sequence:	Fig. 4.39.
Synthesis and target:	GIP is mainly a product of endocrine GI tract cells and acts on insulin or glucagon secreting pancreatic islet cells.
Function:	GIP triggers (dependent on glucose) insulin release, inhibits glucagon synthesis, and blocks glucagon-induced lipolysis.
Receptor:	GIP receptor is a GPCR from the secretin protein family.

4.10.5.1 Introduction

The glucose-dependent insulinotropic polypeptide (aka gastroinhibitory polypeptide (GIP)), originally identified as a regulator of gastric acid release, of the major gastric enzyme pepsin and of gastrin, is first of all a major insulin formation and release stimulus.

4.10.5.2 Structure and Genes

The *GIP* gene with its six exons is located on chromosome 17 close to an IFG2 binding protein (IGF2BP1). After removal of the signal peptide the remaining 132 amino acid long precursor is cleaved at two furin recognition sites by PC1 giving rise to the mature hormone. The cleavage is, surprisingly, affected by PC1, as has been demonstrated in PC1-defective mice (Ugleholdt et al. 2006).

```
                  1                   2                   3                   4                   5
1 2 3 4 5 6 7 8 9 0 1 2 3 4 5 6 7 8 9 0 1 2 3 4 5 6 7 8 9 0 1 2 3 4 5 6 7 8 9 0 1 2 3 4 5 6 7 8 9 0
                                                          m v a t k t f a l l l l s l f l a v g l g   -21 -  -1
e k k e g h f s a l p s l p v g s h a k v s s p q p r g p r Y A E G T F I S D Y S I A M D K I H Q Q     1 -  50
D F V N W L L A Q K G K K N D W K H N I T Q r e a r a l e l a s q a n r k e e e a v e p q s s p a k    51 - 100
n p s d e d l l r d l l i q e l l a c l l d q t n l c r l r s r                                      101 - 132
```

Fig. 4.39 Primary gastroinhibitory peptide (GIP) sequence. The 153 amino acid long protein bears a 21 amino acid long signal peptide. By the prohormone convertase 1 (!see text!) the precursor is cleaved at the *inverted* arginine residues and the mature polypeptide (*in uppercase*) released (Source: GenBank NP_004114)

4.10.5.3 Physiology

GIP is classified as an incretin, that is, a stimulator of insulin production. It is rapidly inactivated in serum by the dipeptidyl peptidase IV. In diabetes mellitus type 2 patients, a prolongation of GIP (or GLP1, another incretin) presence in blood is tried as an option to enhance insulin release.

4.10.5.4 Phylogeny

Until now, the protein library only contains vertebrate GIP sequences: fish, frogs, birds, and mammals. The GIP from *D. melanogaster* appears to be an isomerase with no relation to vertebrate GIP.

4.10.6 PNP, NPY, PYY

Fact sheet 4.39: Pancreatic polypeptide (PNP)

Gene:	Chromosome 17; locus 17q21.31; four exons.
Sequence:	Fig. 4.40: PNP.
Synthesis and target:	PNP is the product of pancreatic islet PP cells. PNP receptors are present on pancreatic acinar cells.
Function:	PNP acts on exocrine pancreatic secretion. It is furthermore involved in blockade of food intake.
Receptor:	The PNP receptor is a heptahelical membrane GPCR.

Fact sheet 4.40: Peptide tyrosine tyrosine (PYY)

Gene:	Chromosome 17; locus 17q21.31; five exons.
Sequence:	Fig. 4.40: PYY.
Synthesis and target:	PYY is released from endocrine cells in the ileum and colon. It binds to PNP receptor 1 (PNP-R1, aka NPY-R4) which is expressed in brain, coronary arteries, and gut and to NPY-R2 expressed preferentially in the nervous system.
Function:	PYY has a role in signaling between the enteric nervous system and the central nervous system. PYY signals decrease appetite.
Receptor:	Receptors belong to the heptahelical membrane GPCR class.

Fact sheet 4.41: Neuropeptide Y (NPY)

Gene:	Chromosome 7; locus 7p15.3; four exons.
Sequence:	Fig. 4.40: NPY.
Synthesis and target:	NPY is secreted by neurosecretory cells. In the hypothalamus NPY signals appetite. NPY receptors are expressed in brain, spleen, ileum, kidney, testis, placenta, and in aortic smooth muscles.
Function:	Apart from its appetite-regulating function NPY acts anxiolytically. NPY defects impair development of the olfaction. NPY can control growth of neuron precursors in adults.
Receptor:	Receptors are heptahelical membrane GPCR.

PNP

1234567891 1234567892 1234567893 1234567894 1234567895 (positions 1–50)	
m a a a r l c l s l l l l s t c v a l l l q p l l g a q g	-29 - -1
A P L E P V Y P G D N A T P E Q M A Q Y A A D L R R Y I N M L T R P R Y g K R h k e d t l a f s e w	1 - 50
g s p h a a v p r e l s p l d l	51 – 66

PYY

1234567891 1234567892 1234567893 1234567894 1234567895 (positions 1–50)	
m v f v r r p w p a l t t v l l a l l v c l g a l v d a	-28 - -1
Y P I K P E A P G E D A S P E E L N R Y Y A S L R H Y L N L V T R Q R Y g K R d g p d r l l s k t f	1 - 50
f p d g e d r p v r s r s e g p d l w	51 – 79

NPY

1234567891 1234567892 1234567893 1234567894 1234567895 (positions 1–50)	
m l g n k r l g l s g l t l a l s l l v c l g a l a e a	-28 - -1
Y P S K P D N P G E D A P A E D M A R Y Y S A L R H Y I N L I T R Q R Y g K R s s p e t l i s d l l	1 - 50
m r e s t e n v p r t r l e d p a m w	51 – 79

Fig. 4.40 Primary sequences of pancreatic polypeptide (PNP), protein tyrosyl-tyrosine (PYY), and neuropeptide-Y (NPY): Precursor proteins contain a 29 or 28 amino acid long signal peptide (on *gray* background). Prohormone convertase-1 cleaves the precursors at **KR** (position 38/39; shown *inverted*); endopeptidases shorten the peptides till **G** (37) which is finally amidated. The mature peptide is shown highlighted *yellow*. The four prolines determining the polyprolin (PP)-fold are shown on *green* background. The associated peptide in the PNP precursor highlighted (*light blue*) is called pancreatic icosapeptide (Sources: GenBank NP_002713 (PNP), NP_004151.3 (PYY), NP_000896.1 (NPY))

4.10.6.1 Introduction

Pancreatic polypeptide (PNP), neuropeptide Y (NPY), and peptide tyrosine tyrosine (PYY) share a common three-dimensional structure, the so-called PP-fold a polyproline coil of amino acids 1–8 (*lower part* in Fig. 4.41), an amphipathic helix with a polar and a hydrophobic side (aa 15–30; *upper part* of Fig. 4.41. Both parts are linked by a β-turn called sequence (*right side* in the figure). The three-dimensional structure is tightened by hydrophobic interactions between prolines and the hydrophobic side of the helix.

4.10.6.2 Structure and Genes

Two peptides (PNP and PYY) are coded for by closely linked genes on the long arm of chromosome 17 (17q21). The *NPY* gene, however, is located on chromosome 7 (7p15). *PNP2* and *PYY2* (arisen by gene duplication) are on chromosome 17 in 17q11.2. Products of these genes are shortened due to frameshift mutations and are not active as hormones.

The three precursors are cleaved by the signal-peptidase and furthermore by PC1 and finally amidated, giving rise to 36 amino acid long hormones. Characteristic features are the prolines (*boxed* in Fig. 4.40 at the start of the sequence causing the proline coil and the particular PP-fold (Fig. 4.41). Whether these peptides are always present in the PP-fold has been debated (Bettio et al. 2002).

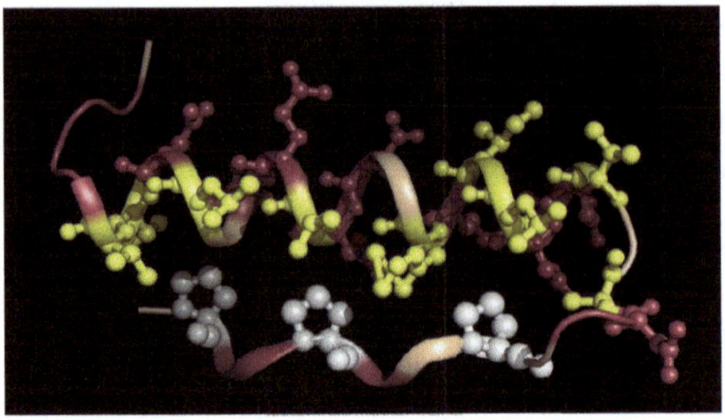

Fig. 4.41 Common structure of pancreatic polypeptide (PNP), protein tyrosyl-tyrosine (PYY), and neuropeptide-Y (NPY). The polyproline (PP) -fold is composed of an amphipathic helix (*upper part*) with polar (*red*) amino acids on one side (*facing upwards and to the back*) and hydrophobic (*yellow*) amino acids on the other side (*facing to the front and downwards*), paired with a proline coil (*lower part*). The hydrophobic helical side interacts with the equally hydrophobic proline residues *white* of the proline coil which due to the inflexible structure of the proline rings form an extended helix (Source: Lerch et al. 2004; PDB: 1PPT (RasMol))

4.10.6.3 Physiology: PNP

Pancreatic islet PP cells form PNP (see Sect. 10.5). PNP acts back on the pancreas blocking its activity, in addition to gut contractions. PNP influences gluconeogenesis and decreases fat levels. PNP receptors are present in the brain, too.

4.10.6.4 Physiology: NPY

NPY is one of the most abundant neuropeptides in the human organism. NPY once released signals hunger. For this action NPY is expressed in the brain (arcuate nucleus and nearby areas; see Sect. 4.3), in the GI tract, and other organs (see also Sect. 11.5). In the brain NPY acts as neurotransmitter, usually together with noradrenaline (see Chap. 7).

In addition, NPY is required for normal development of the olfactory sense: postpartum NPY supports cellular division and development of olfactory neurons. NPY-deficient mice have only half of the olfactory cell precursors as wildtype mice and develop fewer mature olfactory neurons.

NPY neurons in the arcuate nucleus interact with POMC neurons and their MSH products for regulation of fertility and energy homeostasis. In older postmenopausal women NPY expression is enhanced compared to the POMC expression.

Because NPY is anxiolytically active, individuals with the ability to release larger amounts of NPY deal better with stress. Moreover, NPY counteracts the usage of alcohol. Persons with a Leu→Pro exchange at amino acid 7 of the NPY signal peptide (in Fig. 4.40 *bold* and highlighted *blueish*) consumed significantly more alcohol than those with the prominent sequence.

Table 4.5 Affinities of neuropeptide-Y (NPY) receptors for pancreatic polypeptide (PNP), protein tyrosyl-tyrosine (PYY), and NPY (Berglund et al. 2003)

Receptor	Relative affinities of different ligands
NPY-R1	NPY = PYY > NPY[2-36][a] > NPY[3-36][b] ≥ PNP
NPY-R2	NPY ≥ NPY[2-36] = NPY[3-36]
NPY-R4	PNP > PYY ≥ NPY > NPY[2-36]
NPY-R5	NPY = PYY = NPY2-36 > hPNP[c] > rPNP[d]
NPY-R6	??[e]

From Berglund et al. (2003)
[a]NPY[2-36] lacks the N-terminal amino acid of NPY (at the proline coil).
[b]NPY[3-36] lacks 2 N-terminal amino acids of NPY (at the proline coil)
[c]hPNP: human PNP
[d]rPNP: rat PNP
[e]NPY-R6 is (in humans, swine, or guinea pigs) due to mutations C-shortened and thus afunctional as receptor. Because NPY-R6 is strongly expressed in heart- and skeletal muscles, a new function is supposed

4.10.6.5 Physiology: PYY

PYY was originally isolated from the small intestine. Its actions are comparable to those of PNP. In addition, PYY acts on salt release and blood support by regulating the diameter of capillaries. PYY is part of the communication between the ENS and the CNS: PYY receptors are present on vagus nerve endings in the GI tract.

4.10.6.6 Physiology: Receptors for PNP, PYY, and NPY

The three peptides bind with similar, but divergent, affinity to the different NPY receptors (Table 4.5). As a result, the NPY-R2 appears to be NPY specific; NPY-R4, however, to be PNP specific. PYY does not preferentially bind to any of these receptors. PYY binds both to NPY-R1 and to NPY-R5 with equal affinity as does NPY. Therefore, PYY may induce similar actions as NPY.

4.10.6.7 Phylogeny

The most ancient PYY and NPY genes and peptides were discovered in *Lampetra fluvialis*, *Petramyzon marinus*, and related species. PYY and NPY genes are closely linked in the vicinity of a so-called hox gene; hox genes are responsible for time and spatially controlled gene expression. In fish these hox genes are duplicated, however, without duplicated NPY and PYY genes. As the zebrafish genome is almost completely sequenced, the duplicates appear to have been lost.

In molluscs and insects NPY peptides have been described. The peptide homology is only scarce. Using a ClustalW analysis these peptides are less related to vertebrate NPY than are, for example, human galanin or FSH sequences which usually are not regarded as NPY related (Fig. 4.42). A proline coil is only partially present; the feature appears accidentally.

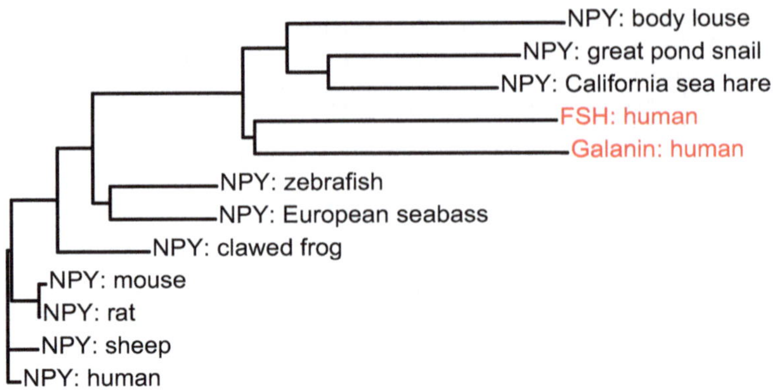

Fig. 4.42 Potential relation of NPY peptides of vertebrates and nonvertebrates

The *PNP* gene located in the direct vicinity of *PYY* is most probably derived from a local gene duplication shortly before or at the time of amphibian evolution. PNP is only active in the tetrapod pancreas. In fish a PY peptide is expressed, for example, in European seabass *Dicentrachus labrax* **YPPKPESPGS NASPEDWAKY HAAVRHYVNL ITRQRY** (amino acids identical to the human PYY are *underlined*). Such a peptide is expressed in the pancreas and brain. The gene might stem from a local duplication of the *Pyy* gene (Conlon 2002).

4.11 Neuropeptides in the Enteric Nervous System

4.11.1 Endorphins and Enkephalins

Endorphins and enkephalins are endogenous opioids. They act via opioid receptors. β-Endorphin is a processed peptide of POMC (see Sect. 4.4). Neoendorphins are cleaved from the prodynorphin, and enkephalins from the proenkephalin A and from prodynorphin (Figs. 4.43 and 4.44).

These neuropeptides equally undergo maturation by prohormone convertases. Thus far we got used to dibasic peptide motifs as recognition sites for these enzymes. However, here, an additional enzyme, cathepsin L is active, which cleaves after furin-like recognition sites too. Similarities between proenkephalin A and prodynorphin A are not only restricted to similar peptide sequences and active prohormone convertases. Whereas the *PENK* gene possesses two exons and the *PDYN* gene four, the exon 1 still codes for the signal peptide and the region with the six cysteines equally spaced in both sequences. This analogy is shown by drawing similar disulfide bridges in prodynorphin without formal proof in the literature (Fig. 4.44).

In the GI tract, endorphins and enkephalins increase the chyme's staying in the stomach and intestine by inhibiting motor activity on the gut musculature after binding to opioid receptors.

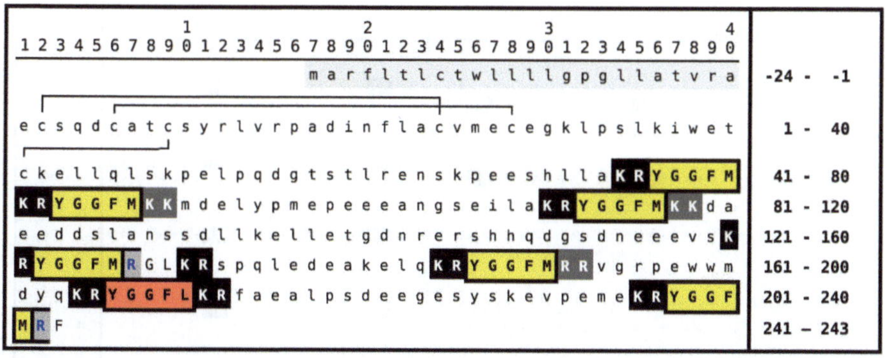

Fig. 4.43 Primary sequence of proenkephalin A. The precursor protein comprises six copies of met-enkephalin and a single copy of leu-enkephalin. According to Yasothornsrikul et al. (2003) cathepsin L is the active prohormone convertase recognizing dibasic dipeptides (such as prohormone convertase 1 and 2) but singular amino acids **R**, too; met-enkephalin is equally excised from the **YGGFMRGL** (162–169) and **YGGFMRF** (237–243) (Source: GenBank P01210)

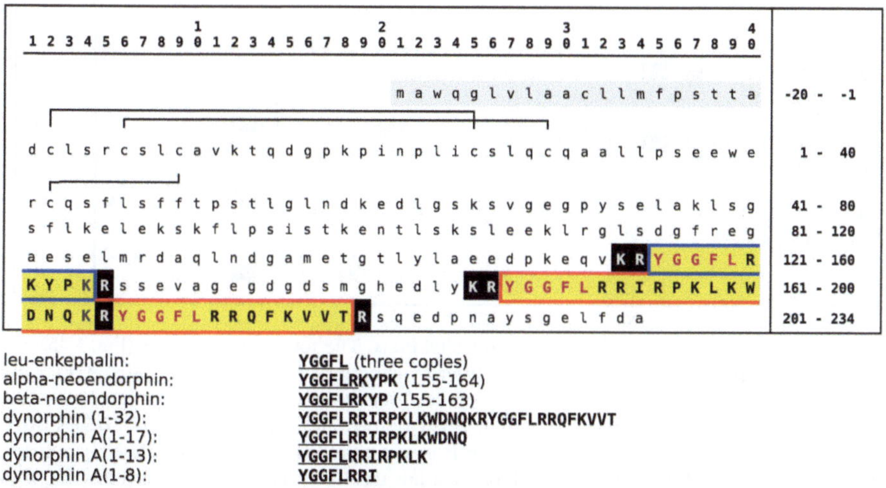

Fig. 4.44 Primary sequence of prodynorphins (proenkephalin B). Those products of peptide maturation isolated are shown below the primary sequence. Due to the mixture of utilized dibasic and monobasic motifs, a cathepsin L endopeptidase or a similar enzyme must be assumed to be acting as prohormone convertase (Source: GenBank P01213)

4.11.2 Tachykinins: Substance P, Neurokinin, and Endokinin

Tachykinins are characterized by a C-terminal pentapeptide: **FxGLM**-NH_2, with x = **V** or **F** (Figs. 4.45 and 4.46). Among others neurokinin A, neurokinin B, and substance P belong to these neuropeptides.

These peptides, excluding endokinins, all 10 to 12 amino acids long, are neurotransmitter acting via neurokinin receptors. In the GI tract substance P stimulates

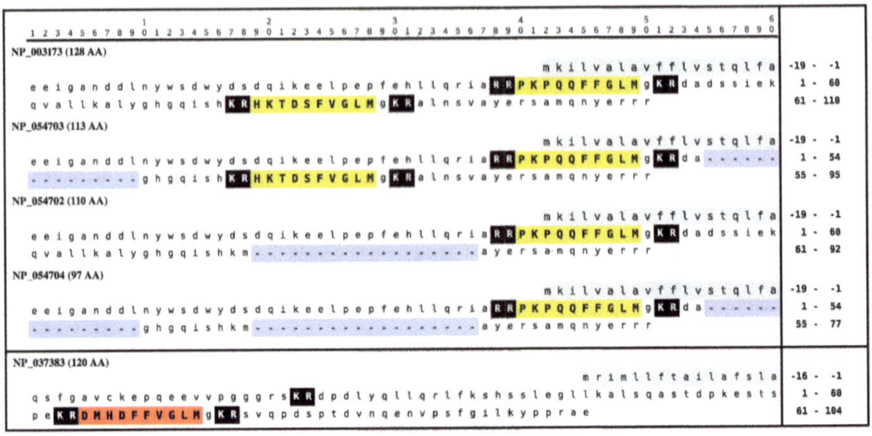

Fig. 4.45 See Fig. 4.46

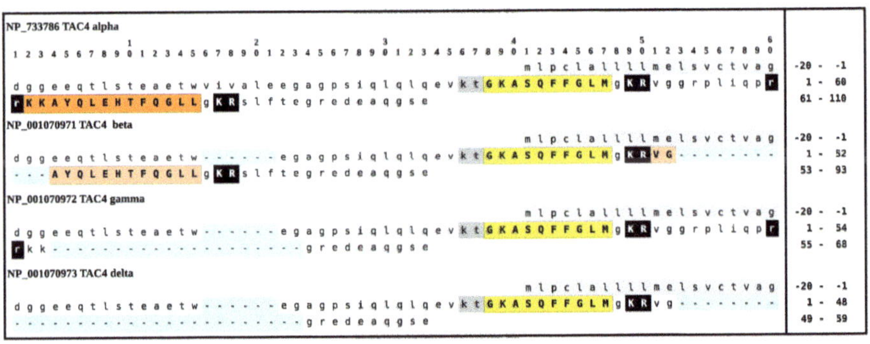

Fig. 4.46 Primary sequence of different protachykinins. The four tachykinin 1-3 (Fig. 4.45) and the four tachykinin 4 isoforms (this figure) differ by exon loss due to alternate splicing. Lacking amino acid from missed exons is indicated by single *dashes on light blue*. Protachykinin 1 gives rise to substance P (**PKPQQFFLGM**-NH_2; highlighted *yellow*) and neurokinin A (**HKTDSFVLGM**-NH_2; *green*), the sole product from tachykinin B is neurokinin B (**DMHDFFVLGM**-NH_2; *orange*; *above* NP_733786, fifth sequence). PC1 or PC2 are the likely prohormone convertases. From tachykinin 4 endokinin A (**GKASQFFGLM**; *yellow*) and endokinin C (**KKAYQLEHTFQGLL**-NH_2; *orange*) or endokinin D (**VGAYQLEHTFQGLL**-NH_2; *light orange*) are derived. The endokinin peptides have be isolated, however, the enzyme that recognizes **KT** has still to be identified (Source: GenBank *no. in the first column*)

muscle contraction and saliva production (see Sect. 11.5). Tachykinins stimulate vascular muscles, stomach secretion, and kidney functions.

Endokinin is the product of the tachykinin 4 gene (Fig. 4.46). The murine hemokinin 1 is the analogue of human endokinin A. In contrast to the other vertebrates analyzed, the human tachykinin 4 gene lacks the dibasic endopeptidase motif in front of the endokinin A obscuring the cleavage site.

Receptors for all tachykinins are the neurokinin receptors (NKR 1–3). Tachykinin 1 expression was preferentially found in the brain, marrow, and GI tract, where tachykinin 4 was not found to be expressed; the latter is preferentially found in placenta. The different splice variants of tachykinin 4 exhibit differential expressions (Page et al. 2003). In contrast to neurokinins and substance P which act as neurotransmitters, Page et al. (2003) found that endokinins act in an endocrine manner. They stimulate mostly lymphocyte development.

4.11.3 Gastrin-Releasing Peptide/Bombesin

Fact sheet 4.42: Gastrin-releasing peptide (GRP)/Bombesin)

Gene	Chromosome 18; locus 18q21; three exons.
Sequence	Fig. 4.47.
Synthesis and target	GRP is made in the lungs and brain and acts on pancreatic islet cells.
Function	GRP stimulates insulin and glucagon release and influences anxiety behavior.
Receptor	GRP receptor is a heptahelical GPCR.

12345678901234567890123456789012345678901234567890 (positions 1–50)	
mrgrelplvllalvlclaprgrav	-23 - -1
plpagggtvltkmyprGNHWAVGHLMgKKstgesssvsergslkqqlrey	1 - 50
irweeaarnllglieakenrnhqppqpkalgnqqpswdsedssnfkdvgs	50 - 100
kgkvgrlsapgsqregrnpqlnqq	101 - 124

Fig. 4.47 Primary sequence of gastrin-releasing peptide (GRP)-bombesin. From GRP (*boxed on yellow*) the decapeptide neuromedin C (*uppercase*) can be cleaved by an endopeptidase cutting after arginine. Because splice-variants do not change the GRP sequence, they are omitted here (Source: NP_001012531)

4.11.3.1 Introduction

Gastrin-releasing peptide (GRP) is the human analogue of the toad bombesin. Bombesin had originally been isolated from toads and frogs (Erspamer et al. 1972) (hence the name; for an early review, see Erspamer 1971).

4.11.3.2 Structure and Genes

GRP is released from a precursor by the action of the signal-peptidase and PC2. A furin-like enzyme can excise the neuromedin-C (Fig. 4.47).

4.11.3.3 Physiology

GRP is expressed mostly in neuroendocrine cells of the lungs. Injected into humans it enhances gastrin, GIP, PNP, glucagon, and insulin levels in blood similarly to bombesin (**PQRLGNQWAVGHLM**-NH_2) from the toad *Bombina bombina*.

The C-terminal decapeptide neuromedin C (amino acids 18–27) is active as an neurotransmitter. This peptide exhibits strong homology to neuromedin B (Fig. 4.47).

In mice it was found that GRP is expressed in the lateral nucleus of the amygdala where Pavlovian learned fear associations are formed. The GRP receptor was found in interneurons of the lateral nucleus (Shumyatsky et al. 2002). The results demonstrated a negative feedback function of GRP regulating fear and a relationship between the GRP receptor, long-term potentiation, and the amygdala-dependent memory for fear.

Studying the angiogenic effect of adrenomedullin and its associated peptide PAMP, which had been shown to bind to GRP receptor, Martínez et al. probed GRP for angiogenic activity and could confirm that GRP has potent angiogenic capacity, for example, in tumors (Martínez 2006; Martínez et al. 2005).

The latest addition to the bouquet of GRP features stems from Sakamoto (2011), who has shown expression of GRP in the spinal cord and its relation to male sexual functions (for a review, see Sakamoto 2011).

4.11.3.4 Phylogeny

A GRP receptor analogue has already been found in the California purple sea urchin (*S. purpurata*; echinoderms), however, GRP or bombesin could not be detected in the fully sequenced genome (Burke et al. 2006). Thus, GRP is most probably restricted to vertebrates.

4.12 Nonsteroidal Regulators of Reproduction

4.12.1 Activin/Inhibin

Fact sheet 4.43: Activin/Inhibin

Gene:	Inhibin-α: Chromosome 2; locus 2q35; two exons . Inhibin-β-A: Chromosome 7; locus 7p14.1; three exons . Inhibin-β-B: Chromosome 2; locus 2q14.2; two exons.
Sequences:	Fig. 4.48.
Synthesis and target:	Dimerization of two inhibin-β-chains gives rise to activins (A,B,AB); association of a α- to a β-chain generates inhibin. Activin-A binds to different serine/„‘threonine kinase receptors, that is, on pituitary gonadotropic cells: inhibin binds to an inhibin-binding protein and to β-glucan.
Function:	Activins and inhibins are paracrine regulators in endocrine glands and in organs of reproduction; in other tissues and organs they act as cytokines
Receptor:	Activin-A receptors are serine/threonine kinases. Cell surface receptors of inhibins have not yet been found; inhibins bind to β-glucan and to inhibin-binding proteins.

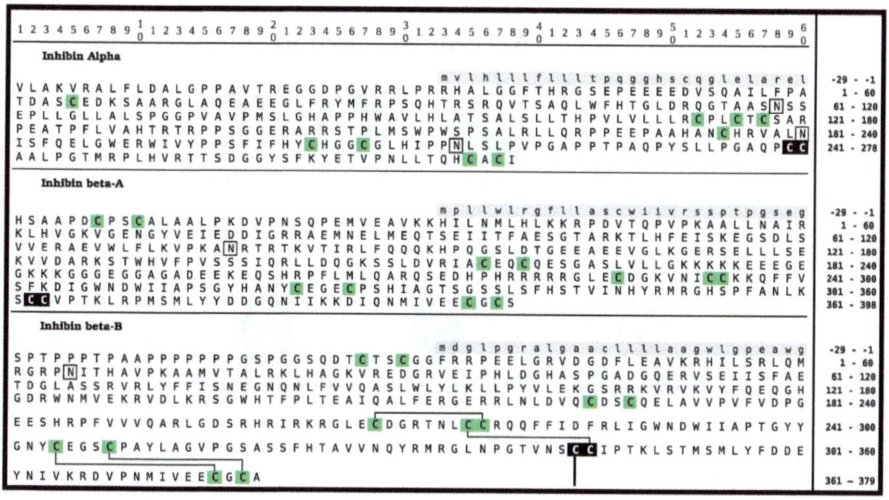

Fig. 4.48 Primary sequence of activin/inhibin. Three polypeptides are cysteine knot proteins as can be seen in the figure for inhibin-β B. Every β chain binds to another β chain generating thus an activin dimer or to an α chain giving rise to an inhibin. The dimers are stabilized by a single or multiple cysteine disulfide bridges. At least the first cysteine of last CC motif (*inverted*) is involved. (Source: inhibin alpha: NP_002182; inhibin beta A: NP_002183; inhibin beta B: NP_002184); disulfide bonds of inhibin dimers have not yet been provided for by analysis or X-ray structure; only activins have been crystallized thus far. The cysteine knot of the beta A chain has been crystallized (2ARV)

4.12.1.1 Introduction

As with follistatin the activins and inhibins are not only required for feedback regulation from gonads to the pituitary; activin-A at least is a regulator of dental development and a potential autocrine stimulator in placenta.

4.12.1.2 Structure and Biochemistry

The three different polypeptides give rise to five dimeric proteins: activin A, AB, or B, and inhibin A or B (Fig. 4.49). The chains contain, as does any TGF-β family member, a cysteine knot. There is a differential expression of β-A and β-B genes. Defects in β-A are neonatally lethal; those of β-B are not (Thompson et al. 2004).

Receptors for activins are serine/threonine membrane kinases signaling via SMAD proteins, a common TGF-β and bone morphogenic factor (BMF) signal transduction pathway. There are activin receptors and activin-like receptors, the latter having been stimulated by other TGF family proteins.

Receptors for inhibins are largely unknown. Today β-glucan and an inhibin-binding protein are regarded as mediators of inhibin action. Together with β-glucan, inhibin binds to the activin receptor type-II (ActRII).

With β-glucan binding to activin, activin can no longer interact with an activin receptor. Its function is blocked. Likewise the actions of follistatin are blocked by activin (Florio et al. 2004).

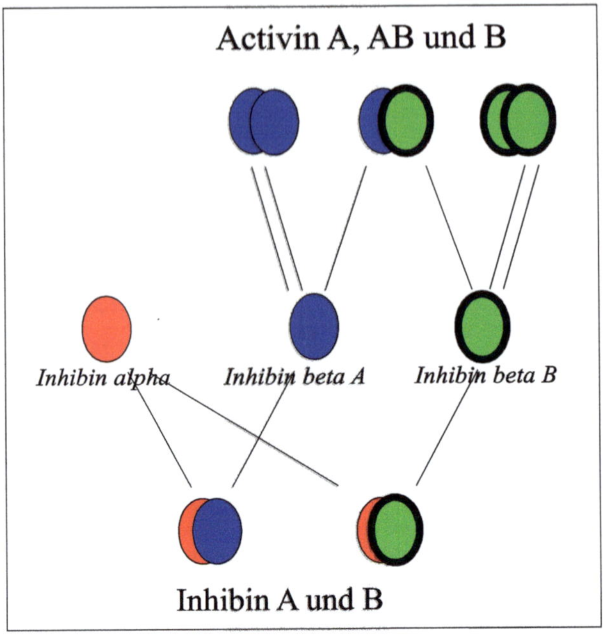

Fig. 4.49 Model of inhibin chain dimerization. Three proteins might dimerize to five distinct proteins. Beta chains dimerizing to each other give rise to activins; a beta chain binding to an alpha chain generates an inhibin

4.12.1.3 Physiology

Inhibins and activins were discovered for their control on FSH release. Additionally they act as regulators in ovary and testis. Furthermore they are called cytokines. Each chain is expressed separately and the expression is tissue, sex, and time controlled. In adult rhesus monkeys, only inhibin-B is present in serum.

Differences in β-A and β-B expression found in animals during uterine development could not be confirmed in humans: both subunits were equally present in brain, spinal cord, liver, kidney, or adrenal (Thompson et al. 2004).

During the estrous cycle β-chain expression is modulated in any animal species analyzed. The α-chain is expressed in any follicular stage: in antral follicles the β-B chain is more abundantly found; in later stages the β-A chain is prevalent, and thus there is an apparent switch in β-chain expression. The potentially formed activin-A is, however, not present in the serum, but mostly bound to follistatin. Activin-A controls expression of the FSH receptor on granulosa cells and activates steroidogenesis.

In the pituitary both beta chains are made in gonadotropic cells: the A-chain also in somatotropic and lactotropic cells, and the B-chain additionally in thyrotropic cells. A paracrine (possibly autocrine) stimulation of FSH by activin-B in rats has been confirmed. Cultured pituitary cells secrete activin-B, but not activin-A. Once activin-B is neutralized by antibodies in these cell cultures, FSH secretion is stopped. These antibodies blunt, *in vivo* applied, the preovulatory FSH surge in rats (Thompson et al. 2004, and references).

In his review on paracrine effects Denef (2008) summarizes the paracrine actions of inhibins/activin as well as follistatin on gonadotropin secretion.

4.12.1.4 Phylogeny

Inhibin genes and activin receptors were found together with other TGF-β family members in insects which demonstrates their early metazoic origin.

4.12.2 Follistatin

Fact sheet 4.44: Follistatin (FST)

Gene:	Chromosome 5; locus 5q11.2; six exons.
Sequence:	Fig. 4.50; follistatin has been crystallized: 2B0U.
Synthesis and target:	Follistatin is preferentially made in folliculostellate pituitary cells, in ovarian follicles, and in the kidney. It binds to proteins of the TGF-β family.
Function:	Follistatin controls *i.a.* by binding to activin/inhibin FSH biosynthesis. In addition it controls the activity of other TGF-β family members.
Receptor:	Follistatin binds to heparan sulfate proteoglucans.

4.12.2.1 Introduction

Since it could be established that FSH and LH are made in the same gonadotropic cells, regulators that provide for differential expression of LH and FSH synthesis get into the focus. Follistatin is one of such regulators. Being first isolated from the ovarian follicular fluid, today the folliculostellate pituitary cells are also known to secrete follistatin as well. Follistatin is considered to be an endocrine as well as paracrine regulator of FSH synthesis.

4.12.2.2 Structure and Biochemistry

Follistatin is member of the transforming growth factor-β family of proteins. By alternative splicing two isoforms are generated with similar, only C-terminally differing, sequences: FS-288 and FS-315 (Fig. 4.50).

Follistatin inhibits action of other TGF-β proteins by forming heterodimers with these proteins; among those are bone morphogenetic protein (BMP), activin and inhibin, as well as α_2-macroglobulin. Two follistatin molecules bind to one activin dimer and the inhibin-α/β-dimer binds only one follistatin polypeptide. This leads to the conclusion that follistatin bind to the β-activin subunit. Whether the follistatin–inhibin interaction is biologically relevant, is not yet conclusively analyzed.

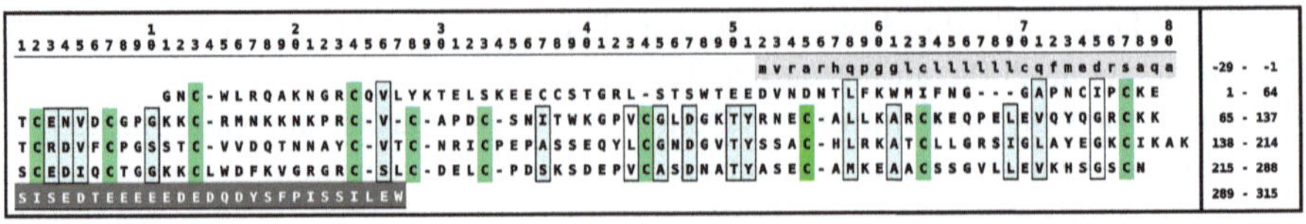

Fig. 4.50 Primary sequence of follistatin. Following the signal peptide (*light gray*) the protein exhibits four domains (*one per row*), characterized by identical cysteine positions and conservative amino acid exchanges. The (*inverted*) C-terminal sequence distinguishes the two follistatin isoforms (Source: NP_037541, Shimasaki et al. 1988)

4.12.2.3 Physiology

Follistatin blocks in a paracrine manner the action of activin and other proteins of the TGF-β family. It is made by pituitary folliculostellate and ovarian granulosa cells. Whereas activins stimulate the maturation of FSH-dependent antral follicles and block the generation of the corpus luteum, follistatin is expressed only in the final state follicle and facilitates corpus luteum generation as an activin antagonist (Lin et al. 2003).

It is not yet evident whether follistatin is endocrine active, that is, whether it is transported via the blood to distant target organs. As a paracrine regulator it is active not only in gonads, but in a variety of other organs, too (Schneider et al. 2000).

4.12.2.4 Phylogeny

Like inhibins, follistatin is an "ancient" metazoic protein found in common precursors of vertebrates and arthropods.

4.12.3 Antimüllerian Hormone

Fact sheet 4.45: Antimüllerian Hormone (AMH), Müllerian Inhibitory Factor (MIF)

Gene: Chromosome 19; locus 19p13.3; five exons.

Structure: Glycoprotein of the TGF-β family.

Synthesis and target: AMH is expressed in fetal and postnatal Sertoli cells; it acts on the AMH receptor in fetal müllerian duct cells. In women AMH is made by granulosa cells of preantral and antral follicles up to the moment when its synthesis is blocked by FSH.

Function: AMH acts as a testicular hormone on the müllerian ducts and induces their regression. A postnatal role has yet to be found. AMH in serum or seminal fluid is used to determine functional reproduction capacity of men and women in assisted reproductive technology.

Receptor: AMH receptor is a membrane serine/threonine kinase.

5 10 15 20 25 50 35 40 45 50	
mrdlpltslalvlsalgal	-19 - -1
lgtealRAEEPAVGTSGLIFREDLDWPPGSPQEPLCLVALGGDSNGSSSP	1 - 50
LRVVGALSAYEQAFLGAVQRARWGPRDLATFGVCNTGDRQAALPSLRRLG	51 - 100
AWLRDPGGQRLVVLHLEEVTWEPTPSLRFQEPPPGGAGPPELALLVLYPG	101 - 150
PGPEVTVTRAGLPGAQSLCPSRDTRYLVLAVDRPAGAWRGSGLALTLQPR	151 - 200
GEDSRLSTARLQALLFGDDHRCFTRMTPALLLLPRSEPAPLPAHGQLDTV	201 - 250
PFPPPRPSAELEESPPSADPFLETLTRLVRALRVPPARASAPRLALDPDA	251 - 300
LAGFPQGLVNLSDPAALERLLDGEEPLLLLLRPTAATTGDPAPLHDPTSA	301 - 350
PWATALARRVAAELQAAAAELRSLPGLPPATAPLLARLLALCPGGPGGLG	351 - 400
DPLRALLLLKALQGLRVEWRGRDPRGPGRAQRSAGATAADGPCALRELSV	401 - 450
DLRAERSVLIPETYQANNCQGVCGWPQSDRNPRYGNHVVLLLKMQVRGAA	451 - 500
LARPPCCVPTAYAGKLLISLSEERISAHHVPNMVATECGCR	501 - 541

Fig. 4.51 Antimüllerian hormone (AMH). The protein is cleaved from a precursor. The three-dimensional structure has not yet been determined, thus disulfide bridges cannot be assigned. The asparagine (**N**; on *blue background*) is glycosylated (Source: Swiss-Prot 03971)

4.12.3.1 Introduction

In the nineteenth century Johannes Müller (1801–1858) described the müllerian ducts as the female analogue to the male wolffian ducts found by Caspar E. Wolff (1733–1794). Experiments by A. Jost in the middle of the twentieth century could only be interpreted that in males a protein induces regression of müllerian ducts. Thus male fetuses do not develop ovaries and ovarian ducts. The development of wolffian ducts into seminiferous tubules requires testosterone that female fetuses do not yet have.

Therefore, in the female sex development the müllerian ducts evolve which due to the missing AMH do not regress, whereas wolffian ducts are not developed due to the lack of testosterone.

The hormonal nature of AMH was proven by A. Jost while analyzing the calf of Freemartin: two calf fetuses of different sexes share a placenta. This situation induces in the female calf the occurrence of male sex characteristics. The effect had to be induced via the placenta and the blood; thus AMH has been proven to act in an endocrine manner (Teixeira et al. 2001).

4.12.3.2 Structure and Gene

AMH is a member of TGF-β growth factors, the *AMH* is coded for in the short arm of chromosome 19 in locus 19p13.3 and contains five exons. It has not yet been crystallized. Which enzyme releases the N-terminal hexapeptide is not known.

The structure cannot be found in the Protein Data Bank, however, there is a presumably deduced structure on the Internet when you look for "AMH structure".

4.12.3.3 Physiology

AMH is already expressed in (male) Sertoli cells during fetal development; in women AMH is expressed in preantral and antral follicles. AMH in male fetuses induce the müllerian ducts to regress via apoptosis. Postpartum AMH is measurable until puberty and is downregulated when testosterone levels increase. The expression in girls and women is arrested by LH of the menstrual cycle (Teixeira et al. 2001).

The AMH receptor, AMHRII, is a serine/threonin kinase of type 2 in the cell membrane. The protein recruits a type 1 receptor kinase. Together these proteins phosphorylate SMAD proteins that induce further signals and gene activations (Teixeira et al. 2001).

AMH defects leave the müllerian ducts intact and their development while concomitantly male sex characteristics develop (müllerian duct persistence syndrome) characterized first by maldescendant testes. Mutations of AMH and AMH receptor can both be responsible for the syndrome (Josso et al. 2005)

4.12.3.4 Phylogeny

AMH has been identified in mammals, birds, reptiles, amphibians, and fish. Although female birds show the heterogamete type ZW and males are homogamete ZZ, genes such as *sry*, *sox9*, *sf-1*,and *dax1* together with *amh* have a sex-determining role. In reptiles sex determination depends more on environmental temperature (Thurston and Korn 2000). In fishAMH is a growth factor of early germ cell without influence on any müllerian ducts because they are lacking in fish (Shiraishi et al. 2008)

4.13 Angiotensins and Renin

Fact sheet 4.46: Angiotensin-II

Gene:	Chromosome 1; locus 1q42.2; five exons.
Sequence:	`DRVYIHPF` (Fig. 4.52).
Synthesis and target:	Angiotensinogen is a liver protein, renin a renal enzyme, and angiotensin-converting enzyme (ACE) a lung enzyme. Renin cleaves from the angiotensinogen the angiotensin I (highlighted in Fig. 4.52) which is later converted by ACE into the active angiotensin-II.
Function:	Angiotensin-II stimulates in the adrenal the synthesis (and release) of aldosterone. Angiotensin- is a major regulator of blood pressure, water, and electrolyte homeostasis.
Receptor:	The angiotensin receptor (AGT-R) is a heptahelical GPCR with two splice variants: AGT receptor 1 (AGT-R1) signals stimulation of aldosterone synthesis; in contrast, AGT-R2, a heptahelical GPCR, too, acts to reduce aldosterone.

4.13.1 Introduction

Angiotensin is released by renin from the angiotensinogen precursor. This precursor is a liver protein. In the juxtaglomerular renal cells (Fig. 10.12) blood pressure is estimated. With decreasing blood pressure or with increasing osmolarity, that is, increasing ion content in the blood, these juxtaglomerular cells release renin. The only known substrate of renin is angiotensinogen (Sect. 11.8.7).

5 10 15 20 25 30 35 40 45 50	
m r k r a p q s e m a p a g v s l r a t i l c l l a w a g l a a g	-31 - -1
D R V Y I H P F H L v i h n e s t c e q l a k a n a g k p k d p t f i p a p i q a k t s p v d e k a	1 - 50
l q d q l v l v a a k l d t e d k l r a a m v q m l a n f l q f r i y q m h s e l w q v v h q a t v	51 - 100
l s p t a v f g t l a s l y l g a l d h t a d r l q a i l g v p w k d k n c t s r l d a h k v l s a	101 - 150
l q a v q g l l v a q g r a d s q a q l l l s t v v g v f t a p g l h l k q p f v q g l a l y t p v	151 - 200
v l p r s l d f t e l d v a a e k i d r f m q a v t g w k t g c s l m g a s v d s t l a f n t y v h	201 - 250
f q g k m k g f s l l a e p q e f w v d n s t s v s v p m l s g m g t f q h w s d i q d n f s v t q	251 - 300
v p f t e s a c l l l i q p h y a s d l d k v e g l t f q q n s l n w m k k l s p r t i h l t m p q	301 - 350
l v l q g s y d l q d l l a q a e l p a i l h t e l n l q k l s n d r i r v g e v l n s i f f e l e	351 - 400
a d e r e p t e s t q q l n k p e v l e v t l n r p f l f a v y d q s a t a l h f l g r v a n p l s	401 - 450
t a	451 - 452

Fig. 4.52 The angiotensinogen precursor. Renin cleaves angiotensin I (highlighted *yellow*) from the angiotensinogen precursor. The Angiotensin-converting enzyme removes two C-terminal amino acids and generates the active angiotensin II (**DRVYIHPF**) (Source: NP_000020.1)

4.13.2 Structure and Genes

Renin cleaves from the precursor the angiotensin-I: **DRVYIHPFHL** (highlighted *yellow* in Fig. 4.52). Angiotensin-I circulates and is converted by angiotenin-converting enzyme (ACE) into Angiotensin-II: **DRVYIHPF**.

Angiotensin-II can additionally be formed (without renin and ACE) by the endopeptidase chymase, which is relevant for the pathophysiology of hypertension, atherosclerosis, and diabetic renal insufficiency (Miyazaki and Takai 2006).

In the brain, however, renin and ACE are expressed and the local neuronal angiotensin-II formation appears in the classical way (Saavedra 1992).

4.13.3 Physiology

Angiotensin-II triggers in the adrenal synthesis of the mineralocorticoid aldosterone (see Fig. 11.15). This action is signaled by AGT-R1.

Pathophysiological actions of angiotensin-II are mostly related to reactive oxygen radicals, so-called ROS. In addition, metalloproteinases, the PDGF receptor, and an EGF receptor set off a cascade that leads to vasoconstriction, fibrosis, hypertrophy, or inflammation. However, with the AGT-R2 receptor expressed, the tissue is protected against these symptoms.

The central actions of angiotensin-II target baroreflex, blood pressure control, and control of water intake/thirst are discussed in Sect. 11.8.

4.13.4 Phylogeny

Angiotensin-I has been found in fish and later vertebrates. ACE is a very old enzyme with an existence reported in insects and bacteria, as well. Renin, however, is not found earlier than fish. This leaves angiotensin-II formation restricted to vertebrates. In *Ciona intestinalis* and later in fish the AGT-R1 has been found.

4.14 Atrial Natriuretic Peptides

Fact sheet 4.47: ANP — BNP — CNP

Gene:	ANP; Chromosome 1; locus 1p36.2; three exons.
Gene:	BNP; Chromosome 1; locus 1p36.2; three exons.
Gene:	CNP; Chromosome 2; locus 2q37; two exons.
Sequence:	Fig. 4.53.
Synthesis and target:	ANP is formed in atrial myocytes in reaction to elevated blood pressure. It acts on renal NP receptors. BNP is likewise made in the heart (and in the brain). It acts on cardiac NP receptors to prevent fibrosis in heart muscle cells. CNP is a brain peptide.
Function:	ANP is an endocrine regulator of water homeostasis and a major regulator of cardiac development. BNP appears to be a paracrine organizer of cardiac tissue, whereas CNP participates in bone formation and prolongs ossification.
Receptor:	The ANP and BNP receptor is a membrane guanylate cyclase. The CNP receptor is a similar guanylate cyclase. CNP binds additionally to the NP clearance receptor NP-R3.

4.14.1 Introduction

Since 1984 the hormone from the myocytes of the right atrium has been known which controls volume and electrolyte balance. Its origin helped to name it atrial natriuretic peptide (ANP). A second related peptide was later found in the brain (and after that in the heart, too) which was labeled brain natriuretic peptide (BNP). When the third peptide was found in the brain, it was simply named CNP.

4.14.2 Structure and Genes

The genes for ANP and BNP are closely linked on the short arm of chromosome 1 (1p36.2).[33] CNP is coded for on chromosome 2, on the long arm (2q36).

ANP is synthesized in right atrial myocytes, BNP in the ventricle and in brain. CNP, apart from being made in the brain, is formed in circular endothelia.

The characteristic feature is the intramolecular disulfide bridge that generates a cyclic peptide. In this cyclic part the homology of the three peptides is obvious (Fig. 4.54).

[33]http://www.ncbi.nlm.nih.gov/projects/mapview/maps.cgi?TAXID=9606&CHR=1&MAPS=genes-r&QSTR=600295[MIM]&QUERY=uid%2812721970%29&BEG=11%2C780K&END=11%2C900K&oview=default

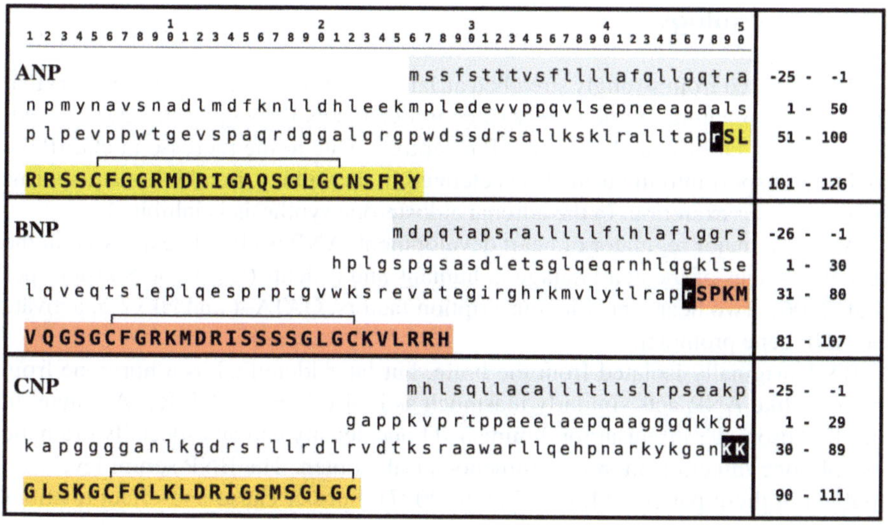

Fig. 4.53 Primary sequences of the three atrial natriuretic peptides ANP, BNP, and CNP. The three peptides, ANP, BNP, and CNP, are generated from precursors, which are cleaved by the signal peptidase and different endopeptidases to release the C-terminal natriuretic peptides. The cyclic nature of these peptides due to disulfide bridges is indicated by connecting the cysteine residues (Source: GenBank AAA35529.1 (ANP), P16860.1 (BNP), BAA14351.1 (CNP))

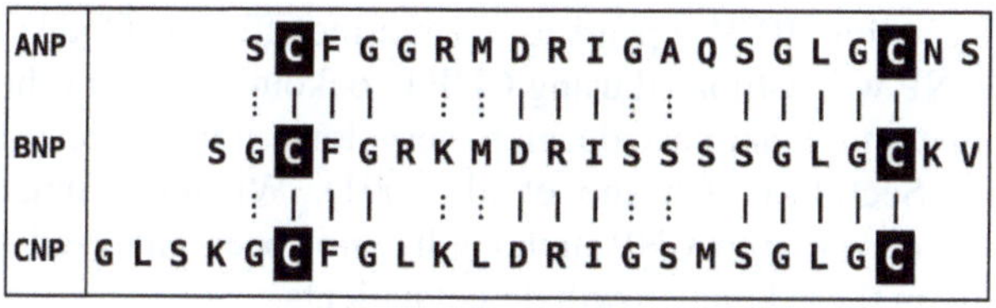

Fig. 4.54 Sequence homologies of atrial natriuretic peptides. Identical amino acids are labeled by |, similar amino acids by ⁝, cysteines are shown *inverted*

The prohormone convertase for ANP is corin, a type II membrane endopeptidase with several characteristic regions: trypsin-like peptidase domain, LDL receptor domain, scavenger receptor domain, and two cysteine-rich FZ domain (for frizzled involved in signal transduction of the Wnt protein). Defects of corin are similar to those of ANP. The CNP precursor is cleaved by furin (Wu et al. 2003a). The CNP precursor has a PC2 motif, too (Fig. 4.53; aa 88–89).

The receptor for natriuretic peptides are membrane-located guanylate cyclases NP-R1 and NP-R2.

4.14.3 Physiology

ANP is released from strongly stretched atrial myocytes. The mechanoreceptors that signal this stretching are nonselective cation channels (Zhang et al. 2008b). ANP acts on renal, cardiac, and adrenal NP receptor 1. Due to the increase in cGMP the sodium transport into the urine is accelerated in the kidney. In addition aquaporins augment water excretion. In the adrenal aldosterone synthesis is inhibited.

ANP is a major regulator of heart development. ANP is already expressed in the very first stages of the heart anlage in humans and rodents (Chuva de Sousa Lopes et al. 2006). Two heart-specific transcription factors, GATA-4 and Nkx2-5, activate the ANP gene promoter.

BNP, originally isolated from the brain, but later identified as a hormone from the heart like ANP, acts similarly inasmuch as it also binds to NP-R1. A mutant in the BNP promoter is causal for diminished bone density and an especially dramatic loss of bone mineralization in postmenopausal women. The BNP serum levels are doubled in these patients (Takeishi et al. 2007). Further cause-and-effect relations have not been identified.

In BNP knockout mice pathological changes in heart tissue were evident: increase and hardening of connective tissue (Tamura et al. 2000). $BNP^{-/-}$ mice did not exhibit–compared with wildtype mice–alterations of potassium serum levels, aldosterone levels, urine sodium or potassium excretion, which led to the conclusion that BNP does not or only marginally affects metabolic parameters. In contrast to ANP's endocrine action, BNP appears as a paracrine effector (Ogawa et al. 2001).

The role of CNP was estimated using CNP knockout mice: Chusho et al. (2001) demonstrated that these animals develop very long bones because ossification is retarded (see Sect. 11.6; Chusho et al. 2001). Whether human variants of achondrodysplasia are due to CNP mutants has not been analyzed thus far. CNP might act as drug in these bone growth defect patients.

4.14.4 Phylogeny

Although in hagfish (*Eptatretus burgeri*) only a single NP gene has been found (Kawakoshi et al. 2003), cartilaginous and bony fish and later vertebrates have three and more NP genes (Houweling et al. 2005). There are up to four CNP genes. CNP-4 is regarded as the primordial gene of cyclostomes (Kawakoshi et al. 2006).

As before for other hormones there are some reports that antibodies against human or rodent NP bind to snail, crab, or even protozoa specifically. Any biochemical support for a hormone or cloning and sequencing has, however, not been provided. Takei thus rejected in 2001 these reports as lacking proof. Since that time any further confirmation for expression of NP in invertebrates is missing.

Guanylate cyclase receptors have been found in many divergent vertebrates and invertebrates, however, the NP-R1 and NP-R2 are vertebrate specific (Fitzpatrick et al. 2006). NP-R3, the NP-clearance receptor is equally restricted to vertebrates.[34]

4.15 Additional Neuropeptides

4.15.1 Parathormone

Fact sheet 4.48: Parathormone (PTH)	
Gene:	Chromosome 11; locus 11p15.2; three exons.
Sequence:	Fig. 4.55.
Synthesis and target:	PTH is synthesized and released by parathyroid glands as a response to low serum calcium concentrations. PTH acts on the PTH receptor on bone cells and in the kidney.
Function:	PTH induces decalcification of bone. Simultaneously it inhibits renal calcium excretion.
Receptor:	PTH receptor is a heptahelical GPCR.

4.15.1.1 Introduction

Parathormone (PTH; Fig. 4.55) is the product of the parathyroid gland (pars parathyroidea or glandula parathyroidea). Very sensitive calcium sensors estimate the free calcium concentration ($[Ca^{2+}]_{free}$) in blood. When $[Ca^{2+}]_{free}$ rises parathormone release is stopped and calcitonin release initiated; when $[Ca^{2+}]_{free}$ declines PTH is released and calcitonin release inhibited.

4.15.1.2 Structure and Genes

The *PTH* gene is located on chromosome 11, about 1.5 megabase pairs (MBp) away from *CALCA*. After removing the signal peptide the furin endopeptidase acting, for example on the somatostatin precursor (Sect. 4.3.5), removes the N-terminal

```
                  1                   2                   3                   4                   5
1 2 3 4 5 6 7 8 9 0 1 2 3 4 5 6 7 8 9 0 1 2 3 4 5 6 7 8 9 0 1 2 3 4 5 6 7 8 9 0 1 2 3 4 5 6 7 8 9 0
                                                  m i p a k d m a k v m i v m l a i c f l t k s d g   -25 - -1
k s v k k r S V S E I Q L M H N L G K H L N S M E R V E W L R K K L Q D V H N F V A L G A P L A P R    1 - 50
D A G S Q R P R K K E D N V L V E S H E K S L G E A D K A D V N V L T K A K S Q                        51 - 90
```

Fig. 4.55 Sequence of human parathormone. The signal peptide is highlighted *light gray*, the mature peptide highlighted *yellow* and in *uppercase* (Source: GenBank NP_000306.1)

[34]http://www.ncbi.nlm.nih.gov/sites/entrez?Db=homologene&Cmd=ShowDetailView&TermToSearch=699

hexapeptide. Although a **KR**-motif is present, PC1 is not used; these cells obviously do not express PC1.

A parathormone-related peptide (PTHrP) is not made in the parathyroid gland, but in bone by chondrocytes (see Sect. 11.6.3) and by tumor cells.

4.15.1.3 Physiology

PTH is required for control of free calcium ion concentration ($[Ca^{2+}]_{free}$) in the blood. In the PTH forming chief cells in the glandula parathyroidea and in the kidney the calcium-sensing receptor (CASR[35]) is expressed, a heptahelical membrane GPCR. Its expression is *i.a.* stimulated by calcitriol (dihydroxy vitamin D_3). The signal transduction cascade from CASR is directed to the PTH promoter.

Sufficiently high $[Ca^{2+}]_{free}$ is required, for example, to achieve a fast transport in signals with calcium influx. The function of many proteins requires complexes with calcium, such as integrins that do not bind to their ligands on cells or in the extracellular matrix without calcium.

Several hereditary diseases are linked to mutants in the PTH protein/gene or in the PTH receptor (see NCBI:OMIM:PTH[36]): chondrodysplasia or enchondromatosis, as well as hyper- and hypoparathyroidisms.

4.15.1.4 Phylogeny

The phylogeny of PTH is related to calcium sensing by CASR. In fish there are PTH-like polypeptides. These are expressed but in gills. Okabe and Graham (2004) demonstrated that gills and parathyroid glands are functionally related: the controlling factor for expression of PTH and CASR in gills and parathyroid glands is in fish, chicken, and mammals GCM2 (glial cell missing) already found in *D. melanogaster* and which is the only transcription factor whose expression is restricted to the parathyroid gland. Whereas fish take calcium from the water, tetrapods got independent of this source by developing the parathyroid gland but use the same hormone for calcium homeostasis (Okabe and Graham 2004).

No PTH could be identified in invertebrates. There are anecdotal descriptions of PTH actions on neurons and ganglia of snails without molecular proof. Antimammalian-PTH antibodies could bind to snail tissue, but the molecule(s) recognized have thus far escaped identification (Hull et al. 2006).

4.15.2 Stanniocalcin

4.15.2.1 Introduction

Stanniocalcin (STC) is the hormone of the corpuscles of Stannius, that is, endocrine fish glands associated with teleost kidney. At elevated $[Ca^{2+}]_{free}$ STC is released which blocks calcium uptake in the gill. Thus fish regulate their calcium levels. In

[35]http://omim.org/entry/601199

[36]http://www.ncbi.nlm.nih.gov/entrez/dispomim.cgi?id=168468

Fact sheet 4.49: Stanniocalcin (STC)	
Sequence:	Fig. 4.56 (Coho salmon).
Synthesis and target	STC is formed and released in the corpuscles of Stannius in fish. It acts on the calcium gill transporter and kidney phosphate transporters.
Function:	STC blocks calcium transporters in gills to reduce calcium uptake. Kidney phosphate transporters, in contrast, are stimulated.
Receptor:	(unknown; July 2008/October 2011).

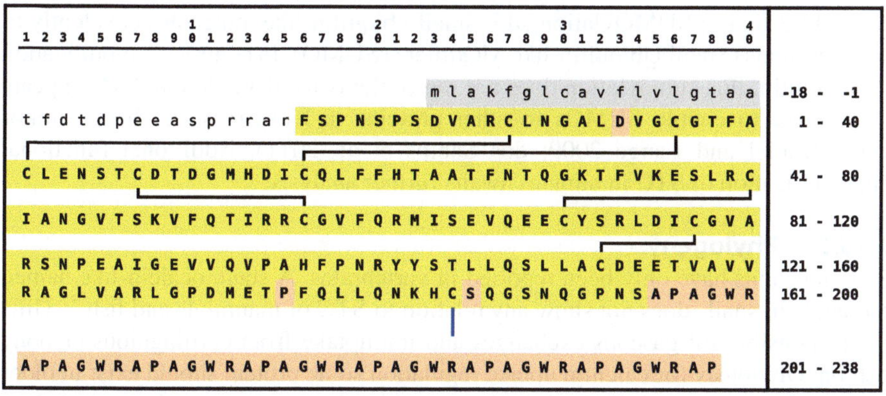

Fig. 4.56 Sequences of stanniocalcin from *Oncorhynchus kisutch* (Coho salmon/silver salmon) and *Oncorhynchus keta* (chum salmon). The precursor from −18 to 238 is from *O. kisutch*. The *O. keta* sequence (highlighted *yellow*) differs from the *O. kisutch* sequence in three amino acids (highlighted *light orange*). We superimposed these two sequences because for *O. keta* the disulfide bonds have been determined (shown by *black* lines connecting the cysteine residues). The cys residue 184 serves as a intermolecular disulfide bond linking two stanniocalcins in a homodimer (Source: Swiss-Prot P43647.2 (*O.keta*) and GenBank AAB26419.1 *O.kisutch*)

addition STC stimulates renal resorption of phosphate which leads to more calcium incorporated into bone and reduces $[Ca^{2+}]_{free}$ as well.

4.15.2.2 Structure and Genes

Stanniocalcin is generated from a precursor that is cleaved by the signal peptidase and a furin-like enzyme, inasmuch as PC1 or PC2 would degrade the STC peptide further at their recognition motifs (Fig. 4.56; PC1: amino acids 227–228) and PC2: (amino acids 12–13; 94–95; 226–227). The molecule exists as a disulfide-linked homodimer.

In humans (as well as in rodents) two stanniocalcin genes have been found: on chromosome 8 (8p21) and on chromosome 5 (5q35.2).

4.15.2.3 Physiology

The calcium sensor receptors CaSR in gills and corpuscles of Stannius induce at elevated $[Ca^{2+}]_{free}$ STC release from the corpuscles, but according to recent findings, from gills and other tissues, too. Calcium transport from water through the gills into the circulation is blocked by STC. STC augments, in addition, renal phosphate resorption. By reducing uptake and forcing bone deposition as calcium phosphate serum calcium concentrations are reduced.

Recently STC have been found in rodents and humans. Apart from some speculation on the role of STC in mammals, it has been shown that STC is a SUMO E3 ligase: SUMOylation adds small ubiquitin-like molecules covalently to other proteins (Small Ubiquitin-like MOdifier = SUMO). In contrast to modification by ubiquitin, this sumoylation does not tag proteins for degradation. STC appears as a ligase to couple SUMOs to other proteins in leukemias and breast tumor cells (Daniel and Lange 2009; dos Santos et al. 2011). Additional functional characterization of STC in humans awaits further analysis.

4.15.2.4 Phylogeny

Stanniocalcins have only been reported in vertebrates. A stanniocalcin-like named molecule in snails does not show any relation to STC of mammals and fish.[37] Gills are responsible for gaseous exchanges and ion uptake from cartilaginous or bony fish on. Obviously for calcium uptake regulation some protein should exist in older classes; the presence of STC-related polypeptides, however, cannot be excluded.

4.15.3 Erythropoietin

Fact sheet 4.50: Erythropoietin (Epo)

Gene:	Chromosome 7; locus 7q21; five exons.
Sequence:	Fig. 4.57; the protein has been crystallized: 1EER; there is structural homology to GH, PRL or leptin.
Synthesis and target:	Epo is synthesized in oxygen-sensing cells of juxtaglomerular kidney cells.
Function:	Epo stimulates the proliferation of reticulocytes which mature into erythrocytes.
Receptor:	The receptor belongs to the family of interleukin growth factors such as GHRH-R or PRL-R.

[37] Our ClustalW analysis

```
                  1                   2                   3                   4                   5
1 2 3 4 5 6 7 8 9 0 1 2 3 4 5 6 7 8 9 0 1 2 3 4 5 6 7 8 9 0 1 2 3 4 5 6 7 8 9 0 1 2 3 4 5 6 7 8 9 0
                                                      m g v h e c p a w l w l l l s l l s l p l g l p v l g    -27 -  -1
A P P R L I C D S R V L E R Y L L E A K E A E N I T T G C A E H C S L N E N I T V P D T K V N F Y A            1 -  50
W K R M E V G Q Q A V E V W Q G L A L L S E A V L R G Q A L L V N S S Q P W E P L Q L H V D K A V S           51 - 100
G L R S L T T L L R A L G A Q K E A I S P P D A A S A A P L R T I T A D T F R K L F R V Y S N F L R          101 - 150
G K L K L Y T G E A C R T G D R                                                                              151 - 166
```

Fig. 4.57 Primary sequence of erythropoietin. The signal peptide is in *lowercase* highlighted *gray*. The protein is highlighted *yellow*; helical regions are underlined in different colors

4.15.3.1 Introduction

The endocrine role of erythropoietin has somehow escaped our attention. Although not being part of central control via the hypothalamus and pituitary, it is a sensu stricto hormone made in the kidney, transported via the blood, and acting in the bone marrow on reticulocytes. Even structurally it is closely related to other hormones, for example, GH, PRL, or leptin.

4.15.3.2 Structure and Genes

The *EPO* gene on chromosome locus 7q21 has five exons. The gene is activated in kidney cells when the oxygen concentration in the medullary cortex drops. While in normoxic cells hypoxia-inducible factor (HIF) is oxidized by reactive oxygen species (ROS); in hypoxic cells, HIF is stable and can bind to the *EPO* promoter and stimulate transcription.

The translated protein is a tetrahelical glycoprotein hormone such as GH, PRL, or leptin. The helical regions are underlined with different colors in Fig. 4.57, the N-glycosylation sites are boxed. There are additional O-glycosylation sites not indicated.

The receptor is very similar to the GH receptor belonging to the CSF-receptor family (see Sect. 8.5. It is expressed in small numbers (some 1,000 molecules per cell) on erythroid cells. The EPO-R is the earliest molecule to distinguish the erythroid from the myeloid lineage.

4.15.3.3 Physiology

EPO transcription is controlled by the oxygen content in the renal medullary cortex. The REPOS cells (Wenger and Hoogewijs 2010) are characterized by HIF (hypoxia-induced factor) which is under normoxic situation oxidized and thus degraded. When the oxygen content drops due to reduced oxygen partial pressure in blood, the production of reactive oxygen species (ROS) is blocked and HIF remains stable. Together with co-factors, HIF induces transcription of the *EPO* gene. After transport via the blood, EPO acts in the bone marrow on erythropoiesis by stimulating cell division of reticulocytes.

The receptor interaction of EPO and the EPO-R has been crystallized. Figure 8.6 exemplifies such an interaction for the highly homologous GH and GH receptor. A picture for EPO and its receptor appeared very similar.

The available literature does not discuss a mechanism whereby EPO would be released by a release triggering signal. On the contrary, there is no evidence for storage of EPO in secretory granules (personal communication by R. Krstić, author of Human Microscopic Anatomy, see Chap. 10). In an experiment to analyze EPO secretion as a requirement for autocrine growth stimulation of EPO transfected cells, Villeval et al. (1994) transfected the UT-7 pluripotent cell line with an EPO vector and got an EPO secretion that stimulated UT-7 growth. Anti-EPO antibodies blocked UT-7 proliferation, likewise an additional retention signal at the C-terminal end of the EPO sequence. This shows that normally EPO is not retained, at least in these cells.

Because EPO belongs to the cytokine family of proteins, it should be noted that cytokines and protein/peptide hormones use different secretory pathways. Cytokines are constitutively secreted whereas secretion of hormones is regulated. The signals whether a peptide/protein is sorted into either of the two pathways seem to be generated by the protein/peptide sequence itself as has been recently shown in the case of gastrin (Bundgaard and Rehfeld 2008). Comparing EPO, GH, PRL, and leptin and, for example, generating hybrid sequences should answer the question of which part of a molecule is required for the regulated pathway.

EPO, once released, can enter the bloodstream using the fenestrated peritubular ascending arterial vas rectum (Krstić 1991, plate 154).

EPO acts on cells of the erythroid lineage. These are present in the bone marrow. The EPO receptor (EPO-R) has been found to be the earliest marker distinguishing the erythroid and the myeloid lineage. The stimulation of erythroid precursor cells enhances reticulocyte and erythrocyte production. Absence of EPO results in lack of erythrocyte renewal and thus in anemia. Injection of EPO can substitute for the failure of endogenous EPO production.

Being available as a drug, EPO has been used for nonnormal stimulation of erythropoiesis. Such an enhancement of erythrocytes might facilitate thrombosis which may be fatal. Because not all EPO preparations are biosimilars, detection of doping is possible but difficult. EPO doping with the unchanged human sequence is impossible to track. For this reason, athletes with an abnormally increased hematocrit are prevented from competition for a certain period, but not challenged for doping (see Chap. 15).

4.15.3.4 Phylogeny

GenBank contains erythropoietin entries from primates, humans, orangutans, guenons, and rodents. There is more work required to clarify the phylogeny of EPO.

Invertebrate Protein and Peptide Hormones 5

Contents

5.1 Metabolically Active Peptide Hormones 135
5.1.1 Crustacean Hyperglycemic Hormone 135
5.1.2 Bombyxin and Insulin-Like Peptides (ILP) 137
5.1.3 AKH, RPCH, and HrTH 139
5.2 Regulation of Heart Frequency and Pressure by Neuropeptides 141
5.2.1 Cardioacceleratory Peptides: CAP 141
5.2.2 Cardioexcitatory Peptide, NDWF-Amide 143
5.2.3 Enterins 144
5.2.4 *Mytilus* Inhibitory Peptides (MIP; AMrP) 145
5.2.5 Diuretic Hormones 147
5.3 Kinins 148
5.3.1 Pyrokinin/Myotropin and Pheromone Biosynthesis Activating Neuropeptide (PBAN) 149
5.3.2 Orcokinins 151
5.3.3 Leucokinins/Lymnokinins 153
5.3.4 Tachykinin-Related Peptides (TRP) 155
5.3.5 Sulfakinins 157
5.4 Neuropeptides of Reproduction 158
5.4.1 PTTH 158
5.4.2 PTSH; MIP 160
5.4.3 Pheromonostatic Peptide, Sex Peptides 161
5.4.4 GIH, VIH 163
5.4.5 TMOF 164
5.4.6 Nebcolloostatin 165
5.5 Peptide Hormone of Metamorphosis and Molting 166
5.5.1 Molt-Inhibiting Hormone (MIH) 166
5.5.2 Corazonin 168
5.5.3 Ecdysis-Triggering Hormone (ETH) 169
5.5.4 Eclosion Hormone (EH) 170
5.5.5 Bursicon 172
5.6 Regulators of Food Intake: RFamide and FMRFamide 174
5.6.1 Introduction 174
5.6.2 Biochemistry and Structure 175
5.6.3 Physiology 177

B. Kleine, W.G. Rossmanith, *Hormones and the Endocrine System*,
DOI 10.1007/978-3-319-15060-4_5

5.6.4 Phylogeny 178
5.7 Neuropeptide Regulators of Juvenile Hormone Metabolism 178
5.7.1 Allatotropins 178
5.7.2 Allatostatins 179
5.8 Peptide Hormones of Skin: Pigment-Dispersing Hormone 181
5.8.1 Introduction 181
5.8.2 Biochemistry and Structure 182
5.8.3 Physiology 182
5.8.4 Phylogeny 183
5.9 Other Neuropeptides 184
5.9.1 Neuropeptide-F—Two Peptide Genes 184
5.9.2 Proctolin 187
5.10 Summary and Overview 188

When analyzing invertebrates in the 1930s and 1940s, Berta and Hans Schaller developed the idea of neurosecretion using the large oceanic snail *A. californica*[1] (Califonia sea hare) with its few and relatively large neurons. Neuropeptides have been found in all metazoans. Where whole genome sequences are available such as in bees *A. mellifera*[2]) sequence motifs (signal peptides, cleaving sites of prohormone convertase, terminal glycine) could be investigated. In bees, 200 different potential neuropeptides were identified and the majority of those could be confirmed by chemical analysis (Hummon et al. 2006). In further insects, that is, *D. melanogaster* or *C. elegans*,[3] the two model organisms of developmental biology, the potential inventory of neuropeptides is identified.

Within deuterostomes some tunicates (e.g., *C. intestinalis*[4]) and echinoderms (*S. purpurata*) genomes have been fully sequenced. In these species, the focus has been on vertebrate hormone homologues. Since 2008, when we first looked this up, several peptides have been found related to RF-amides, or to GnRH in echinoderms and tachykinins and GnRH in tunicates (Kawada et al. 2011; Roch et al. 2014; Rowe et al. 2014). These hormones have been discussed in chapter 4. Therefore, when we talk about invertebrates in the following, mostly this is about protostomes.[5]

Table 5.1 summarizes structural motifs and other characteristics of invertebrate hormones. Individual hormones are described in the following sections.

[1]Aplysia californica.

[2]Apis mellifera.

[3]Caenorhabditis elegans.

[4]Ciona intestinalis.

[5]As far as PubMed is concerned; Aug. 2008

Table 5.1 Sequence motifs of invertebrate neuropeptides

Name	Motif
Adipokinetic hormones	p**ELNFx**$_{4/5}$-NH_2
Allatotropins	**TARGF**-NH_2
Allatostatins type-A	**Y/F-x-FG-L/I**-NH_2
type-B	**Wx**$_6$**W**-NH_2
type-C	**PISCF**-OH
Bombyxin—insulin-like peptide	B chain → C-peptide → A chain
CAP	nonapeptides with intramolecular disulfide-bridge like oxytocin/vasopressin
Enterins	**PxxxHxxFV**-NH_2
FMRF/RFamide	**FMRF**-NH_2 or **RF**-NH_2
Cardioexcitatory peptide	**N**D**WF**-NH_2
Leucokinins	**FxxWG**-NH_2
Short neuropeptide-F	**xxxxRLRF**-NH_2
MIP	**PxFY**-NH_2 (y is F, I, V)
Orcokinins	H_2N-**NxDEI**
PTTH	Cysteine knot like gonadotropins/TSH/NGF
Pyrokinins	**FxPRL**-NH_2
Sulfakinins	Tyrosine-sulfated **YGHMRF**-NH_2
Tachykinin	**FxGLM**-NH_2

5.1 Metabolically Active Peptide Hormones

5.1.1 Crustacean Hyperglycemic Hormone

Fact sheet 5.1: Crustacean hyperglycemic hormone (CHH)

Sequence: Fig. 5.1.

Synthesis and target: CHH and related peptides are formed in neurosecretory cells of the X-organ and-after axonal transport into the neurohemal organ sinus gland-are released into the hemolymph.

Function: CHH controls carbohydrate metabolism.

Receptor: Transmembrane guanylate cyclase (mGC).

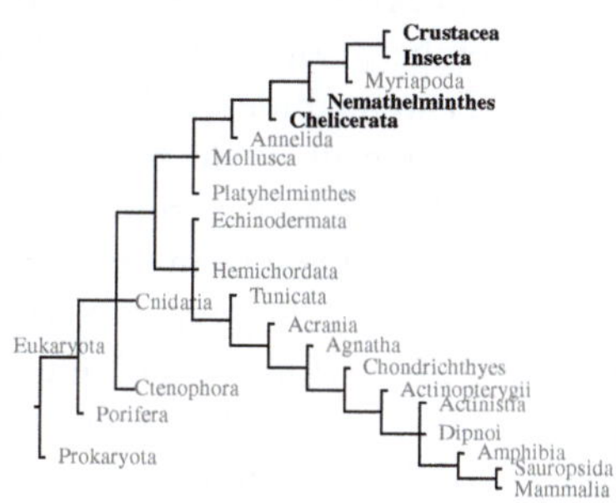

5.1.1.1 Introduction

CHH of crustacea is a hormone of the X organ in the eye-stalk released in the sinus gland. CHH is the prototype of a peptide family: CHH/MIH/MOIH/GIH/

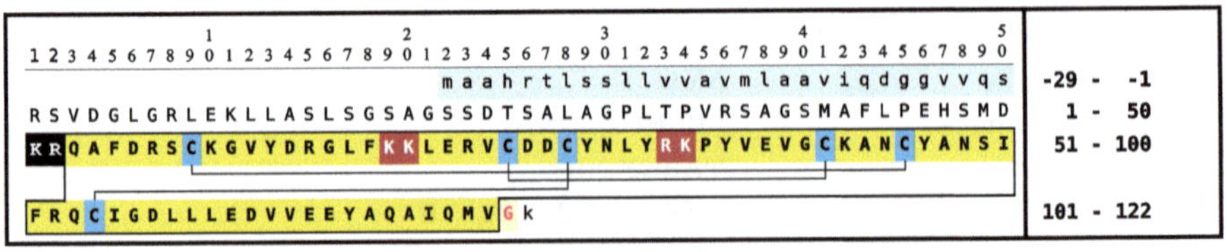

Fig. 5.1 The crustacean hypoglycemic hormone (CHH) of the common European hermit crab (*P. bernardus* [Pagurus bernhardus]): signal peptide (first line, *light blue*); the precursor CHH amino acids 1–126 (*uppercase*) is cut by prohormone convertase (PC) 1 after the motif KR (inverse) to release CHH amino acids 53–126 which is then processed to the amidated CHH (amino acids 53–124 *boxed*; *bold* on *yellow*); unused PC2 motifs are shown *white* on *dark red*; *black lines* between cysteine residues (*black* on *blue*) are analogous to Yasuda et al. (1994). The phenylalanine (amino acid 55) may be epimerized; the N-terminal glutamine in the secreted peptide is modified to pyroglutamate (Source: GenBank DQ450960)

Table 5.2 CHH family members

Abbreviation	Name	Alternative name
CHH	*Crustacean hyperglycemic hormone*	Ion transport peptide (ITP)
MIH	*Molt-inhibiting hormone*	
MOIH	*Mandibular organ-inhibiting hormone*	
GIH/VIH	*Gonad/vitellogenesis-inhibiting hormone*	

VIH. Crustacean hyperglycemic peptide stimulates the carbohydrate metabolism; molting-inhibiting hormone (MIH) suppresses the molt required for growth in ecdysozoans; mandibular-organ (MO)–inhibiting hormone blocks methylfarnesoate synthesis in the mandibular organ; gonad-inhibiting/vitellogenesis-inhibiting hormones (GIH/VIH) modify gonadal functions.

The CHH homologue of insects, ion transport protein (ITP), is formed and released in the corpora cardiaca.

5.1.1.2 Biochemistry and Structure

The members of the CHH/MIH/MOIH/GIH/VIH-family (abbreviations in Table 5.2) are peptides with about 80 amino acids with six characteristic cysteines and thus three disulfide bridges (see also Fig. 5.21). The CHH precursor bears another, CHH precursor-related peptide (CPrP; amino acids 1–50) whose function remains yet unknown. In crustaceans CHH is formed in the X organ (XO) and released in the sinus gland (SG). The CHH homologue of insects, ion transport protein (ITP), is formed and released in the corpora cardiaca (CC). An alternative ITP was identified by Dircksen et al. (2001) in the pericardial organ. By alternative splicing of the same RNA a C-terminal modified peptide was generated.

5.1.1.3 Physiology

The CHH receptors in crustaceans are transmembrane guanylate cyclases (mCG), which after CHH-binding stimulate intracellular cGMP production. CHH stimulates amylase release in the midgut. This in turn increases sugar content in the hemolymph. CHH (as MIH) may inhibit ecdysteroid biosynthesis which delays molting. ITP similarly binds to an mGC and regulates diuresis in flies.

5.1.1.4 Phylogeny

CHH/ITP and related peptides have been found in chelicerates, nematodes, crustaceans, and insects.

5.1.2 Bombyxin and Insulin-Like Peptides (ILP)

Fact sheet 5.2: Bombyxin and insulin-related peptides (ILP)

Sequence: Fig. 5.2.

Synthesis and target: Bombyxins/ILPs are synthesized by neurosecretory cells of the CNS, as well as in the gut (in *D. melanogaster*).

Function: Bombyxin/ILPs have important functions for growth.

Receptor: ILP/bombyxin receptor is (like the insulin receptor) a transmembrane tyrosine kinase. An insulin receptor substrate (IRS) is needed for signal induction.

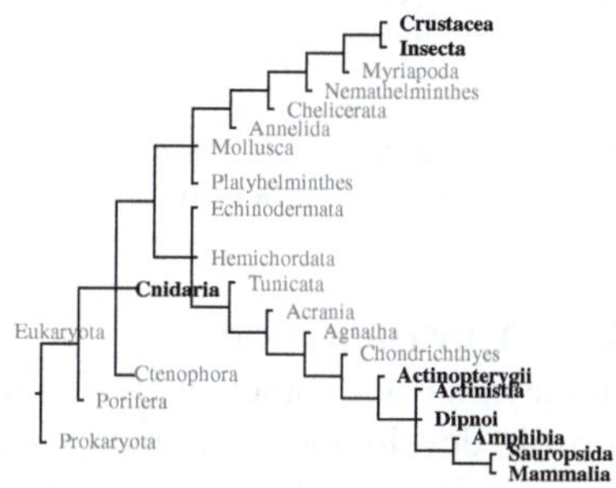

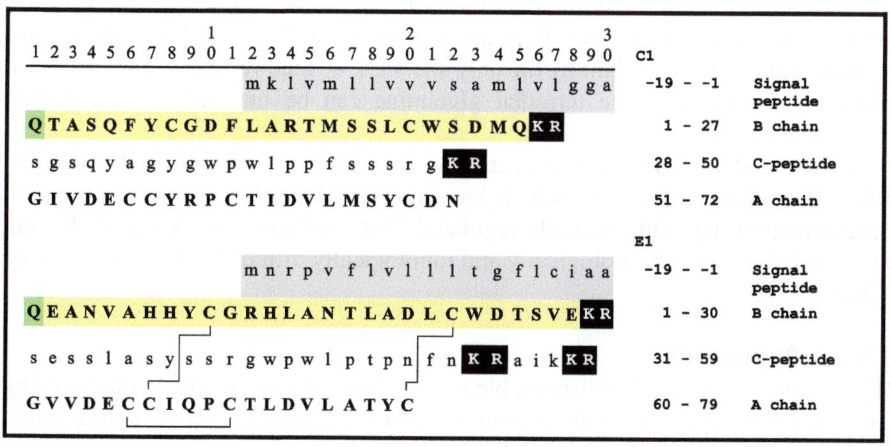

Fig. 5.2 Primary bombyxin sequences from *B. mori* (Bombyx mori): signal peptide amino acids −19 to −1 in the first row with *grey background*. Prohormone convertase-1 (PC1) cuts off the C-peptide (*lowercase*) (**KR** are shown *white on black*); *black lines* between cysteine residues of the A- and B-chains indicate disulfide bonds according to Maruyama et al. (1992); N-terminal glutamine (**Q**) cycled pyroglutamates (p**E**, on *green*) (Source: GenBank P21808, P15410)

Table 5.3 Differential expression of ILP in *D. melanogaster* (From Brogiolo et al. 2001)

	Embryo	Larva
dIlp1	Not expressed (NE)	Not analyzed
dIlp2	**in midgut,** in mesodermal stages 12–16	In imaginal disks[a], **in 7 neurosecretory cells in *partes intercerebrales* (PI) and in salivary glands**
dIlp3	NE	**In 7 neurosecretory PI cells** (see dIlp2)
dIlp4	**in mesodermal stages 2–6,** in the anterior midgut rudiment	**In midgut**
dIlp5	NE	**In 7 neurosecretory PI cells** (s. dIlp2), in gut
dIlp6	NE	in gut
dIlp7	Ubiquitous but yolk sack, **in midgut**	**In 10 neurons of the ventral nerve cord**

Strong, moderate, low expression
[a]Give rise to wings, gonads, limps, eyes, and antennae

5.1.2.1 Introduction

Bombyxins and other ILPs represent neuropeptides involved in regulation of growth, development, fecundity, metabolic homeostasis, and longevity. Nagasawa et al. (1986) identified the small neuropeptide of the prothoracic gland as an insulin of insects.

5.1.2.2 Biochemistry and Structure

The insulin-like peptides show a similar structure to vertebrate insulin. The precursor bears a signal peptide, B-chain, C-peptide, and the A-chain. In contrast to vertebrate insulin, in bombyxin only the PC1 is required for cleaving of the precursor. In addition, the terminal glutamine can be internally closed to the pyroglutamate ring.

B. mori and *D. melanogaster* each have several ILP genes (Kawakami et al. 1989; Brogiolo et al. 2001), which are expressed in different cells and due to developmental stage differentially regulated. In *D. melanogaster* seven dILP were identified; these are chronologically and topologically differentially expressed (see Table 5.3).

5.1.2.3 Physiology

ILPs/Bombyxins control indispensable growth functions of insects. Synthesized by a few neurosecretory cells in the pars intercerebralis, ILPs are released in the corpora cardiaca. Release depends on available food.

The insulin receptor of flies (coded for by a singular gene and alternatively spliced to different proteins) is related to the vertebrate IR. In contrast to this IR, an analogous insulin–receptor substrate (IRS) phosphorylated upon ligand binding and thus initiating signal transduction is integrated into the receptor protein chain.

However, a separate IRS gene exists in *D. melanogaster* which codes for chico, which when expressed and phosphorylated initiates further signal cascades. The insulin receptor (InsR) is indispensible in insects. Without the InsR irreparable developmental defects of the fly nervous system occur (Fernandez et al. 1995). To which degree insulin signaling influences growth and differentiation is discussed later. Receptor knockout is lethal for the embryo.

A special function has been reported for the Ilp1 of bees. Expression of Ilp1 is due to feeding of the larva by royal jelly thus fed as a queen to be. Only when royal jelly has been fed, Ilp1 is expressed with the consequence that special checkpoints in development are taken that ultimately lead to the raising of a new queen (Wheeler et al. 2006).

5.1.2.4 Phylogeny

ILPs appear ubiquitously in eumetazoans.

5.1.3 AKH, RPCH, and HrTH

Fact sheet 5.3: Adipokinetic hormone (AKH), red-pigment–concentrating hormone (RPCH), hypertrehalosemic peptide (HrTH)

Sequence: Fig. 5.3.

Synthesis and target: AKH is formed in the corpora cardiaca by neurosecretory cells. It acts on GPCR on adipocytes of the fat body and stimulates protein kinase A.

Function: AKH controls metabolism pathways providing energy for flying.

Receptor: Heptahelical G-protein-coupled transmembrane receptor.

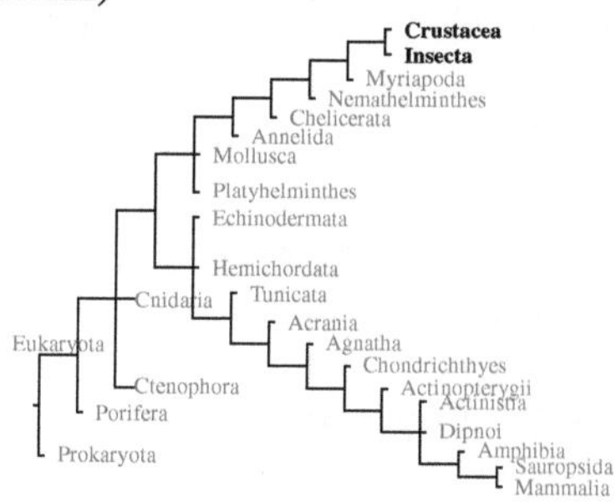

5.1.3.1 Introduction

Adipokinetic hormones (AKH) are released in the corpora cardiaca of insects and target the fat body. They induce release of sugars and lipids that can be used by equally AKH-stimulated wing muscle cells providing the required forces for flying.

5.1.3.2 Biochemistry and Genes

AKH originates from a preproprotein precursor. At first the signal peptide is cleaved off by the signal peptidase, afterward, by PC1 or PC2, the AKH associated peptide(AAP) (Fig. 5.3); finally, the C-terminal glycine is oxidized to amide by peptidylglycine alpha-hydroxylating monooxygenase (PHM) and the N-terminal glutamine transformed into Pyroglutamate. Characteristic for AKH are

```
                  1                   2                   3
1 2 3 4 5 6 7 8 9 0 1 2 3 4 5 6 7 8 9 0 1 2 3 4 5 6 7 8 9 0
- - - - - - - - - m g w v l k a l v v i a a l i a v m c e a   -21 - -1
Q L T F T P N W G K R s g l q d g p c k l s t e v l m h i y     1 - 30
k l v e t e a q k l v e c g k f g g n                          31 - 49
```

Fig. 5.3 Primary sequence of the adipokinetic hormone (AKH) of *Periplaneta americana*. The signal peptide is shown on *gray background*, the AKH peptide *boxed* and on *yellow* background, and the PC1 motif *black on white* (Source: GenBank AAV41425)

Table 5.4 Several adipokinetic hormones in insects

Species	Hormone	Sequence	Dibasic motif	GenBank entry
L. migratoria	AKH I	p**ELNFTPNWGT**-NH_2	GKR	P55319
	AKH II	p**ELNFSAGW**-NH_2	*GRR*	P08379
	AKH III	p**ELNFTPWW**-NH_2	GKR	P19872
Schistocerca gregaria	AKH I	p**ELNFTPNWGT**-NH_2	GKR	P18829
Schistocerca nitens	AKH II	p**ELNFSTGW**-NH_2	*GRR*	P53807
M. sexta[a]	AKH	p**ELTFTSSWG**-NH_2	GKR	P67788.1
D. melanogaster	AKH	p**ELTFSPNW**-NH_2	GKR	P61855
C. maenas[b]	RPCH	p**ELNFSPGW**-NH_2	GKR	Q26324

[a]Manduca sexta
[b]Carcinus maenas

the N-terminal **ELFN** sequences. Some locusts have been shown to express several AKH peptides (Table 5.4).

5.1.3.3 Physiology

After AKH binding to the heptahelical G-protein coupled transmembrane receptor in adipocytes of the fat body the protein kinase A is stimulated by cAMP. This enzyme phosphorylates the lipid storage droplet protein-1 (LSDP-1) and the triglycerol lipase (TG lipase). Both events induce rapid lipid release into the hemolymph (Patel et al. 2005).

Red-pigment–concentrating hormone (RPCH) of crustaceans is built like AKH. Some X organ neurons secrete RPCH in the sinus gland of the eye stalks. In addition, RPCH/AKH is expressed in the stomatogastric ganglion. Chung and Webster (2004) have shown that RPCH was initially found in X-organs/sinus glands, but was formed in other neurons, presumably in the postcommissural organ. RPCH seems relevant for the rhythmic movements in the gastrointestinal tract (the pyloric rhythm, gastric mill) controlled by 28 neurons of the stomatogastric nerve system (Thirumalai and Marder 2002).

Recently RPCH was also found to be a modulator of the heart ganglion in *Cancer borealis* (Cruz-Bermudez and Marder 2007). Because different peptides and further substances influence the pulses of heart beating the author suggests that due to expression in the heart ganglion an accelerated distribution of hormones in a given animal appears feasible and thus a shortening of the response time to hormonal release.

5.1.3.4 Phylogeny

Thus far AKH/RPCH have been found in crustaceans and insects.

5.2 Regulation of Heart Frequency and Pressure by Neuropeptides

5.2.1 Cardioacceleratory Peptides: CAP

Fact sheet 5.4: Cardioacceleratory peptide (CAP)

Sequence:	Fig. 5.4.
Synthesis and target:	CAP are neuronally synthesized in insects and crustacea and released in neurohemal organs.
Function:	CAP stimulates heart frequency.
Receptor:	CAP receptor is a heptahelical GPCR.

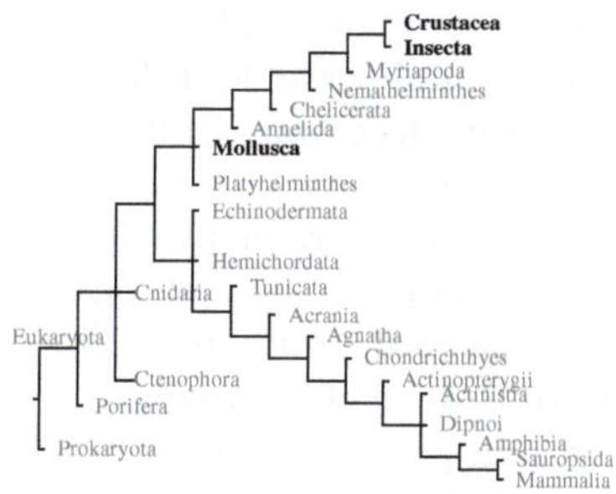


```
                                    1                   2                   3                   4                   5
1 2 3 4 5 6 7 8 9 0 1 2 3 4 5 6 7 8 9 0 1 2 3 4 5 6 7 8 9 0 1 2 3 4 5 6 7 8 9 0 1 2 3 4 5 6 7 8 9 0
                                    m k m s s t s w l g r t w l v t a g s l l l l v f l v t n a q a  Cm   -32 -  -1
                                          m s t i s t h g r a g v m v l t a l l l l v l a a h a h a  Oi   -29 -  -1
g p v a K R d i d s l l d g k i K R P F C N A F T G C g K K R s d p e l e g l a s g s e l n d i t k  Cm     1 -  50
g p l v K R d i g d l l e g k d K R P F C N A F T G C g K K R s d p g l e g v a s s s e l d a l a k  Oi     1 -  50
h v l a e a r l w e q l q s k m e a m r m l a s r m d s r p v f R R K R s l i h p q h d r v h s v t  Cm    51 - 100
h v l a e a k l w e q l q n k m e v m r s l a a r m e n h p l y R R R R s a p q q p r h h l t t t p  Oi    51 - 100
t l n h k g d a e k q                                                                                     101 - 111
k q k v e s e k q                                                                                         101 - 109
```

Fig. 5.4 Primary sequences of crustacean cardioacceleratory peptide (CCAP) of Carcinus maenas and Orconectes immunis. The signal peptides are shown on *gray background*; CAP are *boxed* and in *yellow*. Dibasic peptide motifs are shown *white on black* (Source: GenBank ABB46291 and ABB46293)

5.2.1.1 Introduction

The first cardioacceleratory peptide of crustaceans (CCAP) has been identified by Stangier et al. (1987) in the pericardial organ. Injections of this peptide increased the heart frequency. CAP has also been found in insects and molluscs. These molecules resemble the posterior pituitary hormones such as oxytocin or vasopressin.

5.2.1.2 Biochemistry and Structure

CCAP are nonapeptides with an intramolecular disulfide bridge. At the C-terminus these peptides are amidated. With the genome of *D. melanogaster* fully sequenced Park et al. (2002) expressed vasopressin receptor-like GPCR in cell lines and observed calcium mobilization or cAMP enhancement when using CCAP as ligand. However, the CCAP concentrations required for these stimulations were very high which made a specific interaction of CCAP with these receptors unlikely and suggested an unspecific stimulation.

5.2.1.3 Physiology

CCAP in crustaceans and insects has not only been found in the pericardial organ, but in several brain neurons (e.g., in the tobacco hornworm (*M. sexta*) in the subesophageal, thoracic, abdominal, and terminal ganglia (Loi et al. 2001). In crustaceans Trube et al. (1994) found CCAP in neurons and neurosecretory cells of the ventral nerve cord. Veelaert et al. (1997) observed neurosecretory cells of the pars intercerebralis of the locust brain secreting CAP into the corpora cardiaca, therein stimulating AKH release, an analogy to the hypothalamic–pituitary axis.

CCAP is preferentially expressed during late stages of molting when the CCAP content in the hemolymph increases (Phlippen et al. 2000; Gammie and Truman 1999; Loi et al. 2001). In order to meet the enhanced water uptake during molting CCAP is thought to provide the required stimulation of heart frequency and thus increased pumping. Gammie and Truman (1997) have additionally demonstrated (again in *M. sexta*), that CAP induces those essential movements during the molt, resulting in breaking of the old cuticle and in its stripping off. In *D. melanogaster* Davis et al. (2007) provided evidence of CAP activating those enzymes coloring the newly built skeleton: tyrosine hydroxylase (TH) and DOPA carboxylase, which are dealt with in more detail in the chapter on vertebrate catecholamine biosynthesis.

Table 5.5 Cardioacceleratory peptides (M-CAP) of molluscs, compared to a CCAP (Source Vehovszky et al. 2005)

Species	Hormone	Sequence
Helix pomatia	M-CAP I	PF C NSYG C YNS-NH_2
	M-CAP II	LF C NGYGG C QNL-NH_2
Orconectes immunis	CCAP	PF C NAFTG C -NH_2

CAP of molluscs (M-CAP) are related to the CCAP: cyclic peptides with some amino acid exchanges (Table 5.5). Vehovszky et al. (2005) have shown that M-CAP influences the central motor of ingestion.

5.2.1.4 Phylogeny

So far CAP have been observed in molluscs, crustaceans, and insects. The presence of a GPCR of the vasopressin family does not allow extrapolating an expression of CAP/vasotocin/vasopressin-like molecules in all bilateria.

5.2.2 Cardioexcitatory Peptide, NDWF-Amide

Fact sheet 5.5: Cardioexcitatory peptide NDWF-NH_2)

Sequence	**NDWF**-NH_2; unknown precursor protein.
Synthesis and target	**NDWF**-NH_2 positive brain neurons target cardiovascular muscles in molluscs.
Function	**NDWF**-NH_2 stimulates blood flow and increases blood pressure in molluscs.
Receptor	(unknown).

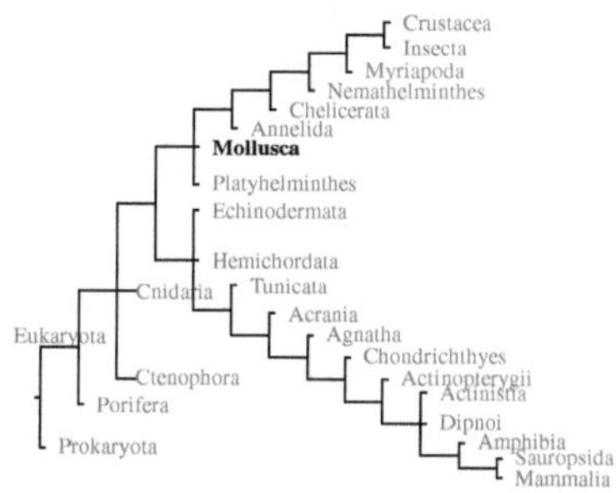

The molluscan cardioexcitatory peptide is a tripeptide with a D-amino acid: D-tryptophan (DW): asparaginyl-D-trytophanyl-phenylalanyl-amide (**N**D **WF**-NH_2). Morishita et al. (1997) used it to stimulate in *A. kurodai*[6] heart contraction, but not heart frequency. Later it was found in *A. californica* that not only in the heart, but in all tissues with **NDWF**-positive neurons muscle contractions can be induced by this peptide. The peptide was further identified in snails (Morishita et al. 2003a,b).

[6] Aplysia kurodai.

D-Amino acids

In ribosomes amino acid incorporation is restricted to L-amino acids. The observation of D-isomers presupposes a posttranslational isomerization by a racemase or isomerase. D-Aspartate has been shown to be an neurotransmitter, as well as D-serine. Further D-amino acids incorporated into proteins were found in the animal kingdom. Isomerase acts mainly close to the N-terminus. The spider enzyme (from *Agelenopsis aperta*) uses an **LxF** motif as a substrate and isomerizes the x amino acid; the **NxF** of molluscs is closely related to the spider motif.

Neither the gene coding for **N**D**WF**amide, nor the enzyme isomerizing the tryptophan have thus far been identified (see also the box "D-amino acids"). **N**D**WF**amide has thus far been found only in molluscs.

5.2.3 Enterins

Fact sheet 5.6: Enterins

Sequence:	Fig. 5.5.
Synthesis and target:	Enterins are formed in cerebral and buccal ganglia of molluscs.
Function:	Enterins are effective in aorta and gut and inhibit muscle contraction.
Receptor:	(thus far unknown).

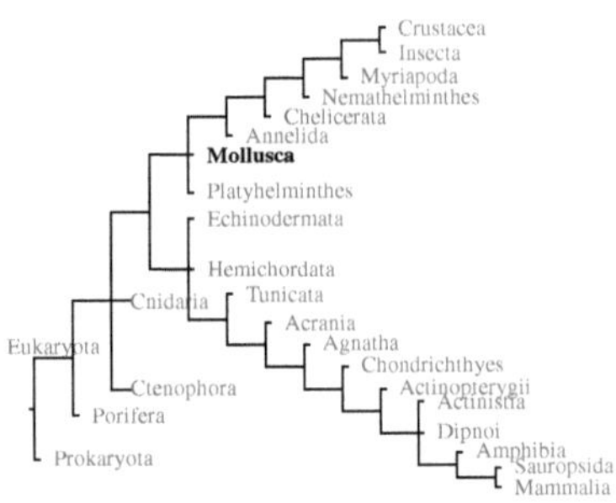

5.2.3.1 Introduction

Enterins were identified as characteristic neuropeptides of aplysia (*A. californica* and *A. kurodai*) (Furukawa et al. 2001). No homologue has ever been found in other phyla. Enterins in concert with AMrP are antagonists of **N**D**WF** of molluscs and inhibit contractions of the aorta as well as of the gut.

5.2.3.2 Biochemistry and Structure

The enterin-precursor protein is relatively large. Dibasic and monobasic cleavage sites for prohormone convertases are present. The peptides shown in Fig. 5.5 *boxed* have actually been isolated. The enterin peptide motif is **PxxxHxxFV**-NH_2. The nonactive additional peptides have been isolated as well, which support the model of precursor cleavage as shown in Fig. 5.5 (Furukawa et al. 2001).

```
                  1                   2                   3                   4                   5
1 2 3 4 5 6 7 8 9 0 1 2 3 4 5 6 7 8 9 0 1 2 3 4 5 6 7 8 9 0 1 2 3 4 5 6 7 8 9 0 1 2 3 4 5 6 7 8 9 0
                                          m a k h d v t v m t l l l v v c a l h v f d a q g    -25 -  -1
t d v k l n d g f l r s g i m n i p f q R R V S P K Y G H N F V g K R s g f q s p v s p s d y f i p    1 -  50
d e l n d e d e s p n m d t y a l f r e l l g e y p s s e l s d d d v i r p y p v v p w a w d r f e   51 - 100
n K R f s k e n e K R G S P G F S H S F V g K R m d l s a l e k e l i a k l k a a d l l s p l e t e  101 - 150
a p g K R R L P S Y G H S F L g K R m p v d v f p R A I P S Y S H N F V g K R s g n g e n y f d d l  151 - 200
d t f g d i s q R A D L G F T H S F V g K R g n t d f s r n p l a r l s q v q n R A I P N F V H K F  201 - 250
V g K R s v h n i v K R S S P F Y G H N F V g K R d f e d a s e g l d e e e g d i d g y s d d l d v  251 - 300
K R V P G Y S H S F V g K R I P G Y S H S F V g K R T P G Y S H S F V g K R V P G Y S H S F V g K R  301 - 350
V P G Y S H S F V g K R V P G Y S H S F V g K R V P G Y S H S F V g K R V P G Y S H S F V g K R T P  351 - 400
G Y S H S F V g K R A P G Y S H S F V g K R T P G Y S H S F V g K R A P G Y S H S F V g K R A P G Y  401 - 450
S H S F V g K R A P G Y S H S F V g K R V P G Y S H S F V g K R A P G Y S H S F V g K R V P G Y S H  451 - 500
S F V g K R A P G Y S H S F V g K R e l g e d e i n f l k e v d a a d i s r q l a e e d e k e a m v  501 - 550
s v d d k e t l s n e e d a s e d d f e K R E L N F Q H A F V g K K d e e g d m g v e m e e e e m e s  551 - 600
e K G Q P G Y G N A F L g K R Q P S Y G H S F V g K R Q P G Y S H S F V g K R V P S F G H S F V g K  601 - 650
R Q P S Y T H A F V g K R D P G F N H A F V g K R Q P S F G H S F V g K R Q P S F T H A F V g K R Q  651 - 700
P S F G H S F V g K R s g a e d i d n t f t d n f v v k g s d p l e d v l d f y s d s q l i g q p a  701 - 750
a a n e e e l q q e a a e e s e n q t g t k q l d d k s n v p a s s l g d d l p s p v s l s k q e d  751 - 762
a e d s h i m a t s s t
```

Fig. 5.5 Primary sequence of the enterins-prepropr otein from *A. californica*. The signal peptide is shown on *gray background*; the enterins are *boxed* and with *gray background*. Monobasic and dibasic peptide motifs are shown *white on black*. C-terminal glycines are usually oxidized to amides. N-terminal glutaminyl amino acids are internally cycled to pyroglutamate (Origin: Furukawa et al. 2001 GenBank Q95P23)

5.2.3.3 Physiology

Enterins inhibit gut contractions either as neurotransmitters or in an endocrine manner. Their presence in cerebral and buccal ganglia suggests a role in food ingestion. Whether enterins exhibit additional regulatory functions is an open question (Furukawa et al. 2001).

5.2.3.4 Phylogeny

Enterins have only been identified in molluscs.

5.2.4 *Mytilus* Inhibitory Peptides (MIP; AMrP)

5.2.4.1 Introduction

MIP were first found in the pedal ganglia of the clam *Mytilus edulis* (Hirata et al. 1988). In the meanwhile related peptides were observed in other mussels and in snails. The characteristic feature of MIP is inhibition of muscle contractions..

5.2.4.2 Biochemistry and Structure

The characteristic motif of MIP and related, "aplysia-MIP–like peptides" (AMrP) is **PxFF/I/V**-NH_2. The precursor protein of *A. californica* (Fig. 5.6) encloses 24 peptides, six or seven amino acids long; 21 of these contain this motif, and two others bear a methionine or leucine at the last position. **GSPRFF**-NH_2 is found elevenfold.

Fact sheet 5.7: *Mytilus* Inhibitory Peptides

Sequence:	Fig. 5.6.
Synthesis and target:	MIP are neuropeptides in several ganglia, especially the pleural and abdominal ganglia.
Function:	MIP inhibit muscle contractions.
Receptor:	(thus far unknown).

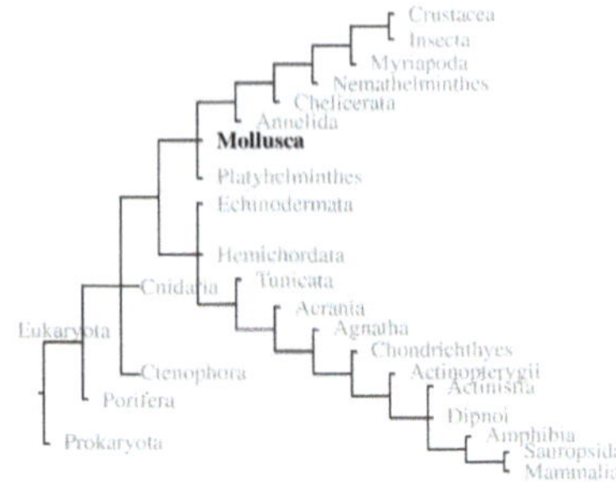

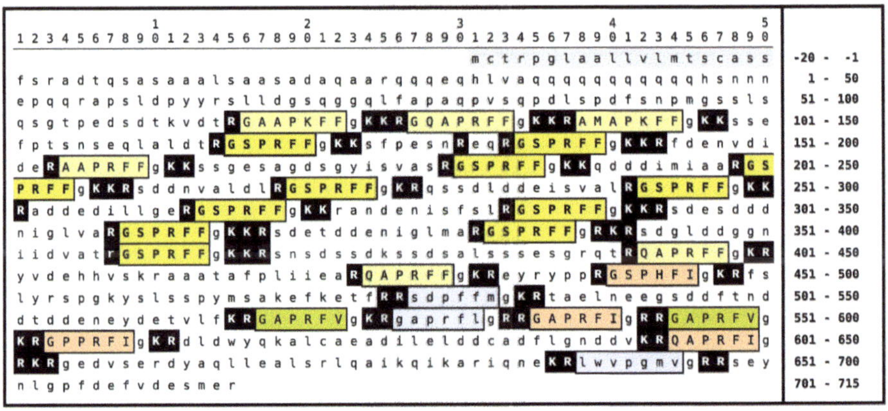

Fig. 5.6 Primary sequence of mytilus inhibitory peptide (MIP, AMrP) from *A. californica*. The signal peptide is shown on *gray background*; the MIP are *boxed* and with *colored background*. The prevalent peptide **GSPRFF**-NH_2 is shown in *uppercase*. Monobasic and dibasic peptide motifs are shown *white on black*. C-terminal glycines are oxidized to amides. N-terminal glutaminyl amino acids may be internally cycled to pyroglutamate (Origin: Fujisawa et al. 1999; GenBank AAF80382)

5.2.4.3 Physiology

MIP/AMrP neurons were observed in several ganglia, most prominent in pleural and abdominal ganglia, additionally in cerebral, buccal, and pedal ganglia. MIP/AMrP inhibit muscle contractions: contractions of the esophagus, of the Penis retraction muscle, or body wall muscle were differentially inhibited by several *A. californica*-AMrP. Most active were **GAPFRV**-NH_2, **GAPFRI**-NH_2, and **GPPFRI**-NH_2 (Fujisawa et al. 1999). In *A. kurodai* Sasaki et al. (2004) analyzed stimulation of the aorta vasoconstrictor muscle by **N**D**WF**amide and its inhibition by enterins and AMrP (**GSPRFF**-NH_2). In cultivated pleural ganglion neurons McDearmid et al. (2002) identified a 4-aminopyridine-sensitive potassium channel that could be stimulated by **GAPRFV**-NH_2 and **GSPRFF**-NH_2.

5.2.4.4 Phylogeny

To date MIP/AMrP seem to be restricted to molluscs. So-called myoinhibitory peptides (MIP = leucomyosuppressin) from the prothoracic gland of insects are structurally unrelated to the molluscan MIP.

5.2.5 Diuretic Hormones; (DiuH)[7]

Fact sheet 5.8: Diuretic Hormones (DiuH)

Sequence:	Fig. 5.7.
Synthesis and target	DiuH are formed by neurosecretory cells of the brain and of the corpora cardiaca and act on GPCR in malpighian tubules.
Function:	DiuH stimulate sodium chloride secretion and water transport.
Receptor:	A GPCR of the secretin receptor family.

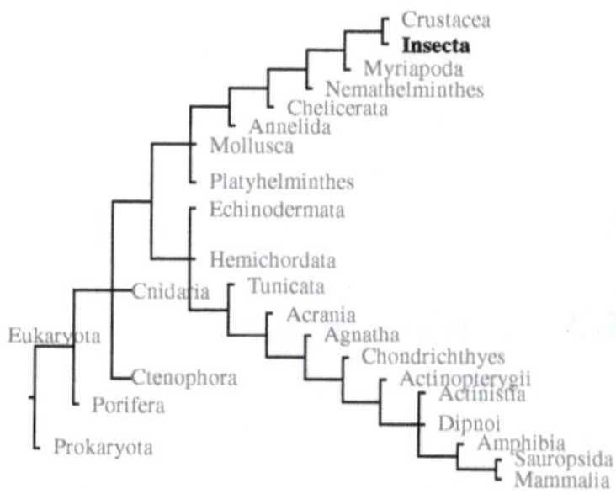

5.2.5.1 Introduction

Diuresis in insects takes place in the malpighian tubules (MT), appendices to the mid gut. When a mosquito or a bug has fed on a blood meal, the animal or human blood is hypotonic compared to the insect hemolymph: approximately 280 mOsm in blood

```
                  1                   2                   3                   4                   5
1 2 3 4 5 6 7 8 9 0 1 2 3 4 5 6 7 8 9 0 1 2 3 4 5 6 7 8 9 0 1 2 3 4 5 6 7 8 9 0 1 2 3 4 5 6 7 8 9 0
                                                              m m w w a i w c v m v v v s s a a s a    -19 -  -1
a p a p d s a p m d l v q i d s a g p d d e s l g y a v s s l e g r y g a e a p w l y l l a e m p r      1 -  50
d s q i g r a a v K R R M P S L S I D L P M S V L R Q K L S L E K E R K V H A L R A A A N R N F L N     51 - 100
D I g K R g l q w s r s e q p s a y y                                                                 101 - 119
```

Fig. 5.7 Primary sequence of DiuH from Manduca sexta.The signal peptide is shown on *gray* background; the diuretic hormone is *boxed* and on *yellow background*. Dibasic peptide motifs are shown *white on black*. The C-terminal glycine will be oxidized to amide (Origin: Kataoka et al. 1989 GenBank P21819)

[7]Because there are two hormones that are abbreviated DH (diuretic hormone and diapause hormone) we abbreviate diuretic hormone as DiuH.

and about 370 mOsm in the hemolymph. By secreting sodium and chloride ions into the lumen of the MT and simultaneous uptake of water the food becomes isotonic for the insect and can be ingested and digested. These processes are stimulated by diuretic hormone.

Vasopressin/oxytocin-like peptides as well as CRH-related peptides have been found active as DiuH,[8] furthermore there exist some calcitonin-related diuretic peptides. Because they are analyzed in great detail, CRH-like insect peptides and calcitonin-related peptides are regarded as *the* DiuH being in vitro much more active as inotocin or similar peptides (see Gaede 2004).

5.2.5.2 Biochemistry and Structure

DiuH are formed by the action of the PC1 (and PC2) from precursor peptides. They are C-terminally amidated. Whether the associated peptides (see Fig. 5.7) are physiologically active is not known. In some species some genes were identified as coding for active peptides with amino acid chain lengths of 30 to 50. A differential expression has not been described in the literature. Apart from DiuH, mainly kinins and CAP are stimulators of diuresis.

The DiuH-receptor in *A. aegyptii*[9] has been identified as a GPCR of the secretin receptor family, expressed on the surface of the MT (Jagge and Pietrantonio 2008).

5.2.5.3 Physiology

DiuH are formed in neurosecretory cells of the pars intercerebralis and of the corpora cardiaca and released thereof. In thoracic and abdominal ganglia an however much reduced expression could be observed (Audsley et al. 1997). DiuH acts via the hemolymph on receptors at the malpighian tubules which in turn increase intracellular cAMP. A similar enhanced intracellular cAMP formation was observed in DiuH receptor expressing cell lines (Gaede 2004).

Because the DiuH content in adult locusts was much reduced on the day after the final molt, as after a feeding, a role of DiuH in the postmolting processes was suggested.

5.2.5.4 Phylogeny

Diuretic hormones have been found in 20 different insect species, but not in other phyla.

5.3 Kinins

Named as kinins we describe neuropeptides and neurohormones isolated due to their myotropic or gut-stimulating properties. Different kinin families can be differentiated by characteristic sequence motifs: pyrokinins/pheromone-

[8]Diuretic hormone.

[9]Aedes aegyptii.

biosynthesis–activating neuropeptide (**FxPRL**-NH_2[10]), leucokinins/lymnokinins (**FxxWG**-NH_2), sulfakinins (sulfated **RF**-NH_2), or orcokinins (N-terminal **NxDEI** peptides.

5.3.1 Pyrokinin/Myotropin and Pheromone Biosynthesis Activating Neuropeptide (PBAN)

Fact sheet 5.9: Pyrokinins/PBAN

Sequence:	Fig. 5.8.
Synthesis and target:	PBAN is formed in the subesophageal gland and released in the corpora cardiaca.
Function:	PBAN stimulates pheromone biosynthesis.
Receptor:	(yet unknown).

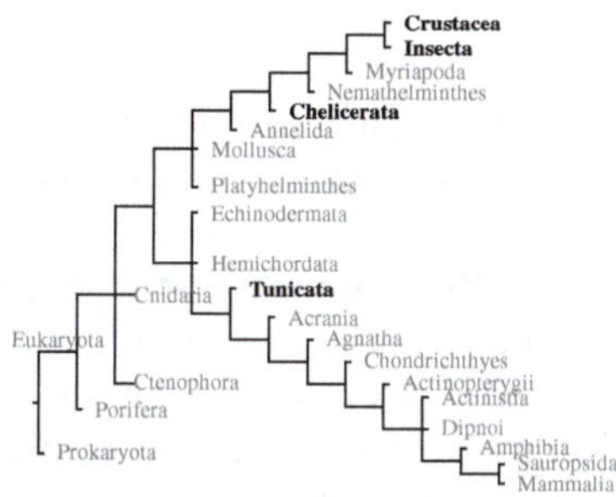

5.3.1.1 Introduction

PBAN is formed by singular neurosecretory cells of the subesophageal gland and released in this gland or, after axonal transport, in the corpora cardiaca. PBAN acts on the enzymes of the pheromone glands. The presence of PBAN has been proven in

1 2 3 4 5 6 7 8 9 0 1 2 3 4 5 6 7 8 9 0 1 2 3 4 5 6 7 8 9 0 1 2 3 4 5 6 7 8 9 0 1 2 3 4 5 6 7 8 9 0	
m y k t n i v f n v l a l a l f s i f f a s c	-23 - -1
T D M K D E S D R G A H S E R G A L W F G P R L g K R s m k p s t e d n r q t f l r l l e a a d a l	1 - 50
k f y y d q l p y e r q a d e p e t k v t k k i i f t p k l g R s v a n p r t h e s l e f i p r l g	51 - 100
R R L S E D M P A T P A D Q E M Y Q P D P E E M E S R T R Y F S P R L g R t m s f s p r l g R e l s	101 - 150
y d y p t k y r v a r s v n k t m d n	151 - 169

Fig. 5.8 The pheromone-biosynthesis–activating neuropeptide (PBAN) and diapause hormone (DiaH) primary sequence from Bombyx mori. The signal peptide is shown on *gray background*, the diapause hormone [1–24] and PBAN [102–132] are *boxed* on *yellow/orange background*. Monobasic and dibasic peptide motifs are shown *black on white*. C-terminal glycines are oxidized to amides. Three additional neuropeptide (α-SG [74–80], β-SG [83–99], and γ-SG [108–115]) are equally released by prohormone convertases. N-terminal glutamine amino acids are internally cycled to pyroglutamate (Origin: Sato et al. 1993; GenBank P09971)

[10]x represent any amino acid.

```
                  1                   2                   3                   4                   5
1 2 3 4 5 6 7 8 9 0 1 2 3 4 5 6 7 8 9 0 1 2 3 4 5 6 7 8 9 0 1 2 3 4 5 6 7 8 9 0 1 2 3 4 5 6 7 8 9 0
                                  m i g f a v f s s f n r f t t i f v c v l l c v v y l l s y a s g    -33 -   1
e y d g R d s s s g s n n d R a p s n e f g s c t d g k c i K R T S Q D I T S G M W F G P R L g R R      1 -  50
R R a d R K p e i n s d i e a f a n a f e e p h w a i v t i p e t e K R Q I T Q F T P R L g r e s g     51 - 100
e d y f s y g f p k d q e e l y t e e q i y l p l f a s r l g R R v p w t p s p r l g r q l h n i v    101 - 150
d K p R q n f n d p r f                                                                                151 - 162
```

Fig. 5.9 The pyrokinin precursor of the honeybee (Apis mellifera). The signal peptide is shown on *gray background*; pyrokinins are *boxed*, in *uppercase* and on *yellow background*. Whether the peptides [111–130] and [134–142](in *lowercase*) are active as pyrokinins is not reported in the literature; Audsley and Weaver (2006) do not mention them. Monobasic and dibasic peptide motifs are shown *white on black*. Glycines at the C-terminus are always oxidized to amide. N-terminal glutamines may be internally cycled to Pyroglutamate (Origin: GenBank NP_001104182)

lepidopterans, however, in other insects there are PBAN-like peptides; the analogous peptide of *D. melanogaster* is called "hugin."

5.3.1.2 Biochemistry and Structure

Pyrokinins constitute a group of peptides characterized by a C-terminal **FxPRL**-NH_2 motif and exerting muscle contracting (myotropic) activity. Particular pyrokinins are the PBAN where precursor sequences have only been identified in lepidopterans (Figs. 5.8, 5.9).

The precursor of the silk moth (*B. mori*) contains in addition to PBAN the DiaH[11], a pyrokinin, too. Whereas in the silk moth diapause hormone and PBAN are coded for by the same gene, *D. melanogaster* pyrokinins are found within the *hugin*, the *capa* and the cardioacceleratory peptide *Cap2b* genes (Kean et al. 2002; Baggerman et al. 2002; Meng et al. 2002). In the mosquitoes *A. aegyptii* and *A. gambiae*[12] there are three short FxPRL-amide and a single PRL-amide peptide found in a pyrokinin precursor (GenBank Q16N80 and Q7PTL2). The PBAN neuropeptide proprotein of the moth *Agrotis ipsilon* includes a diapause hormone, where the leucine-amide is replaced by isoleucine-amide, PBAN, and two additional pyrokinins; in the oriental tobacco budworm (*Helicoverpa assulta*) DiaH and PBAN and two further pyrokinins are present in the precursor.

5.3.1.3 Physiology

Originally pyrokinins were found due to their myotropic activity. They further influence contractions of the locust oviduct, diapause of the eggs of the silk moth (diapause hormone; Fig. 5.8), acceleration of pupation in larvae of flesh flies, or melanization and reddening in larvae of moths (Torfs et al. 2001). Ectopic (over-) expression of the *hugin* gene in *D. melanogaster* resulted in over 50 % of the larvae in lethality during the second pupation; only 5 % of the flies reached the adult stage.

---

[11]Diapause hormone.

[12]Anopheles gambiae.

There were unreparable damages and erratic behavior during the molt (Meng et al. 2002).

PBAN stimulates via the PBAN receptor, a heptahelical GPCR; calcium influx and cAMP increase in cells of the pheromone gland. Thus the acetyl CoA carboxylase activity is enhanced in some species, and in other ones the reduction of fatty acids to aldehydes or alcohols is stimulated, which in turn induced pheromone biosynthesis and release (Tillman et al. 1999). The pyrokinin-1 receptor of the fruit fly is a heptahelical GPCR as well (Cazzamali et al. 2005).

5.3.1.4 Phylogeny

Although PBAN thus far[13] has been observed in insects, homologous genes were found in ticks and spiders as well as in echinoderms without proof of their expression. Pyrokinins were detected in insects, but also in crabs (Torfs et al. 2001).

5.3.2 Orcokinins

Fact sheet 5.10: Orcokinins

Sequence	Fig. 5.10.
Synthesis and target	Orcokinins are neuropeptides, for example, from the stomatogastric ganglion; they act as neurotransmitters and hormones.
Function	Orcokinins contract the gut muscles and regulate the rhythm in the pylorus. They participate in the circadian regulation of motor activity.
Receptor	(not yet known).

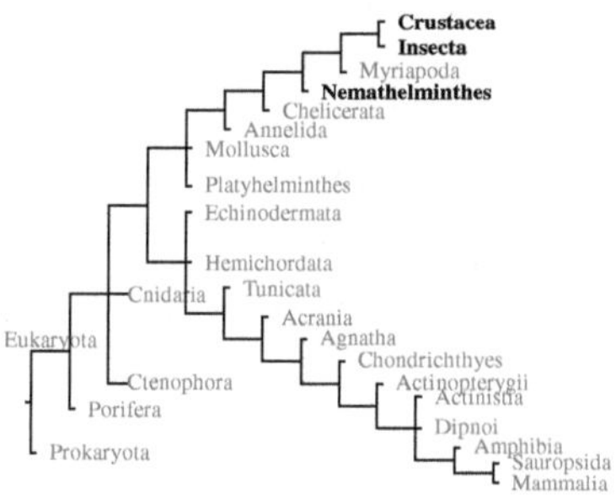

5.3.2.1 Introduction

From *O. limosus*[14] Stangier et al. (1992) reported a gut-active neuropeptide: orcokinin. Later they identified similar peptides in other crabs. Orcokinins have now been found in crustaceans, insects, and nematodes; potential mussel or snail orcokinins are not specified as sequences in GenBank.

[13](in Sept 2014)

[14]Orconectes limosus.

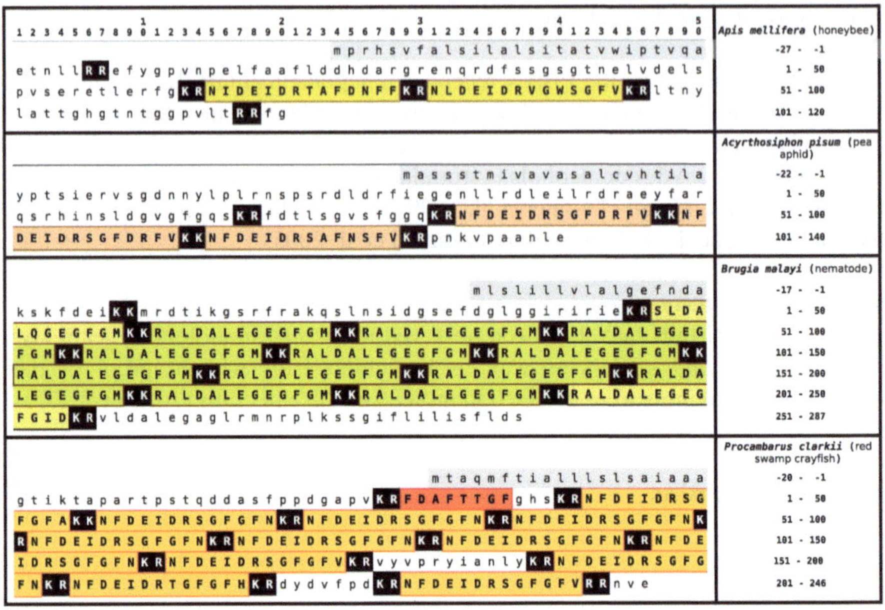

Fig. 5.10 Primary sequences of orcokinins from the arthropods (honeybee), pea aphid (*Acyrthosiphon pisum*), of the nematode *Brugia malayi*, and the red swamp crayfish *Procambarus clarkii*. Signal peptides are on *gray background*, orcokinins are *boxed* and on *colored background*. Dibasic peptide motifs are shown *white on black* (Origin: GenBank P85832, XP_001947462, EDP37605, Q9NL82)

5.3.2.2 Biochemistry and Structure

In some crustaceans and nematodes there are 10 and more identical or very similar neuropeptides on the same orcokinin precursor to be cleaved and released by prohormone convertase 1 or prohormone convertase 2 (**KR** or **KK** motif). In insects there are only few orcokinins on the precursor, however, the N-terminal propeptide sequences are longer (Fig. 5.10). Bees only need PC1; the pea aphid (*Acyrthosiphon pisum*) has to use PC1 and PC2 for orcokinin release (Fig. 5.10).

The *red boxed* **FDAFTTGF**amide peptide in Fig. 5.10 from the red swamp crayfish (*P. clarkii*[15]) was first isolated and analyzed as orcomyotropin from *O. limosus*. That precursor of *O. limosus* has not been sequenced.

Until now no orcokinin receptor has been described. Because in *C. elegans* orcokinins and all the GPCR are known due to total genome sequencing, an orcokinin receptor might most probably be found in this species.

[15] Procambarus clarkii.

5.3.2.3 Physiology

The evolutionary conservation of the neuropeptide structure within the prepropolypeptide in nematodes and crustaceans may point to essential functions of orcokinins. Orcokinin from *O. limosus* acts on gut contractions: contraction amplitude and frequency are enhanced. Orcokinin forming neurons were detected in the stomatogastric ganglion and other ganglia. They exhibit stimulatory activity on rhythm and activity of the pylorus. Whether orcokinins act as neurotransmitters or in an endocrine fashion via the hemolymph is not yet decided. In the hemolymph constant concentrations of orcokinins were observed. On the other hand the middle gut is innervated by several orcokinin neurons.

Finally another orcokinin function in the circadian control of locomotor activity in cockroaches has recently been found. Injections of orcokinin into the accessory medulla of the lobus opticus resulted in stable phase shifts in locomotor activity (Hofer and Homberg 2006).

5.3.2.4 Phylogeny

To date, orcokinins were found in insects, crustaceans, and nematodes.

5.3.3 Leucokinins/Lymnokinins

Fact sheet 5.11: Leucokinins, Lymnokinins

Sequence:	Fig. 5.11.
Synthesis and target:	Leucokinins are brain neuropeptides that act via GPCR on muscle cells as well as on malpighian tubules.
Function:	Leucokinins stimulate gut musculature, but most of all they stimulate diuresis.
Receptor:	The lymnokinin receptor belongs to the heptahelical GPCR family.

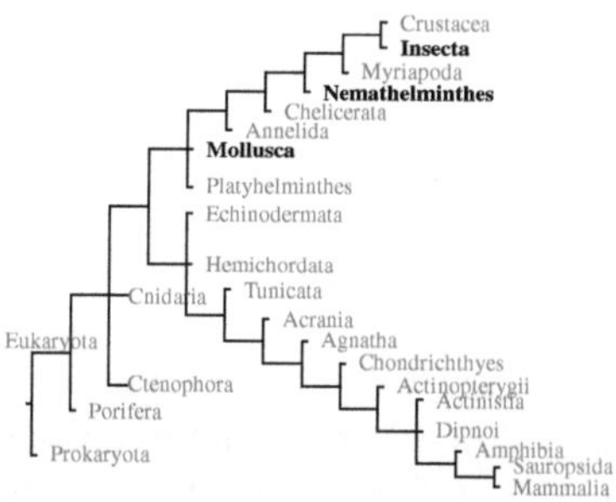

5.3.3.1 Introduction

Leucokinins, locustakinins, and lymnokinins constitute the **FxxWx**amide neuropeptide family, found with similar structure in insects, molluscs, and nematodes. At first leucokinins were identified due to their gut contracting activity. In the meanwhile regulation of diuresis in malpighian tubules appears as the characteristic feature of leucokinins and related peptides.

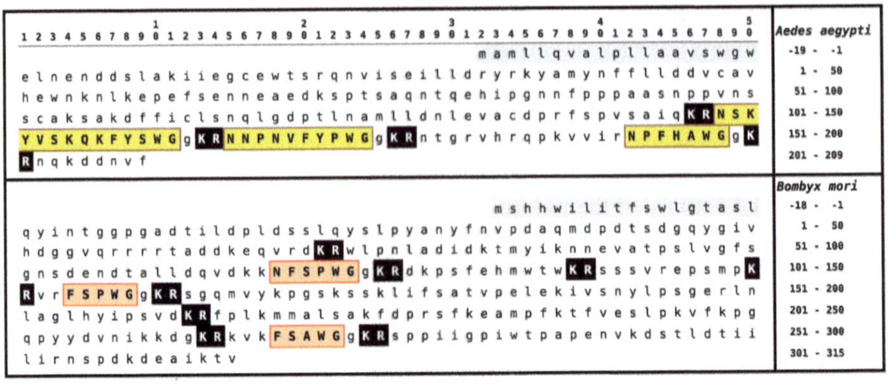

Fig. 5.11 Primary sequence of leucokinins from arthropods: silk moth (*lower*) and yellow-fever mosquito (*upper panel*). The signal peptide is shown on *gray background*, leucokinins are *boxed* and on *yellow* or *red background*. Dibasic peptide motifs are shown *white on black* (Origin: GenBank O02036, BAG50367)

Table 5.6 Kinins from the cockroach *Leucophaea maderae*

	Sequence
I	DPA F NS W G-NH_2
II	DPG F SS W G-NH_2
III	DQG F NS W G-NH_2
IV	DAS F HS W G-NH_2
V	GSG F SS W G-NH_2
VI	pESS F HS W G-NH_2
VII	DPA F SS W G-NH_2
VIII	GAD F YS W G-NH_2

5.3.3.2 Biochemistry and Structure

Leucokinins were originally isolated by Holman et al. (1986b,a) in the cockroach *L. maderae*[16] (Table 5.6). Precursor proteins have been identified in the silk moth, the yellow-fever mosquito, and *D. melanogaster*. The three peptide sequences of the yellow-fever mosquito precursor (Fig. 5.11) have been isolated. In the silk moth the precursor has been determined by genome sequencing; the kinins themselves are not yet proven. They might be N-terminally prolonged. The leucokinin of *D. melanogaster*, **NSVVLGKKQR F HS W** G-NH_2 isolated by Baggerman et al. (2002), contains a dibasic peptide motif where prohormone convertase might cleave the precursor.

The lymnokinin from *L. stagnalis*[17] bears a C-terminal serine-amide: **PS F HS W** S-NH_2 (Cox et al. 1997). In *C. elegans* a leucokinin-like precursor

[16]Leucophaea madera.

[17]Lymnaea stagnalis.

protein has been identified (GenBank AAM22049) containing twice the sequence **QFYAWAG** wherefrom a p**EFYAWA**-NH_2 might result.

The lymnokinin receptor has been first characterized by Cox et al. (1997). Other GPCR were later observed in insects: honeybees, fruit flies, and yellow fever mosquitoes. In ticks (i.e., in chelicerates) a very similar receptor was sequenced although a leucokinin has not been found in ticks. In the placozoan *Trichoplax adhaerens*, thus in an very basic metazoan species, a leucokinin-like receptor was sequenced together with the entire genome. Three *A. aegyptii* leucokinins (Fig. 5.11) were found to bind to the identical receptor, but with different affinities (Pietrantonio et al. 2005).

5.3.3.3 Physiology

Holman et al. (1986b) called the leucokinins originally cephalomyotropins, *that is*, peptides of the brain eliciting muscle contractions. When additional leucokinins were isolated, for example, in locusts, gut contractions served as the bioassay for testing of active fractions.

Leucokinins became important for insect research because they promote diuresis of sodium and potassium chloride. In blood-feeding insects, for example, *A. aegyptii*, a lot of salt and water has to be removed postprandially. This happens in the malpighian tubules. Therein leucokinins stimulate salt export. By inhibiting this export it might be possible to block reproduction of these dangerous disease vectors.

5.3.3.4 Phylogeny

Until now leucokinins and similar peptides of the **FxxFx**amide family have been found in molluscs, insects, and nematodes. A potential receptor might have already existed since the origin of metazoan evolution.

5.3.4 Tachykinin-Related Peptides (TRP)

Fact sheet 5.12: Tachykinin-related peptides

Sequence: Fig. 5.12.

Synthesis and target: TRP is found in the brain, in several head ganglia, in the thorax and abdomen, as well as in neuroendocrine cells of the midgut.

Function: TRP activates gut musculature and olfactory nerves.

Receptor: The TRP receptor belongs to the rhodopsin family of heptahelical GPCR.

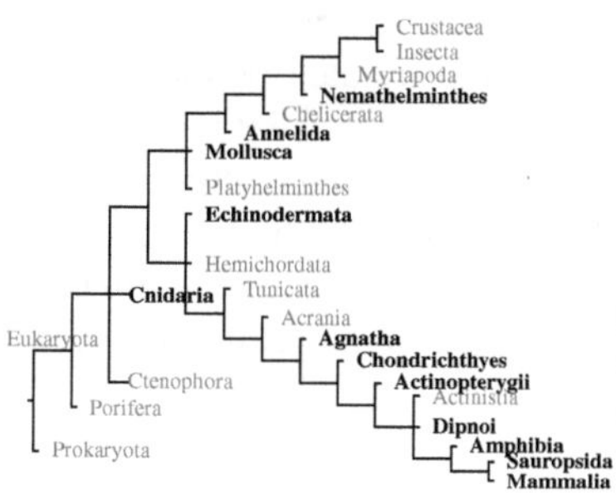

```
                  1                   2                   3                   4                   5
1 2 3 4 5 6 7 8 9 0 1 2 3 4 5 6 7 8 9 0 1 2 3 4 5 6 7 8 9 0 1 2 3 4 5 6 7 8 9 0 1 2 3 4 5 6 7 8 9 0
                                                        m a f p c p r s r v g a l i l v t l s l i a v v l c   -26 -  -1
a p e e s p K R A P S G F L G V R g K K d n v i p e f n s d e f n e a a d K R A P A M G F Q G V R g   1 -  50
K K d q d e d l g y d K R G P S M G F H G M R g K K e q q d f l r d l v d K R G P N M G F M G M R g   51 - 100
K K d p l d f d y y e K R A P S M G F Q G M R g K K d q w e e d a d m y K R A P S M G F Q G M R g K   101 - 150
K d y f d d e d e y f K R M G F M G M R g K K e g d f e v d d y p e d g l w s e d g e g e e m d K R   151 - 200
A P A A G F F G M R g K K V P A S G F F G M R g K K G P S V G F F A M R g K K A P S A G F M G M R g   201 - 250
K K A P S G F M G M R g K R e d l d g e d l d s l l q y l d s a y q h g r d K R n g e r a p g s K K   251 - 300
A P S G F L G T R g K K d w p t q q g a e a g t e p d i h t s l s d                                     301 - 334
```

Fig. 5.12 Primary sequences of tachykinin-related peptide from *L. maderae* (LemTRP). The signal peptide is shown on *gray background*, the TRPs are *boxed* on *yellow background*. Dibasic peptide motifs are shown *black on white*. LemTRP-2 and LemTRP-3 have been found as long and short peptides that contain in their long form an intramolecular dibasic peptide motif (Origin: GenBank AAX11211; Nässel 1999; Predel et al. 2005)

5.3.4.1 Introduction

In this book, tachykinins such as substance P have already been described among vertebrate hormones. These peptides bear a common **FxGLM**-NH_2 motif. Invertebrate tachykinin-related peptides belong to the **FxGxR**-NH_2 family (Nässel 1999; Review). They were found in several invertebrate phyla: molluscs, annelids, insects, and crustaceans.

5.3.4.2 Biochemistry and Structure

Up to 14 different tachykinin-related peptide sequences are present in the precursor gene of the TRP in cockroaches, for example, *L. maderae*, where the peptides could equally be identified by chemical analysis. However, in other species, only a few TRP peptides were found.

From different insects and spoon worms (annelids), presumable TRP receptor sequences have been published. All of these proteins belong to the rhodopsin family of GPCR.

5.3.4.3 Physiology

Tachykinin-related peptide (of *L. migratoria*) was found in several brain regions, furthermore in head, thorax, and abdominal ganglia as well as in the midgut. Tachykinin-related peptides stimulate gut musculature as do sulfakinins or allatotropins. When TRP was inactivated by RNA interference in *D. melanogaster*, it could be observed that olfaction was impaired and the locomotion was hyperactive (Winther et al. 2006). Crustaceans form their TRP in endocrine gut cells. These cells secrete TRP when potassium chloride concentrations are increased. A role of TRP in the control of feeding has been assumed (Christie et al. 2007).

5.3.4.4 Phylogeny

Tachykinins and tachykinin-related peptides might be present in all metazoans. A precursor gene has existed at least before protostomes and deuterostomes developed separately.

5.3.5 Sulfakinins

Fact sheet 5.13: Sulfakinins

Sequence: Sulfated **DYGHMRF**-NH_2 (Fig. 5.13).

Synthesis and target: Sulfakinins are formed in individual brain neurons. The molecules act on other neurons and on gut muscle cells.

Function: Sulfakinins appear to inhibit feeding.

Receptor: Sulfakinin receptor is a GPCR analogous to gastrin/cholecystokinin receptors.

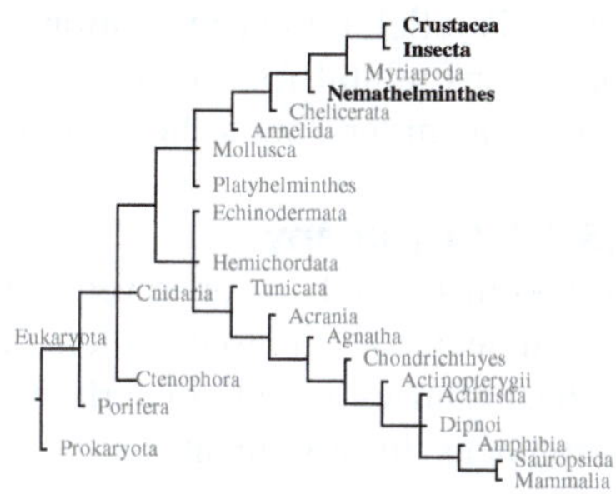

5.3.5.1 Introduction

The first sulfakinin from the cockroach *L. maderae* was isolated using a bioassay that determined frequency and intensity of gut contractions.

5.3.5.2 Biochemistry and Structure

Sulfakinins are cleaved from a precursor protein. They are characterized by a C-terminal **DYGHMRF**-NH_2 sequence (Fig. 5.13). Some sulfakinins were isolated with a sulfated tyrosine (**Y**); when the sequence has only been identified by RNA/DNA analysis it cannot be assumed that the final peptide is indeed sulfated. Admittedly, nonsulfated sulfakinins were equally active (Nichols 2007, 2003). Tyrosylprotein sulfotransferase (TPST), the enzyme that adds a sulfate group to

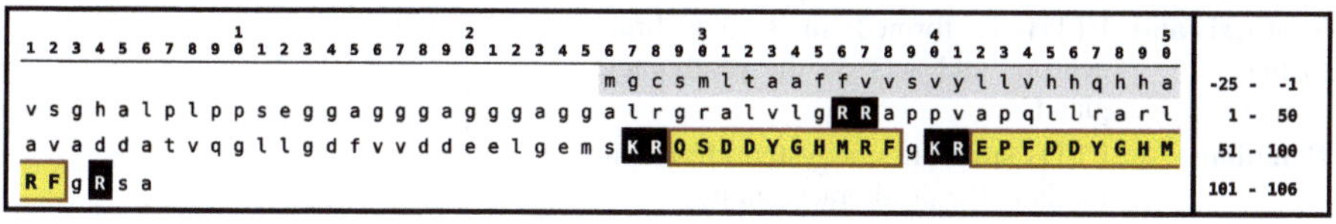

Fig. 5.13 Primary sequence of the sulfakinin from *Grillus bimaculatus*. The signal peptide is shown on a *gray background*; the sulfakinins are *boxed*, *uppercase*, and on *yellow background*. Dibasic or furin peptide motifs are shown *black on white* (Origin: GenBank CAL48349)

tyrosine, has been found in all vertebrates analyzed, in insects, nematodes and annelids, and, too, in several bacteria (for a review, see Moore 2003).

Sulfakinins resemble the equally sulfated C-terminal fragments of gastrin and cholecystokinin. The cionin from *C. intestinalis* also seems to be related.

Sulfakinin receptors of *D. melanogaster* were identified due to their analogy to vertebrate gastrin/CCK receptors and characterized. In other insects and nematodes sulfakinin receptor genes could be established as well.

5.3.5.3 Physiology

Originally sulfakinins were identified for their gut muscle stimulation. In larvae very few brain neurons secrete sulfakinins. Their number increases in pupae and adults. Parallel to their regulation of muscle contractions it was recently found that sulfakinins inhibit food intake (Wei et al. 2000). Thus sulfakinins not only act as neurotransmitters, but as hormones in sensu stricto.

5.3.5.4 Phylogeny

Sulfakinins have been observed in insects and crustaceans. Their relation to gastrin and cholecystokinin, to caerulein, and to cionin as well, argues for the presence of such substances already in early metazoans. The homology of the sulfotransferases strengthens this argument.

5.4 Neuropeptides of Reproduction

5.4.1 PTTH

5.4.1.1 Introduction

Prothoracicotropic hormone (PTTH) is formed in a few neurosecretory cells of the insect brain. It is secreted into the corpora allata and stimulates in the prothoracic gland the synthesis of ecdysteroids. These are indispensable for growth and development.

Fact sheet 5.14: Prothoracicotropic hormone (PTTH)

Sequence: Fig. 5.14.

Synthesis and target: PTTH is formed in a few brain neurons and acts on the prothoracic gland.

Function: PTTH stimulates ecdysteroid biosynthesis in the prothoracic gland.

Receptor: Tyrosine kinase receptor *torso*.

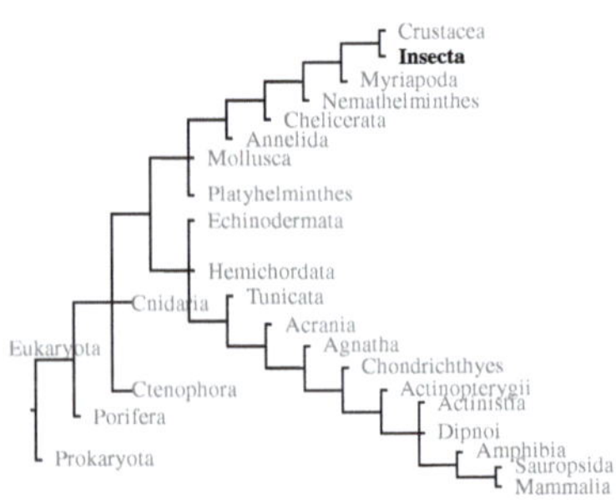

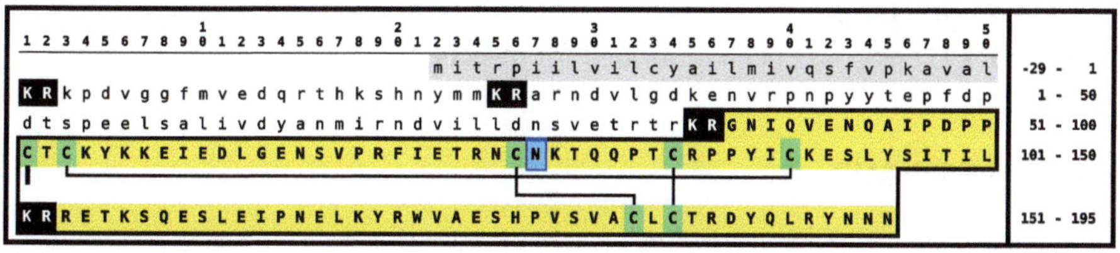

Fig. 5.14 Primary sequence of the prothoracicotropic hormone (PTTH) from silk moth. The signal peptide is shown on a *gray background*, the PTTH is *boxed* on *yellow background*, and the N-glycosylated arginine (**N**) is boxed, too. Dibasic peptide motifs are shown *black on white* (Origin: P17219)

5.4.1.2 Biochemistry and Structure

PTTH is built as a homodimer from two identical peptide chains (Ishizaki and Suzuki 1994 Fig. 5.14); the subunits are linked by a disulfide bridge between the two cysteine residues at position 130 (in Fig. 5.14 indicated by a *gray bar*).

The singular genes (or cDNAs) for PTTH from *B. mori*, *D. melanogaster* and from several *Helicoverpa* species (Noctuidae) have been cloned. Characteristic features are the dibasic peptide motif in front of the PTTH peptide, the positioning of seven cysteine residues, and a glycosylated arginine. Sequence homology is between 40 and 98 % of amino acids (Sauman and Reppert 1996).

A PC1 that recognizes **KR** would internally cleave PTTH again, thus the recognition site might be **RKR** or **RxRK**. The latter motif would also be cleaved by furin. The PTTH-releasing endopeptidase has not been described in the literature thus far.

In 2009, the torso receptor was identified as the PTTH receptor. Its role in early fly embryo-genesis had already been found some time ago (for a review, see Li 2005). The early ligand for torso is trunc. Later in larval development, torso expression is restricted to the prothoracic gland where PTTH exerts its effects (Rewitz et al. 2009)

5.4.1.3 Physiology

PTTH triggers ecdysone biosynthesis. The torso receptor triggers a cAMP- and calcium-mediated stimulation in cells of the prothoracic gland. The restricted expression to a few laterodorsal neurosecretory cells in the insect's brain points to a very specific function of PTTH.

PTTH is released in a circadian rhythm. The pacemaker neurons (expressing *per*) are placed in close vicinity to the PTTH neurons and connected to these synaptically (Sauman and Reppert 1996).

The amount of PTTH mRNA has been found to be constant during larvae development, however, PTTH pulses were secreted into the hemolymph before eclosion indicating a regulation of the secretion and not of the synthesis of PTTH. One transcription factor necessary for expression of PTTH is myocyte enhancer factor 2 (MEF2) which binds to the PTTH promoter.

5.4.1.4 Phylogeny

Thus far PTTH has only been described for insects.

5.4.2 PTSH; MIP

Fact sheet 5.15: Prothoracicostatic Hormone (PTSH); Myoinhibitory Peptide (MIP)

Sequence: Fig. 5.15.

Synthesis and target: PTSH is synthesized in the brain of larvae and acts on the prothoracic gland.

Function: PTSH inhibits ecdysone synthesis in the prothoracic gland.

Receptor: The sex peptide receptor is also the receptor for PTSH.

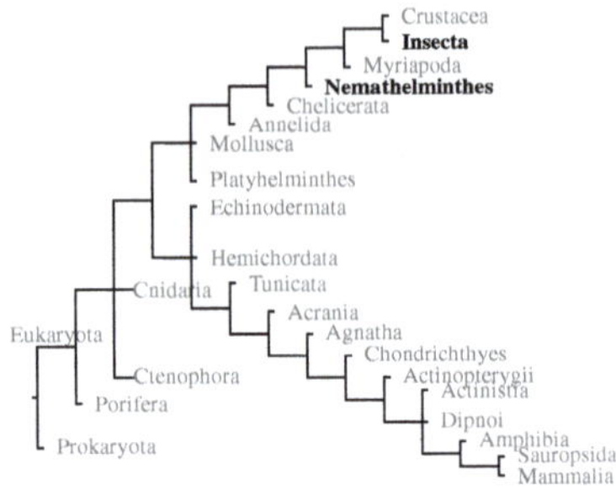

5.4.2.1 Introduction

Although PTTH stimulates ecdysone biosynthesis in the prothoracic gland, the hormone inhibiting this biosynthesis had not been identified for some time. Either Neb-TMOF or several allatostatins had been described as prothoracicostatic; finally, Hua et al. (1999) identified an additional neuropeptide capable of inhibiting PTTH stimulation.

5.4.2.2 Biochemistry and Structure

The consensus sequence of *B. mori* PTSH contains six invariable amino acids (Fig. 5.16). Known PTSHs all possess an allatostatin type B motif **Wx$_6$W**-NH$_2$.

```
                  1                   2                   3                   4                   5
1 2 3 4 5 6 7 8 9 0 1 2 3 4 5 6 7 8 9 0 1 2 3 4 5 6 7 8 9 0 1 2 3 4 5 6 7 8 9 0 1 2 3 4 5 6 7 8 9 0
                                                              m r w c l f a l w v f g v a t v v t a    -19 -  -1
a e e p h h d a a p q t d n e v d l t e d d K R a w s s l h s g w a K R A W Q D M S S A W g K R A W      1 -  50
Q D L N S A W g K R G W Q D L N S A W g K R A W Q D L N S A W g K R G W Q D L N S A W g K R d d d e     51 - 100
a m e K K S W Q D L N S V W g K R A W Q D L N S A W g K R A W Q D L N S A W g K R g w n d i s s v w    101 - 150
g K R A W Q D L N S A W g K R A W Q D M S S A W g K R a p e k w a a f h g s w g K R s s i e p d y e    151 - 200
e i d a v e q l v y q q a p n e e h i d a p e K K a w s a l h g t w g K R p v k p m f n n e h s a t    201 - 250
t n e a                                                                                                251 - 254
```

Fig. 5.15 Primary sequence of prothoracicostatic hormone (PTSH) from the silk moth. The signal peptide is shown on a *gray background*; the PTSH is *boxed*, in *uppercase* on *yellow background*. Dibasic peptide PC1 motifs are shown *black on white*. The peptide amino acids 25–33 cannot be amidated due to a G→A substitution and might thus be inactive (Source: NP_00103689)

Fig. 5.16 Consensus sequence of prothoracicostatic hormone (PTSH) from the silk moth. The larger the letter the more frequent the respective amino acid (Max. = 10/10; Min. = 1/10) (Origin: NP_001036890)

In addition, Hua et al. (1999) observed that the PTSH consensus sequence **AWQDLNSAW**-NH_2 is identical to the sequence of myoinhibitory peptide 1 (MIP1) from *M. sexta* where a prothoracicostatic activity had not been found before.

5.4.2.3 Physiology

MIP/PTSH peptides not only inhibit ecdysone biosynthesis, but also block gut peristaltic and possibly juvenile hormone biosynthesis . However, since the discovery of a potential B-type allatostatin receptor in the fruit fly (Johnson et al. 2003) no other study in particular about topological and temporal expression of this protein in the course of fly development has been published.

Apart from MIP/PTSH peptides FRMF-like peptides additionally regulate the prothoracic gland and its ecdysone synthesis (Yamanaka et al. 2006). The questions of the influence exerted by the combination of different peptides together with PTTH on the hormone production, how the regulation is progressing in the course of development, and how gene defects influence this development cannot be answered yet. In 2010 the sex peptide receptor was identiefied by Yamanaka et al. (2010).

5.4.2.4 Phylogeny

Myoinhibitory peptides have only been found in flies thus far, prothoracicostatic peptides, however, are found in moths (Lepidoptera) and beetles (Coleoptera). In addition to insects, *C. elegans* possess a protein homologous to the PTSH precursor with a **Wx$_6$W**-NH_2-allatostatin-type-B peptide/PTSH to be cleaved by PC1.

5.4.3 Pheromonostatic Peptide, Sex Peptides

5.4.3.1 Introduction

With the release of pheromones female insects signal a sexual mature state and mating eagerness. During copulation the females receive together with the sperms peptide hormones inhibiting pheromone synthesis and release. Thus these females become temporarily unattractive to other males. After egg positioning pheromones are again released and further copulation enabled.

5.4.3.2 Biochemistry and Structure

The pheromone synthesis inhibiting peptide from *Helicoverpa zea* (Fig. 5.17) is a relatively long peptide compared to the sex peptides from the accessory gland of

Fact sheet 5.16: Pheromone synthesis inhibiting (pheromonostatic) peptide (PSP); sex peptide (SP)

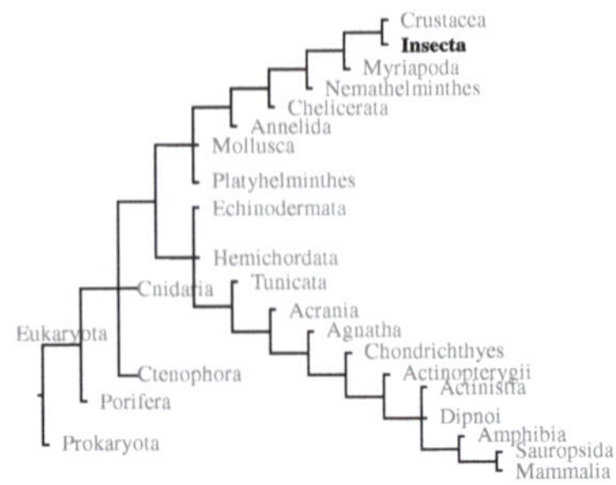

Sequence: Fig. 5.17.

Synthesis and target: PSP/SP are formed in the male genitalia and are transferred together with the sperms into the female genitalia.

Function: These peptides inhibit pheromone biosynthesis.

Receptor: The receptor is a heptahelical membrane GPCR.

									1										2										3		
1	2	3	4	5	6	7	8	9	0	1	2	3	4	5	6	7	8	9	0	1	2	3	4	5	6	7	8	9	0		
pE	I	I	N	N	N	D	Y	H	D	D	H	H	G	D	Q	P	T	L	L	L	R	S	Q	P	E	R	M	R	P		1 - 30
L	L	L	A	R	D	A	F	G	G	E	C	P	P	S	S	F	H	K	L	K	N	W	C	H	I	V	-amide				31 - 57

Fig. 5.17 Peptide sequence of pheromonostatic peptide-1 (PSP-1) from *Helicoverpa zea* (cotton bollworm aka corn earworm aka tomato fruitworm). PSP-1 shown *boxed, on yellow background* and in *uppercase*. It is N-terminally modified to pyroglutamate and C-terminally amidated. The intramolecular disulfide bridge is indicated (Origin: AAB35024)

fruit flies which are much shorter. A common feature of both types of molecules is a disulfide bridge close to the C-terminus.

5.4.3.3 Physiology

Formation of sex peptides/PSP in the male genitals and their reception by the female are a most interesting neuropeptide transmission. This is neither a neurotransmission, nor an endocrine interaction via the hemolymph. It is pheromone-like inasmuch as it happens between two individuals. The closest relationship appears to be feeding of royal jelly to the larvae creating bee queens and thus influencing ILP1 expression.

PSP/SP are synthesized in the male genitalia more precisely in the ejaculatory duct; there is, however, an expression in the heart in both sexes.. With the seminal fluid as vector they are transferred into the female. Shortly after copulation a reduction of PBAN synthesis in brain neurons could be observed. Whether PSP/SP receptors on cells of the female genitalia and neuronal contacts to the brain or a transport of PSP/SP into the brain are required has been analyzed in only a few examples. In female *D. melanogaster* increasing amounts of SP could be found in the brain after copulation (Nagalakshmi et al. 2004); and still, in order to respond to SP functional neurons are necessary that should express the egghead protein (Soller et al. 2006).

PSP/SP not only influences the PBAN synthesis and release; they additionally stimulate egg deposition and juvenile hormone release. The male thus ensures that its sperms and not those of a competitor are fertilizing the eggs. In several species the PSP/SP action is reduced after several days. The duration of effects is dependent on the presence of sperms onto which the PSP/SP adhere with their N-terminal end and from where they diffuse after fertilization has occurred (Kubli 2003).

When the PSP receptor from the *D. melanogaster* (a GPCR) had been cloned its expression in the genital tract and the CNS of females fruit flies was observed (Yapici et al. 2008). In flies with receptor mutants egg disposition was not seen and these flies copulated again as virgin females.

5.4.3.4 Phylogeny

PSP/sex peptides have only been found in insects.

5.4.4 GIH, VIH

Fact sheet 5.17: Gonad-inhibiting hormone (GIH); vitellogenesis inhibiting hormone (VIH)

Sequence:	Fig. 5.18.
Synthesis and target:	GIH are synthesized in sinus glands in the crustacean eye stalks. They act on the gonads.
Function:	GIH inhibits molting and facilitates sexual maturation.
Receptor:	(not yet identified).

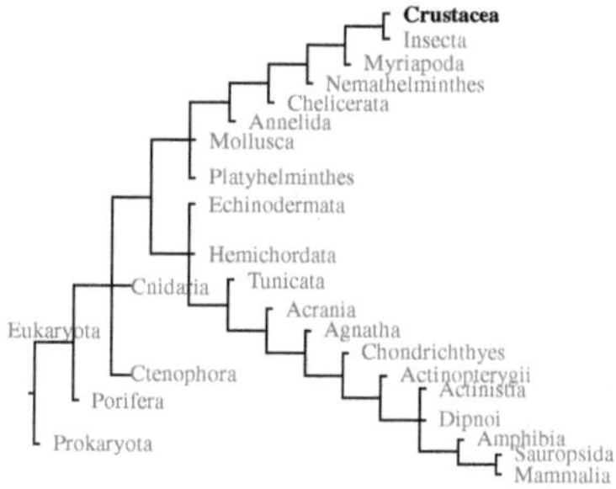

5.4.4.1 Introduction

As with the related CHH GIH/VIH is a peptide from crab synthesized in the X organs in eye stalks and released in the nearby sinus gland.

5.4.4.2 Biochemistry and Structure

The GIH precursor has the signal peptide and the GIH peptide with a C-terminal **GKR** to be oxidized to amide. Six cysteines are similarly spaced as in CHH. In contrast to CHH there is no associated N-terminal peptide in GIH/VIH.

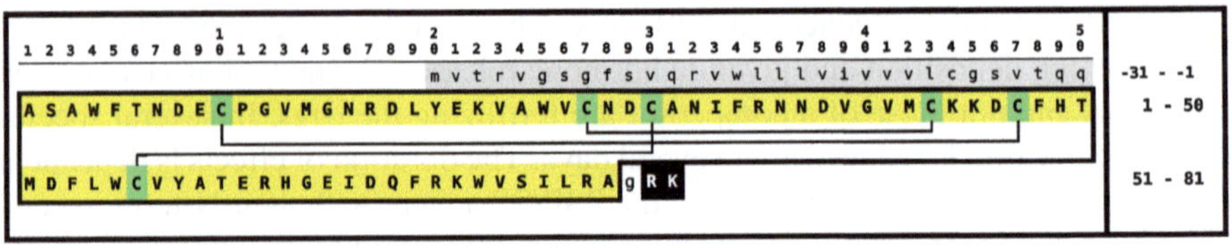

Fig. 5.18 Peptide sequence of gonad-inhibiting hormone (GIH) from the American lobster (*H. americanus* [Homarus americanus]). GIH is *boxed*, shown on *yellow background* and in *uppercase*. It is C-terminally modified to amide. The intramolecular cysteine bridges are indicated (Origin: P55320)

5.4.4.3 Physiology

GIH is a hormone controlling the life cycle of the crustacean. As long as GIH is present, molting is inhibited. After a decrease of GIH levels maturation occurs and vitellogenesis is activated. Finally molting occurs.

5.4.4.4 Phylogeny

GIH/VIH are thus far only known in crustaceans.

5.4.5 TMOF

Fact sheet 5.18: Trypsin-modulating oostatic factor (TMOF)

Sequence: Fig. 5.19.

Synthesis and target: TMOF is released by gonads and acts on gut enzymes.

Function: TMOF inactivates enzymes.

Receptor: TMOF binds to an unknown receptor in the gut epithelium and inhibits trypsin generation.

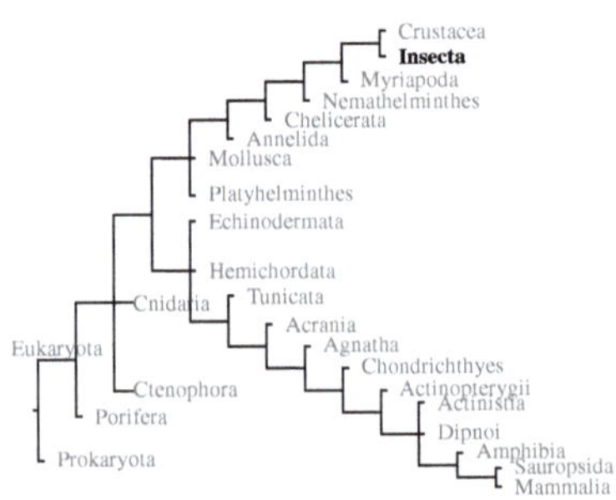

5.4.5.1 Introduction

A trypsin-modulating oostatic principle was first observed 75 years ago (Iwanov and Mescherskaya 1935, additional references at Borovsky 2003). With the sequence yet unknown it had already been found that an ovarian extract could inhibit vitellogenesis and digestion of blood meal in *A. aegyptii*. Borovsky et al. (1990) identified the sequence of TMOF. TMOF might be a pesticide against the yellow fever mosquito.

1 2 3 4 5 6 7 8 9 10 1 2 3 4 5 6 7 8 9 20 1 2 3 4 5 6 7 8 9 30 1 2 3 4 5 6 7 8 9 40	
m n k i i a a l v l f t a v i g a l a	-19 - -1
D Y P A P P P P P P k p y h a p p p p p y h a p p h h a p a p l h p v v h t y p	1 - 40
v k a p a a k c g a n l l v g c a p s v a h v p c v p v h p h p p p p a h y	41 – 78

Fig. 5.19 Peptide sequence of trypsin-modulating oostatic factor (TMOF) from yellow fever mosquito (*A. aegyptii*). TMOF is *framed, on gray background* and in *uppercase letters* (Source: P19425; Borovsky et al. 1990)

5.4.5.2 Biochemistry and Structure

With its seven proline residues TMOF is an unusual peptide. For TMOF receptor activity the N-terminal four amino acid residues are critical. An **DYPR** analogue was found to be fourfold more active than TMOF or natural **DYPA**. TMOF is cleaved from the precursor protein by the signal peptidase and another endopeptidase cleaving after lysine, that is, a trypsin or chymotrypsin-like enzyme.

From the grey flesh fly, a similarly active TMOF was isolated whose sequence **NPTNLH** differs strongly from the one from *A. aegyptii* (Bylemans et al. 1994).

5.4.5.3 Physiology

The name "oostatic factor" is misleading. TMOF inhibits synthesis and release of gut enzymes. Thus it inhibits gonads only in an indirect way. Released from gonads, TMOF can diffuse without any transporter protein across the gut wall. In adult mosquitoes TMOF is released about 30 h after a blood meal. At this time the blood is digested and TMOF might inhibit excess digestion enzymes. Larvae treated with TMOF die from starvation. TMOF might also influence the prothoracic gland (Gilbert et al. 2002)

5.4.5.4 Phylogeny

The original findings of oostatic or antigonadotropic factors include as a source a decapod crustacean (compare Borovsky 2003), but a sequence has not been published. Only in insects has TMOF been cloned and sequenced.

5.4.6 Nebcolloostatin

5.4.6.1 Introduction

While isolating TMOF a second peptide was identified with oostatic activity which was named colloostatin due to its closeness to collagen. It has not been analyzed thus far in great detail.

5.4.6.2 Biochemistry and Function

The sequence **SIVPLGLPVP IGPIVVGPR** of colloostatin resembles preprocollagen-IV from fruit flies, however, the collagen-IV from the grey flesh fly has not yet been sequenced. The enzymes releasing colloostatin are not known.

Fact sheet 5.19: Collagen-like oostatin from the grey flesh fly *Neobellieria bullata* (Neb-Colloostatin)

Sequence:	SIVPLGLPVP IGPIVVGPR.
Synthesis and target:	Potentially from preprocollagen IV by unknown endopeptidase(s).
Function:	Blocks vitellogenesis.
Receptor:	(not yet known).

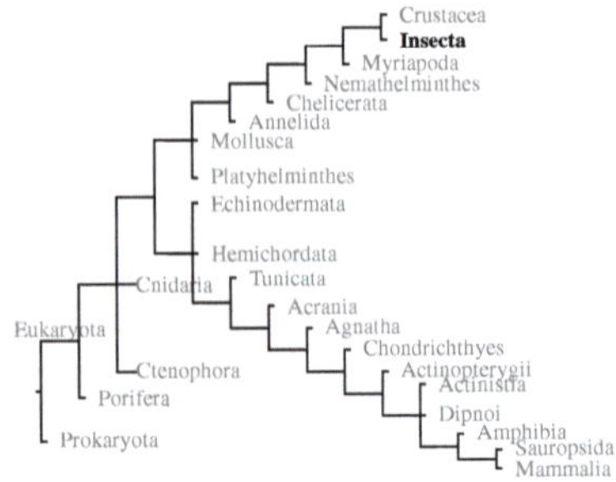

The peptide has been isolated from the abdomen of *Neobellieria bullata* and inhibits vitellogenesis at nanomolar concentrations.

5.4.6.3 Physiology

Recently Wasielewski and Rosinski (2007) observed that in the mealworm (*Tenebrio molitor*) colloostatin as well as TMOF inhibit vitellogenesis, slows down ovary development, delays ovulation and egg deposition, and finally reduces the number of eggs deposited.

5.4.6.4 Phylogeny

Colloostatins have only been found in insects.

5.5 Peptide Hormone of Metamorphosis and Molting

5.5.1 Molt-Inhibiting Hormone (MIH)

Fact sheet 5.20: Molt-inhibiting hormone (MIH)

Sequence:	See, for example, Fig. 5.20.
Synthesis and target:	MIH is synthesized in the X organ of crustaceans and released in the sinus gland.
Function:	MIH delays molting (this view is challenged).
Receptor:	Possibly a membrane guanylate cyclase.

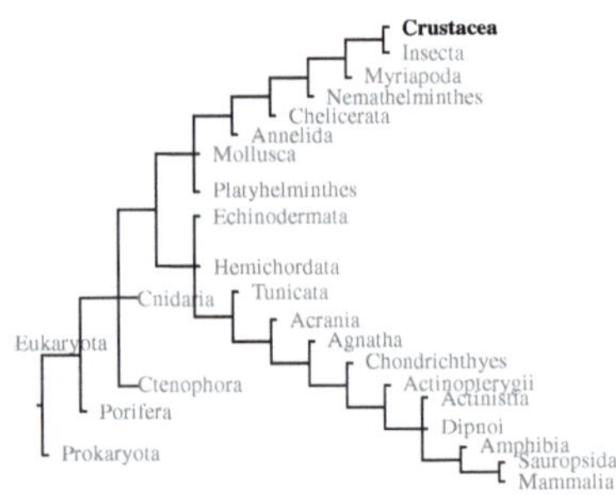

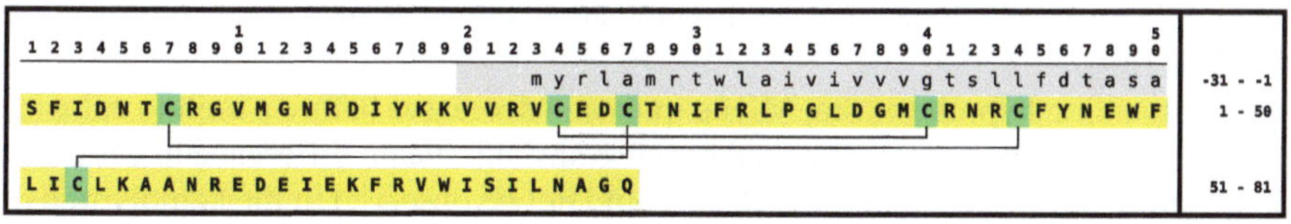

Fig. 5.20 Peptide sequence of molt-inhibiting hormone (MIH) from the Kuruma prawn (*Marsupenaeus japonicus*). The signal peptide is *highlighted gray*; MIH is *boxed, highlighted yellow and in uppercase*. The intramolecular cysteine bridges are indicated (Source: P55847)

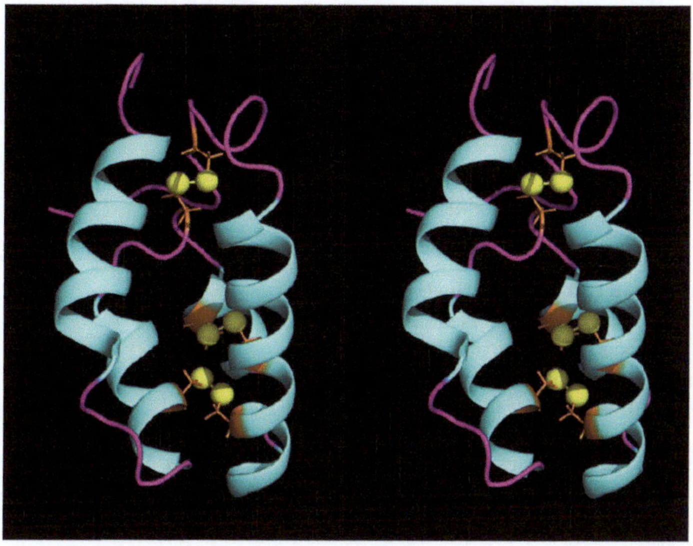

Fig. 5.21 Stereo view of the structure of molt-inhibiting hormone (MIH) from the Kuruma prawn (*Marsupenaeus japonicus*). Helices are shown in *light blue*, and other sequence parts in *pink*; cysteines are in *orange* and disulfide bridges drawn *yellow* (Source: Katayama et al. 2003)

5.5.1.1 Introduction

MIH belongs to the family of CHH/GIH/VIH peptides. Like those it is made in a few neurosecretory cells of the crustacean eye stalks and released in the sinus gland therein.

5.5.1.2 Biochemistry and Structure

As with GIH/VIH the MIH precursor protein consists of the signal peptide and the hormone alone (Fig. 5.20). The cysteine positions and the three disulfide bridges are similar to CHH and GIH (Fig. 5.21).

5.5.1.3 Physiology

MIH is like CHH a hormone controlling ecdysone biosynthesis via the sinus gland. The actions of MIH are also mediated by guanylate cyclase. NO synthase seems to play a role. A specific receptor has not yet been found, only one candidate gene (Zheng et al. 2008, and an unpublished sequence of a guanylate cyclase (AGB51125)).

The standing hypothesis that MIH expression decreases before a molt to enhance ecdysone levels has been challenged because any significant difference in ecdysone concentrations before and after molting could not be observed in the hemolymph or the eye stalks (Chung and Webster 2005).

5.5.1.4 Phylogeny

MIH is a crustacean hormone.

5.5.2 Corazonin

Fact sheet 5.21: Corazonin

Sequence: pETFTYSHGWTN-NH_2.

Synthesis and target: Corazonin is released in the CNS and acts on abdominal Inka cells.

Function: Corazonin triggers in Inka cells release of ecdysis-triggering hormone (ETH).

Receptor: An heptahelical GPCR.

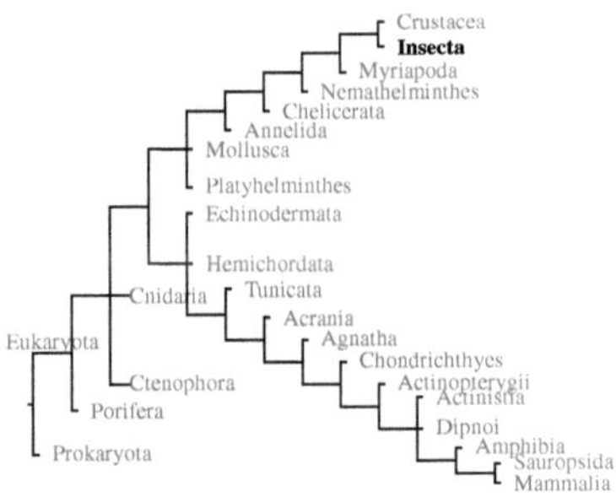

5.5.2.1 Introduction

Corazonin was originally found and characterized by Veenstra in cockroaches and thereafter in other insects (Veenstra 1989). A corazonin gene in beetles could not be found. Corazonin's role has only recently been identified by Kim et al. (2004b). It is the earliest peptide acting in the molting cascade.

5.5.2.2 Biochemistry and Structure

The corazonin peptide has 11 amino acids, a N-terminal pyroglutamate, and a C-terminal amide. Variations between different insects are mostly due to residue 7: histidine in bees and arginine in cockroaches and in flies (Fig. 5.22). The corazonin receptor is a GPCR of the rhodopsin family (Cazzamali et al. 2002).

5.5.2.3 Physiology

Corazonin is a brain neuropeptide. It is synthesized by one pair of dorsomedial neurons, three pairs of dorsolateral cells in the brain, and by eight pairs of cells in the ventral cord. The dorsolateral corazonin neurons survive metamorphosis, however,

Sequence	Positions
`         1         2         3         4         5`	
`12345678901234567890123456789012345678901234567890`	
`                             mvnsqililfilsltitivmc`	-21 - 1
`QTFTYSHGWTNgKRstsleelanrnaiqsdnvfancelqklrlllqgnin`	1 - 50
`nqlfqtpcellnfpKRsfsenmindhrqpaptnnny`	51 - 86

Fig. 5.22 Peptide sequence of corazonin from the honeybee (*A. mellifera*). The signal peptide is highlighted *gray*, and the corazonin peptide *blocked, highlighted yellow and in uppercase*. Dibasic motifs are *inverted* (Source: Q5DW47)

the others undergo apoptosis therein effected by ecdysone. After the decay of the pre-ecdysial ecdysone peak corazonin neurons secrete corazonin which triggers pre-ecdysis-triggering hormone (PETH) and ETH in abdominal Inka cells.

Corazonin has originally been described as a cardioacceleratory peptide. This function appears of minor importance now. Additionally corazonin acts as dark color inducing hormone (DCIN) because it induces dark pigmentation in certain albino locusts.

5.5.2.4 Phylogeny

As of today, corazonin has been found in insects, but not in every species analyzed.

5.5.3 Ecdysis-Triggering Hormone (ETH)

5.5.3.1 Introduction

More than 100 years ago, Ikeda postulated a role of the silk moth's epitracheal glands in air uptake of tracheae (Ikeda 1913). Zitnan et al. (1996) isolated ETH from Inka cells and demonstrated in additional studies its role in ecdysis and metamorphosis.

Fact sheet 5.22: Ecdysis-triggering hormone (ETH)

Sequence: Table 5.7.

Synthesis and target: ETH is the product of epitracheal Inka cells. It acts on the CNS.

Function: By ETH pre-ecdysis and ecdysis are triggered. Moreover stimulated by ETH tracheae fill with air.

Receptor: An heptahelical GPCR.

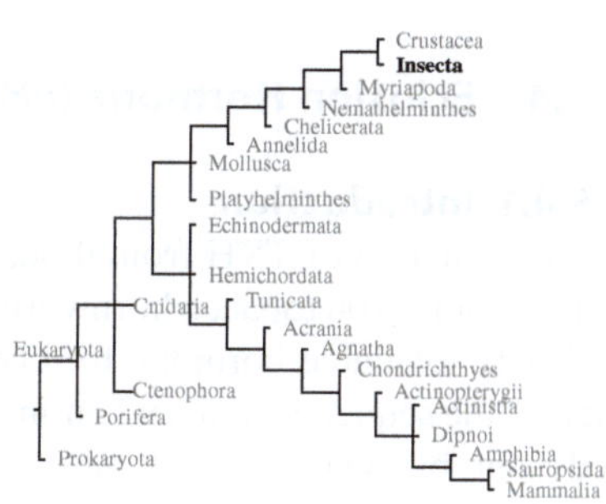

Table 5.7 Ecdysis-triggering hormones

Species	Short name	Sequence
B. mori, M. sexta	**PETH**	**SFIKPN.NVPRV**-NH_2
B. mori	**ETH**	**SNEA FDEDVMGYVIKSNKNIPRM**-NH_2
M. sexta	**ETH**	**SNEAISPFDQGMMGYVIKTNKNIPRM**-NH_2
D. melanogaster	**ETH1**	**DDSSPGFFLKITKNVPRL**-NH_2
D. melanogaster	**ETH2**	**GE..NFAIKNLKTIPRI**-NH_2
A. aegyptii	**ETH1**	**DETPGFFIKLSKSVPRI**-NH_2
A. aegyptii	**ETH2**	**GDFENFFLKQSKSVPRI**-NH_2
B. mori	**ETH-AP**	**NYDSGNHFDIPKVYSLPFEFYGDNEKSLNNDDAEE...YYAKKMGSM**-OH
M. sexta	**ETH-AP**	**NYDSENRFDIPKLYPWRAENTELYEDDAQPTNGEEINGFYGKQRENM**-OH

5.5.3.2 Biochemistry and Structure

The silk moth (*B. mori*) precursor bears—as in the tobacco hornworm—three different peptides: the pre-ecdysis—triggering hormone (PETH), ETH, and associated peptide (ETH-AP). In fruit flies (*D. melanogaster*) and in the yellow fever mosquito (*A. aegyptii*) there are two ETH peptides. The structural motif is **I/LKxxKxI/VPRx**amide.

5.5.3.3 Physiology

ETH and PETH are hormones of the molting cascade: In Inka cells the neuropeptide corazonin stimulates an increase of intracellular cGMP. This induces ETH synthesis and release. ETH in turn acts locally on tracheae which fill with air. ETH also acts in an endocrine way on EH and CAP neurons and together with these neurons and their hormones induce the behavioral patterns of ecdysis.

5.5.3.4 Phylogeny

ETH has only been found in insects. An ETH receptor like protein was identified in echinoderms

5.5.4 Eclosion Hormone (EH)

5.5.4.1 Introduction

As the isolation of TRH from thousands of sheep hypothalami before, the isolation of EH from 1,700 tobacco hornworm larvae was a heroic task. This isolation resulted in 3.5 μg available hormone for chemical analysis. Marti et al. (1987) determined thereof the sequence of the 62 amino acids of EH. Following PTTH which stimulates ecdysone biosynthesis, EH was the second hormone of the molting cascade to be identified.

Fact sheet 5.23: Eclosion hormone (EH)

Sequence:	Fig. 5.23.
Synthesis and target:	EH is the product of ventromedial neurons that acts on Inka cells.
Function:	EH stimulated the ecdysis program *i.a.* by activation of CAP release.
Receptor:	EH receptor is a receptor guanylate cyclase.

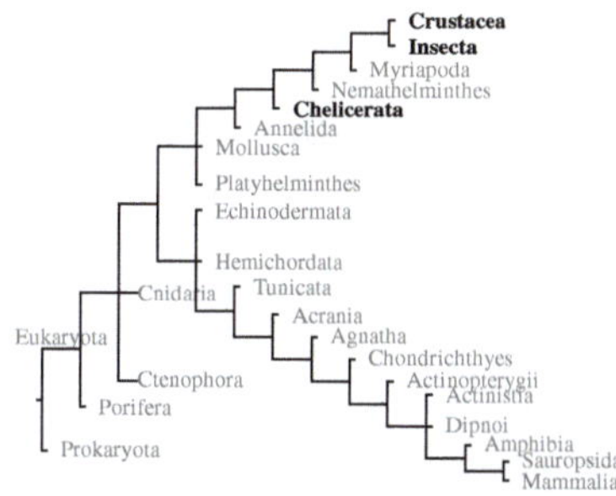

Sequence	Position
m a g k v t v a f f m f a m i a f l a n f g y v e c	-26 - -1
N P A I A T G Y D P M E I C I E N C A Q C K K M L G A W F E G P L C A E S C I K F K G K L I P E C E	1 - 50
D F A S I A P F L N K L	51 - 62

Fig. 5.23 Primary sequence of the eclosion hormone (EH) of the tobacco hornworm. The signal peptide is shown on *gray background*, and the EH polypeptide in *bold* and *uppercase* letters. Disulfide bridges are indicated (Source: P11919)

5.5.4.2 Biochemistry and Structure

Eclosion hormone is cleaved from the precursor by the signal peptidase. The characteristic features are three disulfide bridges which let the peptide resemble the CHH/MIH/ion transport protein family although any significant sequence homology has not been established. Very recently Chang et al. (2009) identified in the *Bactrocera dorsalis* (oriental fruit fly) the membrane guanylate cycles BdmGC-1 as the receptor for EH.

5.5.4.3 Physiology

EH is made in ventromedial neurosecretory cells of the brain and released from neurosecretory cells of the CNS. It is released either centrally or in the periphery: for the latter release axons of EH neurons project to the abdominal proctodeal nerve and EH is released within a neurohemal organ there. Central release takes place in the corpora cardiaca. For the ecdysis program central release is required. The abdominal release stimulates several skin glands (Copenhaver and Truman 1986a,b; Hewes and Truman 1991; Truman 1992, 2005; Gammie and Truman 1999; Truman and Copenhaver 1989).

As shown in the section on ETH EH participates in the ecdysis program. It acts *i.a.* on tracheal filling with air, stimulates ETH release in Inka cells and activates CAP release. EH-defective mutants are unable to perform ecdysis in the normal pattern, but sometimes they eventually escape the old skin.

5.5.4.4 Phylogeny

In GenBank there are insect and crustacean EH sequences and one from a spider.

5.5.5 Bursicon

Fact sheet 5.24: Bursicon

Sequence: Fig. 5.24.

Synthesis and target: The bursicon dimer is made in a few neurosecretory cells of thoracic and abdominal ganglia. It acts on epidermal and wing cells.

Function: Bursicon enables hardening of skin and wings after ecdysis.

Receptor: A heptahelical GPCR.

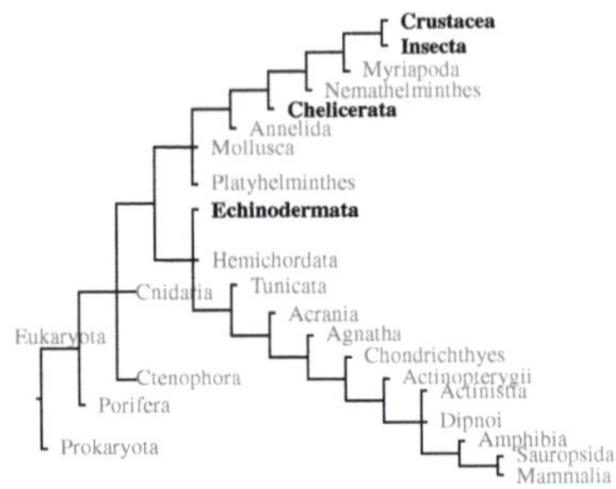

5.5.5.1 Introduction

Some decades ago, it had already become obvious that to finish a molt a longtime unknown hormone stimulated hardening and coloring of the coat. This hormone was called bursicon. Only recently in 2005, the complete bursicon molecule as a dimer of

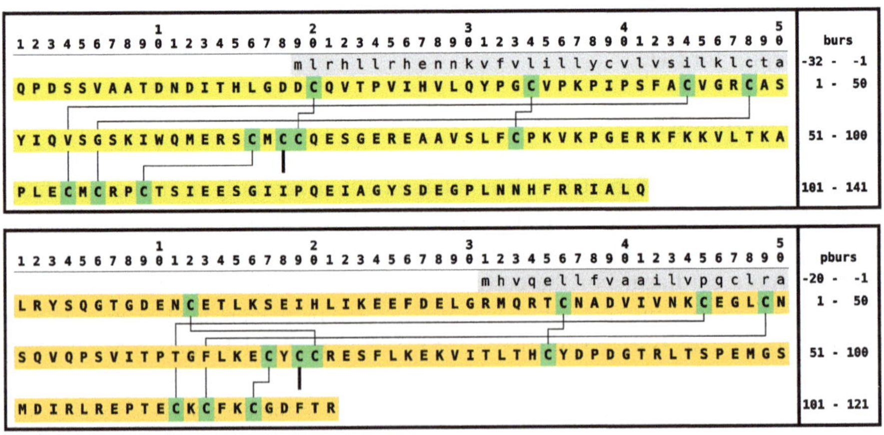

Fig. 5.24 Peptide sequences of the two bursicon subunits from the fruit fly (*D. melanogaster*). The signal peptide is highlighted *gray*; the peptides are in *uppercase*. The intramolecular disulfide bridges taken from the source are indicated. The intermolecular disulfide bridge coupling the two subunits is indicated by *short bold lines* (Source: Q9VD83 (burs) and Q9VJS7 (pburs))

two cysteine knot polypeptides was identified, together with the GPCR specifically binding the dimer of both proteins (Luo et al. 2005).

5.5.5.2 Biochemistry and Structure

Bursicon is formed by the two subunits burs and pburs (partner of burs), with similar composition: after the signal peptide follows a cysteine knot polypeptide. The organization of the cysteine bridges is the same: Cys1 → Cys7[18]; Cys2 → Cys8; Cys3 → Cys9; Cys4 → Cys10; Cys5 intermolecular; and Cys6 → Cys11. Homology exists between the respective subunits in different insect species analyzed. In the fruit fly and the red flour beetle the receptor for the bursicon heterodimer (burs/pburs) is a GPCR with characteristic analogies to the glycoprotein hormone receptors of vertebrates, the glycoprotein hormones equally heterodimers of cysteine knot proteins.

Burs and pburs have been found expressed in the same and in different cells (Honegger et al. 2008). A homodimer has been shown to be active different from bursicon to activate immune and stress genes (An et al. 2012).

5.5.5.3 Physiology

The known functions of bursicon are the induction of the stabilization of wings and coat coloring. The bursicon dimer is synthesized in some cells of the subesophageal ganglion, and thoracic and abdominal ganglia (SEG, TG, AG), and released by neurosecretion. Pburs is additionally expressed in an intrinsic cell of corpora cardiaca. For the tobacco hornworm, this has been recently analyzed in detail (Table 5.8; Dai et al. 2008).

During larval stages, bursicon is expressed in the SEG and TG; in the pupa all CNS ganglia bear some bursicon-expressing neurons, in the adult animal a pair of

Table 5.8 Differential expression of bursicon in the nerve system of the tobacco hornworm (abbrev.: CAP: cardioaccelatory peptide; CC: corpora cardiaca; SEG: subesophageal ganglion; TG: thoracic ganglion; AG: abdominal ganglion; TAG: terminal abdominal ganglion; St-3 larvae: third instar larvae (Source Dai et al. 2008)

	3rd instar larvae	Cells	Cotranslated	Pupae	Cells	Cotranslated	Adult	Cells	Cotranslated
CC	pburs	›10		Not shown			pburs	5–7	AKH
SEG	Bursicon	2 × 2	CAP 1 pair	Bursicon	2 × 2		None		
TG1	Bursicon	2 × 2	CAP 1 pair	Bursicon	2 × 2		None		
TG2/3	Bursicon	1 × 2		Bursicon	2 × 2		None		
AG1	Bursicon	1 × 2		Bursicon	2 × 2	CAP 1 × 2	Bursicon	3 × 2	CAP 1 × 2
AG2	Bursicon	1 × 2		Bursicon	2 × 2	CAP 1 × 2	Bursicon	3 × 2	CAP 1 × 2
AG3-5	None			Bursicon	2 × 2	CAP 1 × 2	Bursicon	2 × 2	CAP 1 × 2
AG6	None			Bursicon	1 × 2		Bursicon	3 × 2	CAP 1 × 2
TAG	None			Bursicon	1 × 2		Bursicon	3 × 2	CAP 1 × 2

[18]Counted from the N-terminus: Cys1 first cysteine, Cys2 second cysteine …

cells in the SEG and in seven pairs of cells in abdominal ganglia still maintain some bursicon expression. Bursicon's role has been reviewed very recently (White and Ewer 2013): it is responsible for postecdysial behavior which is governed by a pair of subesophageal cells and for wing expansion and hardening under the endocrine control of seven pairs of abdominal neurosecretory bursicon cells.

5.5.5.4 Phylogeny

Bursicon dimers have been found in insects, crustaceans, and chelicerates. A bursicon-like protein has been sequenced in echinoderms.

5.6 Regulators of Food Intake: RFamide and FMRFamide

Fact sheet 5.25: FMRFamides

Sequence:	**FMRF**-NH_2 and N-terminally extended analogues.
Synthesis and target:	FMRFamides originate in neurons and act via synapses on GPCR or ion channels.
Function:	FMRFamides were identified for their muscle contraction activity.
Receptor:	FMRF receptors in insects and nematodes are heptahelical GPCR.

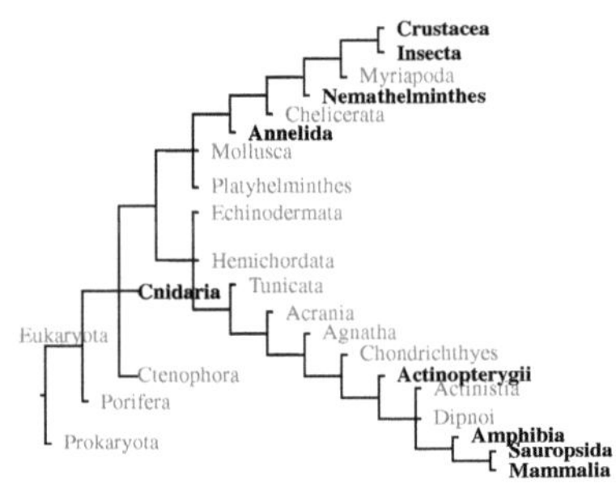

5.6.1 Introduction

Thirty years ago Price and Greenberg (1977) described the isolation of **FMRF**-NH_2 as a cardiostimulatory peptide from the sunray Venus clam (*Macrocallista nimbosa*), equally active inducing contraction of the radula muscle of the lightning whelk *Busycon contrarium*. In the meantime many vertebrate and invertebrate RFamide peptides have been isolated, that is, kisspeptins, prolactin releasing peptide. Neuropeptide FF (NPFF) (**FLFPQPQRF**-NH_2) and the cotranslated NPAF or NPSF have a role on nociception in human and other mammals (Vilim et al. 1999 and references). Other RFamide peptides have been found expressed in the mammalian and human brain, for example, the RFamide-like peptide in the hypothalamus, but their function has escaped elucidation thus far (Bechtold and Luckman 2007).

1 2 3 4 5 6 7 8 9 0 1 2 3 4 5 6 7 8 9 0 1 2 3 4 5 6 7 8 9 0 1 2 3 4 5 6 7 8 9 0 1 2 3 4 5 6 7 8 9 0 (1, 2, 3, 4, 5 above each 0)	
m g i a l m f l l a l y q m q s a i h s e i	-22 - -1
i d t p n y a g n s l q d a d s e v s p s q d n d l v d a l l g n d q t e r a e l e f r h p i s v i	1 - 50
g i d y s k n a v v l h f q k h g R K p r y k y d p e l e a K R r s v q d n f m h f g K R q a e q l	51 - 100
p p e g s y a g s d e l e g m a K R a a m d r y g R D P K Q D F M R F g R D P K Q D F M R F g R D P	101 - 150
K Q D F M R F g R D P K Q D F M R F g R D P K Q D F M R F g R T P A E D F M R F g R T P A E D F M R	151 - 200
F g R S D N F M R F g R s p h e e l r S P K Q D F M R F g R P D N F M R F g R s a p q d f v r s g K	201 - 250
m d s n f i r f g K s l k p a a p e s k p v k s n q g n p g e r s p v d k a m t e l f K K q e l q d	251 - 300
q q v k n g a q a t t t q d g s v e q d q f f g q	301 - 325

Fig. 5.25 Primary sequence of the FMRF precursor from the fruit fly. The signal peptide is highlighted *gray*, peptides discovered chemically are *boxed*, FMRF peptide *bold*, on *yellow background* and in *uppercase*, and additional peptides in *lowercase*; within the precursor a CRH-like peptide is found: 96–116. Monobasic and dibasic peptide motifs for prohormone convertase are shown *inverted*; all those glycine residues in front of these monobasic or dibasic motifs are oxidized to NH_2 by PHM (Source: P10552)

In invertebrates some species have multiple FMRFamide-like genes; for example, in *C. elegans* there are 22 RFamide genes (*flp-1* to *flp-22*) that can be translated into 50 peptides.[19]

5.6.2 Biochemistry and Structure

Whereas in molluscs the tetrapeptide **FMRF**-NH_2 is found, in insects the FMRF-amides are N-terminally extended, for example, in *D. melanogaster* (Fig. 5.25).

In *C. elegans* only FLP have been found (see Table 5.9), in toto 63 different RFamides in 23 genes whereas these genes, however, code for additional neuropeptide without RFamide C-termini. Mostly the precursor proteins contain a single FLP type, however, in multiple copies: flp1, flp3, flp7, or flp18. Sometimes a precursor harbors different FLP types: flp14 or flp17.

In addition to PC1 or PC2 motifs, there are **RxxR**, **RxxK**, and monobasic cleavage sites for prohormone convertases, for example, in the great pond snail *L. stagnalis* (Fig. 5.26). A furin-like enzyme (cleaving **RxxR** or **RxxK**) has been described in *L. stagnalis* (Smit et al. 1994). Southey et al. (2008) evaluated the usage of potential cleavage sites in insects and showed that in those organisms analyzed, an endopeptidase cutting after **RxxK** or after a single **K** was not observed. Cleavage required either dibasic **RK** or **KK** or **RxKK** motifs. A similar analysis in molluscs had been published before by Spijker et al. (2004). It appears, however, fairly impossible with eight FMRFamides at hand to identify those that had not been liberated. Three of seven FMRFamides are cleaved by an RxxR recognizing PC, for example, furin. One of the seven by PC1 (**KR**) and three others by an

[19] For reasons of conciseness we call FMRFamides those peptides with N-terminally extended **FMRF**amide, and variants, for example, **FIRF**amide are labeled FLP for FMRFamide-like peptides

Table 5.9 FMRFamide-like peptides in *C. elegans*

Gene	Example	Consensus	Peptide count
flp-1	**SADPNFLRF**-NH_2	PNFLRF	6
flp-2	**SPREPIRF**-NH_2	EPIRF	2
flp-3	**SAEPFGTMRF**-NH_2	GTRMRF	9
flp-4	**PTFIRF**-NH_2	FIRF	2
flp-5	**AKFIRF**-NH_2	KFIRF	3
flp-6	**KSAYMRF**-NH_2	YMRF	1
flp-7	**SPMQRSSMVRF**-NH_2	MVRF	4
flp-8	**KNEFIRF**-NH_2	FIRF	1
flp-9	**KPSFVRF**-NH_2	FVRF	1
flp-10	**QPKARSGYIRF**-NH_2	YIRF	1
flp-11	**AMRNALVRF**-NH_2	L/FVRF	2/1
flp-12	**RNKFEFIRF**-NH_2	FIRF	1
flp-13	**APEASPFIRF**-NH_2	PFIRF	6
flp-14	**KHEYLRF**-NH_2	YLRF	1
flp-14	**SLLDYRF**-NH_2	DYRF	1
flp-14	**EIVFHQISPIFFRF**-NH_2	FFRF	1
flp-15	**GGPQGPLRF**-NH_2	GPLRF	2
flp-16	**AQTFVRF**-NH_2	FVRF	2
flp-17	**KSQYIRF**-NH_2	YIRF	1
flp-17	**KSAFVRF**-NH_2	FVRF	1
flp-18	**EMPGVLRF**-NH_2	PGVLRF	6
flp-19	**WANQVRF**-NH_2	Q/SVRF	1/1
flp-20	**AMMVRF**-NH_2	MVRF	1
flp-21	**GLGPRPLRF**-NH_2	PLRF	1
flp-22	**SPSAKWMRF**-NH_2	WMRF	1
flp-23	**TKFQDFLRF**-NH_2	FLRF	3

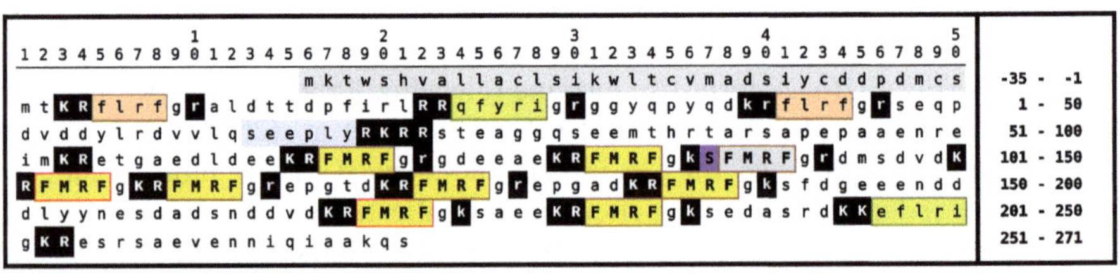

Fig. 5.26 Primary sequence of the FMRF precursor from the great pond snail (*L. stagnalis*). The signal peptide is highlighted *gray*. FMR- like peptides are *bold* and in *uppercase*; additional peptides are in *lowercase*. Monobasic and dibasic peptide motifs for prohormone convertase are shown *inverted*; all those glycine residues in front of these monobasic or dibasic motifs are oxidized to NH_2 by PHM. Whether a **SFMRF**amide (137–141) has ever been isolated is not described (Source: P19802)

unknown RxxK recognizing PC. In flies, in contrast, all 10 FLP can be cleaved by RxxR- recognizing furin, whereas the associated peptides exhibit a **RxxK** motif. It might well be that the degree of FMRF release and of associated peptides is regulated by the differential expression of prohormone convertase, or that although the FMRFamides are conserved, they are not released.

The thus far identified receptors for FMRFamides are GPCR: for example, in *D. melanogaster* (genbank AAF47700) or in *C. elegans* (GenBank ACG61342). Of the many FLP in *C. elegans* only *flp10* and *flp17* coded peptides bind to this receptor.

5.6.3 Physiology

As far as can be said today, (FM)RFamides act preferentially as neurotransmitters, and less as hormones *in sensu proprio*. Immunohistology in vertebrates with fluorescent antibodies most often generates an image of the entire nervous system, because in some areas more than 40 % of all neurons are labeled (Pernet et al. 2004). There is, however, for example, in *C. elegans* differential expression of different *flp* genes pointing to special functions of individual peptides or peptide types. Only for a few of them have specific receptors been discovered. Some hints to flp functions may come from the analysis of defect mutants: *flp-1* defect nematodes do not sense hyperosmolarity any further; they do not react to contacts with the nose, but react to contacts with the remaining body. They exhibit a variant movement coordination compared to wildtypes (Li et al. 1999).

FMRFamide was originally found as a peptide inducing muscle contraction in snails. A similar role has been observed in earthworms (*Eisenia fetida*, *Lumbricus terrestris*; Csoknya et al. 2005). In anthozoans (Cnidaria), too, muscle cells were contracted by antho-RFamide neurons (Pernet et al. 2004). It has been supposed that food intake and reproduction are controlled by antho-RFamides. Tessmar-Raible et al. (2007) have argued that RFamide neurons are multifunctional sensory cells that present an ancient developmental pattern in invertebrates (annelids, molluscs, hemichordates, and cephalochordates) as well as in fish (at least in teleosts).

Lingueglia et al. (1995) identified apart from the GPCR in garden snail *Helix aspersa*, a FMRF-controlled sodium channel. This amiloride sensitive channel assembled by four identical subunits is largely analogous to the renal human epithelial sodium channel sodium channel (eNaC) required for sodium resorption. This molluscan channel is preferentially opened by the tetrapeptide **FMRF**amide. **FLRF**amide is partial agonist and **FKRF**amide is an antagonist. N-terminally elongated peptides are less active than the original tetrapeptide (Cottrell 1997).

Similar sodium channels have been found in *C. elegans*; they were called degenerines because single point mutation led to neuronal degeneration. In this species these channels have a role in tactile sensing; they do not act as FMRFamide receptors.

5.6.4 Phylogeny

RFamides and FMRFamides have been found in many eumetazoans. Although in protostomes peptides are present in multiple copies on the precursor, the neuropeptide FF in humans exists as a single copy. Phylogenetic evaluations have not been performed; some vertebrate RFamides have been listed by Osugi et al. (2006). Apart from neuropeptide FF, kisspeptin and prolactin-releasing peptide are among the RFamides.

5.7 Neuropeptide Regulators of Juvenile Hormone Metabolism

5.7.1 Allatotropins

Fact sheet 5.26: Allatotropin

Sequence:	GFKNVEMMTARGF-NH_2.
Synthesis and target:	Allatotropins are released by neurosecretory brain neurons in the corpora cardiaca and act on the corpora allata.
Function:	Allatotropins are stimulators of juvenile hormone synthesis.
Receptor:	The allatotropin receptor is a heptahelical GPCR.

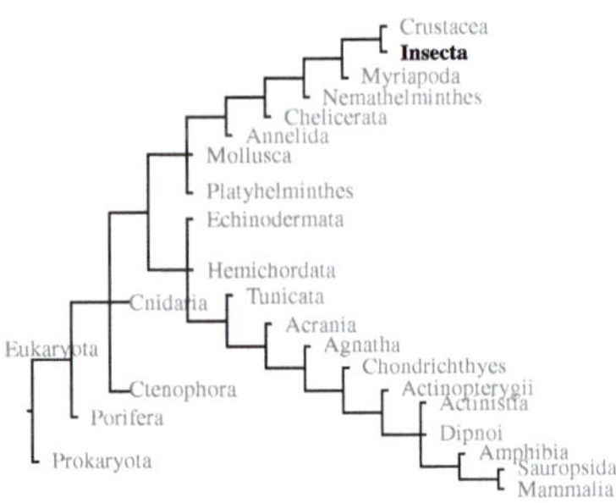

5.7.1.1 Introduction

Allatotropin (Fig. 5.27) is produced in neurosecretory brain cells, released in the corpora cardiaca and acts via the hemolymph in the corpora allata stimulating juvenile hormone biosynthesis.

5.7.1.2 Biochemistry and Genes

The *M. sexta* allatotropin (MAS-AT) is generated by PC1 and a furin-like protease from the precursor. It is C-terminally amidated.

The allatotropin receptor has recently been discovered as GPCR (Yamanaka et al. 2008).

5.7.1.3 Physiology

The balance between ecdysone and juvenile hormone is critical for the outcome of molting. If JH is missing, the imago develops. An additional dose of JH, in contrast,

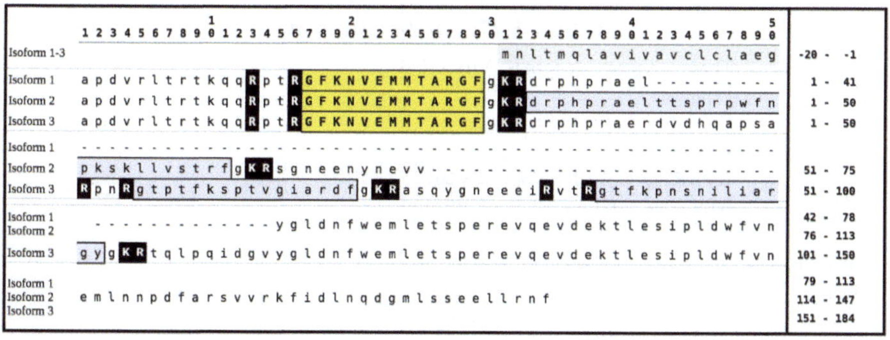

Fig. 5.27 Primary sequence of the allatotropin precursor of the tobacco hornworm (*M. sexta*). Three isoforms derived by alternative splicing give rise to the allatotropin and to additional peptides. The signal peptide is highlighted *light gray*, allatotropin *boxed, bold* and in *uppercase*, and other peptides are *boxed*. RxxR- and dibasic peptide motifs for prohormone convertases are *inverted*; glycine residues in front of **KR** motifs are oxidized to amides (Source: P21786)

blocks metamorphosis and leads to an additional instar. Allatotropin regulates the JH production in a thus far unknown way.

Allatotropin is made in a few neurosecretory cells of the insect brain. The control of its release in the corpora cardiaca has not been sufficiently analyzed.

In addition, allatotropin is expressed during the development of the antennal lobe in *M. sexta* (Utz et al. 2008) and further acts on gut contractions.

5.7.1.4 Phylogeny

The known allatotropins were discovered in butterflies, flies, and beetles. The five C-terminal amino acids **TARGF**-NH_2 are conserved, a synthetic **TARGF**-NH_2 has been found to stimulate juvenile hormone biosynthesis, too. In further insects (e.g., in locusts) the allatotropin activity is present: brain peptides stimulate JH synthesis and the peptide has not yet been isolated. The *B. mori* allatotropin is identical to that of *M. sexta*. For *D. melanogaster* an investigation in the protein database looking for the peptide **TARGFGKR** did not succeed. **TAMRGF** could be found, but the prohormone convertase motifs were lacking.

In two annelids allatotropin-like precursors have been found, however, from these precursors no peptides can be cleaved that might resemble allatotropin (GenBank P46978 and P46980).

5.7.2 Allatostatins

5.7.2.1 Introduction

With use of isolated ıcorpora allata in vitro, three peptide types inhibiting JH biosynthesis have been isolated from different insect species: those from cockroaches were labeled type A, those from crickets type B, and lepidopteran allatostatins as type C.

Fact sheet 5.27: Allatostatins (AST)

Sequence: Three types: A-type: Y/F-X-FG-L/Iamide
B-type: WX_6Wamide
C-type:
EVRFRQ[C]YFNPIS[C]F-OH
with an intramolecular disulfide bridge.

Synthesis and target: The action of allatostatins on the juvenile hormone biosynthesis takes place at the axon ends of AST neurons in the corpora allata.

Function: AST inhibit juvenile hormone biosynthesis and are involved in additional regulations.

Receptor: Type-A receptors are GPCR of the galanin family, a type-B GPCR belongs to the bombesin family, and type-C receptor are members of the somatostatin/opioid GPCR family.

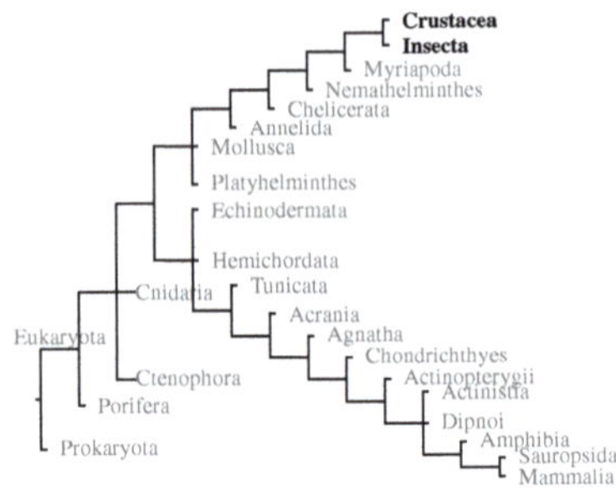


```
                  1                   2                   3                   4                   5
1 2 3 4 5 6 7 8 9 0 1 2 3 4 5 6 7 8 9 0 1 2 3 4 5 6 7 8 9 0 1 2 3 4 5 6 7 8 9 0 1 2 3 4 5 6 7 8 9 0
                                                              m l y s s l p v c f l v l g a a l c a   -19 -  -1
p e r m q n e a e p h d l q p h e a e p h s d h v a p l a K R S P H Y D F G L g K R a y s y v s e y     1 -  50
K R L P V Y N F G L g K R S R P Y S F G L g K R s v d e d q s n d e q q l t t s d l d q a a l a e l    51 - 100
f d q y d d a e K R A R P Y S F G L g K R f a d d e t s e e K R A R A Y D F G L g K R L P M Y N F G   101 - 150
L g K R A R S Y N F G L g K R Y S K F N F G L g K R E R D M H R F S F G L g K R s g d d v s a d d s   151 - 200
d n y f d v                                                                                          201 - 206
```

Fig. 5.28 Primary sequence of the *Helicoverpa armigera* allatostatin precursor (type-A): Allatostatins (pattern **F/Y-x-FGL**-NH_2) are on *yellow background* and in *uppercase*; additional peptides are *in lowercase*. Dibasic prohormone convertase peptide motifs are *inverted*; glycines in front of those are all oxidized to amides (Source: O44314)

In vivo, however, not all of these peptides are functional or not always (see Audsley et al. 2008).

5.7.2.2 Biochemistry and Structure

Type A allatostatins are characterized by the *Y/F-X-FGL*-amide motif and by their origin from large precursors with more than 30 single peptides (Fig. 5.28). Only PC1 acts on the (**KR** motif) to release AST, whereas other peptides from the same precursor are cleaved by PC1 plus a furin-like enzyme.

Type B allatostatins with their $\mathbf{Wx_6W}$-NH_2 have only been found in crickets thus far. Although type A AST act in additional species and insect families, type B AST are only active in crickets. They are related to other, myoinhibitory peptides in other families (Stay and Tobe 2007). It is only in crickets that these type B peptides act allatostatically (Audsley et al. 2008).

Type C allatostatins from different species possess in each case one single sequence: for example, pyro-**EVRYFRQ C YFNPIS C F**-OH in *D. melanogaster* with an intramolecular disulfide bridge and an N-terminal pyroglutamate. Such a peptide is active as allatostatin in moths and mosquitoes.

It is noteworthy that the allatotropin-2 from the moth *Spodoptera frugiperda*, **RVRGNPISCF**, exhibits remarkable C-terminal sequence homology to C-type AST.

The AST receptors are GPCR from different families: type A—galanin receptor family; type B—bombesin receptor family; type C—somatostatin/opioid receptor family.

5.7.2.3 Physiology

In the tobacco hornworm the allatostatic effect has been discovered for a type-C peptide: the JH biosynthesis in isolated corpora allata of fifth instar larvae could be inhibited by the isolated peptide 0–4 h, as well as 24 h after molt and in 3-day-old adults, too (Kramer et al. 1991; Audsley et al. 2008). JH synthesis was inhibited in the same organs of other species, however, not as completely as in *M. sexta*.

The focus on the larval stage where inhibition could be observed is necessary and the pointer to in vivo or in vitro is important because any effect could rarely be observed in vivo. Whereas the *M. sexta* type-C peptide showed activity at concentrations of 10 nM, in the organs of other species levels of 1 μ M were required for inhibition.

It would be relevant, too, to note whether constitutive JH biosynthesis had been inhibited or (only) an experimental increase by allatotropin. In summary, only few species have been analyzed in detail and the findings do not yet appear conclusive (compare the recent reviews by Audsley et al. 2008 and Stay and Tobe 2007).

The immunological characterization of allatostatin receptor is also far from being finished.

5.7.2.4 Phylogeny

Allatostatins have thus far only been found in insects and crayfish, whereas allatostatin receptor-like proteins have been discovered in placozoans, nematodes, and echinoderms.

5.8 Peptide Hormones of Skin: Pigment-Dispersing Hormone

5.8.1 Introduction

Adaption to light in eyes and skin is mediated in crustaceans by two antagonistic peptide hormones: red-pigment-concentrating hormone (RPCH/AKH) and pigment-dispersing hormone (PDH) called pigment-dispersing factor (PDF) in insects. The discovery of an endocrine influence on dark/light adaption stems from Koller and Perkins (reviewed by Rao 2001). In insects PDF expression is directly coupled to neurons of the circadian pacemaker.

Fact sheet 5.28: Pigment-dispersing hormone (PDH)

Sequence: GFKNVEMMTARGF-NH_2.

Synthesis and target: The PDH is a hormone of the eye stalks and acts on the skin of crayfish.

Function: Influenced by PDH the crustacean skin decolorizes.

Receptor: The PDH receptor is a GPCR of the secretin family.

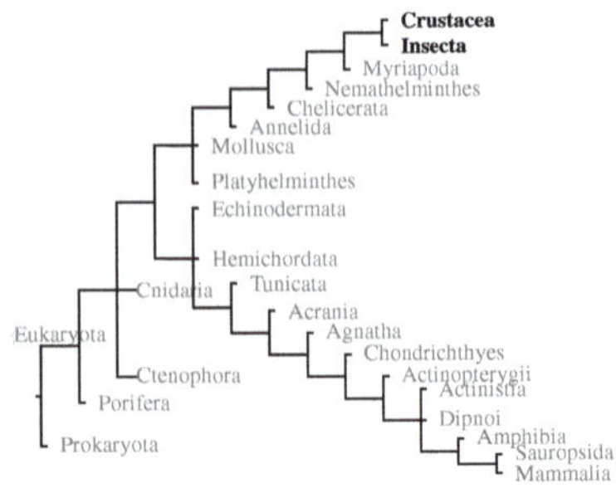


```
                  1                   2                   3                   4                   5
1 2 3 4 5 6 7 8 9 0 1 2 3 4 5 6 7 8 9 0 1 2 3 4 5 6 7 8 9 0 1 2 3 4 5 6 7 8 9 0 1 2 3 4 5 6 7 8 9 0
                                                              m r s a m v v l v l v a m v a v f t r a    -20 - -1
q e l k y p e r e v v a e l a a q i y g w p g s l g t m a g g p h K R N S E L I N S I L G L P K V M    1 - 50
N E A g R R                                                                                              51 - 56
```

Fig. 5.29 Primary sequence of pigment-dispersing hormone (PDH) from spiny-cheek crayfish *O. limosus*. The PDH is *boxed* and in *uppercase*. Dibasic peptide motifs for prohormone convertase are *inverted*; the glycine placed in front is oxidized to amide (Source: P37085)

5.8.2 Biochemistry and Structure

Pigment-dispersing hormones (PDH; Fig. 5.29) are peptides of 18 amino acids with conserved N- and C-termini as well as other conserved amino acids. The consensus sequence for β-PDH found in crustaceans *and* insects is **NSELINSxLxxSxxxxxA**-NH_2. α-PDH has only been found in a few crustacean species; its consensus sequence is **NSGMINSILGIPxVMxxA**-NH_2. It is noteworthy that in the ocean shrimp *Pandalus jordani* in addition to two different α-PDH a β-PDH had been sequenced, too (see Table 5.10).

PDF receptors in nematodes and insects are GPCR of the secretin family; in crustaceans no receptors have been deposited in GenBank.[20]

5.8.3 Physiology

PDH are products of the eye stalk. Most PDH neurons secrete into the neurohemal organ sinus gland; some are, however, not connected to the sinus gland.

In crustaceans PDH serves light/dark adaption. Influenced by PDH and its antagonist PCH pigments in the eye's ommatides are moved and light adapted.

[20] As of September 2014.

Table 5.10 Sequence comparison of PDH from crustaceans and insects

	Sequence
β-PDH	
Crustaceans:	
Uca puligator, Cancer magister, C. maenas, Pastifastacus leniusculus, Callinectes sapidus I	**NSELINSILGLPKVMNDA**-NH_2
Callinectes sapidus II	**-------L--ISAL--E**--NH_2
P. clarkii, O. immunis, O. limosus	**---------------E**--NH_2
Penaeus aztecus, Penaeus vannamei I/II	**----------I-------**-NH_2
Penaeus vannamei III	**-------L---------**-NH_2
Penaeus japonicus I	**----------I----T--**-NH_2
Penaeus japonicus II	**-------------F-I--**-NH_2
Pandalus jordani I	**---------------T--**-NH_2
Armadillidium vulgare	**----------A-R-L-N-**-NH_2
PDF	
Insects:	
Periplaneta americana	**--------------L---**-NH_2
Acheta domesticus	**---I----------L---**-NH_2
Romalea microptera	**---I---------LL---**-NH_2
Carausius morosus	**---------A----L---**-NH_2
D. melanogaster	**---------S---N----**-NH_2
α-PDH	
Crustaceans:	
Pandalus borealis, Pandalus jordani II	**--GM---I--I-R--TE-**-NH_2
Pandalus jordani III	**--GM---I--I----A--**-NH_2
Macrobrachium rosenbergii	**--GM---I--I----AE-**-NH_2

In insects it was shown that PDF neurons are involved in the regulation of the circadian pacemaker. They control daylight-dependent fly activity. PDF has been localized to lateral brain neurons found in larvae in the Bolwig organ and in adult flies in the extraretinal eyelet (Helfrich-Forster et al. 2002).

5.8.4 Phylogeny

GenBank contains PDH/PDF sequences only from crustaceans and insects.

Fact sheet 5.29: Neuropeptides F (NPF)

Sequence: Fig. 5.30.

Synthesis and target: NPF is made in a few neurosecretory cells of the CNS.

Function: NPF is a neurotransmitter and involved in circadian control of foraging in larvae. It plays in additional role in reaction to stressors.

Receptor: The NPF receptor is a heptahelical GPCR of the rhodopsin family.

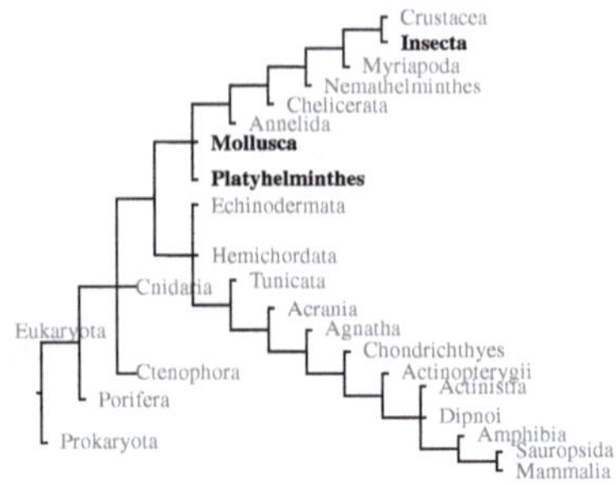

5.9 Other Neuropeptides

5.9.1 Neuropeptide-F—Two Peptide Genes

5.9.1.1 Introduction

Neuropeptide-F is considered as the invertebrate homologue of neuropeptide Y. The sequence homology, however, is five of 36–40 amino acids, one proline residue close to the N-terminus, one or two tyrosines with a conserved spacing, and two arginines at the C-terminal end (highlighted *red* in Fig. 5.30). The analogy, however, extends to NPF/NPY receptor homology and to functional activities. It has been, for example, possible to use anti-PYY antisera (against the vertebrate PYY) to identify neurons that later turned out to express NPF.

There is another "short neuropeptide-F" (sNPF) not to be mixed up with neuropeptide-F (NPF). The four sNPF1..4 peptides in *D. melanogaster* are coded for by a different gene and differentially expressed compared to NPF. Although NPF has been discovered in platyhelminthes, that is, in an original metazoan species, sNPF has only been found in arthropods.

1234567890123456789012345678901234567890	
m a s g t f t q r l l v a l m i f a l i a d l s t l v a a	-29 - -1
R P Q D S D A A S V A A A I R Y L Q E L E T K H A Q H A R P R F g K R g g y l n	1 - 40
p a i f g q d e q e v d w q d s t f s r	41 - 60

Fig. 5.30 Primary sequences of neuropeptide F from *A. aegyptii*. The signal peptide (−29 to −1) is highlighted *light gray*. NPF (*boxed*) is cleaved by PC1. The mature peptide is C-terminally amidated; the [*red highlighted*] amino acids are identical with mammalian neuropeptide Y

Fact sheet 5.30: Short neuropeptide F (sNPF)

Sequence: Fig. 5.31.

Synthesis and target: sNPF is made in neurosecretory cells of the CNS and acts preferentially as a neurotransmitter.

Function: sNPF is expressed in the olfactory sensory cell and is involved in circadian control of larval foraging.

Receptor: sNPF receptor is a heptahelical GPCR.

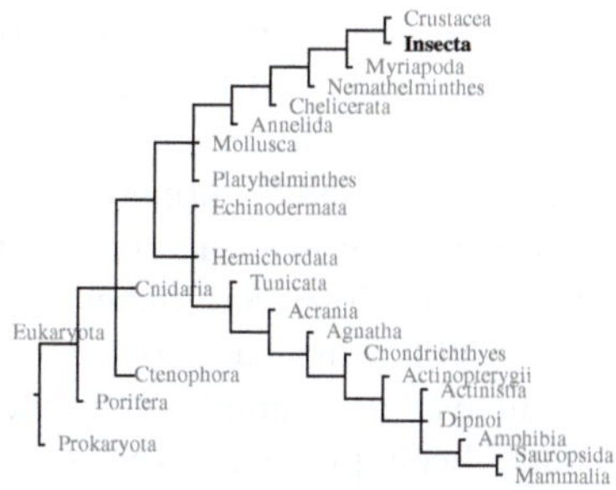

5.9.1.2 Structure and Genes

The NPF peptide is cleaved from the precursor by the signal peptidase and a PC1. It is amidated C-terminally by PHM.[21]

sNPF was originally isolated as NPF-like peptide from beetle brain (Led-NPF; Spittaels et al. 1996). Further sNPF have been identified in several other insect families. sNPFf of *A. gambiae* are cleaved from the precursor by the signal peptidase, a PC1/PC2, and a furin-like peptidase (see Fig. 5.31). C-termini are amidated. It is not known which enzyme(s) release sNPF2 or sNPF4. The NPF and the sNPF receptors are heptahelical GPCR of the rhodopsin family with homology to NPY receptors.

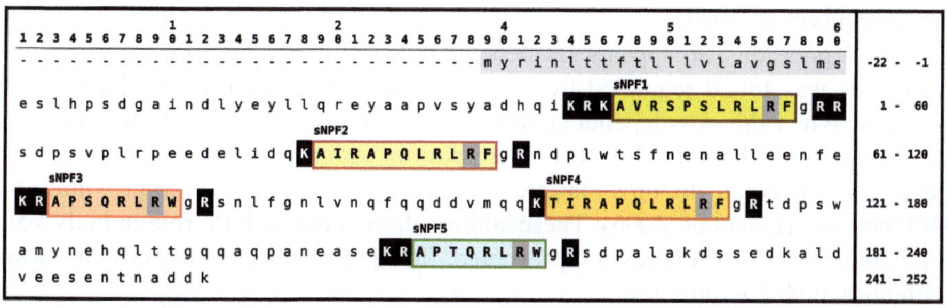

Fig. 5.31 Primary sequence of short neuropeptide F (sNPF) from *A. gambiae*. The signal peptide is highlighted *light gray*. Five sNPF are *boxed*. They are C-terminally cleaved by a furin-like peptidase (motif **RxxR**) and N-terminally by prohormone convertases or other enzymes. Mature peptides are C-terminally amidated

[21] Peptidylglycine alpha-hydroxylating monooxygenase.

5.9.1.3 Physiology

NPF is a neuropeptide found in arthropods, molluscs, and flatworms. In *D. melanogaster* there are four NPF-forming neurons with axons projecting in other brain areas and the ventral nerve cord. Due to its presence in neurohemal organs, for example, in the corpora cardiaca NPF is regarded as a neurohormone. In turbellarians NPF supports organ regeneration: these animals are capable if, for example, the head has been removed to generate fully with the help of NPF (Kreshchenko 2008; Kreshchenko et al. 2008).

The sNPF gene has, however, only been found in insects: flies, beetles, and mosquitoes, for example. It is made in many neurons. A detailed analysis has recently been published (Nässel et al. 2008). Because of its presence in the hemolymph, sNPF is regarded as a neurohormone and as a neurotransmitter, too. The majority of sNPF neurons have been positive for other neurotransmitters. The conclusion thus is that sNPF acts preferentially as a neurotransmitter.

In *D. melanogaster* additional NPF-releasing endocrine cells were detected in the middle gut (Veenstra et al. 2008), sNPF secreting cells, however, could not been found. NPF-secreting cells contained tachykinins, as well; sNPF was discovered in hypocerebral ganglia which innervate the foregut.

NPF as well as sNPF are considered involved in foraging and food intake, in analogy to NPY. sNPF has an additional role in olfaction in flies. NPF and NPF receptor expression studies using transgenes or RNA interference, have demonstrated that without NPF signals flies reject food, whereas an enhancing NPF induced signals resulted in prolonged food intake (Nässel et al. 2008). Yamanaka et al. (2008) have argued that sNPF are coupling food intake to juvenile hormone synthesis. During diapause, NPF had not been expressed in potato beetles (*Leptinotarsa decemlineata*) (Huybrechts et al. 2004).

NPF expression also appears sex dependent: some particular NPF expression in male flies was found regulated by the *transformer* gene with an involvement of *fruitless*. When this sex-dependent NPF expression is disabled, male flies change their courtship behavior. This NPF expression is additionally controlled by genes generating circadian rhythms that are expressed in neurons in close proximity to NPF neurons (Lee et al. 2006). These authors discovered an NPF role in body size regulation: with NPF expression prevented, flies got remarkable larger than those with normal NPF expression.

Looking for additional analogies of NPF and NPY, Dierick and Greenspan (2007) observed that NPF knockout flies were more aggressive than wildtype flies, independently of the effect mediated by an enhanced serotonin level.

5.9.1.4 Phylogeny

NPF has been found in different invertebrates, from flatworms on, in molluscs and in insects. The structural homology, including that of the NPF receptor, to the NPY/PYY proteins and their receptors in vertebrates suggests that this molecule might be present in the entire bilaterian subregnum.

As of today, sNPF is a peptide of insects.

5.9.2 Proctolin

Fact sheet 5.31: Proctolin (PCT)

Sequence: Fig. 5.32.

Synthesis and target: Proctolin is made by neurons in the CNS and other parts of the insect and crustacean nervous system.

Function: Proctolin acts on muscle activity and enhances the frequency of muscle contractions.

Receptor: The *D. melanogaster* proctolin receptor is heptahelical GPCR.

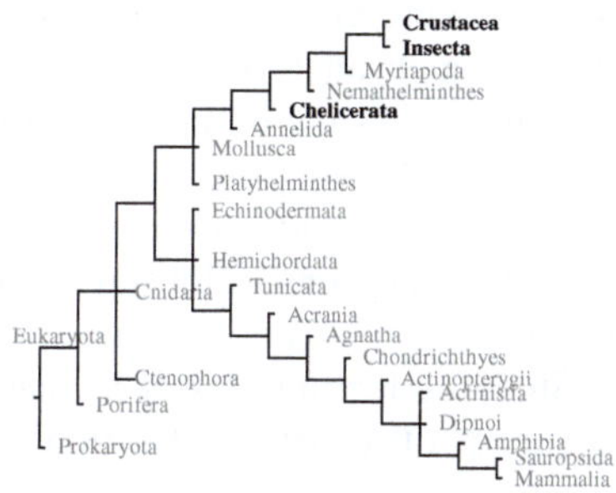

1 2 3 4 5 6 7 8 9 0 1 2 3 4 5 6 7 8 9 0 1 2 3 4 5 6 7 8 9 0 1 2 3 4 5 6 7 8 9 0 1 2 3 4 5 6 7 8 9 0 (tens: 1, 2, 3, 4, 5)	
mgvprshgtgigcgsghrwllvwmtvlllvvpphlvdg	-28 - -1
RYLPTRshgddldklrelmlqilelsnedpqqqqqqqqqqhpqlrlhne	1 - 60
atggsssssninnprvsngnsnaawlqklsamgaldelggdgarfgpnyg	60 - 102
ry	

Fig. 5.32 Primary sequence of proctolins from *D. melanogaster*. The signal peptide is highlighted *light gray*. Proctolin (on *yellow* background and in *uppercase*) is cleaved C-terminally by a furin-like peptidase (motif **RxxxxR**) and N-terminally by the signal peptidase. (source: GenBank CAD30643; the cleavage site of the signal-peptidase was estimated using the program Signal 3.0 (http://www.cbs.dtu.dk/services/SignalP/))

5.9.2.1 Introduction

Cockroach proctolin was the first neuropeptide ever sequenced (Starratt and Brown 1975). Using anti-PCT antisera PCT neurons were identified first in insects, later in crustaceans and even in the horseshoe crab *Limulus polyphemus*, a living fossil.

5.9.2.2 Structure and Genes

Proctolin is a simple pentapeptide without protection at the C-terminus or N-terminus. The molecule is instead by itself an inhibitor of enkephalinases. PCT is cleaved from the precursor by the signal peptidase and a furin-like peptidase (recognition motif **RxxxxR**). The last years have seen an increase in sequenced proctolin precursors.

5.9.2.3 Physiology

PCT acts via a GPCR from the rhodopsin family (in *D. melanogaster*). The attribute of PCT most analyzed is its role in muscle contractions. Using PCT and anti-PCT antibodies the immunohistology to identify PCT neurons has been developed (Bishop et al. 1981; Eckert et al. 1981). PCT neurons have been found in the brain and in thoracic and abdominal ganglia. Siwicki et al. (1985) identified more than 1,400 PCT neurons in lobster. Apart from its activity on muscles, PCT act on the oviduct, too.

Studying leg muscles of locusts Evans (1984) observed that PCT enhances the frequency of muscle contractions by a blockage of potassium influx and the thus-enhanced membrane resistance.

5.9.2.4 Phylogeny

GenBank contains PCT sequences from limulus (Chelicerata), crustaceans, and insects.

5.10 Summary and Overview

Table 5.11 summarizes several important features of invertebrate neuropeptides. All these neuropeptides are active as hormones; they are secreted by neurosecretory cells into the hemolymph and act on distant cells. Those neuropeptides synaptically active have at least endocrine active analogues in vertebrates, the reason to present them here.

Table 5.11 Neuropeptides in insects, crustaceans, and other protostomes

Name	Site of synthesis	Synaptically/endocrine	Vertebrate analog
CHH	X organ	Endocrine	
Bombyxin/ILP	CNS/GI tract	Endocrine	Insulin
AKH	CC	Endocrine	
CAP	Neurons	Paracrine/synaptically	Oxytocin/vasopressin
NdWF	Neurons	??	
Enterins	Neurons	Endocrine/synaptically	
MIP/AMRP	Neurons	Synaptically	
DiuH	Neurons	Endocrine	
Pyrokinin/PBAN	Neurons	Endocrine	
Orcokinins	Neurons	??	
Leucokinins	Neurons	Endocrine?	
TRP	Neurons		Tachykinin
Sulfakinins	Neurons	Synaptically/endocrine	Gastrin, CCK
PTTH	Neurons	Endocrine	
PSP/SP	Male accessory gland	Pheromonal	
GIH/VIH			
TMOF	Gonads	Endocrine	
Nebcolloostatin	??	??	
MIH	X organ	Endocrine/paracrine	
Corazonin	CNS		
ETH	Inka cells	Endocrine	
EH	CNS	Endocrine	
Bursicon	Neurons	Endocrine	
RFamide/FMRFamide	Neurons	Synaptically	Neuropeptide FF
Allatotropin	Neurons	Endocrine	
Allatostatin	Neurons	Synaptically/paracrine?	
PDH	Neurons	Endocrine/synaptically	

Hormones from Mevalonate: Juvenile Hormone and Steroid Hormones 6

Contents

6.1 Introduction 192
6.2 Pathways to Juvenile Hormone and Steroids 192
6.2.1 JH Synthesis 192
6.2.2 From Farnesyl Pyrophosphate via Squalene to Cholesterol 194
6.3 Juvenile Hormones 196
6.3.1 Introduction 196
6.3.2 Biochemistry and Structure 196
6.3.3 Physiology 199
6.4 Steroids 201
6.5 Steroidogenic Enzymes 203
6.5.1 CYP11A1 203
6.5.2 HSD3B 204
6.5.3 CYP17 205
6.5.4 HSD17B 206
6.5.5 5α Reductase 211
6.5.6 CYP21 213
6.5.7 CYP11B1 214
6.5.8 CYP11B2 215
6.5.9 CYP19 216
6.5.10 Concluding Remark 220
6.6 Progesterone 221
6.7 Sex Hormones 223
6.7.1 Gonadal Development 224
6.7.2 Gonadal Hormone Production 225
6.7.3 Androgen Production in the Adrenal Glands 225
6.8 Corticoids 226
6.8.1 Cortisol 226
6.8.2 Aldosterone 228
6.9 Brain Steroids 229
6.10 Additional Steroid Hormones 230

The original version of this chapter was revised.
An erratum to this chapter can be found at DOI 10.1007/978-3-319-15060-4_17

B. Kleine, W.G. Rossmanith, *Hormones and the Endocrine System*,
DOI 10.1007/978-3-319-15060-4_6

6.10.1 Insect and Crustacean Steroid Hormones........ 230
6.10.2 Steroids in Plants and Fungi........ 230
6.11 Dihydroxyvitamin D_3 (Calcitriol)........ 232
6.11.1 Vitamin D_3 Synthesis........ 233
6.11.2 Vitamin D Uptake from Food........ 235
6.11.3 Synthesis of 25-Hydroxyvitamin D_3........ 235
6.11.4 Synthesis of Calcitriol........ 235
6.11.5 Calcitriol Degradation........ 236

6.1 Introduction

With mevalonic acid as the common origin in vertebrates and invertebrates, a variety of hormones is synthesized from isoprenyl oligomers, squalene and cholesterol. Most relevant for vertebrates are steroid hormones based on the polycyclic sterane frame. In arthropods, however, noncyclic juvenile hormones are derived from farnesoate or farnesoic acid methyl esters and ecdysteroids are based on cholesterol.

All the metabolic steps involved depend on a variety of enzymes. These enzymes are not present in the same cells, in the same cellular compartment; steroidogenic enzymes are expressed in different cells and targeted individually to predetermined cellular compartments. This is exemplified in the hormone interplay of mother and fetus where some precursors have to cross the placental barrier in order to become fully metabolized.

The identification of steroid hormone in, for example, some mussel extract does not at all argue for a biosynthesis of this hormone by the mussel itself; it could have been taken up with the food. Likewise is the finding of modification by testosterone using mussel tissue no argument for testosterone formation by the mussel. This finding only says that steroids such as testosterone or related molecules might be modified by an enzyme from the mussel. For testosterone biosynthesis from cholesterol there are at least five enzymes necessary, at least in vertebrates, and none of these has ever been identified in molluscs. This said, there is no formal proof of testosterone or estradiol biosynthesis apart from vertebrates. The necessary enzymes are missing in invertebrates at least in those fully sequenced thus far and the isolation of biosynthetic intermediates has not been published.

To give the full image, we present the juvenile hormone and steroid-forming enzymes, their topological distribution, the hormone structures, and the actions of individual hormones. In addition ecdysones and some plant and fungal steroids are listed.

6.2 Pathways to Juvenile Hormone and Steroids

6.2.1 JH Synthesis

The starting molecule for juvenile hormone (JH) and steroid biosynthesis (Schooley and Baker 1985) is acetyl-coenzyme A (acetyl-CoA) which is converted by the acetoacetyl-CoA-thiolase into acetoacetyl-CoA and by the action of 3-hydroxy-3-methylglutaryl-CoA synthase into (HMG-S) hydroxymethylglutaryl-CoA. From

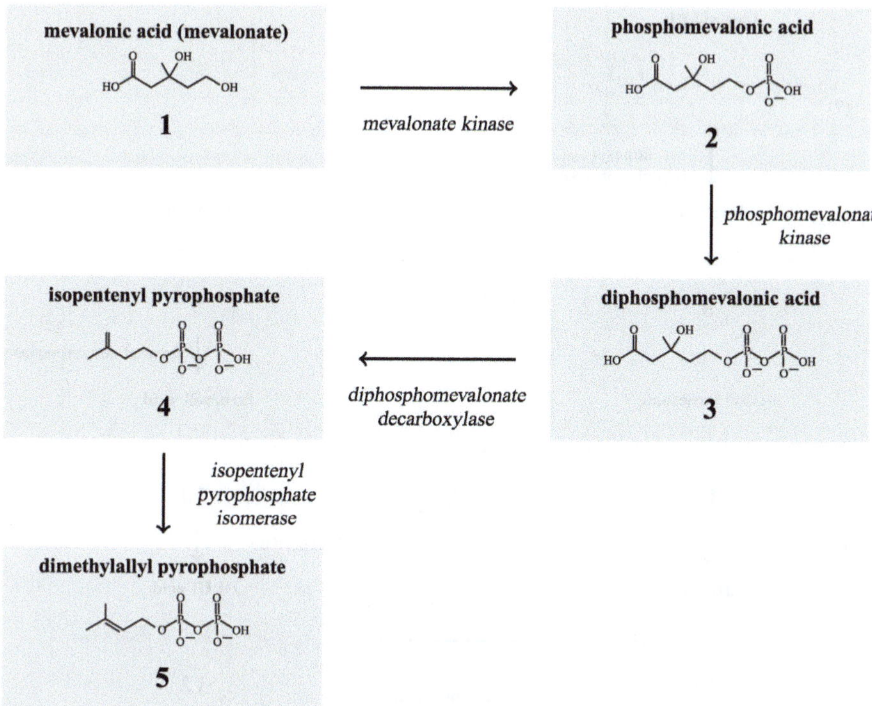

Fig. 6.1 Synthesis of isopentenyl pyrophosphate and dimethylallyl pyrophosphate

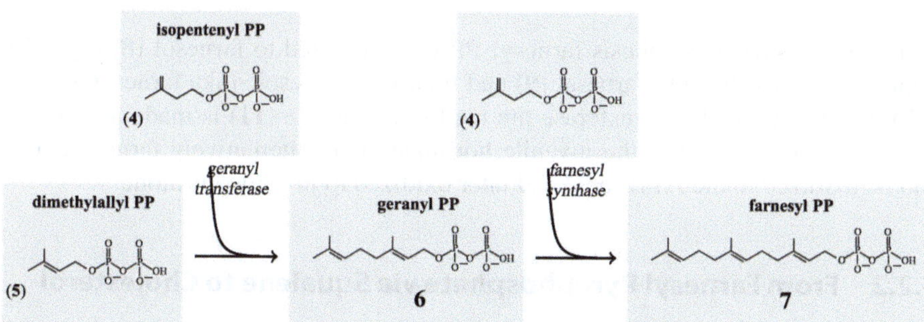

Fig. 6.2 Stepwise addition of isopentenylpyrophosphate generates farnesyl pyrophosphate

this substance the 3-hydroxy-3-methylglutaryl-CoA reductase (HMG-R) generates mevalonate (**1**; Fig. 6.1). With the mevalonate kinase, the phosphomevalonate kinase and the diphosphomevalonate decarboxylase isopentenyl-pyrophosphate is generated (isopentenyl-PP; **4**).

The isopentenyl-isomerase generates finally dimethylallyl-PP (**5**, Figs. 6.1 and 6.2). From dimethylallyl-PP plus isopentenyl-PP geranyl-PP (**6**) is derived. Using this geranyl-PP plus isopentenyl-PP the farnesyl synthase builds farnesyl-PP (**7**; Fig. 6.2).

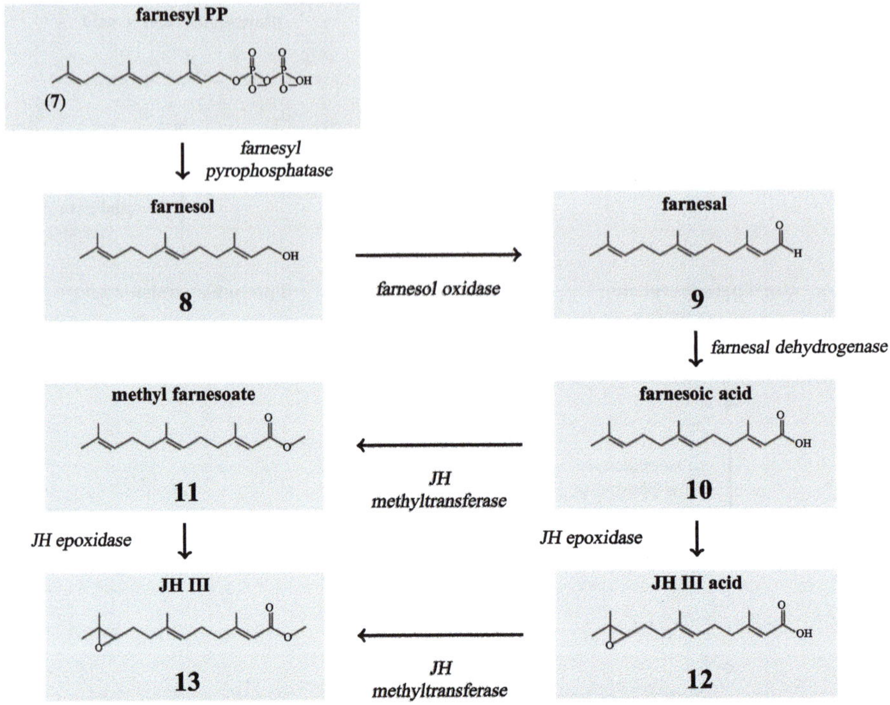

Fig. 6.3 Synthesis of juvenile hormone (*JH*) III from farnesyl pyrophosphate

During the JH biosynthesis farnesyl-PP is hydrolyzed to farnesol (**8**; Fig. 6.2), which is then oxidized to farnesal (**9**) and finally to farnesoic (aka farnesenic) acid (**10**). By the JH-methyl transferase the methyl farnesoate (**11**) is made and by the JH-epoxidase oxidized to the juvenile hormone (**13**). Alternatively farnesoic acid can be oxidized to the JH III acid and later oxidized to juvenile hormone.

6.2.2 From Farnesyl Pyrophosphate via Squalene to Cholesterol

For steroid generation two farnesyl-PP are coupled by the squalene synthase to squalene (**14**; Fig. 6.4). After epoxidation this squalene is rearranged by the oxido-squalene synthase to lanosterol, which now has the four cyclic rings characteristic for steroids. Lanosterol is oxidized by several oxidation steps to cholesterol, the base of all steroids, or to another compound-like vitamin D_3. Dimethylcholestatrienol (**17**), made in the ovarian follicle by the lanosterol-14α-demethylase (CYP51), is a paracrine stimulator of ovum maturation. From 7-dehydrocholesterol (**18**) branches off the biosynthesis of vitamin D_3 and related compounds (Fig. 6.4).

The 14α-demethylase is the cytochrome P450 enzyme (CYP51) which has been found with identical function in all living phyla: in bacteria, plants, and animals.

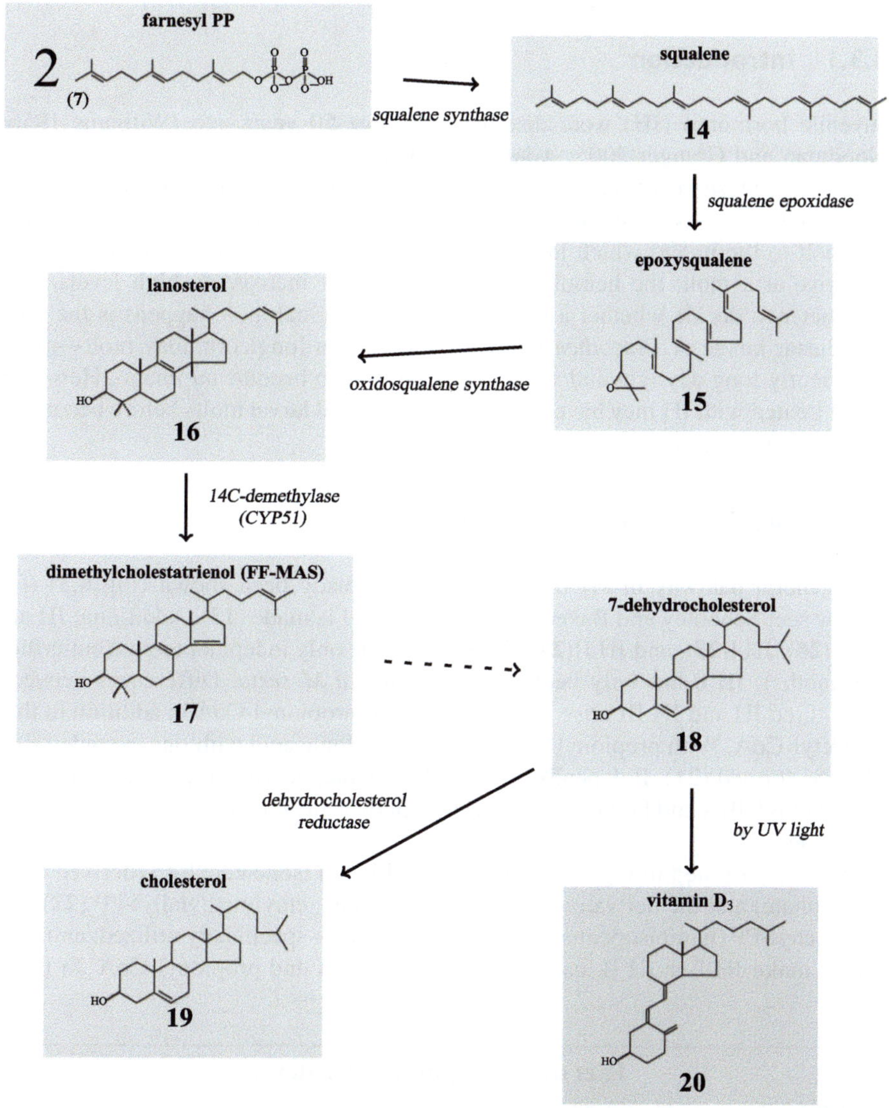

Fig. 6.4 Synthesis of cholesterol from squalene. *FF-MAS* follicular fluid meiosis-activating sterol

JH and steroids are terpenes: monoterpenes are derived from two isopentenyl building blocks, sesquiterpenes[1] such as farnesol of 15 carbon atoms; squalene or lanosterol are triterpenes with 30 carbon atoms.

[1]Sesqui: one and a half.

6.3 Juvenile Hormones

6.3.1 Introduction

Juvenile hormones (JH) were discovered almost 50 years ago (Williams 1956; Goodman and Granger 2005; Adams 2003). They are the product of the corpora allata. All ecdysozoans[2] have to molt in order to grow. In insects there are, however, two different molts: first the molt between two larval stages (instars) and then the molt to the imago which involves in holometabolic insects pupation. In order to arrive at a molt, the hemolymph ecdysone must increase to high levels. The parameter to decide whether a larval molt or a imaginal molt happens is the JH[3]: fifth instar larvae of *Psacothea hilaris*, an east Asian longicorn, may molt—given sufficiently long daylight and warm temperature—to become an imago. However, when treated with JH they have to pass two additional larval molts before becoming an imago (Munyiri and Ishikawa 2004).

6.3.2 Biochemistry and Structure

The general pathway of JH biosynthesis has already been shown (Fig. 6.3) (for review see: Schooley and Baker 1985). Thus JH III is made (**13**). Additional JH are JH 0 (**26**), JH I(**25**), and JH II(**24**). These are made only in lepidopterans (butterflies and moths); JH 0 has only been found in eggs of *M. sexta*. Differences between these three JH and JH III arise from the usage of propionyl-CoA in addition to that of acetyl-CoA. With propionyl-CoA a homomevalonic acid with one extended side chain is formed (**21**). It depends on a single, double, or triple usage of the derived ethylmethylallyl- and homoisopentenyl pyrophosphate whether JH II, JH I, or JH 0 are made.

As demonstrated in Fig. 6.5 dimethylallyl-PP and isopentenyl-pp (derived from mevalonate) and the derivatives of homomevalonate, ethylmethylallyl-PP (**22**) and isohexenyl-PP (homoisopentenyl-PP; **23**), are species-specifically utilized: animals, which make JH I or JH II, can use both acetyl-CoA and propionyl-CoA. In those

Fact sheet 6.1: Juvenile hormone

Structure:	Fig. 6.3: **13**.
Synthesis and target:	JH is synthesized in the corpora allata from mevalonate. Its targets are all tissues.
Function:	In larvae JH ensures molting to an additional larval stage. In adults it supports vitellogenesis and the reproductive cycle.
Receptor:	The transcription factor methoprene-tolerant has lately evolved as the JH receptor.

[2]Including the Panarthropods with insects and crayfish, Tardigrada, Onychophora, and the Cycloneuralia with nematodes, Nematomorpha, Kinorhyncha, Priapulida, and Loricifera.

[3]Juvenile hormone.

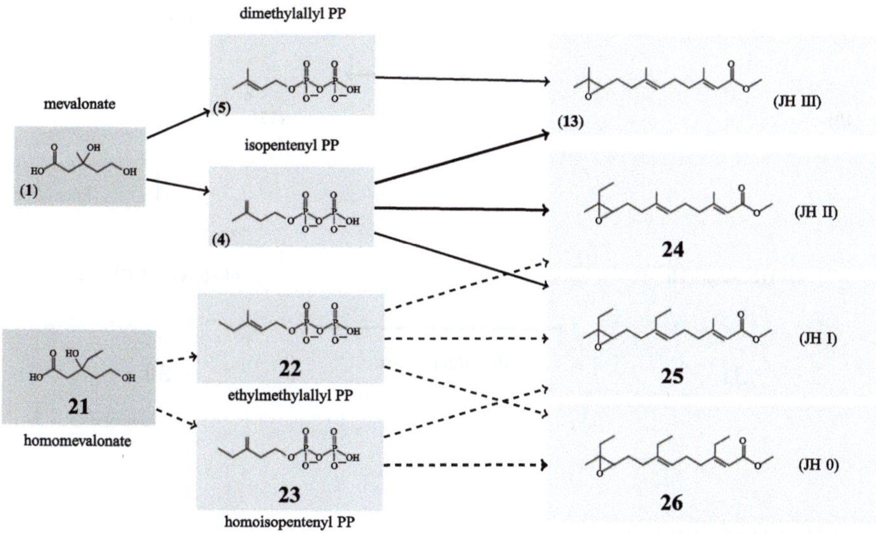

Fig. 6.5 The usage of propionyl-CoA (in Lepidoptera) leads to three additional juvenile hormones

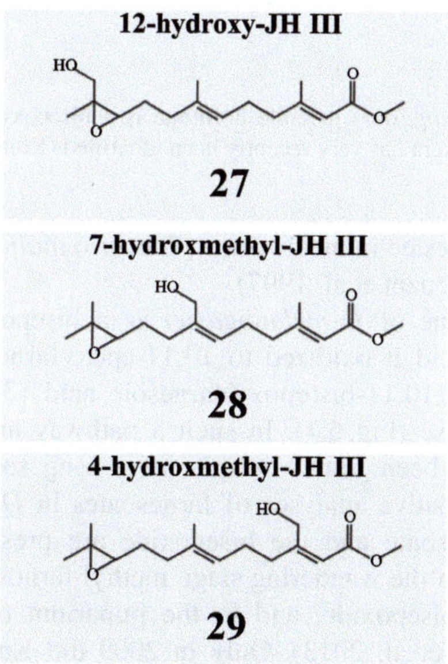

Fig. 6.6 Locusts possess hydroxylated derivatives of JH III

synthesizing JH I, the geranyl transferase adds isohexenyl-PP to ethylmethylallyl-PP, and during JH II synthesis, isoprenyl-PP is added to ethylmethylallyl-PP. These steps distinguish JH I and JH II synthesis.

JH diversity is further augmented by side chain hydroxylation, for example, in migratory locust (Fig. 6.6) where any of the three side-chains can be hydroxylated.

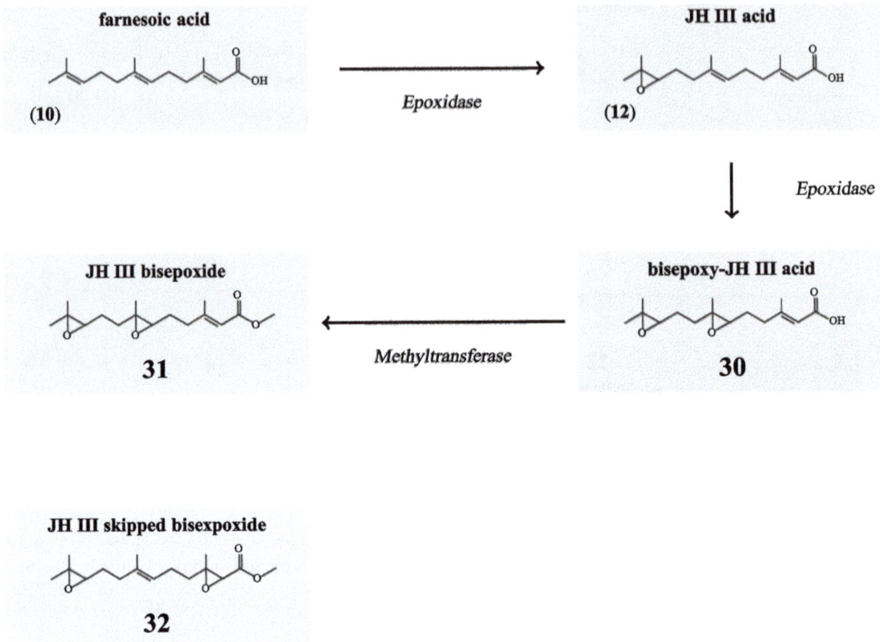

Fig. 6.7 Fruit fly (*D. melanogaster*) juvenile hormone is a bisepoxy-JH III (**31**). The skipped bisepoxide **32** from Heteroptera has very recently been identified (Kotaki et al. 2009)

12-hydroxy-JHIII (**27**) tested in mealworms (*Tenebrio molitor*), was 100-fold more potent than JH III (Darrouzet et al. 1997).

The juvenile hormone of *D. melanogaster* is a bisepoxide (**31**; Fig. 6.7). In a first step farnesoic acid it oxidized to 10,11-epoxyfarnesoic acid (**12**), and in the second step to 6,7;10,11-bis(epoxy)farnesoic acid (**30** before being esterified by methyltransferase (Fig. 6.7). In such a pathway any JHIII would not be produced that had not been found in spite of a long search; very new orders of magnitude more sensitive analyses of farnesoates in *D. melanogaster* showed now that methyl farnesoate and the bisepoxide are present in excess in third instar larvae, whereas in the wandering stage methyl farnoseate in excess of some JHIII and almost no bisepoxide, and in the puparium only methyl farnesoate could be found (Jones et al. 2013). Only in 2009 did Kotaki et al. identify the structure of heteropteran JH as a stereoisomer of the Drosophila JH, and called this bisepoxide juvenile hormone skipped bisepoxide (JHSB$_3$; **32**). The slight difference in the enzyme specificity should help to evaluate the stereospecificity of the epoxidase.

Degradation of JH has two components: hydrolysis by an esterase to the free acid (**12**) and/or hydrolysis by an epoxide hydrolase to the diol derivative (Fig. 6.8). JH III diol (**34**)can be further phosphorylated by a kinase (**35**). Whether diols or diol phosphate are still biologically active has not been analyzed.

In the hemolymph, JH is transported bound to JH binding protein (JHBP).

The juvenile hormone of crustaceans is methyl farnesoate (**11** which is made in the mandibular organ under the control of neuropeptides.

6.3.3 Physiology

Insect juvenile hormones have different functions in the larval and the adult animal. In the larval life JH inhibit precocious molt to the adult imago. Up to 15 molts are required in some species before the imago might arise. In the intensively studied butterflies or moths, for example, metamorphosis happens after the fifth instar. It can be postponed by experimental application of JH. For the regulation of JH synthesis stimulating allatotropins, inhibiting allatostatins, and JH-degrading enzymes such as esterases or epoxide hydrolases play their roles (Fig. 6.8).

The interactions of JH and ecdysons are discussed later.

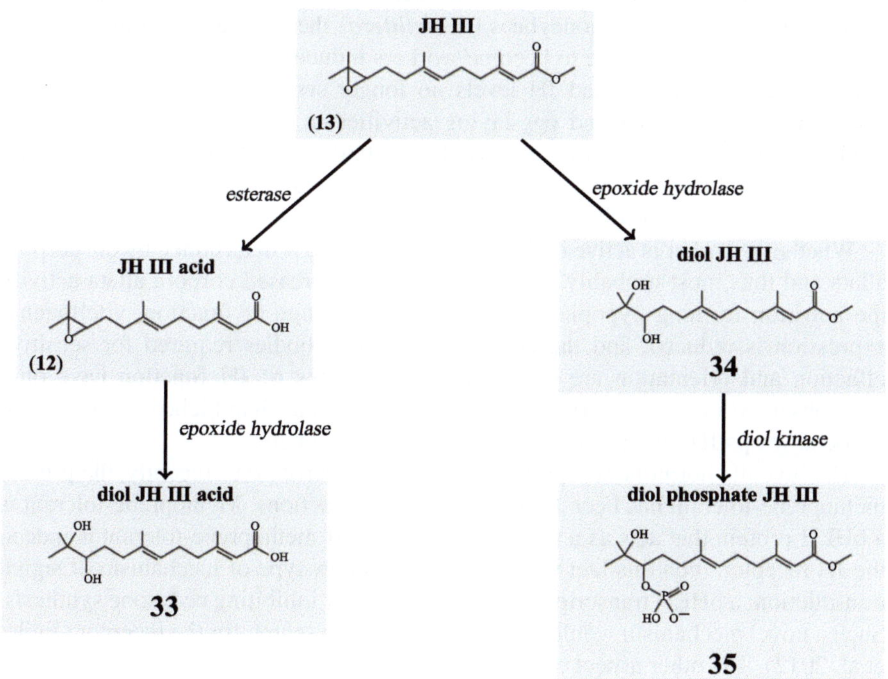

Fig. 6.8 Degradation of juvenile hormone

Beyond their role in molting and pupation, JH exhibits several additional functions in the adult animal. In order to arrive at the final molt and at eclosion, the JH level in the hemolymph must be reduced to zero. After eclosion, however, the corpora allata again start JH synthesis and release. Within one day after eclosion JH levels in the hemolymph are again measurable. This JH simulates vitellogenin transcription, which protein is directly involved in female reproductive cycles. JH acts on the fat body. Furthermore, JH facilitates vitellogenin incorporation into competent follicles (Hartfelder 2000).

Crickets (*Gryllus rubens*) exhibit two polymorphic types: long-winged, flying (LWF) and short-winged nonflying (SWN) crickets; these differ mainly in JH esterase expression (see Fig. 6.8). Thus it could initially be explained how small differences in JH levels in genetically identical crickets lead to two polymorphic phenotypes. In a particular phase JH titers in SWN crickets were lower than in LWF ones. A time-dependent, diurnal JH concentration in the hemolymph of LWF has been observed with fifteenfold to twentyfold JH increase in the light period compared to dark period whereas in SWN animals no difference due to a dark/light cycle could be found. The picture gets more complex by the finding that prolonged elevated JH levels cause the degradation of flight muscles. It is obvious that there is a narrow time window where LWF have to control their JH concentrations without losing their flight capacity (Zhao and Zera (2004).

In social insects the state of an individual within the colony depends on its hemolymph JH levels. In honeybees (*A. mellifera*) the different feeding of larvae to become queens and those to become workers induces enhanced JH levels in the queen larvae. These elevated JH levels no longer serve sex organ shaping. The queen suppresses sexual and egg-laying activities in worker bees by release of a pheromone: (*E*)-9-oxo-2-decenoic acid (Hartfelder 2000). In adult bees, flying activity is stimulated by JH: early in life in queens and drones, later in life for workers to get ready to collect nectar and pollen.

Whether a worker is active in the colony or foraging is determined by the corpora allata and thus most probably JH regulated. With increased corpora allata activity the nutrition forming hypopharynx undergoes a change of function, vitellogenin expression is reduced, and the cerebral mushroom bodies required for sensitive olfaction and orientation are enlarged. Such changes of JH function have only been observed in colony-forming honeybees; in solitary bumblebees, however, or in social wasps JH retains the original gonadotropic functions.

The juvenile hormone receptor has long been evasive. Very recently, the protein methoprene-tolerant has been found to mediate JH actions. Methoprene-tolerant is a bHLH protein that acts as a transcription factor. If methoprene-tolerant is indeed the JH receptor, then this fact would constitute a new type of mechanism of signal transduction: a bHLH transcription factor as receptor inhibiting ecdysone synthesis. Such a novel mechanism would also explain the long search for the receptor (Jindra et al. 2012). The other aspect of this regulation might be the fact that it antagonizes ecdysone-mediated actions and thus is a negative regulator of ecdysone-induced signal cascades.

6.4 Steroids

Steroids are derivatives of hypothetical sterane (**36**; aka gonane in the English literature; Fig. 6.9) a polycyclic molecule with four rings and 17 carbon atoms.

Steroid hormones can be distinguished by the substituents of the 4-ring structure or by modifications of the ring itself: double bonds or one aromatic ring instead of an aliphatic one. Cholesterol is formed from squalene (**14**) which itself is built from mevalonic acid (**1**) via geranyl pyrophosphate (**6**) from farnesyl pyrophosphate (**7**) cyclized into lanosterol (**16**) (Fig. 6.4). Methyl groups in positions 4 and 14 are removed by cytochrome P450 (CYP) monooxygenases described below. The last intermediate is 7-dehydrocholesterol (**18**; see Fig. 6.26), which can be modified to vitamin D_3.

Vertebrate steroid hormones belong to five classes: gestagens such as progesterone with 21 carbon atoms, glucocorticoids such as cortisol and mineralocorticoids such as aldosterone with 21 carbon atoms, androgens such as testosterone with 19 carbon atoms, and estrogens such as estradiol with 18 carbon atoms. All these molecules are derived from common precursors: pregnenolone and progesterone the latter hormone and precursor for other steroids. The first characteristic and common precursor is prenenolone (**37**) which is derived from the membrane lipid cholesterol (**19**).

Numbers in Fig. 6.10 label the carbon atoms of cholesterol. This labeling is identical in all steroids: 17 carbon atoms in four rings, two methyl groups in positions 18 and 19, and the other carbon atoms in the side-chain.

Cholesterol is an important food ingredient for all those searching for healthy nutrition. Elevated cholesterol levels in blood indicate alimentation with a too-high fat content. Without cholesterol the entire steroid hormone anabolism would

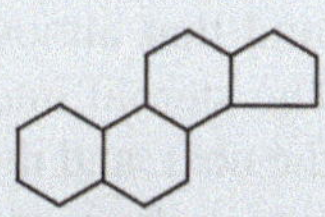

36

Fig. 6.9 Sterane/gonane backbone

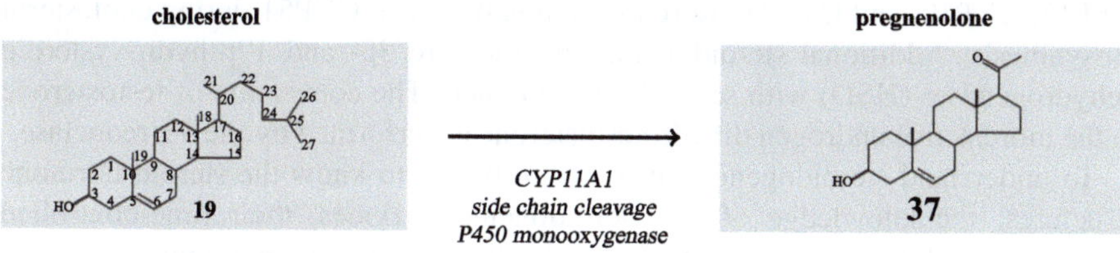

Fig. 6.10 Cholesterol and pregnenolone

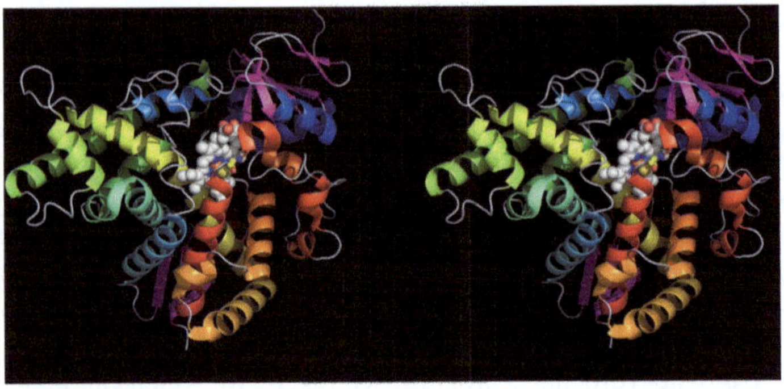

Fig. 6.11 Three-dimensional cytochrome P450 model: the lanosterol demethylase (CYP51). The protein consists of multiple helices (*from blue to red*), two beta sheet areas (*pink*), and the heme structure (*white and red spheres*) complexing a central iron ion (*brown*). Cysteine residue 394 is shown as *spheres*, its *yellow* sulfur atom binding to the iron atom. There are no disulfide bonds stabilizing the structure. CYP51 is widespread in all living organisms; the structure shown is from *Mycobacterium tuberculosis* (Source: Podust et al. (2001); Protein Data Bank entries 1EA1 and PyMOL)

break down, with fatal consequences for metabolic homeostasis and reproduction. In addition cholesterol is indispensable for cell membrane stability.

It has been a long hunt to determine what the characteristic features of steroidogenic cells are. Not long ago, the activator protein StAR (*steroid acute regulator*) was identified as the first prerequisite. StAR itself is very labile: its translation in testis, adrenal, or ovary, as well as in brain is rate limiting for steroid production. StAR transports cholesterol from the cell membrane to the mitochondria, the only compartment where CYP11A1 cleaves the cholesterol side-chain giving rise to pregnenolone.

CYP11A1 is not the sole enzyme of steroidogenesis with such a determined place of action; we have to keep in mind that almost all steroidogenic enzymes act at selected places within the cell. The steroid intermediates thus have to migrate/be transported from one enzyme to the next, and the enzymes retain their positions.

Among the steroidogenic enzymes there are monooxygenase complexed with cytochrome P450 (Fig. 6.11). Such complexes in the liver are involved in detoxification of dangerous compounds. These different monooxygenase are counted CYP1 to CYP21 and further. For steroidogenesis, CYP11A1, CYP11B1, CYP11B2, CYP17, CYP19, and CYP21 are relevant, together with CYP51 during cholesterol biosynthesis. Additional steroidogenic enzymes are 3β- and 17β-hydroxysteroid dehydrogenases (HSD) with several types of each. The conversion of testosterone in the more active androgen dihydrotestosterone is performed by the 5α-reductase.

To understand steroidogenesis it is not sufficient to know the steroid hormone structures: the knowledge of the steroidogenic enzymes, their time-controlled expression, and their compartmental organization are equally important.

6.5 Steroidogenic Enzymes

6.5.1 CYP11A1, Cholesterol Monooxygenase

Fact sheet 6.2: CYP11A1: Cholesterol monooxygenase, 20,22-desmolase

Structure: 3N9Y similar to Fig. 6.11

Gene: Chromosome 15, locus 15:q23, ten exons

Cell type: All steroidogenic cells in adrenal, gonads and brain

Topology: Located in the inner mitochondrial membrane

Function: First steroidogenic enzyme, cleaves cholesterol to pregnenolone

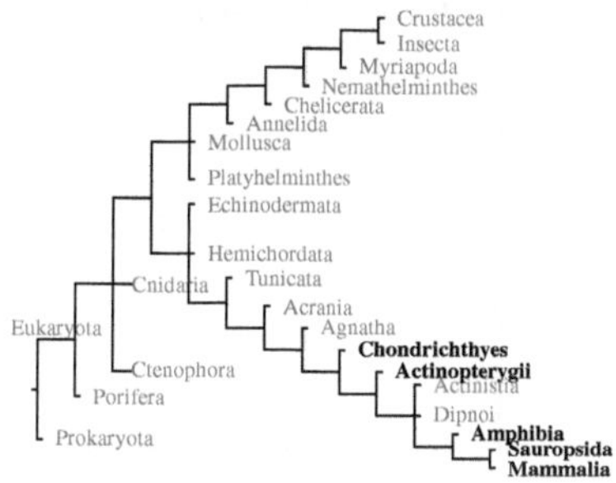

This enzyme (aka P450$_{scc}$ for *side chain cleavage* or 20,22-lyase) cleaves the cholesterol (**19**) side-chain between carbon atoms 20 and 22 (Fig. 6.10). The product of this reaction is pregnenolone (**37**). The earlier name cholesterol monooxygenase indicates that CYP11A1 acts as do all other CYP enzymes by oxidation. The enzymatic complex is assembled by the cytochrome P450 itself, the flavoprotein adrenodoxin reductase, the sulfur–iron-containing adrenodoxin and NADPH as cofactor. The CYP11A1 complex transfers stepwise oxygen molecules onto cholesterol carbon atoms 20 and 22 and takes up the hydrogen atoms, which are stored as water. The necessary electrons for the oxidation stem from NADPH which transfers electrons to the adrenodoxin reductase and in turn reduces adrenodoxin, and then transfers electrons to the cytochrome. From the first oxygen molecule (O_2) one oxygen atom is bound to cholesterol as epoxide, the other atom becomes water. With the second O_2 the epoxide is converted into the diol (plus a water molecule). The third O_2 molecule cleaves the C–C bond and pregnenolone and isocaproic aldehyde are generated, the aldehyde being toxic is removed by the enzyme aldose-ketoreductase 1 B7 (AKR1B7). In very similar ways all CYP enzymes are active. These oxidations are irreversible.

CYP11A1 is an enzyme of the inner mitochondrial membrane (Farkash et al. 1986) as the adrenodoxin reductase. First, StAR has to transfer cholesterol from the cell membrane to the mitochondrium before side-chain cleavage and further steroidogenic conversions can occur.

In GenBank there are sequences from fish, amphibians, reptiles, birds, and mammals. To our surprise two sequences from Trichoplax (an early metazoan precursor) were listed as Cyp11a1. As they are as much related to Cyp24 than to Cyp11a1, the vitamin D_3 hydroxylase (Fig. 6.12) and only have an overall identity of 32 % we do not (yet) consider the putative protein as an active Cyp11a1.

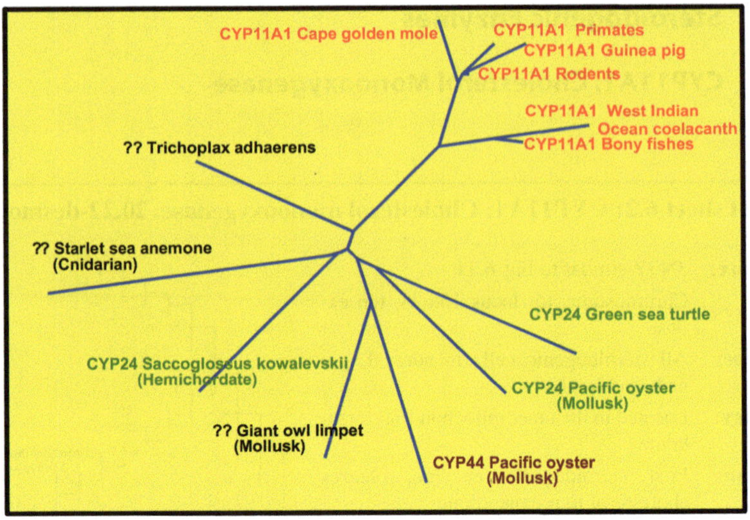

Fig. 6.12 A Cyp11a1 from Trichoplax? The Trichoplax protein TRIADDRAFT_56947 is most probably not Cyp11a1. It is as much related to Cyp11a1 as to different Cyp24.?? denotes a sequence not assigned to any Cyp

6.5.2 3β-HSD

Fact sheet 6.3: 3β-HSD type 1 (HSD3B1) and type 2 (HSD3B2)

Structure: Not yet crystallized.

Gene: Two genes on chromosome 1, locus 1p12, *HSD3B1* four exons, *HSD3B2* five exons.

Topology: In the (smooth) endoplasmic reticulum.

Function: This enzyme converts the 3-hydroxy group of several steroids with a 5,6 double bond into the 3-keto group and moves the double bond to the 4,5 position.

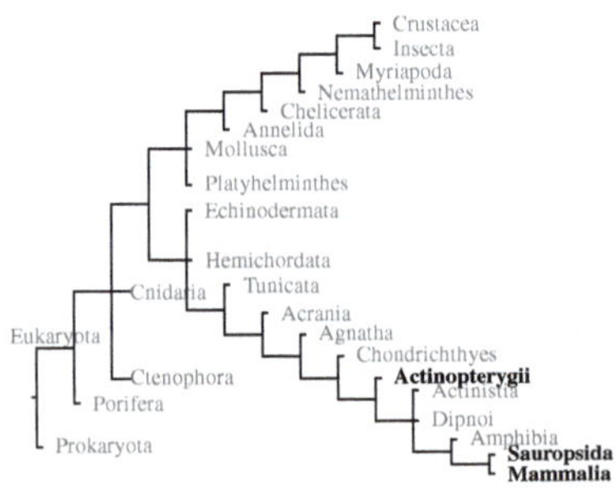

3β-Hydroxy-steroid dehydrogenase-$\delta^{4,5}$-isomerase converts pregnenolone (**37**) to progesterone (**38**) (Fig. 6.13). During this conversion the hydroxyl group on C-3 is oxidized to a keto group and the double bond between C-5 and C-6 is moved to positions 4 and 5.

Both human 3β-HSD isoenzymes are coded for on chromosome 1 and prefer NAD^+ as cofactor. The type 2 enzyme is preferentially expressed in the adrenal and gonads, whereas in placenta and other tissues type 1 is found.

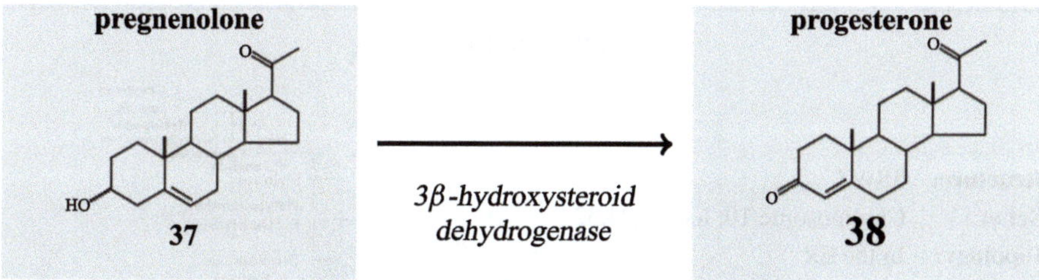

Fig. 6.13 Progesterone formation by 3β-HSD

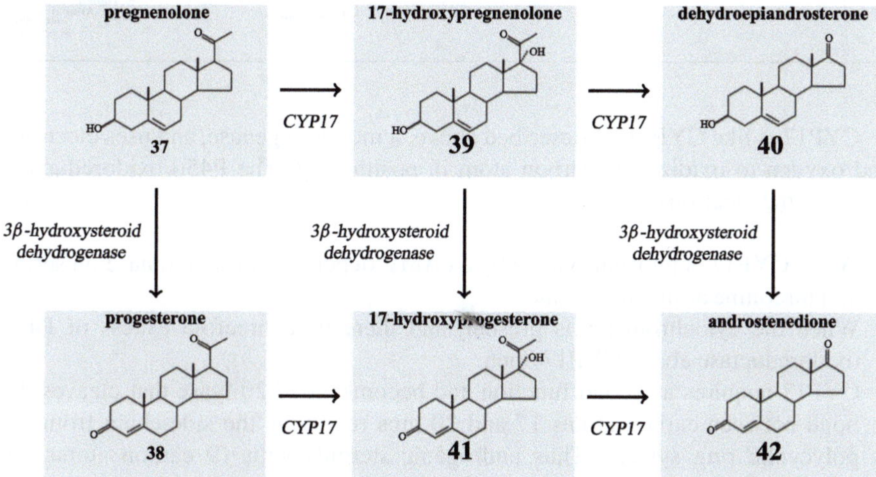

Fig. 6.14 Androgen formation by CYP17 and 3β-HSD

The enzyme is not specific for the conversion of pregnenolone to progesterone. In a similar way, 17-hydroxypregnenolone (**40**) or dehydroepiandrosterone (DHEA (**40**) are concerted 17-hydroxy progesterone (**41**) or androstenedione (43) Fig. 6.14).

3β-HSD is an enzyme of the smooth endoplasmic reticulum.

6.5.3 CYP17: Two Activities in One Enzyme

In the synthesis steps to cortisol and androgens/estrogens pregnenolone and progesterone are oxidized by CYP17. As the name suggests, the target of this enzyme is the C-atom in position 17 of the sterane/gonane backbone (see Fig. 6.10).

The human *CYP17* gene is located on chromosome 10. The enzyme is, like HSD3B, active in the sER. For the conversion of pregnenolone by CYP17 the steroid has to leave the mitochondrion, where it has been generated, and to migrate through the cytosol to the ER.

Fact sheet 6.4: CYP17

Structure: 3SWZ
Gene: Chromosome 10; locus q24.32
Topology: In the ER
Function: Two enzymatic functions: 17-α-hydroxylase and 17,20-lyase

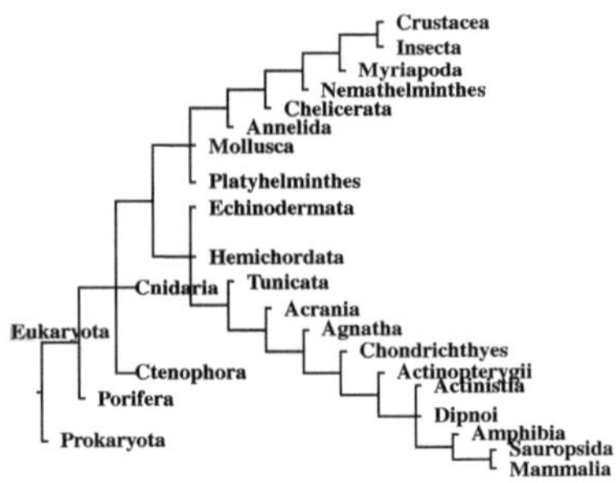

CYP17 is like CYP11A1 described above, a monooxygenase, and uses electrons and oxygen to oxidize the carbon atom in position 17. The P450 oxidoreductase provides the electrons.

- When CYP17 is phosphorylated by a cAMP-dependent protein kinase on serine and threonine amino acids, and
- When the cytochrom B5 is present, and there is a threefold excess of P450 oxidoreductase above CYP17, then
- CYP17 acquires a second function and becomes a 17,20-lyase that cleaves the bond between carbon atoms 17 and 20 thus removing the side-chain from the polycyclic ring system. Thus androgenic steroids with 19 carbon atoms are generated.

In contrast to other vertebrate-specific CYP enzymes, there are CYP17 in almost any deuterostomes and even in bacteria, making it one of the primordial cytochrom P450 monoxygenases such as CYP51.

6.5.4 17β-HSD

17β-Hydroxysteroid dehydrogenases belong with the exception of type 5 to the small dehydrogenase/reductase supergene family (SDR). Type 5 is a member of the aldoketo reductase gene family (AKR). Thus far 14 human enzymes with 17β-HSD activity have been described. Most of these enzymes possess other catalytic activities apart from 17β-dehydrogenase activity. The SDR family of enzymes is of ancient origin and some of the 14 HSD17B genes have homologues in protozoa, plants, and fungi. There are, however, some genes that are not older than the beginning of vertebrate evolution. The SDR enzymes are characterized by the so-called Kallman fold consisting of a bundle of seven parallel sheets, surrounding helices, and binding sites for the NAD/NADP cofactor and steroids.

The intracellular location of these enzymes is diverse. Some of them are cytosolic, some found in microsomes and other vesicles, and others are membrane-bound. It is difficult to see a general rule: whereas 17β-HSD type 5 (HSD17B5)

Fact sheet 6.5: 17β-HSD

Structure:	HSD17B type 1, 4, 5, 8, 10, 11, 14 have been crystallized (see Moeller and Adamski 2009).
Phylogeny:	Type 1: vertebrate; type 2: vertebrate, yeast; type 3: vertebrate; type 4: vertebrate, invertebrate, yeast, plants; type 5: vertebrate, invertebrate, plants; type 6: vertebrate, invertebrate; type 7: vertebrate, yeast; type 8: vertebrate, invertebrate, yeast, plants; type 9: mammalia; type 10: vertebrate, invertebrate, yeast; type 11: mammalia, birds; type 12: vertebrate, invertebrate, yeast, plants, plasmodium; type 13: vertebrate; type 14: vertebrate, invertebrate, plants.
Gene:	See Table 6.1
Topology:	See Table 6.1: subcellular localization and tissue distribution.
Function	17β-HSD catalyze the reduction of the 17-keto group to a 17β-hydroxy group (this is a steroid activating activity, for example, in the case of androstenedione to testosterone or estrone to estradiol conversion) or they catalyze the oxidation of the 17β-hydroxy function to a keto group, which is often the first stage of steroid inactivation. Many of the 14 enzymes have additional catalytic activities.

in the adrenal cortex converts androstenedione (**42**) into testosterone (**44**) in the cytosol, the same conversion in testis is performed by 17β-HSD type 3 in the endoplasmic reticulum.

17β-HSD exert opposite reactions. Some of them (types 1, 3, 5, 7, 12) catalyze the reduction of the 17-keto group to the 17β-hydroxy group. In the case of testosterone and estradiol this is the final reaction to generate these major metabolic and reproductive hormones. Other 17β-HSD (types 2, 4, 6, 8, 11, 12, 13, 14) catalyze the reverse reaction and inactivate testosterone or estradiol by oxidizing the 17β-group. It is worth noting that one enzyme type always catalyzes only one reaction in one direction. Those enzymes that catalyze reductions depend on NADPH as co-factor and those oxidizing the hydroxy group take NAD as co-factor.

The function of 17β-HSD types 1, 3, 5, 7, and 12 is the reduction of the keto group of steroid hormone precursors (Fig. 6.15). In the overall sex hormone generation scheme this is the final or next to final step to arrive at active steroid hormones. The other types catalyze oxidation of the 17β-hydroxy-group which means in almost any tissue inactivating of steroid hormones.

With 14 genes and proteins available in humans, mutations at one gene should be compensated for by one of the other genes/proteins. This is, however, not the case: Defects in *HSD17B3*—the enzyme from testes to convert androstenedione to testosterone—and thus the lack of testosterone in critical phases of sexual development lead to male pseudohermaphroditism (male genotype but external female genitalia). Other HSD17B-linked defects include polycystic ovaries (PCO; HSD17B5; Andersson et al. 1996) and X-linked mental retardation (HSD17B10), the latter defect, however, most probably due to the short-chain 3-hydroxyacyl-CoA

Table 6.1 Human 17β-hydroxysteroid dehydrogenases (Source: Moeller and Adamski 2006; Peltoketo et al. 1999a; Lukacik et al. 2007, PubMed/OMIM, http://www.ensembl.org/Homosapiens

HSD type	Gene name	Other names	Comments	Genomic location	GenBank no.	Subcellular localization	Tissue distribution	Oxidizing/ activating?
1	HSD17B1	E17KSR, EDH17B1, EDHB17, EDH17B2	17β-HSD	17q21.2 (plus one pseudogene)	X13440	Cytosolic	Ovary, placenta, breast	Activating
2	HSD17B2	E2DH, HSD17	17β-HSD, 20α-HSD	16q23.3	L11708	Microsomal	Placenta, liver, GI tract, kidney, uterus, breast, prostate	Oxidizing
3	HSD17B3		17β-HSD	9q22.32	U05659	Microsomal	Testis	Activating
4	HSD17B4	MFP-2, MFE-2, DBP	3-Ketoacyl-DH, 17β-HSD	5q23.1	X87176	Peroxisomal	Widely distributed	Oxidizing
5	AKR1C3	HSD17B5	17β-HSD, 3α-HSD type II	10p15.1	D45850	Cytosolic	Liver, kidney, testis, prostate, adrenal, bone	Activating
6	HSD17B6	HSE, RODH	Retinal reductase/DH	12q13.3	NP_003716.2	(Membrane bound)	(Prostate, liver)	Oxidizing
7	HSD17B7		3-Keto reductase, 17β-HSD	1q23.3, 1q44 (pseudogene) 10p11.2 (short form)	NP_057455.1	(Membrane bound/ associated)	(Ovary, placenta, mammary gland)	Activating
8	HSD17B8		17β-HSD, probably fatty acid CoA-dehydrogenase	6p21.32	D82061	Unknown	Liver, pancreas, kidney, skeletal muscle	Oxidizing
9	RDH5	HSD17B9	Retinal reductase/DH	0135437	12q13.2			

10	HSDH2	HSD17B10, ERAB	17β-HSD, 20β-OH and 21-OH DH, 3α-HSD, 7α-OH- and 7β-OH-HSD	Xp11.22	Q99714	Mitochondrium	Widely distributed	Oxidizing
11	DHRS8	retSDR2, 17β-HSD11, 17β-HSDXI, PAN1B, HSD17B11	17β-HSD	4q22.1	NP_057329.2		Liver, lung, adrenal, steroidogenic cells	Oxidizing
12	HSD17B12	KAR	Ketoacyl-CoA reductase	11p11.2	NP_057226.1		Liver, kidney, heart, skeletal muscle, pancreas, adrenal, pituitary, testis, placenta	Activating
13	HSD17B13	SCDR9, DHRS8	Activity not demonstrated	4q22.1	NP_835236.2	Cytoplasmic	Liver	Oxidizing
14	DHRS10	HSD17B14, retSDR3	17β-HSD	19q13.33	NP_057330.2	Cytoplasmic	Brain, liver, placenta	Oxidizing

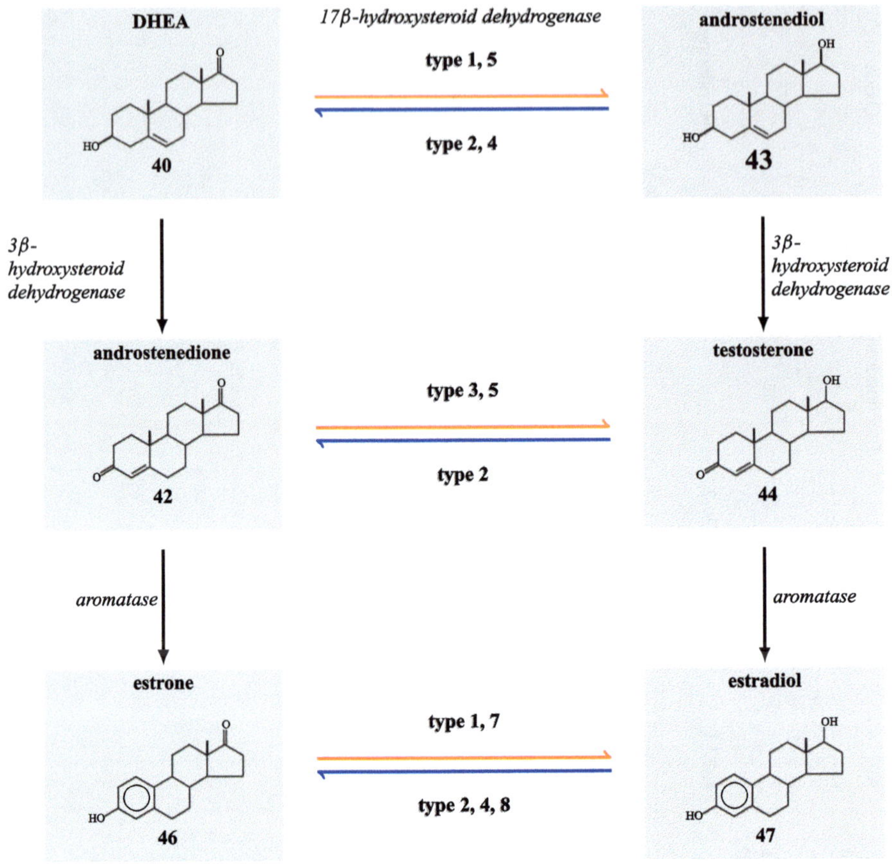

Fig. 6.15 Different activities of 17β-HSD types

dehydrogenase (SCHAD) activity of this gene (Yang et al. 2005).[4] Not yet resolved is the role of HSD17B1 in breast cancer where elevated levels of estradiol are regarded as causal for the disease. Inhibition of HSD17B1 as one critical hormone appears a promising step in disease control (see, e.g., Fournier et al. 2008).

[4]The *HSD17B10* gene is one of a few genes that are maintained active in X chromosome inactivation (Yang et al. 2005), which has, however, recently been rejected (García-Villoria et al. 2010).

6.5.5 5α-Reductase

Fact sheet 6.6: 5α-Reductase

Structure:	no crystallization reported, member of the isoprenyl-cysteine carboxyl methyl-transferase (ICMT) family	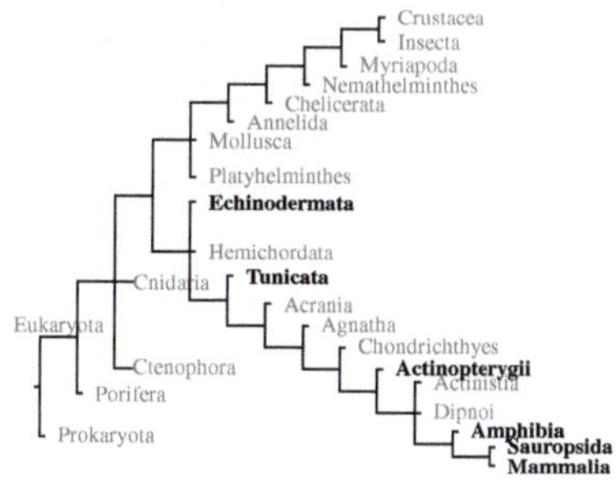
Gene:	SRD5A1: Chromosome 5 locus p15.3, 5 exons SRD5A2: Chromosome 2 locus p23, 5 exons	
Topology:	ER membrane protein; SRD5A1: widely distributed; SRD5A2: preferentially in androgen target tissue	
Function:	reduces testosterone to dihydrotestosterone (type 2), progesterone to dihydroprogesterone and androstenedione to androstanedione	

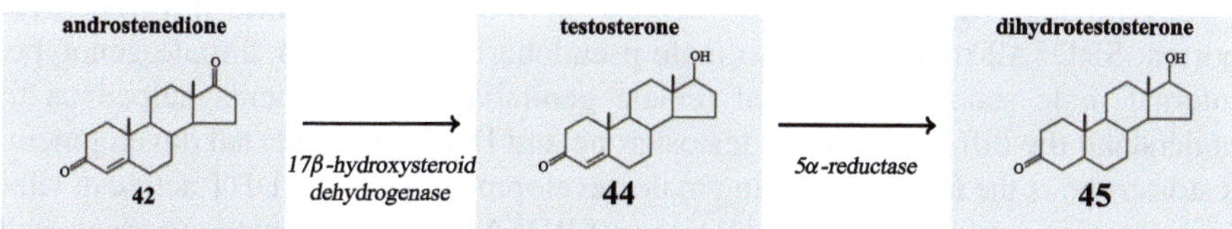

Fig. 6.16 Testosterone: synthesis and reduction into DHT

3-Oxo-5-α-steroid-4-dehydrogenase (5α-reductase; SRD5A) is the enzyme catalyzing conversion of testosterone into dihydrotestosterone (DHT) (Fig. 6.16). Two isoenzymes exist with an overall homology of 46 % of the 254 or 259 amino acids (Fig. 6.17). The phylogenetic separation of the two genes took place when the first echinoderms and tunicates separated; the former having only one Srd5a, the latter having two.[5] Both proteins are members of the isoprenylcysteine carboxyl methyltransferase (ICMT) supergene family (http://www.ncbi.nlm.nih.gov/Structure/cdd/cddsrv.cgi?uid=cl00763; COG3752).

The two enzymes differ by their catalytic properties and by their tissue distribution. The SRD5A2 enzyme almost exclusively uses testosterone to generate

[5]http://www.pantherdb.org/treeViewer/appletTreeViewer.jsp?book=PTHR10556&seq=HUMAN|ENSEMBL=ENSG00000049319|UniProtKB=P31213

```
                    1         1         2         2         3         3         4         4         5
          5         0         5         0         5         0         5         0         5         0
1   M A T A T G V A E E R L L A A L A Y L Q C A V G C A V F A R N R Q T N S V Y G R H - - A L P S H R L    1 -  48
2             M Q V Q C Q Q S P V L A G S A T L V A L G A L A L Y V A K P S G Y G K H T E S L K P A A T    1 -  45

1   R V P A R A A W V V Q E L P S L A L P L Y Q Y A S E S A P R L R S A P N C I L L A M F L V H Y G H R   49 -  98
2   R L P A R A A W F L Q E L P S F A V P A G I L A R Q P L - S L F G P P G T V L L G L F C L H Y F H R   46 -  94

1   C L I Y P F L M R G G K P M P L L A C T M A I M F C T C N G Y L Q S R Y L S H C A V Y A D D W V T D   99 - 148
2   T F V Y S L L N R G R - P Y P A I L I L R G T A F C T G N G V L Q G Y Y L I Y C A E Y P D G W Y T D   95 - 143

1   P R F L I G F G L W L T G M L I N I H S D H I L R N L R K P G D T G Y K I P R G G L F E Y V T A A N  149 - 198
2   I R F S L G V F L F I L G M G I N I H S D Y I L R Q L R K P G E I S Y R I P Q G G L F T Y V S G A N  144 - 193

1   Y F G E I M E W C G Y A L A S W S V Q G A A F A F F T F C F L S G R A K E H H E W Y L R K F E E Y P  199 - 248
2   F L G E I I E W I G Y A L A T W S L P A L A F A F F S L C F L G L R A F H H H R F Y L K M F E D Y P  194 - 243

1   K F R K I I I P F L F                                                                                249 - 259
2   K S R K A L I P F I F                                                                                244 - 254
```

Fig. 6.17 5α-reductases: isoenzyme comparison. SRD5a1 with 259 amino acids is in the *upper* line; (preferentially expressed in adrenal glands and gonads) SRD5a2 is in the *lower* line. For maximal alignment three gaps were introduced indicated by *dashes* (Source: GenBank: NM_001047 und NM_000348)

DHT which binds the androgen receptor (AR) with an enhanced affinity compared to testosterone, that, still, is able to activate AR-dependent gene transcriptions. SRD5A1 prefers progesterone as substrate and can convert androstenedione and testosterone, too. The dihydroprogesterone is an inactive steroid; SRD5A1 thus removes progesterone from the steroid pool.

Several inherited and acquired defects of SRD5A isoenzymes have been reported: SRD5A2 mutants lead to male pseudohermaphroditism, a male genotype, internal male state, but external female genitalia. These patients helped us to understand the different roles of testosterone and DHT during sexual development: testosterone is the hormone driving male development, whereas DHT active at later stages ensures proper male sexual organs. SRD5A1 variants in men are associated with hirsutism, the enzyme active in hair follicles. In mice Srd5a1 variants have been linked to mid-gestational intrauterine death and parturition failure (Mahendroo and Russell 1999). The former fact is explained by increased androgens and therefore estrogens in type 1 knockout females which leads to intrauterine failure as does externally applied estrogen at the gestational stage where the intrauterine death occurred. From day 6 after coitus in mice Srd5a1 is expressed in the decidua; placental androgen increases at about day 10. When Srd5a1 is lacking, the elevated estradiol observed seems causal for fetal death. A fetus surviving this period, however, dies later because parturition fails. Analysis of the hormone levels in the knockout mice points to an additional metabolite of androstandione, 5α-androstane-3α,17β-diol (3α-Adiol) obviously required for the cervical ripening to become elastic. The defect could be overcome by application of reduced androgens (Mahendroo and Russell 1999).

Inasmuch as DHT binds more efficiently to the AR receptor than testosterone, the SRD5A2 has been the target of intensive inhibitor research in order to treat prostate cancer.

6.5.6 CYP21

Fact sheet 6.7: CYP21A2 = CYP21B

Structure:	A CYP protein with 494 amino acids linked to the membrane by 26 N-terminal amino acids; a structure has not been published; Robins et al. (2006) proposed a CYP21A2 model.
Gene:	CYP21A2 on chromosome 6 locus p21.33 within the MHC locus; 10 exons; a second CYP21A1 gene in humans is a pseudogene with a deletion and due to frameshift incomplete, afunctional protein sequence.
Topology:	A membrane enzyme in the ER exclusively in the adrenal cortex.
Function:	Oxidation of the C-21 atom of progesterone or 17α-OH-progesterone.

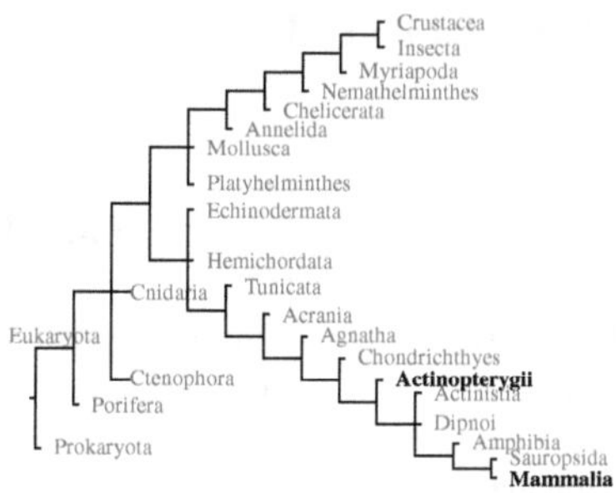

This is the adrenal enzyme to start glucocorticoid and mineralocorticoid biosynthesis. It is only found in vertebrates. In rodents and in humans it is linked to the major histocombatibility complex and occurs in a tandem repeat together with the complement component C4. Whereas in mice Cyp21a1 is the active enzyme and Cyp21a2 is afunctional due to a large deletion, in humans CYP21A2 is the active enzyme and CYP21A1 inactive due to deletion, mutations, and a frameshift. Numerous mutations have been reported because a defect in CYP21A2 gives rise to life-threatening congenital adrenal hyperplasia (CAH).

CYP21A2 is expressed in the two adrenocortical zones: zona glomerulosa and zona fasciculata. In the zona glomerulosa progesterone is converted into 11-deoxycorticosterone (**48**) which is not converted by CYP17, but by the CYP11B2 which generates aldosterone (**51**). In the zona reticularis pregnenolone is converted to 17α-hydroxypregnenolone by CYP17, converted by 3β-HSD to 17α-hydroxyprogesterone, and oxidized further by CYP21 to dihydroxyprogesterone (**52**; aka 11-desoxycortisol) which is terminally converted to cortisol by CYP11B1 in mitochondria (Fig. 6.21).

Numerous mutations of *CYP21A2* have been reported because most of these lead to 21-hydroxylase deficiency, which is one cause of congenital adrenal hyperplasia (see Sect. 14.5.1). 3β-HSD is highly expressed in the zona fasciculata and together with CYP17 ensures conversion of pregnenolone to 17α-hydroxyprogesterone.

CYP21A2 is now required to remove the latter which otherwise will block 3β-HSD. With reduced or lacking CYP21A2 in the zona fasciculata enhanced production of androgens takes place, cortisol is missing, and circulating ACTH—normally controlled by cortisol—promotes adrenal hyperplasia. Near complete CYP21A2 deficiency will also be causal for deficient aldosterone generation and associated diseases (salt-wasting disorders) (Bird et al. 1998; Payne and Hales 2004; White and Speiser 2000).

6.5.7 CYP11B1

Fact sheet 6.8: CYP11B1

Structure:	Not yet crystallized
Gene:	Chromosome 8 locus q21-22; nine exons
Topology:	Inner mitochondrial membrane
Function:	Oxidation of the 11-position of desoxycortisol (=dihydroxyprogesterone) to generate cortisol

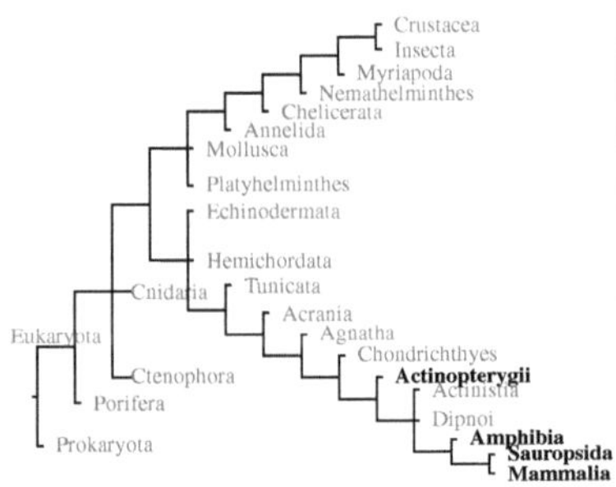

The gene for the enzyme catalyzing the final step of cortisol biosynthesis has been cloned by Chua et al. in 1987. A very similar protein (CYP11B2 aka aldosterone synthase; see below) has been found in humans. These are both mitochondrial proteins and involved in steroidogenesis. The two genes are located close together on chromosome 8. The relatedness of the two human genes is closer than that to any nonprimate Cyp11B gene, where, for example, in cow or mice likewise two separate genes exist for 11β-hydroxylation. Indeed, by changing individual amino acids in the one human enzyme made the enzyme behave as did the other one (Böttner et al. 1998). This is in line with the fact that in other vertebrates 11β-hydroxylation and 18-hydroxylation are performed by the same enzyme. The gene duplication of CYP11B thus appears to be a recent event that happened after primates had evolved.

The enzyme is located in the inner mitochondrial membrane and requires adrenodoxin and NADPH for functional activity. The enzyme acts in the zona fasciculata and converts dihydroxyprogesterone (aka deoxycortisol; **52**) into cortisol (**53**).

In the human fetal adrenal gland CYP21A2 and CYP11B1/CYP11B2 are present in the inner fetal zone and the transitional zone and absent in the outer definitive zone, whereas DHEA is produced in the fetal zone (Coulter and Jaffe 1998). Time-course studies in rat indicated that fetal adrenal corticosterone production depending on the Cyp11A1, Cyp21, and Cyp11B1 (11β-hydroxylase) was already established at gestational day 16 and aldosterone synthesis, which needs additional Cyp11B2 (aldosterone synthase), was enabled by Cyp11B2 expression around gestational day 20 (Mitani et al. 1999).

Human CYP11B1 transcription is stimulated by hypophysal ACTH. Binding of ACTH to its receptor in cells of the zona fasciculata stimulates intracellular cAMP and facilitates the binding of the transcriptional activator ATZ-2. The steroidogenic factor 1 (SF-1) in addition is required for CYP11B1 expression (Wang et al. 2000). Wang et al. have described that the orphan receptor liver receptor homologue 1 contributes to CYP11B1 activation, too (Wang et al. 2001). Downregulation of CYP11B1 is most probably induced by the transcriptional regulator DAX-1 which has been shown to counteract SF-1. Mutations in DAX-1 have been linked to adrenal tumorigenesis.

Overexpression of CYP11B1 might lead to adrenal adenoma. More profound pathologies, however, have been described when CYP11B1 is rendered afunctional due to mutations. This is the case in about 8% of congenital adrenal hyperplasias. The reason for congenital adrenal hyperplasia due to CYP11B1 is similar to CYP21A2 deficiencies. Because cortisol is not or not sufficiently produced by the defective enzyme, ACTH synthesis is not downregulated; it stays elevated and stimulates adrenal growth.

6.5.8 CYP11B2 (Aldosterone Synthase)

Fact sheet 6.9: CYP11B2–aldosterone synthase

Structure:	PDB 4DVQ
Gene:	Chromosome 8 locus q21-22; nine exons
Topology:	Inner mitochondrial membrane in the zona glomerulosa
Function:	Three steps in aldosterone synthesis: 11β-hydroxylation, two-step C-18 oxidation

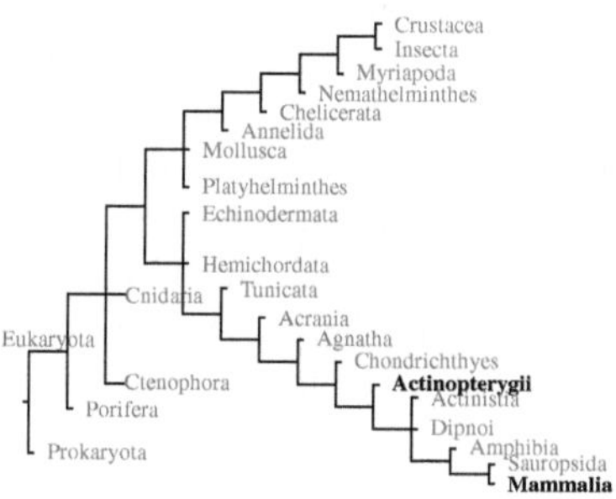

Although CYP11B2 is closely related to the CYP11B1 (see above) with respect to structure and sequence, functionally both enzymes differ considerably: CYP11B1 is part of the hypothalamic—(anterior) pituitary—adrenal axis that leads to cortisol release and CYP11B2 is part of the angiotensin—aldosterone circuit wherein the neuropeptides oxytocin and vasopressin are released from the posterior pituitary and wherein active participation of heart myocytes, juxtaglomerular cells, and lung enzymes occurs (see Sect. 11.8 and Fig. 11.15). It is interesting to note that the promoter of CYP11B2 is far more different from that of CYP11B1 than the coding sequence of the two proteins which are very similar with about 95 % homology at the amino acid level.

CYP11B2 is a mitochondrial enzyme, too, relying on adrenodoxin and NADPH for functional activity.

Exclusively expressed in the zona glomerulosa the CYP11B2 enzyme stepwise catalyzes the last three steps of the aldosterone biosynthesis: oxidation of C11 of deoxycorticosterone (**48**) to corticosterone (**49**), then oxidation of C-18 of corticosterone to 18-hydroxycorticosterone, and again oxidation of the 18-hydroxy group to the aldehyde group of aldosterone (**51**; see Fig. 6.21). The two intermediates have been isolated from normal blood as well as in several patients with either type I or type II aldosterone synthase deficiency (ASD), thus demonstrating the stepwise action of the enzyme (Dunlop et al. 2003). The three enzymatic activities of CYP11B2 were originally attributed to three different enzymes: steroid 11β-hydroxylase, and corticosterone methyl oxidase type 1 and type 2. First in rats (Lauber and Muller 1989) and then in humans (Kawamoto et al. 1992) it was established that a single protein could achieve all three enzymatic functions.

Expression of CYP11B2 is stimulated by extracellular potassium and by angiotensin II. Potassium is sensitively determined in the zona glomerulosa. Elevations of extracellular potassium levels stimulate release of calcium from intracellular stores which binds to calmodulin. Calcium-/calmodulin-dependent activation of protein kinases I and IV leads to phosphorylation and thus activation of several transcription factors including NURR1, NGFIB, ATF-1, or CREB. These transcription factors then stimulate expression of CYP11B2 as well as of StAR, CYP11A1, and CYP21A2 (Dierks et al. 2010, and references). Angiotensin after binding to its G-protein—coupled receptor equally triggers release of calcium and by activation of Ca-/calmodulin-dependent kinases stimulates CYP11B2 transcription, too.

Pathologies of CYP11B2 are either elevated aldosterone production leading to adrenal adenoma or partial to complete CYP11B2 deficiencies. Two types of aldosterone synthase deficiencies are distinguished: no corticosterone methyl oxidase activity, that is, lack of 18-hydroxycorticosterone and of aldosterone (ASD type I) or absence of aldosterone, but not 18-hydroxycorticosterone (ADS type II) (compare Fig. 6.21). Aldosterone synthase deficiencies are still life-threatening diseases and require intensive care and therapy.

6.5.9 CYP19

The key enzyme of estrogen synthesis is the CYP19 or aromatase (Fig. 6.19) which converts either androstenedione or testosterone to estrone, respectively, estradiol. During this aromatization the C19 side-chain gets oxidized in a three-step reaction that requires NADPH and NADPH-cytochrome P450 reductase. The first two steps oxidize the C19 methyl group to hydroxy-methyl and then to an aldehyde; the

Fact sheet 6.10: CYP19–aromatase

Structure:	3EQM
Gene:	Chromosome 15 locus q21.2; ten exons spanning 91 kilobases
Topology:	Mebrane protein in the endoplasmic reticulum
Function:	Conversion of the androgens androstenedione and testosterone into the estrogens estrone and estradiol

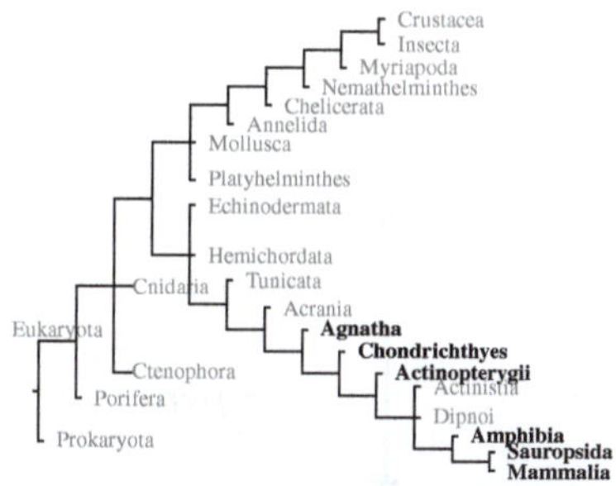

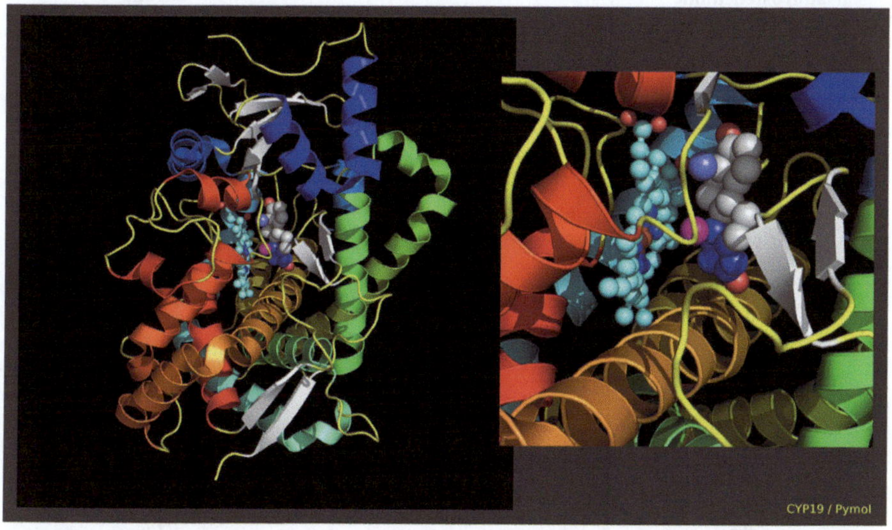

Fig. 6.18 Crystal structure of CYP19 with androstenedione (**42**) as ligand. *Left* side: loops are colored *yellow*, beta sheets in *white*; in the heme group shown in *small spheres* carbon atoms colored *light blue*, nitrogen atoms *dark blue* with a central *red* iron atom; the androstenedione is shown in *larger spheres* with carbon atoms of three rings in *gray*, the first ring—to be aromatized—is shown in *blue*, the side-chain carbon atoms are marked *magenta* (C19) and *light blue* (C18). Oxygen spheres are colored *red*. *Right* side: the center of the *left* side image amplified to demonstrate the close proximity of the heme structure to the C19 atom to be oxidized and thus removed. The *blue* ring of androstenedione will be aromatized

third step involves isomerizations and cleavage of the C19–C10 bond giving rise to the 3-hydroxy aromatized ring of estrogens with 18 carbon atoms and formic acid (compare Simpson et al. 1994). This reaction is irreversible.

The phylogeny of CYP19 has gained new interest in recent years because its sole vertebrate origin has been debated: (Castro et al. 2005) analyzed the chromosome around the CYP19 gene, and confirmed that the chromosomal region already

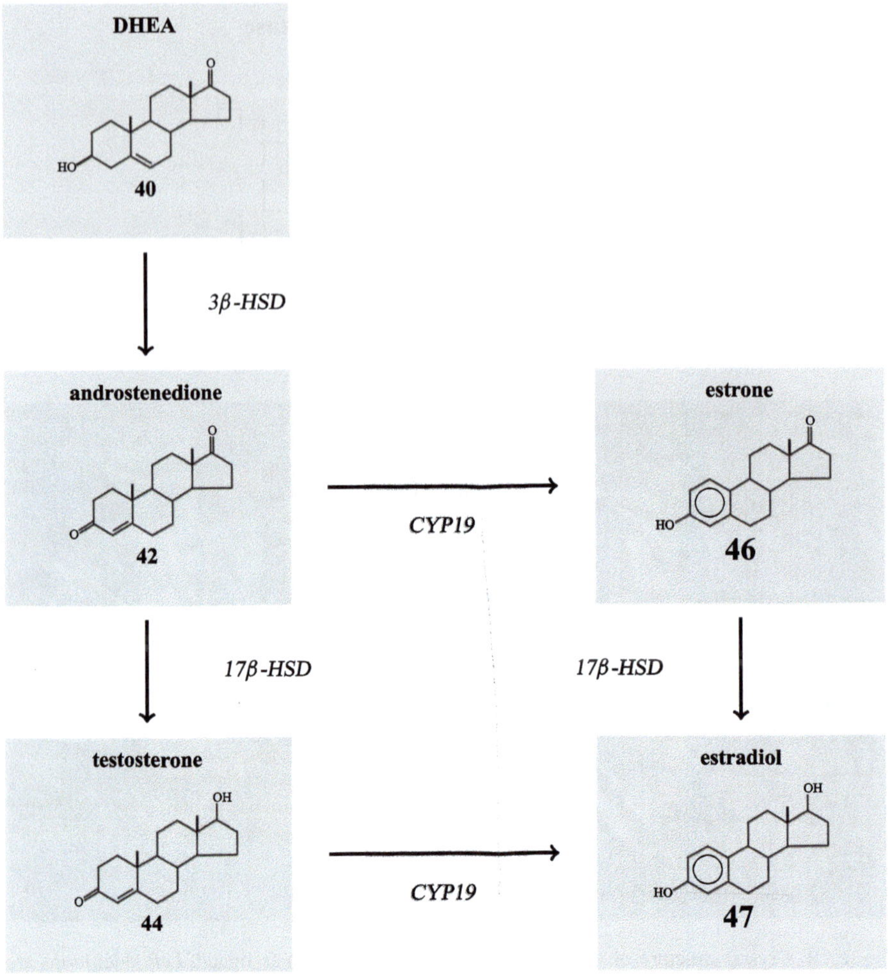

Fig. 6.19 Actions of aromatase (CYP19)

belonged to the MHC before the evolution of vertebrates. They identified a CYP19 gene in lancelets, a prevertebrate species. However, in the sea squirts (tunicates) no CYP19 gene could be found in fully sequenced genomes. In other fully sequenced genomes apart from vertebrates, no CYP19-like gene has ever been observed. There are some strange reports in the literature: (Twan et al. 2006) reported on corals and isolate GnRH from *Euphyllia ancora*. Even more strange is their finding that the coral gonads when stimulated with these and other GnRH compounds secrete sex steroids such as testosterone and estradiol. Given the fact that neither CYP11A1, CYP17, nor CYP19 have ever been cloned outside the vertebrate lineage, such a report awaits confirmation by cloning the three enzymes from the coral species. A similar article on GnRH and steroids in *C. intestinalis* was reported several years before (Fiore et al. 2000). In the light of the lack of any CYP11A1/CYP17/CYP19 analogue in the fully sequenced *C. intestinalis* genome, this paper is very difficult to accept.[6]

The structure of CYP19 has been elucidated by crystallization and x-ray analysis (Fig. 6.18; Ghosh et al. 2009). The N-terminus is on top of the sequence, and the C-terminus in the bottom. In Figure 4 in their article Ghosh et al. propose that integration into membrane is made by N- and C-terminal sequences, the *dark blue* helix next to the N-terminus, and the loops in the lower right side of the *left* image. The channel for androgens to reach the active side and estrogens to leave the enzyme would in their model open between the middle beta-sheet and the green helix behind.[7]

The regulation of CYP19 expression is astonishingly complex. The original complexity is already due to tissue-specific expression. Cells in the ovary, testis, placenta, adipose tissues, brain, bone, skin, and endothelium use their "private" promoters and an individual untranslated exon I (the coding of aromatase starts in exon II). Individual transcriptional activators and thus different transcription factors enable this tissue-specific expression with alternate promoter usage and thus alternative splicing (Table 6.2). The coding region of the aromatase has an additional nine exons, labeled II to X. The entire gene spans more than 90 kbases. In fish (due to the additional genome duplication) there are two aromatase genes. Cyp19a1a from *D. rerio*[8] has nine exons and spans 15.8 kbases; Cyp19a1b has an additional 5′-untranslated exon and spans 13.5 kbases. The promoter of these genes is much shorter than the human one; the two CYP19 variants are differentially expressed, type a in the gonads, and type b in the brain and other tissues (Goto-Kazeto et al. 2004).

Aromatase is expressed in several normal tissues: male and female gonads, placenta, brain, bone, prostate, and adipose tissues. Ovarian aromatase is required to convert the theca-cell–derived testosterone into follicular estradiol. In the testis aromatase has been found in Leydig cells, Sertoli cells, and different sperm precursors. From Cyp19 knockout mice it is known that aromatase deficiency impairs spermatogenesis and the capacity to fertilize oocytes. The placental aromatase is part of the fetal–placental unit that generates estriol (the major estrogen made during pregnancy) from DHEA sulfate involving a 16α-hydroxylation. The placental aromatase fulfills a second role in protecting the fetus in particular the female one from actions of testosterone. The high expression of aromatase ensures that any testosterone does not cross the placenta to masculinize the female fetus.

---

[6]Interestingly, given that such an antidogmatic result should be worth additional efforts the authors did not publish any sequential publication on this subject.

[7]A version of the paper is available freely at http://www.ncbi.nlm.nih.gov/pmc/articles/PMC2820300/pdf/nihms163248.pdf. The PyMOL script to generate the image using the free PyMOL port is presented in the appendix.

[8]Danio rerio.

Table 6.2 The 10 different exon I elements in the aromatase promoter (Source: Santen et al. 2009)

No.	Name	Pseudonym	Transcriptional activator
1	I.1	Placenta major	SP-1, Mash 2, USF-1, USF 2, TSE
2	2.a	Placenta minor 2	
3	I.4	Skin, adipose tissue	Class I cytokines, GRE, GAS, TNFα, glucocorticoids, LRH-1, PKA, PKC, SP-1, SHP
4	I.5	Fetal tissue	
5	I.7	Endothelial cell/breast carcinoma 6	
6	I.f	Brain	Androgens, estrogens
7	I.2	Placenta minor 1	
8	I.6	Bone	GRE, IL-1β, Class I cytokines, TNFα, TNFβ, glucocorticoids, 1,25-dihydroxyvitamin D_3
9	I.3	Adipose/breast cancer	Breast cancer: PGE, cortisol, SF-1, LRH-1, PKA, ERRα, glucocorticoids
10	PII	Ovary, testis – breast cancer – endometriosis	ovary: PGE, SF-1, LRH-1; testis: TGF-α, testosterone, DHT, FSH

Aromatase activity in the brain has been attributed to several functions: during development testosterone's presence primes male behavior in the brain. The acting steroid, however, is not testosterone but estradiol generated locally from testosterone. In addition, recent findings indicate a neuroprotective function of aromatase by converting testosterone and other androgens into estrogens. Furthermore, aromatase might help to deal with brain damage.

In the bone estradiol controls maturation of new bone formation (see Sect. 11.6). The presence of aromatase in bone tissue ensures local production of estrogens. Bone mineral density in aging man has been correlated with estradiol levels and not with testosterone levels which suggests that bone maintenance depends on aromatase.

Major work on aromatase has been performed in the hunt for an efficient treatment for breast cancer. The finding that breast cancer cells depend on estrogen for cell division and that estrogens are locally produced in the breast encouraged studies to use aromatase inhibitors for treatment of this tumor.

6.5.10 Concluding Remark

After this section it should be emphasized that steroids are not per se active. The regulation of steroidogenesis is the regulation of steroidogenic enzymes. These enzymes are tissue-specifically expressed (Table 6.3). Researchers wanting to analyze steroids need to analyze these proteins as well.

Table 6.3 Characteristic enzymes of steroidogenic tissues

Steroidogenic tissue	End product	Characteristic enzyme(s)
Adrenal cortex		
Zona glomerulosa	Aldosterone	CYP11A1, 3β-HSD, CYP21, CYP11B2
Zona fasciculata	Cortisol	CYP11A1, 3β-HSD, CYP17, CYP21, CYP11B1
Zona reticularis	DHEA	CYP11A1, CYP17
Testis	Dihydrotestosterone	CYP11A1, 3β-HSD, CYP17, 17β-HSD, 5α-reductase
Ovary	Estradiol	CYP11A1, 3β-HSD, CYP17, 17β-HSD, CYP19

6.6 Progesterone

Fact sheet 6.11: Progesterone

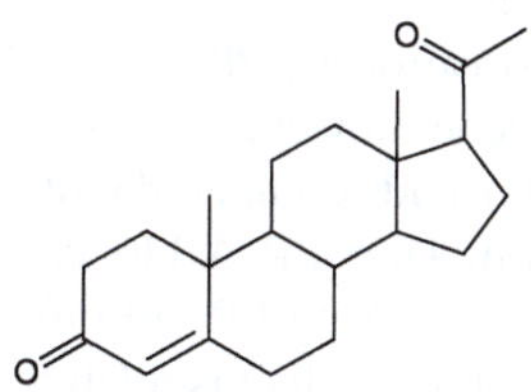

Structure: **38**

Phylogeny: In vertebrates.

Topology: In the smooth ER of luteal cells both 3β-HSD isoenzymes are active; they convert pregnenolone into progesterone.

Receptor: Two progesterone receptors from the same gene on chromosome 11 using different polymerase initiation sites (see http://omin.org/entry/607311).

Function: Progesterone is an intermediate of testosterone, estrogen, cortisol, and aldosterone synthesis. As a hormone it is indispensable for ovulation, nidation of the fertilized egg, and for the maintenance of pregnancy.

In 1934–1935 four groups from Rochester (New York, USA), Basel (Switzerland), Gdansk and Wrocław (now Poland) published independently the progesterone isolation. W.M. Allen named the hormone after *proge*stational *ster*oidal ket*one*.

Progesterone is an intermediate of the sex steroid and the corticosteriod synthesis; on the other hand by itself it is a hormone, which is required for ovulation, the nidation of the fertilized egg, and for the maintenance of pregnancy. Partial loss of progesterone at the end of pregnancy will allow the before-arrested uterus muscles to become active. When due to the birth the progesterone levels drop, milk is injected into the ducts.

In mammals progesterone is formed in most steroidogenic tissues, larger quantities are available after ovulation when the just-ruptured follicle is transformed into

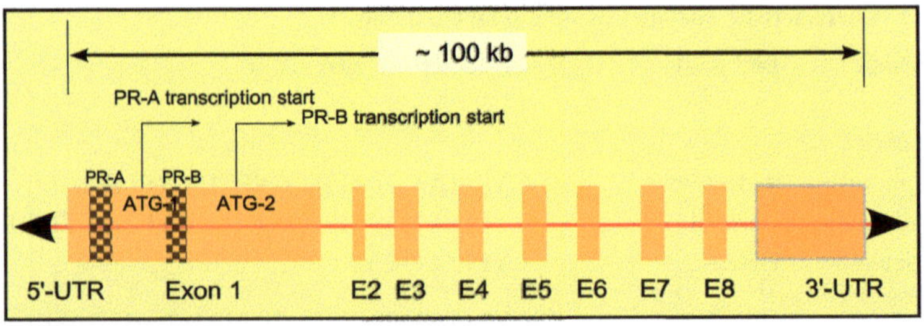

Fig. 6.20 Organization of progesterone receptor gene. The progesterone receptor gene comprises eight exons and spans about 100 kbp. In the same exon 1 there two polymerase start sites, ATG-1 for the isomer PR-A and ATG-2 for the PR-B isomer. Upstream of these two start sites are CG-rich regions (CG islets, *hatched*) whose methylation at least in the case of endometriosis is differentially regulated (Redrawn according to Wu et al. 2006)

the corpus luteum which synthesizes and releases progesterone. Maintaining a high level of progesterone is a characteristic feature of a pregnancy and is ensured by the placental hormone choriogodanotropin.

The progesterone receptor has two isoforms, PR-A and PR-B (Fig. 6.20). Both isoforms are derived from the same gene, however, from different promotors. The A-variant[9] has an additional 165 amino acids compared with the B form. These 165 amino acids form a inhibitory domain. Thus PR-A blocks transcriptions and end PR-B supports them. Mice with defects of both PR are infertile, as are those with only PR-A defects as well. It seems that this defect is related to the lack of nidation.

Differences in the expression of both PR have been noted in the case of endometriosis . In the healthy endometrium both receptors were expressed; in the case of endometriosis only the PR-A was found expressed. This might be at the origin of the disease (Attia et al. 2000). It has also been found that gene silencing at the B promotor due to enhanced methylation is causative (Wu et al. 2006).

In the N-terminal domain of PR there are proline-rich elements . These enable the receptors to bind to SH3 domains of intracellular signal mediators and to initiate fast cell reactions (see Chap. 8) whereas nuclear receptors such as PR normally activate slow reactions (Boonyaratanakornkit et al. 2001).

Progesterone has been repeatedly found in plants in the last 10 years (Iino et al. 2007; Pauli et al. 2010). First hints exist since 1964 (Iino et al. 2007, and citations therein). A side-chain-cleavage enzyme ($P450_{scc}$) in plants capable of cleaving cholesterol or more probable sitosterol (compare Wichtl 1999, S. 473) has never been found. There are in numerous plants cardenolides with a pregnan structure. There is not any evidence that these pregnans are not synthesized from cholesterol or likewise similar 24-substituted cholesterols. There are other steroids in plants (see Sect. 6.10.2); these, however are formed without a side-chain cleavage lyase.

[9]The terms A and B are ambiguously used in the literature. We keep to the use at OMIM (http://omim.org/entry/607311).

The so-called natural progesterone is, however, only a synthetic (*sic!*) derivative of diosgenin present, for example, in yam (*Diocorea*). Diosgenin is a cholesterol with two intramolecular cyclic ether bonds adding two additional rings to the sterane structure. It is necessary to cleave the side-chain either chemically or enzymatically to obtain progesterone. The entire esoteric hype about this product is commercially caused. This product cannot be distinguished by any means from chemically synthesized progesterone or one derived from animals. Yam definitely does not contain any progesterone.

6.7 Sex Hormones

Female and male sex steroids, estrogens, and androgens are major determinants of reproduction. Steroid hormones derived from pregnenolone which itself is generated by cleavage of the cholesterol side-chain using CYP11A1 are characteristic vertebrate hormones (see Sect. 6.5). Invertebrates, fungi, plants, and some aerobic bacteria are able to derive lanosterol (**16**) or ergosterol by cyclization of squalene (**14**). The other steroids derived therefrom, however—are as far as could be found in the literature—not active as sex hormones. There is one singular exception: in fungi (analyzed in *Achlya*) two steroids are generated, antheridiol (**60**) and dehydro-oogoniol (**59**) (see Fig. 6.24), which trigger sexual dimorphism and thus fruit-body formation. In vertebrates, sexual development, maturation of gonads and their activity, is dependent on androgens and estrogens. These hormones are further major regulators of growth and induce sex-specific behavioral priming in the brain.

Fact sheet 6.12: Testosterone

Structure: **44** see Fig. 6.15.

Phylogeny: Most probably only in vertebrates.

Topology: Made by microsomal aromatase and released by diffusion through membranes, in serum bound by steroid-binding globulin (SBG).

Receptor: The androgen receptor is a nuclear receptor preferentially binding dihydrotestosterone which upon ligand binding is transported into the nucleus to drive gene activation. It is expressed in a variety of reproductive and nonreproductive tissues, the latter including brain, bone, skin, anterior pituitary, thyroid, adrenal cortex, liver, kidney tubules, urinary bladder, and cardiac and striated muscle.

Function: Testosterone and the derivative dihydrotestosterone (DHT) bind to the nuclear androgen receptor. During male fetal development testosterone synthesis is activated by the *SRY* gene product. Testosterone then initiates antimüllerian hormone expression which primes the male phenotype. In the brain testosterone triggers male priming in the sexual dimorphic nucleus; the active hormone, however, for this priming is locally converted estradiol. Testosterone stimulates new bone mass formation in epiphyses.

6.7.1 Gonadal Development:

There are a few male-specific genes on the Y chromosome missing on the X chromosome. One of these is the *SRY* gene (sex region Y). If *SRY* is transcribed and translated in the fetus, this is a trigger for synthesis of CYP11A1, HSD3B, CYP17A1, and HSD17B which leads to testosterone synthesis. In the fetus there is no aromatase expressed at this early time. The primordial gonads thus far sexually neutral get male primed. Thereby the characteristic testis cells, Sertoli and Leydig cells, are formed. Antimüllerian hormone (AMH) induces the müllerian tracts to degenerate, the wolffian tracts to develop into the vasa deferentia, and later the testes to descend into the scrotum.

Any of these events might fail which blocks full maturation of the male sexual organs and results almost always in infertility. Similar male infertility arises with androgen receptor (AR) defects or blockage. Inherited AR mutations and the action of anti-androgens inhibit complete development of male reproductive functions. Among the anti-androgenic toxins are stable metabolic derivatives ofDDT,[10] pest management substances such as vinclozolin, as well as some drugs for the therapy of prostate tumors, for example, hydroxyflutamide.

Testosterone triggers male sexual development, however, lack of testosterone in early gestational phases promotes female sexual development. Absence of testosterone and with no antimüllerian hormone present, the müllerian ducts develop into the ovarian ducts, ovaries instead of testes are formed, and the wolffian (mesonephric) ducts degenerate.

Fact sheet 6.13: 17β-Estradiol

Structure: **47** see Fig. 6.15.
Synthesis: By the action of aromatase on testosterone and by 17β-HSD on estrone, stored as estradiol-sulfate.
Receptor: Two human estrogen receptors are both nuclear receptors which on cytoplasmic dimerization due to ligand binding are transported into the nucleus where they act as transcription factors.
Phylogeny: Most probably a steroid exclusively in vertebrates.
Function: Binds to the nuclear estrogen receptor and stimulates gene transcription.

The development of female reproductive organs is not initially dependent on any estrogen production. Female genitalia develop due to the absence of testosterone. In contrast, any inadvertent androgen influx from the mother would masculinize the female fetus. We keep in mind that in humans as in other mammals the presence of testosterone decides the development of sex organs.

This is different in fish where estrogens determine the female phenotype. It is noteworthy that this is why antiandrogens cannot influence fish sex development, whereas in mammals including humans male sexual development is sensitive to anti-

[10]Dichlorodiphenyltrichloroethane.

androgens. In fish, however, any effect of antiestrogens is observable because such substances, for example, DDT and many perchlorated phenol derivatives (PCB and others) occupy the estrogen receptor and inhibit the development of normal female genitalia.

In the rat a sexual dimorphism has been found in special brain areas: in the preoptic *dimorphic nucleus*, male and female rats are differentially primed. It is the testosterone that initiates characteristic male behavior priming in the brain area. When testosterone enters the brain at certain stages during fetal development, this testosterone is converted by the brain aromatase locally and estradiol induces specific gene transcriptions. These effects cannot be mimicked by estrogens which appear to be excluded from this area at the respective time points or which are not yet present. Whether humans have similar sexual dimorphic areas cannot be analyzed due to ethical restrictions.

6.7.2 Hormone Production in the Gonads

In the adult, the major sex steroid synthesis appears in the gonads. Testosterone is made in the Leydig cells of the testis and in the theca cells of the ovarian follicle. Ovarian follicular cells then convert testosterone into estradiol. The same conversion in testis is mainly performed by the Sertoli cells. Estrogen production is under FSH control secreted by the pituitary when stimulated by GnRH. GnRH and FSH secretion, however, are under inhibitory control of testosterone, estradiol, and inhibin. Such a circuit is called feedback inhibition.

6.7.3 Androgen Production in the Adrenal Glands

Outside of gonads androgen synthesis occurs in the zona reticularis of the adrenal cortex. The zonal organization is not only anatomically evident; the functional differences are based on differential expression of steroidogenic enzymes.

After adrenarche the inner cortical zone, the zona reticularis expresses CYP11A1 and CYP17 together with high amounts of the P450 oxidoreductase which leads to 3β-hydroxyandrogens such as DHEA (**40**). The lack of HSD3B in this zone disables progesterone and androstenedione formation. DHEA is further sulfated and released as DHEA sulfate. This product is not especially stored but the serum pool provides androgens to other tissues in need of androgens and where HSD3B is provided as well as HSD17B subtypes to generate testosterone.

Such androgens are causal for masculinization, for example, in postmenopausal women, that is, male facial expressions and beard growth. Inasmuch as no further follicles mature and no ovarian estradiol is made, the androgen-to-estrogen ratio is changed leading to hair growth and skin shrinkage.

CYP21 defects lead to masculinization as well. Owing to the lack or malfunction of CYP21 an excess of androgens is formed and androgenization takes place: hair growth on the torso and legs, low-pitch voice, clitoris increase. Because progesterone and 17α-hydroxyprogesterone are made in excess and the lacking cortisol does not block central and pituitary ACTH stimuli, CRH and ACTH synthesis is not blocked and the large excess of steroids is converted into androgens (see Fig. 14.2).

6.8 Corticoids

The glucocorticoid cortisol (**53**) and the mineralocorticoid aldosterone (**51**) are the products of steroidogenesis in the zona fasciculata and zona glomerulosa, respectively, of the adrenal cortex. The enzymes specifically expressed in these zones are listed in Table 6.3.

6.8.1 Cortisol

The important physiological role of cortisol is most easily recognized by the fact that failure of cortisol synthesis is potentially lethal. In one type of congenital adrenal hyperplasia (CAH) the CYP21 is afunctional. As long as the fetus during pregnancy lives in the uterus while the mother provides cortisol (**53**) and aldosterone (**51**), the defect is not relevant. After parturition, however, the newborn should provide its own cortisol to control ACTH synthesis. Additionally, cortisol is a major regulator of the immune system, particularly in inflammations. If CRH and ACTH regulation fail, the unsuppressed ACTH signals the adrenal glands to produce corticoids. The adrenals react to this permanent stimulus by cellular division and the adrenal increases (hyperplasia). When cytokine mediator release by the immune system is not suppressed cachexia might occur. CYP21 is equally required for aldosterone synthesis (see Fig. 6.21). If the defect is not recognized, the newborn will die. At the origin of CAH might as well be a homozygous defect of CYP11B1, the characteristic enzyme of cortisol biosynthesis.

Fact sheet 6.14: Cortisol

Structure:	(**53** see Fig. 6.21.
Topology:	The final conversion from 11-deoxycortisol (**52**) occurs in the inner mitochondrial membrane. Cortisol can leave the cell by diffusion and is bound in serum by the corticosteroid-binding globulin (CBG); the glucocorticoid receptor is a nuclear receptor in the cytosol of reactive cells: most important liver, adipose tissue, lung, CNS.
Phylogeny:	Cortisol is a glucocorticoid in primates and other mammals, however in birds, reptiles, amphibians, and rodents corticosterone **49** is the major glucocorticoid, not known outside vertebrates.
Function:	Cortisol is synthesized in response to stress; its major function is stimulation of gluconeogenesis. It is also an inhibitor of different circuits: HPA and HPG axis. Inflammatory reactions of the immune system are controlled by cortisol. Cortisol is inactivated by 11β-HSD type 2 oxidation to cortisone.

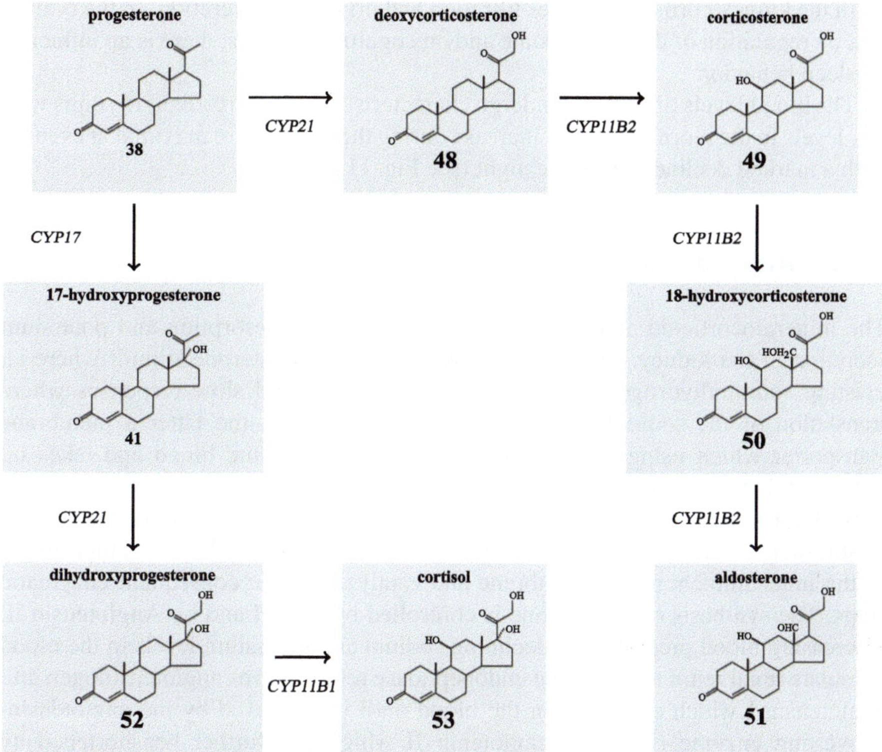

Fig. 6.21 Synthesis of corticoids

Synthesis of cortisol and other adrenal steroids is stimulated by the pituitary ACTH whose release is under hypothalamic CRH control. Cortisol blocks both ACTH and CRH release and vasopressin synthesis as well.

The major role of cortisol is stimulation of gluconeogenesis in the liver: cortisol blocks glucose uptake and its metabolism, for example, in muscle cells. Lipids are metabolized controlled by cortisol; lipid stores are shifted, blood lipoproteins (HDL, LDL, VLDL) are all enhanced to facilitate faster lipid transport.

During stress cortisol counteracts the actions of adrenaline. Adrenaline release empties glycogen stores and cortisol helps to fill up these stores again, at the expense of lipids and amino acids.

Whereas cortisol initiates gene activation in the liver, in other tissues, particularly in muscles, gene activation and translation are shut down. Influenced by cortisol, calcium is no longer taken up in the GI tract and retained in the kidney and formation of new bone is blocked.

Cortisol derivatives are drugs against inflammations. Certain T lymphocytes, that is, cells of the immune system, are driven by cortisol into apoptosis. This suppresses acute inflammation. The origin of the inflammation most often remains.

In the kidney cortisol enhances filtration and drives acid excretion. In the brain it acts on regulation of the mental state and on cognitive abilities; there is an influence on sleep behavior.

The blood levels of cortisol undergo characteristic daily (circadian) rhythms with low levels in the morning, which increase during the day and are maximal at evening with a marked decline during the night (see Fig. 11.2).

6.8.2 Aldosterone

The mineralocorticoid aldosterone(**51**) controls sodium resorption and potassium secretion in the kidney. There are fast responses to aldosterone stimuli where an existing sodium/hydrogenium exchanger is activated and slow responses where translation of the sodium/potassium ATPase is initiated, the latter a membrane transporter which using ATP excretes sodium ions into the blood and takes up potassium ions from there.

The hormone is synthesized exclusively in the zona glomerulosa of the adrenal by the action of CYP11A1, HSD3B, CYP21, and CYP11B2, the latter found in the inner mitochondrial membrane and catalyzing three consequent enzymatic steps. The synthesis of aldosterone is controlled by ACTH and by Angiotensin II. Decreasing blood pressure and declining sodium and potassium levels in the blood stimulate renal renin release. The endopeptidase renin cleaves angiotensinogen into angiotensin I which circulates in the blood until it is cleaved by the angiotensin-converting enzyme (ACE) to angiotenin II which can further be shortened by angiotensinase to angiotensin III which is still active in aldosterone stimulation. Both angiotensins as well as ACTH stimulate in the adrenal transcription of the *CYP11B2* gene.

Aldosterone is inactivated by reduction of first ring to dihydroaldosterone and tetrahydroaldosterone (for a review, see Morris 1993).

Fact sheet 6.15: Aldosterone

Structure:	(**51**) See Fig. 6.21
Phylogeny:	Found in vertebrate tetrapods (plus birds), not in fish
Topology:	Made by CYP11B2 in the mitochondrium
Function:	Stimulates expression of the renal sodium ENaC and of the sodium/potassium ATPase

Aldosterone stimulates transcription of sodium channels and the sodium/potassium ATPase in the kidney adding to the resorption of sodium. The contribution of aldosterone to the homeostasis is described in more detail in Sect. 11.8.7.

Lack of aldosterone due to different mutations of CYP11B2 or CYP21 is the cause of the salt-wasting syndrome which if not treated is lethal in the newborn. Situations with enhanced aldosterone received focus in recent years. Primary aldosteronism as one of the causes of hypertension is actually a prominent object of hypertension research.

6.9 Steroids in the Brain

Since 1995 an independent and particular steroidogenesis and steroid function in the CNS has been established. Synthesis of DHEA, progesterone, testosterone, and estradiol follows those steps already shown (see Fig. 6.22). There are additional enzymes including sulfotransferase and sulfatase that convert cholesterol, pregnenolone, and DHEA into their sulfate esters and the latter back to the sulfate-free forms. A characteristic enzyme of neurosteroidogenesis is the 5α-reductase that has already been presented as a testis enzyme where testosterone is reduced to DHT; however, in the brain an additional 5α-reductase target is progesterone converted to 5α-pregnan-3,20-dione which then can be further reduced to 3α,5α-tetrahydroprogesterone (aka allopregnanolone) by the next reductase, 3α-hydroxysteroid dehydrogenase (HSD3A). From the same 3α,5α-tetrahydroprogesterone the HSD3B generates an isomeric 3β,5α-

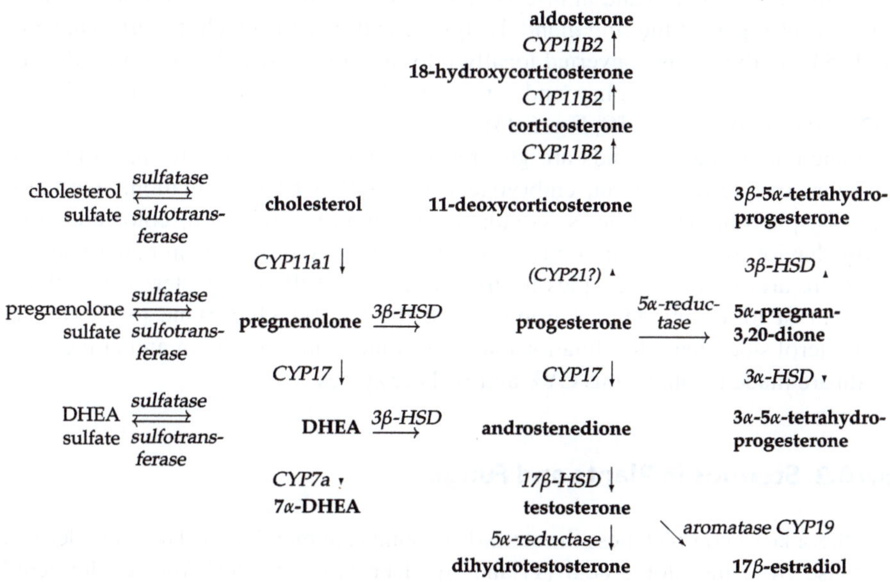

Fig. 6.22 Steroidogenesis in the central nervous system. Those enzymes identified in the brain are indicated. *Arrowheads* are shown where the conversion from biopsies and homogenates is not known; the enzymes and metabolites, however, are observed. *Parenthesis* and *question mark* in the case of CYP21 indicate that the enzyme has not been found thus far. Without CYP21 no 11-deoxycorticosterone can be generated which together with CYP11B2 and corticosterone, however, has been observed (From Stoffel-Wagner 2001; there is some doubt whether the different enzymes found by PCR are indeed active in the brain)

tetrahydroprogesterone. The key enzyme of all steroidogenesis, StAR is equally expressed in the CNS.

The activity of several GABA receptors and glutamate receptors is modulated by allopregnanolone. 7α-Hydroxyprogesterone derived by the action of 7α-reductase on progesterone is known to act on dopaminergic neurons. Neurosteroids are further involved in coping with anxiety: in animal experiments as well as with human volunteers it was shown that neurosteroids act anxiolytically. A connection to Alzheimer disease, to epilepsy, and to the premenstrual syndrome has been postulated, too. It is, however, reality that these three pathologies are linked to any new drug and the future will tell whether neurosteroids are in fact efficient here.

6.10 Additional Steroid Hormones

6.10.1 Insects, Crabs, and Other Early Metazoans

In insects and shellfish the molting hormone ecdysone and its derivatives are generated from cholesterol. These are mostly molting hormones; additionally they do exhibit other activities. Ecdysone is the product of enzymes in the molting gland: in shellfish the Y organ and in insects called the ventral gland, or prothoracic gland which is also part of the ring gland. Ecdysone is derived from cholesterol via keto-diol (**54**). Ecdysone is converted locally into the more active 20-hydroxyecdysone (**56**). Additional active compounds are derived from ketodiol: 25-deoxyecdysone (**57**) and ponasterone A (**58**; Fig. 6.23).

Functions of ecdysteroids are generation of the proper body form, molts, cell division, oocyte maturation, embryogenesis, vitellogenesis, ovulation, spermatogenesis, pheromone synthesis (copulation hormone), translation, color adaption, and behavior. Not all of these processes depend on ecdysteroids in equal manner, but there are, in total, examples in insects for any of these ecdysteroid functions. In contrast to vertebrates, arthropods lack the CYP11A1 enzyme to cleave the cholesterol side-chain. Modulations of the sterane/gonane skeleton and at the side-chain are made by analogous CYP and HSD enzymes.

6.10.2 Steroids in Plants and Fungi

For the sake of completeness the steroids in plants are mentioned. These are derived likewise from mevalonic acid (**1**) and squalene (**14**) and therefrom cycloartenol (**61**) or lanosterol (**16**). There are many different steroid derivatives, however, inasmuch as plants and fungi lack CYP11A1, no pregnenolone- or progesterone-related steroids have ever been found.

The brassinolide from Fig. 6.25 is a major regulator of plant growth. Dwarfism of different plants has been shown to have defective brassinolide-generating enzyme causes. Some fungal steroids are necessary to maintain membrane integrity.

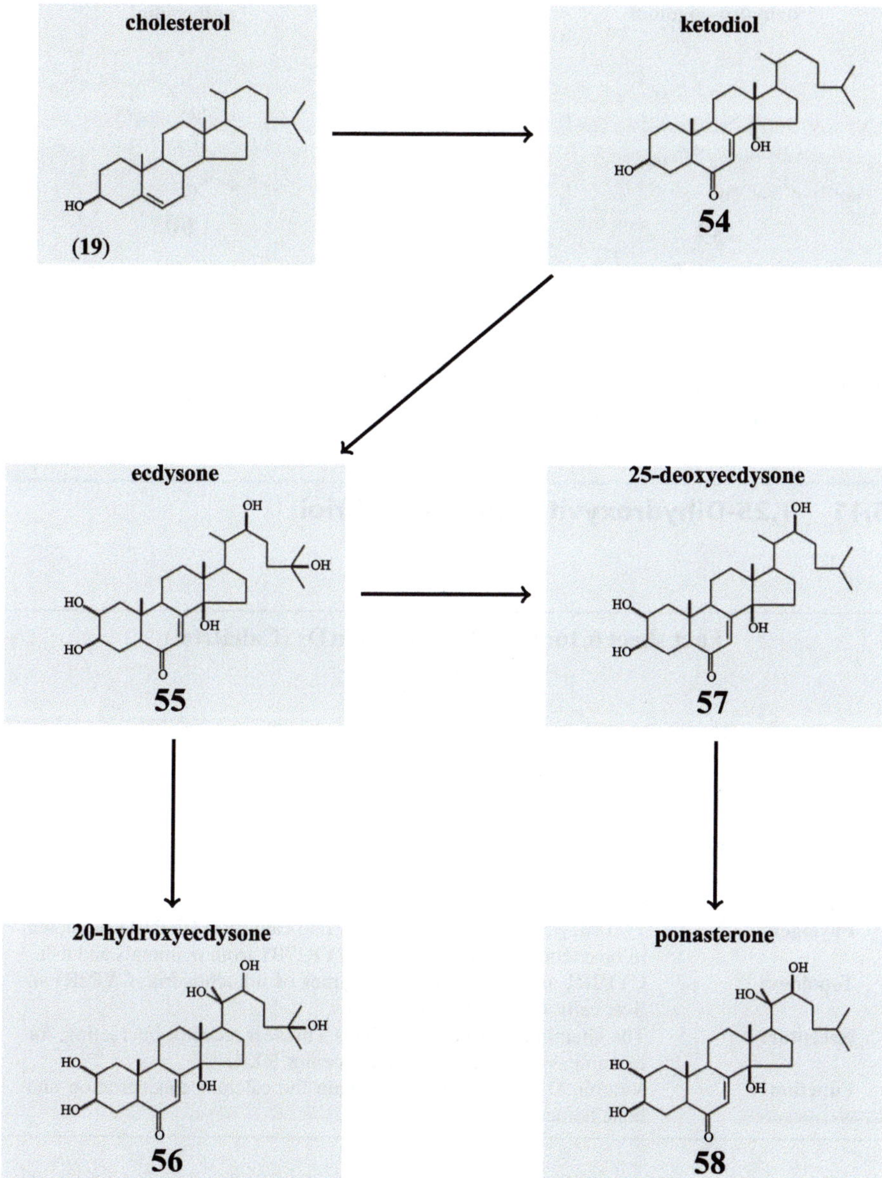

Fig. 6.23 Steroid hormones of insects and shellfish

The steroids antheridiol (**60**) and dehydro-oogoniol (**59**) from the fungi *Achlya ambisexualis* (Fig. 6.24) are differentially expressed in two isoforms called male and female. A female fungus secretes antheridiol which is recognized by a male

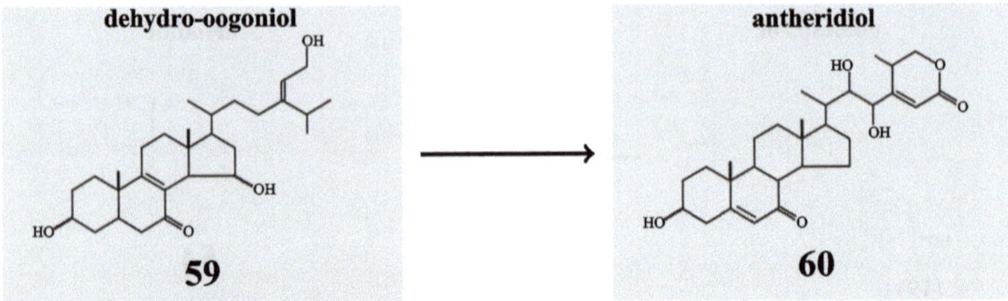

Fig. 6.24 Sex steroids in the fungus *Achlya ambisexualis*

fungus which in turn secretes oogoniol. Mutually attracted by their secreted steroids these fungi together develop fruit bodies.

6.11 1,25-Dihydroxyvitamin D_3 (Calcitriol)

Fact sheet 6.16: Dihydroxyvitamin D_3 (Calcitriol)

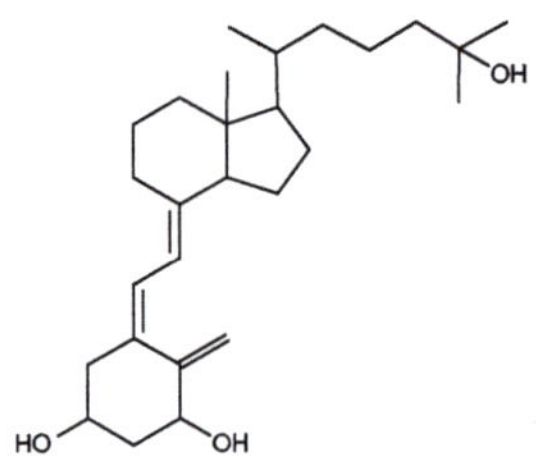

Structure:	**67**
Phylogeny:	Two enzymes convert vitamin D_3 into calcitriol: CYP2R1 is known in invertebrates and vertebrates, CYP27B1 from mammals and fish.
Topology:	CYP2R1 and CYP27B1 are enzymes of mitochondria: CYP2R1 in liver cells, CYP27B1 in the kidney.
Receptor:	The vitamin D3 receptor (VDR) is a nuclear receptor interacting, for example, with the retinoic acid receptor RXR.
Function:	Vitamin D_3 is required to maintain the calcium concentration and bone homeostasis.

Looking at the vitamin D_3 structure (**20**), one would not suppose that is derived from cholesterol. It is called a vitamin, but calcitriol is one hormone to control calcium in the blood (Fig. 6.26): Calcitriol (**67**) is necessary to take up calcium ions in the gut by the gut mucosa. Calcitriol synthesis and all other aspects of calcium homeostasis are then controlled by parathormone and by calcitonin.

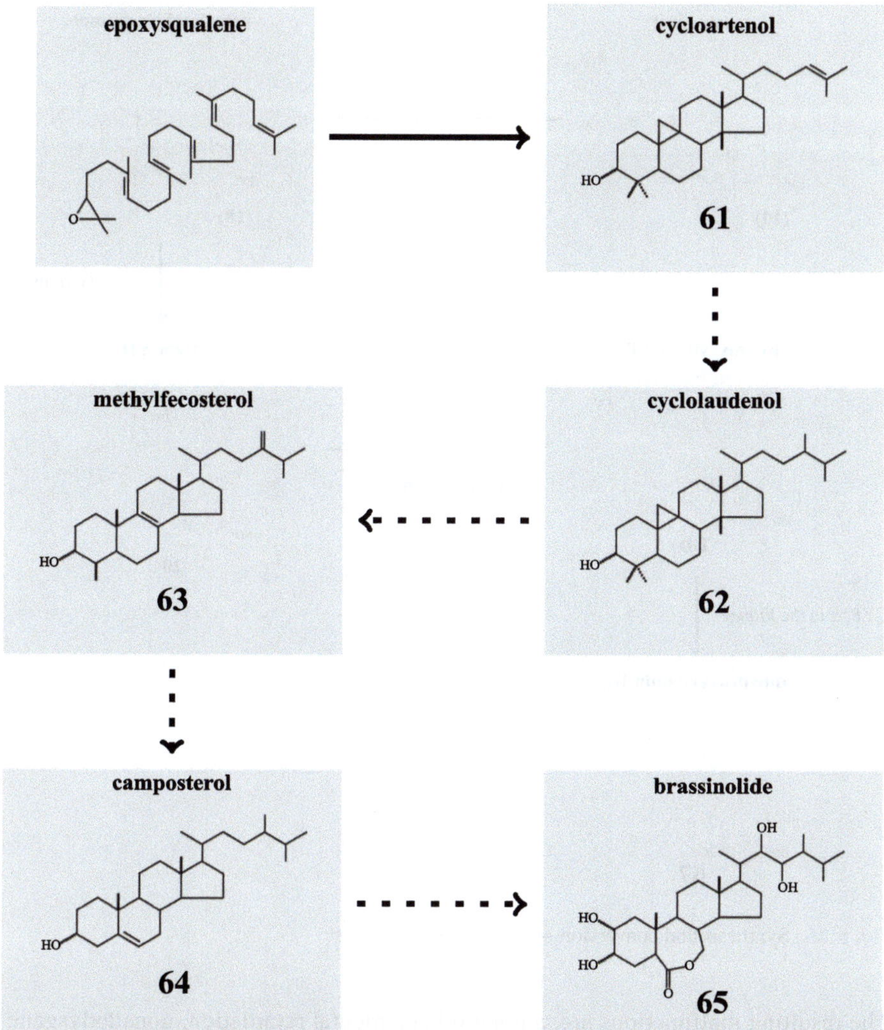

Fig. 6.25 Examples of plant steroids

6.11.1 Synthesis of Vitamin D_3

The synthesis of vitamin D_3 (**20**; cholecalciferol) from 7-dehydrocholesterol (**18**) is initiated by UV light (wavelength between 270 and 300 nm) in the human skin. 7-dehydrocholesterol is the last intermediate of the multistep cholesterol biosynthesis from lanosterol (**16**). The enzyme dehydrocholesterol reductase converting 7-dehydrocholesterol to cholesterol is mutated in Smith–Lemli–Opitz syndrome.

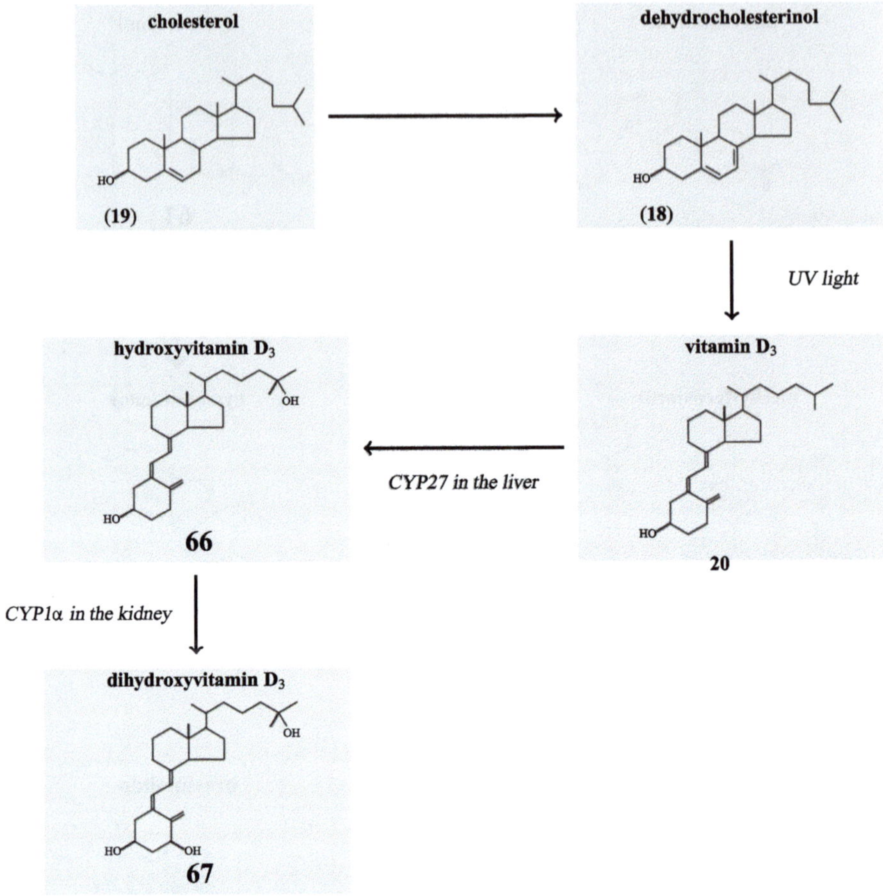

Fig. 6.26 Synthesis and conversion of vitamin D_3

The resulting malfunctions are, among others, mental retardation, gonadodysgenesis, polydactyly, ptosis, and dwarfism.

Stimulated by UV light the bond between carbon atoms 9 and 10 of the 7-dehydrocholesterol (for enumeration, see Fig. 6.10) of the second ring is opened. There is no enzyme necessary for this ring opening. Rotation around the bond between carbon atoms 6 and 7 is freely possible. The vitamin D_3 (**20**) is thus preferentially shown as in Fig. 6.26.

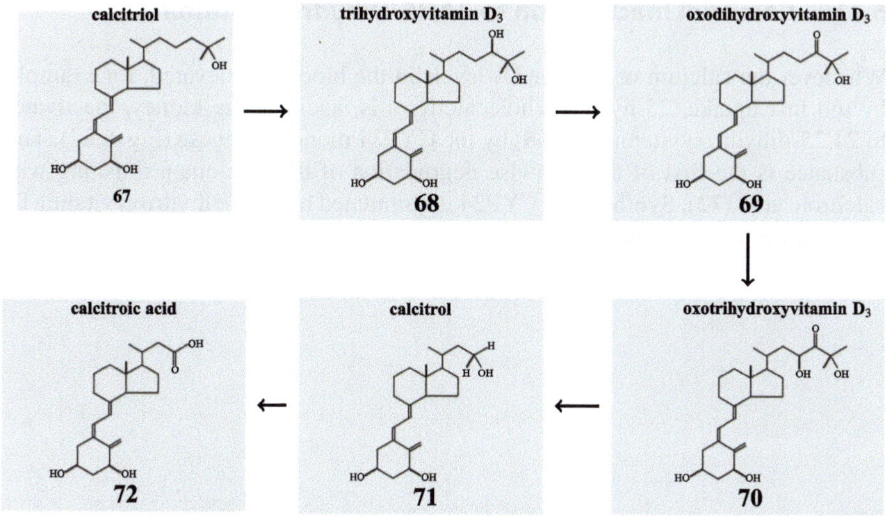

Fig. 6.27 Degradation of 1,25-dihydroxyvitamin D_3 by CYP24

6.11.2 Vitamin D Uptake from Food

Vitamin D_3 might be taken up as vitamin D_2 (24-methyl vitamin D_3, ergocalciferol) or as vitamin D_3 with the nutrition. The liver can utilize all three compounds.

6.11.3 Synthesis of 25-Hydroxyvitamin D_3

From the skin vitamin D_3 is transported to the liver by the vitamin D_3-binding protein. In the mitochondria of hepatocytes vitamin D_3 is oxidized at the C atom 25 by the CYP27 generating 25-hydroxyvitamin D_3 (**66**), or 25-hydroxycholecalciferol. The name CYP27 is due to the activity of the enzyme towards the C atom 27 of bile acids. 25-Hydroxyvitamin D_3 is firmly bound to the vitamin D binding protein (DBP) and circulates in the complex in the blood.

6.11.4 Synthesis of Calcitriol (1,25-Dihydroxyvitamin D_3)

When the calcium concentration in blood declines, 25-hydroxyvitamin D_3 is converted to 1,25-dihydroxyvitamin D_3 (calcitriol; **67**) in the kidney. This conversion is performed by the CYP1α. CYP1α is related to CYP27 and the CYP24 described below.

6.11.5 Calcitriol Inactivation to 24,25-Dihydroxyvitamin D_3

Whenever the calcium or vitamin D_3 levels in the blood are elevated, for example, by too fast uptake, 25-hydroxycholecalciferol is, again in the kidney, inactivated to 24,25-dihydroxyvitamin D_3 (**68**) by the CYP24 monooxygenase (Fig. 6.27). This substance is the first of the stepwise degradation of the side-chain finishing with calcitroic acid (**72**). Synthesis of CYP24 is stimulated by 1,25-dihydroxyvitamin D_3 in the kidney and in bone.

7 Hormones Derived by Amino Acid Conversion

Contents

7.1 Catecholamines 237
7.1.1 Dopamine 239
7.1.2 Noradrenaline and Adrenaline 240
7.2 Thyroxine: The Thyroid Hormone 240
7.3 Melatonin 243

The hormones derived from amino acids include catecholamines, serotonin, melatonin, and the thyroid hormones thyroxine and triiodothyronine.

7.1 Catecholamines

Four hormones are derived from the amino acid tyrosine (Fig. 7.1, **73**): L-3,4-dihydroxyphenylalanine (L-DOPA; Fig. 7.1, **74**), dopamine (Fig. 7.1, **75**), noradrenaline (Fig. 7.1, **76**), and the hormone of the adrenal medulla, adrenaline (Fig. 7.1, **77**); noradrenaline is also released from adrenal medulla cells (Table 7.1). Since the adrenal medulla is derived from neuroectoderm, there are many parallels to noradrenergic neurons.

In American English, the names epinephrine and norepinephrine are commonly used, whereas in British English, the corresponding names are adrenaline and noradrenaline: "epinephrine" is derived from Greek (*epi-nephros*) and "adrenaline" is derived from Latin (*ad-renalis*), but both mean "close to the kidney." In this book, we use the names "adrenaline" and 'noradrenaline" as these are better known internationally.

The original version of this chapter was revised.
An erratum to this chapter can be found at DOI 10.1007/978-3-319-15060-4_17

B. Kleine, W.G. Rossmanith, *Hormones and the Endocrine System*,
DOI 10.1007/978-3-319-15060-4_7

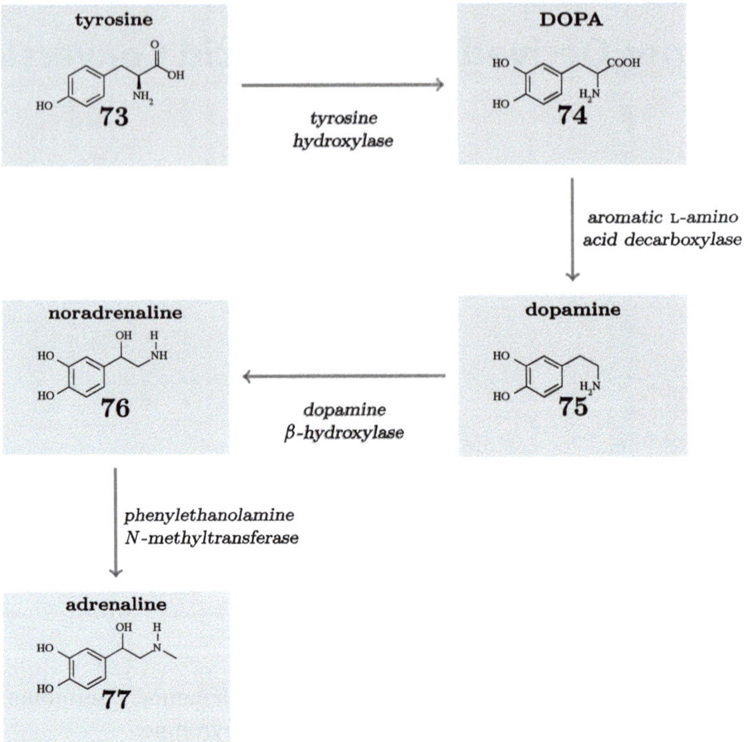

Fig. 7.1 Catecholamine biosynthesis

Table 7.1 Cells secreting catecholamines

Catecholamine	Site of synthesis
DOPA → dopamine	Dopaminergic neurons
DOPA → dopamine →noradrenaline	Noradrenergic neurons
DOPA → dopamine →noradrenaline → adrenaline	Chromaffin adrenal cells

The first two reactions from tyrosine to adrenaline occur in the cytosol. The key enzyme is tyrosine hydroxylase (Fig. 7.1). Its activity determines the amount of noradrenaline and adrenaline produced. Dopamine is active within vesicles. The next enzyme, dopamine β-hydroxylase, is active only inside vesicles. It is expressed in noradrenergic neurons and in the adrenal medulla. The last enzyme of catecholamine biosynthesis, phenylethanolamine *N*-methyltransferase, is *the* characteristic enzyme of chromaffin cells of the adrenal medulla and is stimulated by adrenocorticotropic hormone. Phenylethanolamine *N*-methyltransferase is located in the cytosol; noradrenaline, for it to be converted into adrenaline, has to be transported out of vesicles. Transport into vesicles is performed by vesicular monoamine transporters 1 and 2, which exchange monoamines for hydrogen and maintain a proton gradient across the vesicular membrane (Henry et al. 1994; Erickson et al. 1996; Cartier et al. 2010). How the export of noradrenaline from vesicles is performed is not clear.

Owing to differential expression of the catecholamine-generating hormones, there are some neurons that secrete dopamine (these do not express dopamine β-hydroxylase), many neurons that secrete noradrenaline, and in only in the adrenal medulla cells neurons capable of secreting adrenaline and to a lesser degree noradrenaline. These latter chromaffin cells are of neuroectodermal origin, in contrast to the adrenal cortical cells, which have a mesodermal origin.

The take-home lesson is that noradrenaline:

1. Is a neurotransmitter acting directly from noradrenergic neurons on target cells such as muscle cells via synapses.
2. Is also a hormone from the adrenal medulla acting in an endocrine manner on cells other than those in the adrenal medulla.

Adrenaline, however, is never a neurotransmitter, but is a hormone distributed via the circulation and acting on receptors on a variety of target cells—for example, nerve cells. Some recent findings point to adrenaline synthesis within the brain, but there have been no conclusive functional studies so far.

7.1.1 Dopamine

Dopamine is the hormone controlling prolactin release from lactotropic pituitary cells. These cells would, without stimulation, release prolactin if they were not inhibited by dopamine. Such dopamine stems from hypothalamic neurons releasing dopamine into the median eminence, from where it reaches the pituitary by the portal system of the pituitary stalk in order to inhibit prolactin release

Fact sheet 7.1: Dopamine

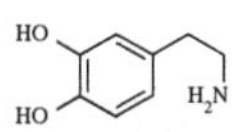

Structure: 75

Function: An intermediate in adrenaline/noradrenaline biosynthesis, a neurotransmitter, and a hormone for the tonic suppression of pituitary prolactin release.

Receptor: Five different dopamine receptors; dopamine receptor D_2 suppresses prolactin release in the pituitary.

The lack of dopamine control results in elevated prolactin levels. Low dopamine and elevated prolactin levels are physiological at the end of gestation. Other brain neurons release dopamine, but these act with dopamine as a neurotransmitter. Degeneration of dopaminergic neurons of the substantia nigra is one of the causes of Parkinson disease.

Prolactin release induced by thyrotropin-releasing hormone can be inhibited by dopamine, whereas stimulation by oxytocin and vasoactive intestinal peptide (see Sect. 4.10) cannot be blocked by dopamine.

7.1.2 Noradrenaline and Adrenaline

Whereas noradrenaline is almost exclusively released at synapses, adrenaline is secreted into fenestrated capillaries as in other neurohemal organs.

Fact sheet 7.2: Noradrenaline (norepinephrine)

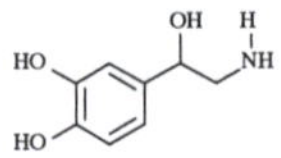

Structure: **76**
Function: Mainly a neurotransmitter and an intermediate in adrenaline biosynthesis.
Receptor: α-Adrenergic and β-adrenergic receptor subtypes.

Noradrenaline is the neurotransmitter of the sympathetic nervous system. On a given heart cell, noradrenaline acts via synapses, whereas adrenaline acts via the circulation. Adrenergic receptors recognize both compounds, but with different affinities.

Fact sheet 7.3: Adrenaline (epinephrine)

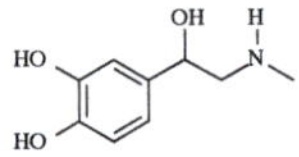

Structure: **77**
Function: Formed in the adrenal medulla from noradrenaline and released as a hormone, accelerates the pulse, contracts blood capillaries, and is the hormone of "fight or flight" reactions in the brain.
Receptor: Binds as noradrenaline to α-adrenergic and β-adrenergic receptors.

In response to stress, coldness, fatigue, shock, or hypoglycemia, for example, adrenaline is secreted with acetylcholine as a neurotransmitter. In the liver, adrenaline stimulates emptying of glycogen stores and gluconeogenesis to cause enhanced glucose consumption. Under the influence of catecholamines, furthermore, lipid catabolism is accelerated and lipid and ketone body levels in the blood are increased.

7.2 Thyroxine: The Thyroid Hormone

Like the catecholamines, the thyroid hormone thyroxine (**78**) and its active derivative triiodothyronine (**79**) are synthesized from tyrosine. In contrast to catecholamines, which are derived from the amino acid monomer tyrosine, the two tyrosine molecules giving rise to thyroxine are part of the polypeptide thyroglobulin,

Fact sheet 7.4: Thyroxine and triiodothyronine

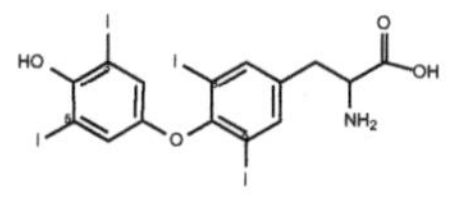

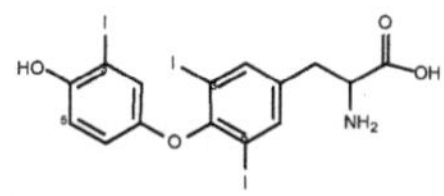

Structure: **78** **79**

Function: Thyroxine and triiodothyronine are required in any cell for metabolism, for development, and for maturation.

Receptor: Two receptor subtypes with three (TR-α) and two (TR-β) splice variants; only TR-α_1, TR-β_1, and TR-β_2 can bind to the hormone; the receptors form homodimers, but preferentially form heterodimers with retinoid X receptor, a retinoic acid receptor.

which—almost exclusively translated in the thyroid gland—contains many tyrosine residues. It is well known that halogens such as iodine, bromine, and chlorine are easily coupled to aromatic six-membered rings such as those of benzene, phenol, estradiol, the amino acids phenylalanine and tyrosine, and the tyrosine of thyroglobulin.

Any iodine taken up with food by the human organism is collected in the thyroid gland by coupling it to the thyroglobulin's tyrosine residues. Iodinated thyroglobulin is afterward stored in the characteristic follicles of the thyroid gland as a colloid (Fig. 7.2).

After its release from the pituitary, thyroid-stimulating hormone (TSH) signals to the thyroid the need for thyroxine. In the thyroid on TSH stimulation, thyroglobulin from the follicles is recruited and hydrolyzed to thyroxine. Thus, there is no thyroxine store, and thyroxine is freshly produced instead (Fig. 7.2).

The iodine-removing enzyme deiodinase converts thyroxine, with four iodine substituents (which is why thyroxine is also called T_4), into triiodothyronine, with three iodine groups (which is why triiodothyronine is also called T_3). This triiodothyronine binds the nuclear thyroid hormone receptor with enhanced activity and is the active hormone in all tissues. Deiodinase is therefore an enzyme in tissues depending on triiodothyronine.

Iodine Coupling to Thyroglobulin (Steps 1 and 2) During the storing of iodine, thyroid peroxidase is active. This enzyme is activated by hydrogen peroxide generated from water by the reduction of the cofactor $NADP^+$. With hydrogen peroxide, iodide is oxidized to molecular iodine (I_2). By the action of thyroid peroxidase, two iodine atoms are successively added to certain tyrosine residues of thyroglobulin (electrophilic substitution of aromatic rings; see Fig. 7.2).

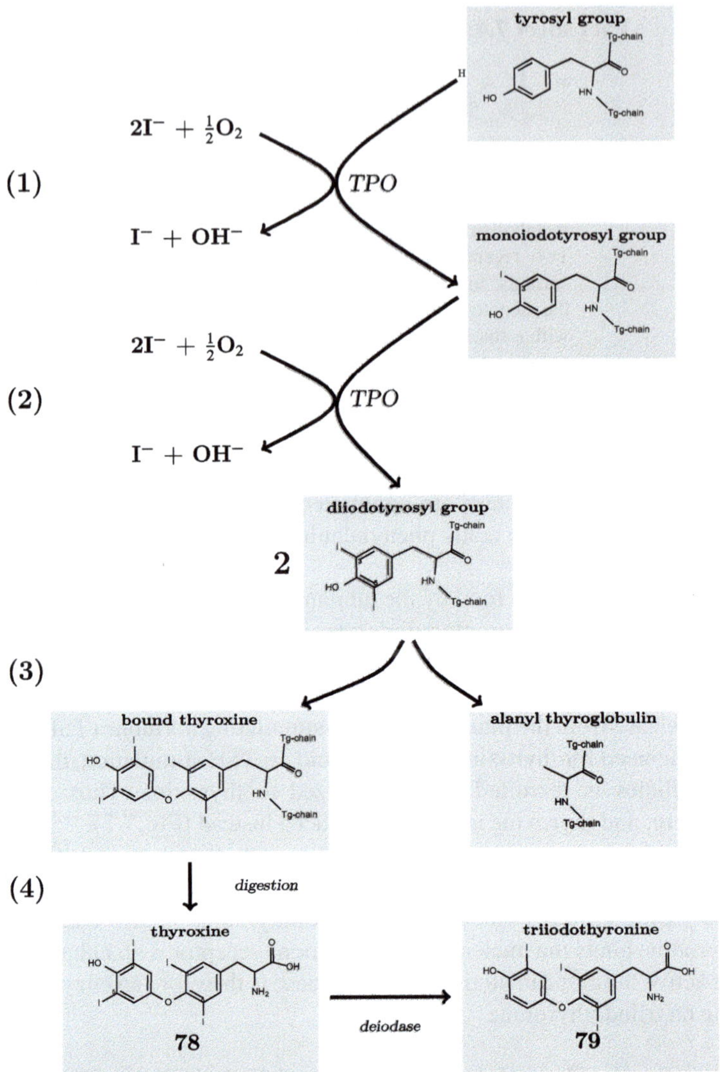

Fig. 7.2 Uptake of iodine and synthesis of thyroxine and triiodothyronine. *Tg* thyroglobulin, *TPO* thyroid peroxidase

Formation of the Thyroxyl Group (Step 3) In the next step of thyroxine biosynthesis, a diiodinated aromatic ring of one tyrosine residue is transfered to another diiodinated tyrosyl group, thus generating the thyroxyl group still coupled to the thyroglobulin. This leaves an alanyl residue (from where the transfered ring comes) and the tetraiodinated thyroxyl residue.

These iodination reactions occur independently of any demand for thyroxine by, for example, TSH and whenever iodine is available. The iodinated thyroglobulin is stored in the follicular colloid.

Release of Thyroxine (Step 4) Stimulated by pituitary TSH, the process of thyroxine release is initiated (Fig. 7.2). By pinocytosis, follicular colloid is taken up into intracellular vesicles. The thyroid follicular cells possess other vesicles, lysosomes filled with endopeptidases. The vesicles with the thyroglobulin are fused to these lysosomes, and the thyroglobulin is degraded by the endopeptidases, generating thyroxine. The free thyroxine molecules diffuse through all membranes and reach the blood to be distributed throughout the circulation.

Deiodinases Thyroxine is converted by deiodinase to triiodothyronine (Fig. 7.2). There are three types of this enzyme. All three are selenoenzymes with a selenocysteine residue. For deiodinase activity, the selenium atom is indispensable. In the case of selenium deficiency, the activity of these deiodinases is maintained longer than that of any other selenoprotein.

Thyroxine and triiodothyronine are major contributors to energy metabolism. They stimulate the basic metabolic rate. Lack of triiodothyronine may be lethal. Reduced triiodothyronine levels have been observed during illness and stress; this has been taken as a caloric reduction and conservation of energy.

Thyroid hormones are bound in blood and are transported by thyroxine-binding globulin (TBG); TBG is encoded by the X chromosome and is translated by hepatocytes. TBG deficiency as well as TBG excess might cause metabolic disturbances.

TBG belongs to the serpin proteinase inhibitors. When TBG is cleaved by an elastase, thyroid hormone is released. Whether cells expressing the triiodothyronine nuclear receptor equally express a surface enzyme to free thyroxine by cleavage of TBG has not yet been analyzed. Such a fact could solve the problem of how hormones target the intracellular receptor. A local release of the hormone by cleavage of the transporter protein would at least be highly feasible.

7.3 Melatonin

Melatonin (Fig. 7.3, **84**) is a product of the pineal gland. This gland with its raspberry-like cells (pinealocytes) and its melatonin has posed many questions for endocrinologists. After a variety of misinterpretations of melatonin's role, this field of endocrine research has lost some focus. Many of the hopes linked to melatonin—for example, its use as an anticancer drug—have not been fulfilled.

Fact sheet 7.5: Melatonin

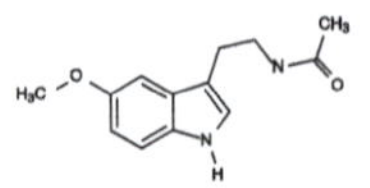

Structure: 84

Function: Melatonin from the pineal gland provides input into the circadian clock in the suprachiasmatic nucleus. Cells in the pars tuberalis measure the circannual rhythm via the nightly melatonin production: a drop in seasonal melatonin level signals springtime and enhanced prolactin production.

Receptor: Two melatonin receptors, MTNR1 and MTNR2, are G-protein-coupled receptors; MTNR1 is modulated by an orphan G-protein-coupled receptor: GPR50.

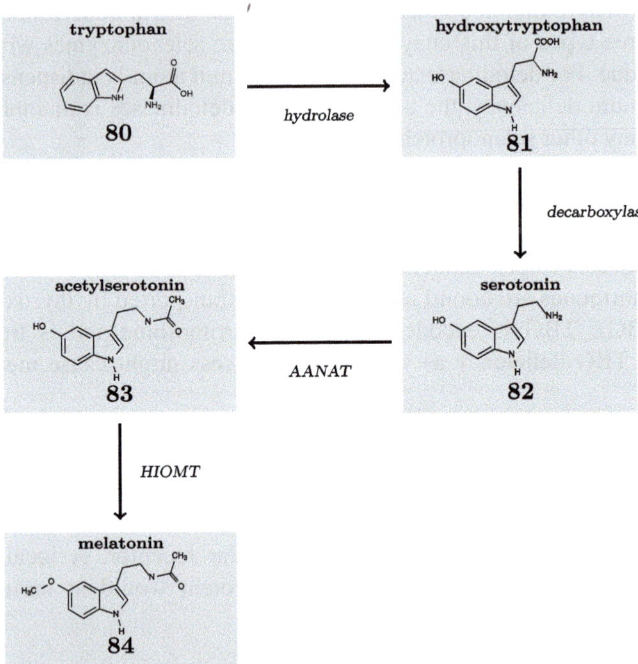

Fig. 7.3 Melatonin synthesis from L-tryptophan. *AANAT* arylalkylamine *N*-acetyltransferase, *HIOMT* hydroxyindole *O*-methyltransferase

Melatonin, 5-methoxy-*N*-acetyltryptamine, is derived from tryptophan (Fig. 7.3, **80**) in a similar way as noradrenaline is derived from tyrosine. An analogous hydroxylation of the aromatic ring (see the synthesis of DOPA) and a following decarboxylation generate 5-hydroxytryptamine (Fig. 7.3, **82**; serotonin). To this serotonin, arylalkylamine *N*-acetyltransferase (AANAT) adds an acetyl group, and hydroxyindole *O*-methyltransferase (HIOMT) (aka acetylserotonin *O*-methyltransferase) transfers a methyl group to the hydroxyl group, which concludes melatonin formation.

Formation of melatonin occurs in a light–dark cycle: in the dark, tenfold more melatonin is released than in brightness. The zeitgeber is located in mammals in the suprachiasmatic nucleus, a hypothalamic region near the optic nerve from where noradrenergic neurons control pineal activity. Mostly β-adrenergic receptors on the pinealocytes mediate the enzymatic activity.

AANAT stimulation is dependent on cyclic adenosine monophosphate. After stimulation of adenylate cyclase by signal transduction from the adrenergic receptors, enhanced cyclic adenosine monophosphate stimulates protein kinase A, which phosphorylates AANAT. Phosphorylated AANAT recruits a 14-3-3 protein. In the complex, the phosphorylation site is blocked from phosphatases and the AANAT is protected from inactivation. The activity of AANAT is thus strongly increased.

The final step of melatonin synthesis, methylation of the 5-hydroxyl group, is catalyzed by HIOMT. The transcription of this enzyme is directly controlled by β-adrenergic receptors and can be inhibited by an adrenergic antagonist. The methyl group cofactor is *S*-adenosylmethionine, whose synthesis is controlled in parallel to the transcription of HIOMT.

Of the many suggested actions of melatonin, only its action as a regulator of sleep has stood the test of time. Whoever has to adapt his or her daily rhythm after traveling across several time zones prefers to take melatonin so as to facilitate adaption to the local light–dark rhythms within the shortest possible time.

Very recent experiments have shown that melatonin helps to control adaption to the yearly seasons. By summing up nightly melatonin amounts, so-called calendar cells in the pituitary stalk estimate differences between light and dark seasons and help the physiological processes to adapt appropriately. This is described in Sect. 12.4.

Hormone Receptors

8

Contents

8.1 Nuclear Receptors 248
8.2 Heptahelical Transmembrane Receptors 249
8.2.1 G Proteins 252
8.2.2 Receptor–G Protein Interactions 253
8.2.3 Target of G Proteins 254
8.2.4 Variability by Differentially Expressed Receptor Subtypes: Somatostatin Receptors 254
8.3 Receptors with Tyrosine Kinase Activity 255
8.4 Membrane Receptors with Serine/Threonine Kinase Activity 257
8.5 Membrane Receptors Without Kinase Activity 257
8.6 Membrane Steroid Receptors: Still Unknown? 258

Hormone receptors belong mainly to two large receptor groups: heptahelical G-protein-coupled membrane receptors and cytosolic nuclear receptors for steroids and thyroid hormones. The exceptions are particularly the insulin and the growth hormone receptors

The signals initiated by the main receptor groups have characteristic differences:

- *Fast-acting receptor signals.* G-protein-coupled receptors (GPCRs) trigger immediate reactions which are converted into long-lasting regulations by other intracellular interaction partners. Ion channel openings, membrane depolarization, release of calcium from intracellular calcium stores, activation of kinases, and lipid conversions by phospholipases are influenced by GPCRs. These reactions occur within seconds or faster. If these immediate reactions are transduced into the cellular nucleus and genes are activated, long-lasting modifications can occur.
- *Slow, genomic changes.* Cytosolic receptors for steroids or the thyroid hormones do not trigger immediate reactions. Hormones which have reached the cytosol by diffusion through the cell membrane initiate hormone receptor dimerization.

B. Kleine, W.G. Rossmanith, *Hormones and the Endocrine System*,
DOI 10.1007/978-3-319-15060-4_8

By this dimerization, a nuclear import signal for the entire complex is generated. After transfer to the cellular nucleus, the dimer associates with characteristic recognition motifs on the DNA. Such an interaction modulates the activity of the associated gene, either activation of transcription or its inhibition. Whereas GPCRs trigger fast signal cascades at the membrane, nuclear receptors act in the cellular nucleus and modify gene activities. These reactions take longer than seconds: they need minutes or hours.

8.1 Nuclear Receptors

In 2004, the Lasker Award was given to three biochemists: Elwood Jensen, Ronald Evans, and Pierre Chambon. They had presented evidence for receptors which are today called nuclear receptors and studied their features (see Fig. 8.1). Forty-eight human nuclear receptors have been identified which recognize steroids, thyroxine, triiodothyronine, vitamin D_3, and vitamin A (Table 8.1). In the liver, many of these nuclear receptors target drugs and poisons.

There are three characteristic features of each nuclear receptor:

1. *Ligand binding.* All nuclear receptors have a domain where the ligand is accepted—for example, the hormone, the vitamin, or the toxin.
2. *Dimerization.* After ligand binding, two ligand-bound receptors form a dimer.

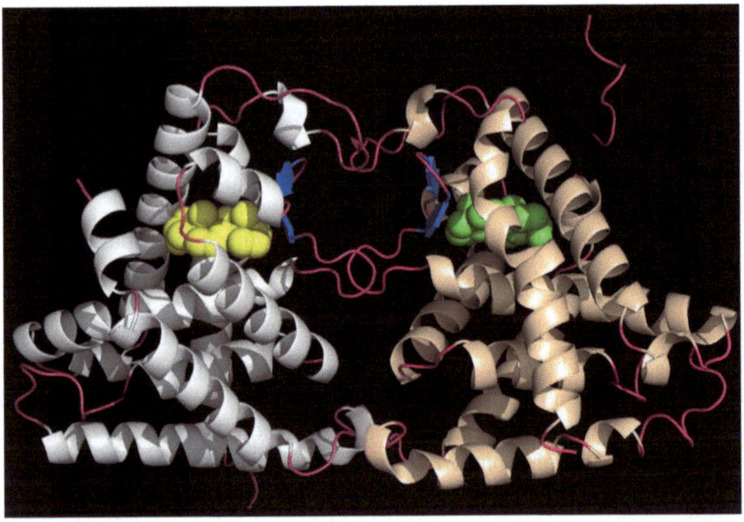

Fig. 8.1 Model of a nuclear receptor: The murine androstane receptor. Androstane (*yellow spheres* and *green spheres*) is embedded into the receptor structure, which consists mainly of helices and a single β sheet (*blue*). Two receptor molecules adhere at the side with the β sheet. This generates the nuclear import signal. The dimer is imported into the cellular nucleus; by binding to recognition motifs, it modulates gene activity (Produced with PyMOL using Protein Data Bank entry 1XNX)

Table 8.1 Nuclear receptors

Hormone/ligand	Receptor	Type	OMIM entries
Thyroid hormone	ErbA (TR)	α, β	190120, 190160
Estrogen	ER	α, β	133430, 601663
Progesterone	PR		607311
Testosterone, DHT	AR		133700
Mineralocorticoid	MR (NR3C2)	α, β	600983
Glucocorticoid	GR		138040
Vitamin D	VDR		601769
All-*trans*-retinoic acid	RAR	α, β, γ	180240, 180220, 180190
9-*cis*-Retinoic acid	RXR	α, β, γ	180245, 180246, 180247
Oxysterol	LXR	α, β	602423, 600380
Bile acids	FXR		603826
Fatty acids	PPAR	α, γ, δ	170998, 601487, 600409
Pregnanes, xenobiotics	PXR (NR1I2)		603065
Xenobiotics	CAR		603881

AR androgen receptor, *CAR* constitutive androstane receptor, *DHT* dihydrotestosterone, *ER* estrogen receptor, *FXR* farnesoid X receptor, *GR* glucocorticoid receptor, *LXR* liver X receptor, *MR* mineralocorticoid receptor, *NR1I2* nuclear receptor subfamily 1, group I, member 2, *NR3C2* nuclear receptor subfamily 3, group C, member 2, *OMIM* Online Mendelian Inheritance in Man (http://www.omim.org/), *PPAR* peroxisome-proliferator-activated receptor, *PR* progesterone receptor, *PXR* pregnane X receptor, *RAR* retinoic acid receptor, *RXR* retinoid X receptor, *TR* thyroid hormone receptor, *VDR* vitamin D receptor

3. *DNA binding.* The dimerization of two nuclear receptors creates a nuclear import signal. This signal triggers transport of the dimer into the cellular nucleus and allows the dimer to bind to its recognition sites. Any functional nuclear receptor has its own recognition sites. After binding of the dimer to the DNA, gene activity in that chromosomal region is modulated. The transcriptional activity might be enhanced or reduced. This will eventually stimulate or suppress cellular functions.

8.2 Heptahelical Transmembrane Receptors

In contrast to steroids, which dock to intracellular nuclear receptors, peptide/protein hormones bind to receptors on the surface of cells. Most of these receptors belong to a protein family where the membrane is spanned sevenfold ("seven" is *hepta* in Greek) by helices, so-called heptahelical receptors (Tables 8.2 and 8.3, Fig. 8.2).

By binding the receptor ligand on the cellular surface, these receptors initiate signal transduction by coupling to intracellular, guanosine triphosphate (GTP)-binding proteins, the G proteins.

Table 8.2 Class A heptahelical receptors: receptors related to rhodopsin and β-adrenergic receptor

Hormone/ligand	Receptor	Subtypes	OMIM entries
Adenosine	Adenosine receptor	A1, A2a, A2b, A3	102775, 102776, 600446, 600445
MSH, ACTH	MC1-R (MSH-R in melanocytes), MC2-R (ACTH-R in the adrenal cortex), MC3-R (in the CNS), MC4-R (AgRP in the hypothalamus), MC5-R (in exocrine glands)		155555, 202200, 155540, 155541, 600042
Noradrenaline, adrenaline	α-Adrenergic receptors	1A, 1B, 1D	104221, 104220, 104219
		2A, 2B, 2C	104210, 104260, 104250
	β-Adrenergic receptors	1, 2, 3	109630, 109690, 109691
Dopamine	DPR	1a, 1b, 2	126449, 126453, 126450
Serotonin (5-hydroxytryptamine)	HTR	1a, 1b, 1d, 1e, 1f	109760, 182131, 182133, 182132, 182134
		2a, 2b, 2c	182135, 601122, 312861
		3a, 3b, 5a, 6, 7	182139, 604654, 601305, 601109, 182137
Acetylcholine	Muscarinic AChR[a]	1, 2, 3, 4, 5	118510, 118493, 118494, 118495, 118496
Angiotensin II	AGTR	1, 2	106165, 300034
Bradykinin	BDKR	B1, B2	600337, 113503
Bombesin	BRS3		300107
Gastrin-releasing peptide	GRPR		305670
Cholecystokinin	CCK-R	A, B	119444, 119445
Neuromedin B	NMBR		162341
Neuromedin U	NMUR	1, 2	604153, 605108
Neuropeptide Y	NPYR	1, 2, (3)[b], 5	162641, 162642, (162643), 602001
Oxytocin	OXTR		167055
Arginine vasopressin	AVPR	1a, 1b	600821, 600264
Galanin	GALR	1, 2, 3	600377, 603691, 603692
Somatostatin	SSTR	1, 2, 3, 4, 5	182451, 182452, 182453, 182454, 182455
GnRH	GnRHR		138850

(continued)

Table 8.2 (continued)

Hormone/ligand	Receptor	Subtypes	OMIM entries
TRH	TRHR		188545
Melatonin	MTNR	1a, 1b	600665, 600804
FSH	FSHR		136435
LH/hCG	LHCGR		152790
TSH	TSHR		603372

AChR acetylcholine receptor, *ACTH* adrenocorticotropic hormone, *ACTH-R* adrenocorticotropic hormone receptor, *AgRP* agouti-related peptide receptor, *AGTR* angiotensin receptor, *AVPR* arginine vasopressin receptor, *BDKR* bradykinin receptor, *BSR3* bombesin receptor subtype 3, *CCK-R* cholecystokinin receptor, *CNS* central nervous system, *DPR* dopamine receptor, *FSH* follicle-stimulating hormone, *FSHR* follicle-stimulating hormone receptor, *GALR* galanin receptor, *GnRH* gonadotropin-releasing hormone, *GnRHR* gonadotropin-releasing hormone receptor, *GRPR* gastrin-releasing peptide receptor, *hCG* human choriogonadotropin, *HTR* 5-hydroxytryptamine receptor, *LH* luteinizing hormone, *LHCGR* luteinizing hormone/human choriogonadotropin receptor, *MC1-R* melanocortin 1 receptor, *MC2-R* melanocortin 2 receptor, *MC3-R* melanocortin 3 receptor, *MC4-R* melanocortin 4 receptor, *MC5-R* melanocortin 5 receptor, *MNTR* melatonin receptor, *MSH* melanocyte-stimulating hormone, *MSH-R* melanocyte-stimulating hormone receptor, *NMBR* neuromedin B receptor, *NMUR* neuromedin U receptor, *NPYR* neuropeptide Y receptor, *OXTR* oxytocin receptor, *SSTR* somatostatin receptor, *TRH* thyrotropin-releasing hormone, *TRHR* thyrotropin-releasing hormone receptor, *TSH* thyroid-stimulating hormone, *TSHR* thyroid-stimulating hormone receptor

[a]Nicotinic AChRs do not belong to the G-protein-coupled receptors

[b]NPYR3 does not bind neuropeptide Y, instead in binds stromal-cell-derived factor; NPYR3, aka CXCR4, is a co-receptor of HIV

Table 8.3 Class B heptahelical receptors: receptors related to the secretin receptor

Hormone/ligand	Receptor	Subtypes	OMIM entries
Glucagon	GcgR		138033
Glucagon-like peptide	GlP-R		138032
Secretin	SctR		182098
Vasoactive intestinal peptide	VIPR	1, 2	192321, 601970
GHRH	GHRH-R		139191
Calcitonin	CTR		114131
Calcitonin-gene-related peptide	CGRPR		114190[a]
Corticotropin-releasing hormone	CRHR	1, 2	122561, 602034[b]

CGRPR calcitonin-gene-related peptide receptor, *CRHR* corticotropin-releasing hormone receptor, *CTR* calcitonin receptor, *GcgR* glucagon receptor, *GHRH* growth-hormone-releasing hormone, *GHRH-R* growth-hormone-releasing hormone receptor, *GlP-R* glucagon-like peptide receptor, *SctR* secretin receptor, *VIPR* vasoactive intestinal peptide receptor

[a]Active only together with receptor-activity-modifying protein 1; OMIM entry 605153

[b]CRHR2 has another specific ligand, urocortin III

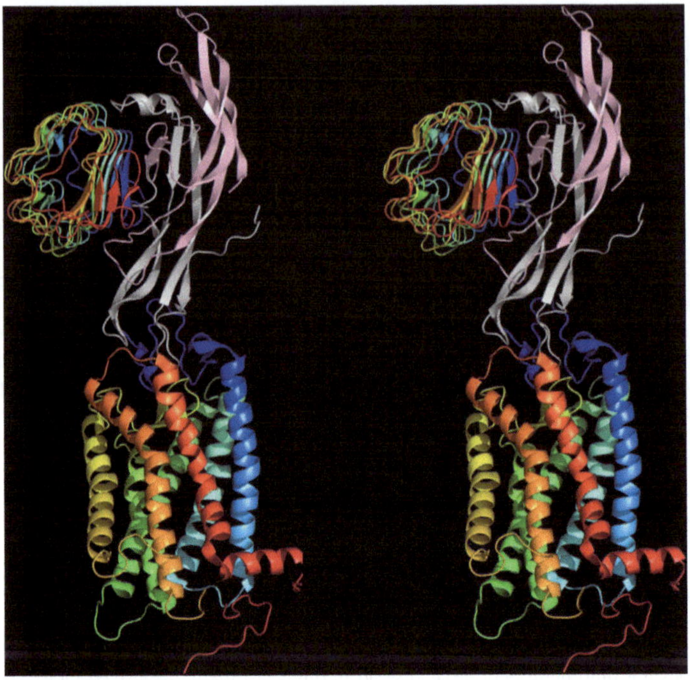

Fig. 8.2 Composite model of a heptahelical receptor. The follicle-stimulating hormone (FSH) receptor has a heptahelical membrane domain coupled to an N-terminal leucine-rich-repeat ectodomain like other glycoprotein hormone receptors and a few other so-called leucine-rich-repeat-containing G protein receptors (LGR). Since no complete LGR structure has been published, we modeled the ectodomain bound by FSH (Protein Data Bank entry 1XWD) to a rhodopsin heptahelical domain (Protein Data Bank entry 1F88) to show receptor–ligand interaction on the surface of cells. The heptahelical domain is inserted into the membrane, and the ectodomain lies on the outside of the cell. FSH binds with its single α-chain helix and the long β-chain loop enclosing the α chain (see Fig. 4.22) to the ten parallel ectodomain β sheets. At the bottom of the FSH dimer, both chains interact with the membrane domain in an as yet unknown stereochemistry. The interaction is testable with FSH mutants (Produced with PyMOL using Protein Data Bank entries 1XWD and 1F88)

8.2.1 G Proteins

G proteins are characterized by changes in conformation depending on whether GTP or guanosine diphosphate (GDP) is bound. The features of G proteins are as follows:

- *GDP/GTP binding.* All G proteins are able to bind either GDP or GTP. The nucleotide is placed into a characteristic ligand pocket.
- *GTP hydrolysis.* In this binding pocket, GTP can be hydrolyzed to GDP and phosphate. There is a faint intrinsic GTPase activity in any G protein;

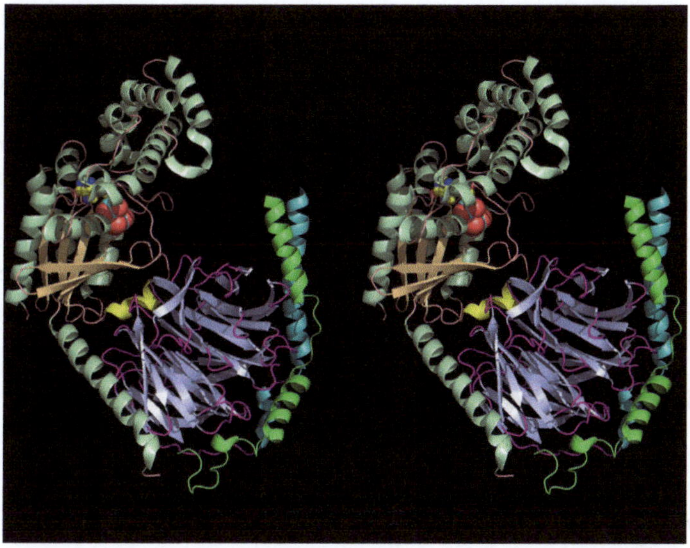

Fig. 8.3 Stereo-model of a heterotrimeric G protein. Three subunits—alpha subunit (*entire upper part* and *lowermost left helix*) with GDP bound, beta subunit (*lower part light blue* β *sheets* and *blue helix and pink loops*), and gamma subunit (*green*)—form the inactive molecule. GDP is shown with *yellow* carbon atoms, *blue* nitrogen atoms, *red* oxygen atoms, and *turquoise* phosphorus atoms. When the receptor binds to the G protein and induces GDP–GTP exchange, the next phosphate group presses on the β-sheet structure (*light orange*), and this in turn forces the contact helix (*yellow*) to press the beta–gamma complex away. An activated G protein consists of only the alpha subunit (From Wall et al. 1995; produced with PyMOL using Protein Data Bank entry 1GP2)

this activity, however, can be enhanced by interaction with GTPase-activating proteins.

- *Guanosine nucleotide exchange*. The hormone receptor with its ligand bound also changes its conformation. These conformational changes allow the interaction of the receptor with a G protein, which triggers GDP–GTP exchange.

Those G proteins binding to heptahelical membrane receptors are formed from three polypeptide chains called alpha, beta, and gamma subunits. The GDP/GTP-binding pocket is within the alpha subunit (Fig. 8.3).

8.2.2 Receptor–G Protein Interactions

The heptahelical membrane receptor changes in conformation after ligand binding. This change makes the intracellular loops between the different membrane domains of the receptor interact with G proteins. The first effect of this interaction is exchange of GDP for GTP. Activated heptahelical GPCRs are GDP–GTP exchange proteins.

With the exchange of GDP for GTP, the alpha subunit of the G protein undergoes structural changes which loosen the interaction of the alpha subunit with the beta–gamma complex and which generate a single alpha subunit with GTP bound and a beta–gamma dimer. The GTP–alpha monomer interacts with a variety of cellular targets. Eventually the intrinsic GTPase activity hydrolyzes GTP to GDP, rendering the GDP–alpha monomer inactive and allowing trimerization with the beta–gamma dimer. GTPase catalyzes hydrolysis, but this can be strongly enhanced by other proteins interacting with the GTP–alpha monomer. The beta–gamma dimer itself can stimulate enzymes within the cells—for example, adenylate cyclase—but its targets are less numerous than those of the GTP–alpha complex.

8.2.3 Target of G Proteins

A variety of intracellular enzymes can be activated by G proteins:

- *Adenylate cyclase.* This enzyme converts adenosine triphosphate to cyclic adenosine monophosphate (cAMP).
- *Guanylate cyclase.* This enzyme converts GTP to cyclic guanosine monophosphate (cGMP).
- *Phospholipases.* These cleave phospholipids. By this mechanism, messengers such as inositol trisphosphate, phosphocholine, and diacylglycerol or eicosanoids such as arachidonic acid and lysophosphatides are generated.
- *Sphingomyelinase.* If this enzyme is activated, then the membrane lipid sphingomyelin is cleaved to ceramide and phosphocholine.
- *Ion channels.* The G-protein alpha subunit subtypes o1 and o2 act on potassium channels.

Inositol trisphosphate, phosphocholine, diacylglycerol, arachidonic acid, ceramide, cAMP, and cGMP are messengers inducing other modifications in cells. Since hormones were called the primary messengers, first cAMP and later all these substances were called second messengers.

8.2.4 Variability by Differentially Expressed Receptor Subtypes: Somatostatin Receptors

Somatostatin is a hormone circulating in the blood. It is made in the hypothalamus as well as in the gastrointestinal tract and the endocrine pancreas, and its target cells are various other cells of the endocrine system. The specific action of somatostatin on the given target cells is achieved not by structural variation of somatostatin (14 versus 28 amino acids) but by variation of the somatostatin receptors.

There are five somatostatin receptors (Fig. 8.4), and they are active as monomers, but also as homodimers or heterodimers; this allows fine-tuned specific reactions (Rocheville et al. 2000b). Furthermore, somatostatin receptors dimerize not only

	Adenylate cyclase inhibition	Phosphotyrosine phosphatase activation	MAPK modulation via G proteins	Potassium channels	Calcium channels	Na/H exchanger	AMPA/kainate glutamate channels	Phospholipase C	Phospholipase A2	Exocytosis inhibition	Cell cycle arrest via PTP, MAPK, RB and p21	Apoptosis via PTP, P53 und BAX	Desensitization of cAMP coupling	Agonist induced endocytosis	Agonist induced enhancement of membrane presence	GH secretion	Insulin secretion	Glucagon secretion
SSTR1	+	+	+		+	+	+			+	+		+		+			
SSTR2	+	+	+	+	+		+	+		+	+		+	+		+		+
SSTR3	+	+	+	+						+		+	+	+				
SSTR4	+	+	+	+					+	+	+			+				
SSTR5	+	+	+	+				+		+	+		+	+		+	+	

Fig. 8.4 Somatostatin receptor (*SSTR*)-mediated signal transduction and functions. *AMPA* α-amino-3-hydroxy-5-methyl-4-isoxazolepropionic acid, *cAMP* cyclic adenosine monophosphate, *GH* growth hormone, *MAPK* mitogen-activated protein kinase, *PTP* protein tyrosine phosphatase, *RB* retinoblastoma protein (Modified from Patel 1999)

with each other, but also with other receptors—for example, dopamine receptor (Rocheville et al. 2000a).

8.3 Receptors with Tyrosine Kinase Activity

The typical example of a tyrosine kinase membrane receptor for a hormone is the insulin receptor (Table 8.4). Insulin receptor consists of two pairs of two polypeptides. The extracellular part is formed by two identical polypeptide chains coupled by disulfide bridges. Each of these chains associates with another polypeptide crossing the membrane and enzymatically active within the cell (Fig. 8.5). This enzymatic activity adds phosphate to tyrosine residues.

Insulin receptors are considered as prototypic class I tyrosine kinase receptors. Other examples are platelet-derived growth factor receptor, vascular endothelial growth factor receptor, and colony stimulating factor 1 receptor. These receptors are class V tyrosine kinase receptors possessing several extracellular immunoglobulin-like domains and a tyrosine kinase domain with an additional 100 amino acids.

In most tyrosine kinase receptors, the receptor tyrosines are autophosphorylated. Other intracellular proteins called "Src" (Rous sarcoma virus protein) bind to these phosphorylated tyrosines. The binding domain of these proteins is called

Table 8.4 Tyrosine kinase receptors

Hormone/ligand	Receptor	Subtypes	OMIM entries
Insulin	Ins-R	α/β	147670
Insulin-like growth factor	IGF-R	I (II)	146370 (147280)
Neurotrophin	NTRK1 (TRKA)		191315
Platelet-derived growth factor	PDGF-R	α/β	173490/173410

IGF-R insulin-like growth factor receptor, *Ins-R* insulin receptor, *NTRK1* neurotrophic tyrosine kinase receptor type 1, *PDGF-R* platelet derived growth factor receptor

[a] IGF-R2 is not a tyrosine kinase receptor!

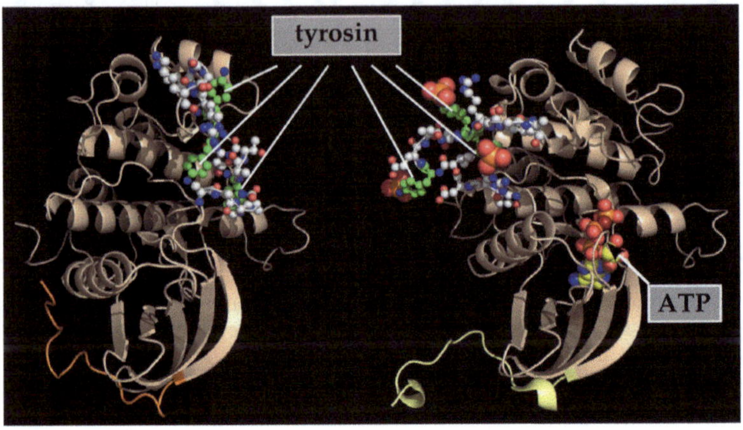

Fig. 8.5 The tyrosine kinase domain of insulin receptor. The same domain is shown on the *left*, in its inactive form, and on the *right*, activated and binding ATP. The loop with the three *green* tyrosines changes its position from the inactive to the active form. Phosphorylation (*enlarged orange/red spheres*) of the tyrosines and loading of ATP change not only the position of this loop, but also the entire structure, as can be seen. The lower β sheet has been kept constant between the two structures, which enables us to identify the subtler changes too. Opening access to the ATP molecules by flipping aside the tyrosine-containing loop activates the enzyme and allows protein phosphorylation to occur (Produced with PyMOL using Protein Data Bank entries 1IRK and 1IR3)

Src homology domain 2 (SH2). Such domains are common to various intracellular signaling proteins.

In insulin receptor, tyrosines are autophosphorylated, but they do not form an SH2 motif. For signal transduction, insulin receptor requires an additional protein, insulin receptor substrate. On phosphorylation, this protein has several SH2 motifs to communicate the insulin signal.

Figure 8.5 exemplifies how phosphorylation and associated structural adaptations modify the molecular topology, demonstrating signal transduction by conformational changes.

Table 8.5 Serine/threonine kinase receptors

Hormone/ligand	Receptor	Subtypes	OMIM entries
Activin	ACVR1	1a, 1b, 1c	102576, 601300, 608981
	ACVR2	2, 2a, 2b	102581, 602730
TGF-β	TGFβ-R	I, II	190181, 190182

ACVR1 activin receptor type 1, *ACVR2* activin receptor type 2, *TGF-β* transforming growth factor β, *TGFβ-R* transforming growth factor β receptor

8.4 Membrane Receptors with Serine/Threonine Kinase Activity

Activin/inhibin receptors (and other receptors for the transforming growth factor β supergene family proteins) are serine kinases (Table 8.5). Induced by ligand binding to the activin/inhibin receptor, the transcription factor SMAD5 is phosphorylated at serine. The phosphorylated SMAD5 is imported into the cellular nucleus, where it binds to SMAD5 motifs to regulate gene activity.

8.5 Membrane Receptors Without Kinase Activity

Leptin, prolactin, erythropoietin, and growth hormone are structurally similar. It is not surprising that their receptors are similar too, and they belong to the same class of membrane receptors (Table 8.6). These receptors, with colony stimulating factor receptor as a prototype, consist of five distinct extracellular domains, a transmembrane region, and a cytoplasmic domain. The genes for such receptors have up to 20 exons. Growth hormone, with its four helices, is bound by two extracellular domains (Fig. 8.6).

These receptors are not enzymatically active themselves. Their activity is mediated by signal transducers and activators of transcription (STAT proteins). These receptors become active when two hormonal ligands and two receptors interact. A STAT protein binds to the intracellular part of the receptors. This STAT protein with its SH2 domain initiates further signal transduction.

Whereas insulin receptor remains enzymatically active, these leptin, prolactin, or growth hormone receptors have no activity by themselves.

Leptin receptor exists in different splice variants. The OB-Rb variant has a 303 amino acid long cytoplasmic domain which transduces signals from the membrane receptor into the cell. Such a variant is expressed in cells where leptin blocks release of neuropeptide Y or agouti-related peptide—for example, on neurons of the arcuate nucleus. Shorter OB-R molecules lack this domain, and they are not capable of inducing signal transduction. Their expression in the choroid plexus or in blood capillaries in the brain suggests that they are active as leptin transporters through the blood–brain barrier. Whether such transport interferes with the in situ expression of leptin in the brain inhibited by hunger and fasting is an important research topic (Tartaglia 1997; Morash et al. 1999).

Table 8.6 Receptors without kinase activity

Hormone/ligand	Receptor	Subtypes	OMIM entries
Growth hormone	GH-R	(Alternative splicing)	600946
Leptin	LEP-R, OB-R	(Alternative splicing)	601007
Prolactin	PRL-R		176761
Erythropoietin	EPO-R		133171
Granulocyte/macrophage colony stimulating factor	GMCSF-R	α, β	306250, 138981

EPO-R erythropoietin receptor, *GH-R* growth hormone receptor, *LEP-R* leptin receptor, *PRL-R* prolactin receptor, *GMCSF-R* granulocyte/macrophage colony stimulating factor receptor

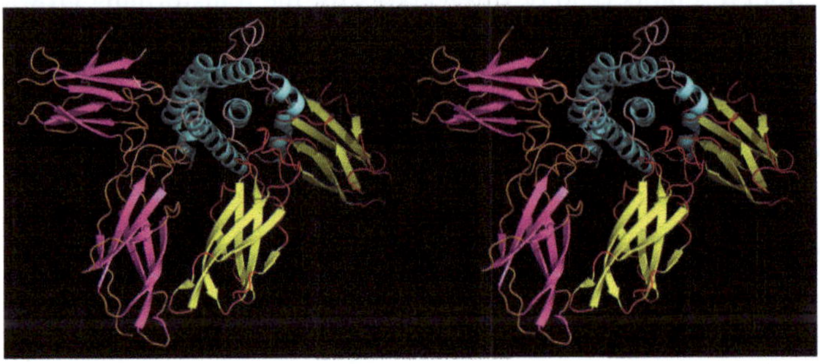

Fig. 8.6 Growth hormone and its receptor. To growth hormone with its *blue* helices bind two different growth hormone receptors with two extracellular domains each, characterized by *yellow* or *magenta* β-sheet structures. The interaction occurs by the *red loops* and *orange loops* and the helices of growth hormone (Produced with PyMOL using Protein Data Bank entry 3HHR)

8.6 Membrane Steroid Receptors: Still Unknown?

The fast action which aldosterone, for example, exerts on lymphocytes or smooth muscle cells cannot be explained by the slow action of nuclear receptors. The hunt for such fast-acting steroid receptors has taken more than 10 years, and is not yet finished. The fast-acting aldosterone receptor is still evasive—its action on the Na^+/H^+ exchanger can be blocked, signal transduction can be measured, but the darned receptor remains unknown.

In the meantime, researchers have extended this search for membrane steroid receptors for cortisol, testosterone, estradiol, vitamin D_3, and thyroid hormones. Since the human genome has been sequenced, candidates for such receptors have become limited. Orphan nuclear receptors, however, do not come into question.

In 2003, Zhu et al. (2003a) reported isolation of a human progesterone-binding GPCR after they had identified an analogous membrane progestin receptor in fish (Zhu et al. 2003b). Two years later, an estrogen-binding receptor was reported, which, on expression in previously negative cells, resulted in an estrogen-induced cAMP level increase—typical for membrane receptors. This molecule was not related to progesterone receptor or any nuclear receptor (Thomas et al. 2005).

An alternative explanation for membrane-receptor-mediated actions of estradiol was presented by Levin (2005, 2008). Membrane actions of estradiol are possible only in a cell having an intact estrogen receptor α gene (*ESR1*). Levin (2005) demonstrated that estrogen receptor α, estrogen receptor β, androgen receptor, and progesterone receptor posses a common **FxxxxxxLL** motif by which, for example, GPCRs are sorted into the plasma membrane. The motif of the nuclear receptors additionally bears a cysteine residue which was shown to be palmitoylated (Pedram et al. 2007). Levin has presented convincing evidence that such palmitoylation inserts the nuclear receptors into membranes, and that mutation of the motif first blocks membrane insertion and inhibits membrane-associated fast reactions as well. This elegant solution avoids the search for additional receptors and provides a new feature for old molecules.

So far this mechanism has not been extended to the membrane aldosterone receptor.

Part II

Endocrine Physiology

Synthesis, Release, and Action

9

Contents

9.1 The Three Steps of Hormone Regulation 263
9.2 Effective Hormone Concentrations of the Hypothalamic–Pituitary–Gonadal Axis 264

9.1 The Three Steps of Hormone Regulation

To understand regulation of hormones, three steps have to be distinguished:

1. Their synthesis
2. Their release or secretion
3. Their action on one or several receptors

We described hormone synthesis while presenting the different hormones: various enzymes by which protein/peptide hormones are correctly cleaved from their precursor or modulated posttranslationally; the two classes of cytochrome P450 (CYP)-dependent monooxygenases and hydroxysteroid dehydrogenases (HSDs), both generating steroids from cholesterol and the intermediates; and those enzymes by which tyrosine or tryptophan is converted into catecholamines, melatonin, or thyroid hormones.

Only steroid release is not blocked by the cell membrane, and steroids leave the synthesizing cell by diffusion. Their release is not deferred, but immediate. The other hormones are not released directly, but are stored in intracellular vesicles and are released on demand. Nerve cells and neurosecretory cells act in a similar way: neurotransmitters and neuropeptides as well as adrenaline and glycoprotein hormones are collected in secretory granules.

These secretory granules are usually located close to the cell surface. When a messenger demands release of a hormone by acting on a particular surface receptor, this ligand binding initiates an increase in intracellular calcium level, either by

B. Kleine, W.G. Rossmanith, *Hormones and the Endocrine System*,
DOI 10.1007/978-3-319-15060-4_9

opening calcium channels or, more often, by release of calcium from intracellular calcium stores. This enhanced calcium level triggers fusion of the granules with the membrane. It has been known for only a short time that soluble *N*-ethylmaleimide sensitive factor attachment receptors (SNAREs) are required for this membrane fusion, by which action two surfaces usually rejecting each other come into contact and fuse (Jena 2004).When the membranes of the vesicle and cell fuse, the inside of the vesicle becomes the outside of the cell and the content of the vesicle is free to leave the cell.

Hormonal release from secretory granules is thus an active process. It is triggered extracellularly by neurotransmitters at synapses, and also by other endocrine or paracrine messengers. Many hormones are released in pulses. For this several cells have to fuse their granules to the cell membrane simultaneously. The regulation of gonadotropin-releasing hormone (GnRH) pulses, for example, is age dependent: postpartum, the GnRH release, which is high in the fetus, is downregulated to almost zero; therefore, luteinizing hormone (LH) and follicle-stimulating hormone (FSH) are hardly released. At puberty, GnRH release increases again. Although the release does not yet occur in steady pulses, mood swings occur in a parallel way. Once the pulse rate has been correctly set at the level of an adult, these mood swings mostly go. During aging, the pulse rate drops again.

Pulsatile GnRH release is required not only for female fertility, it is required for male fertility as well. If by external doses of GnRH or stable agonists a constantly elevated GnRH level is established, pituitary LH and FSH release is blocked and infertility occurs. This is one method of temporary contraception. However, in men constantly elevated GnRH levels maintain elevated testosterone levels, potentially triggering prostate carcinomas.

The action of hormones occurs via receptors, as presented Chap. 8. Since hormones circulate in the blood, specificity of hormone action is exclusively generated by expression of specific receptors in specialized sensitive target cells. Cells react to hormones only when they bear receptors for the given hormone.

9.2 Effective Steroid Concentrations of the Hypothalamic–Pituitary–Gonadal Axis

If the amounts of hormones and their concentrations released in the hypothalamus, the pituitary, and the target organs are measured, an amplification cascade can be observed. Hypothalamic releasing hormones are released in picomolar concentrations: GnRH at 1 ng/l (about 0.6 pM). In the pituitary, the release is in the range of micrograms per liter. In contrast to the decapeptide GnRH, the molar LH release is 25-fold higher. The adrenal release of dehydroepiandrosterone (DHEA) is in the milligram range. Since DHEA is small, about 200,000 DHEA molecules are released per GnRH molecule.

If we assume there are 5,000 GnRH neurons, these release 6 ng GnRH into 6 l of blood—that is, 3.6 pmol, which corresponds to 21×10^{11} GnRH molecules and about 4×10^{8} GnRH molecules per neuron. Since the half-life of GnRH in blood is only a few minutes, most of the pulsatile-released GnRH is metabolized before the next pulse. To reach a concentration of 1 ng/l in blood again, all 4×10^8 molecules have thus to be newly synthesized in the period between the two pulses. At a pulse rate of one pulse every 2 h, each cell has to synthesize and store 41,700 molecules every second. These cells are like a GnRH factory. In the immune system the plasma cells are highly specialized and can produce about 2,000 antibodies per second. With the GnRH precursor (approximately 70 amino acids) being much shorter than an immunoglobulin heavy chain (approximately 400 amino acids), the GnRH neuron synthetic capacity is, however, of the same magnitude.

When GnRH is released at only 0.1 ng/l, the number of molecules drops to 4,100. The exact number is to be treated with caution; however, the magnitude is interesting. All molecules have to undergo the different posttranslational steps described before. With exact knowledge of the enzyme kinetics, we could estimate how many enzyme molecules a GnRH neuron requires to fulfill the GnRH demand.

Figure 9.1 demonstrates additional aspects of quantitative endocrinology: steroids do not exist in their free form, but are preferentially bound to transporter proteins or are sulfated like DHEA. Since steroids are modified only intracellularly, they have to dissociate from their binding globulin before they are able to diffuse

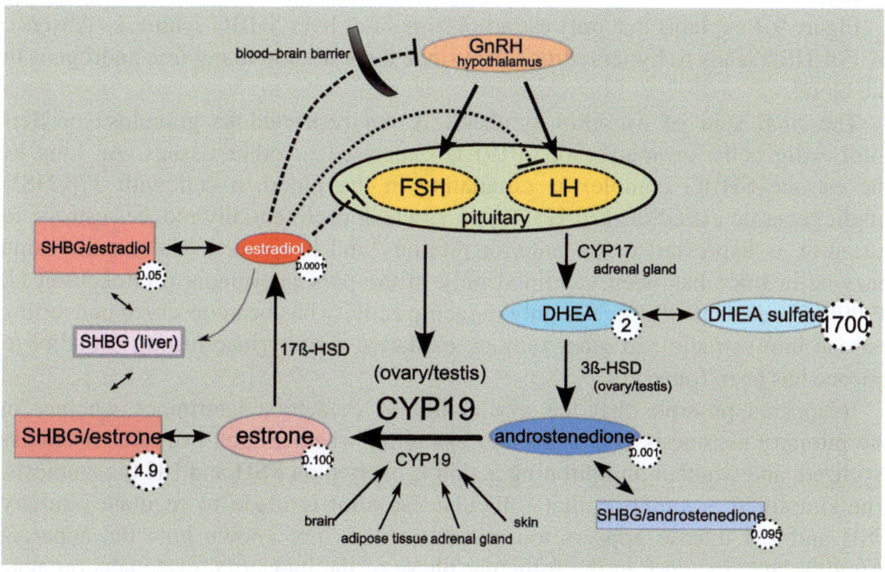

Fig. 9.1 Steroid hormone stores and relative steroid amounts of the hypothalamic–pituitary–gonadal axis. Steroid hormones are poorly water soluble. Complexed to the sex-hormone-binding globulin (*SHBG*), androgens and estrogens can be transported. Dehydroepiandrosterone (*DHEA*), however, is soluble as its sulfate derivatives. The *numbers in the dashed circles* give serum concentrations of the respective hormones or SHGB–hormone complexes in nanograms per milliliter. *CYP* cytochrome P450, *FSH* follicle-stimulating hormone, *GnRH* gonadotropin-releasing hormone, *HSD* hydroxysteroid dehydrogenase, *LH* luteinizing hormone

into reactive cells. Only then can they be modified. The reasons when and why a steroid dissociates from its binding globulin have not yet been identified in the case of androgens or estrogens. There might well be an equilibrium reaction which delivers new free steroids whenever the free steroid concentration in the surroundings of a reactive cell drops. However, from vitamin D_3 and its binding protein (vitamin D_3 binding protein), it is known that the binding protein is cleaved when vitamin D_3 is required. Such a mechanism has not yet been found for sex-hormone-binding globulin (SHBG).

Furthermore, Fig. 9.1 shows that the content of free steroids is considerably lower than that of the complexed or sulfated steroids. For estrogens and SHBG, the ratio is 1:49, for androstenedione it is 1:24, and for DHEA it is 1:1,000. The ratio of progesterone to DHEA to androstenedione to estrone to estradiol is 0.02:2:0.001:0.1:0.0001. These molecules all have similar molecular masses. Thus, DHEA is the steroid made in the largest amounts. The adrenal CYP17 is thus the dominant enzyme. The conversion of androstenedione to estrone by CYP19 in the ovaries and testes, as well as in adipose tissue, skin, brain, and (to a small degree) adrenal gland, reduces the androstenedione concentration to 1/100 that of estrone. This steroid, on its own, is not active, but has to undergo conversion by 17β-HSD into estradiol, which has the lowest serum levels of all steroids, but the highest activity.

Figure 9.1 explains that only estradiol stimulates liver SHBG synthesis. Absence of 17β-HSD leads to hyperandrogenism since there are too many free androgens in the blood.

The final step of estradiol synthesis is not restricted to granulosa or Sertoli/Leydig cells: aromatase (CYP19) is expressed in other tissues. As long as the estrone–SHBG complex is circulating in the blood, a cell with 17β-HSD might generate estradiol. 17β-HSD type 1, which preferentially reduces estrone to estradiol, is expressed in the anterior pituitary and in tumors, whereas the same enzyme in mice has been identified only in the pars intermedia (Peltoketo et al. 1999b; Green et al. 1999). Not only reducing activity has been observed converting estrone into estradiol; in other tumors, oxidative activity inactivating estradiol to estrone has been found.

It appears possible that the availability of cofactors determines whether in the pituitary estrone is converted into estradiol, whether such estradiol is readily oxidized, and whether an inhibiting action is exerted on FSH and LH biosynthesis. The simplified assumption that follicular estradiol is made to regulate pituitary FSH and LH release appears too simplistic. It is not known how the apparent estradiol level increase in the follicular phase of the menstrual cycle acts on gene expression in the pituitary, most notably in gonadotropic cells. Release inhibition by downregulating intracellular calcium and thus inhibiting membrane fusion of secretory granules to the cell membrane does not appear highly probable: estrogen receptors are preferentially transcription factors acting on gene expression and much less on signal cascades.

In Fig. 9.1 it is also shown that estradiol blocks GnRH release. This, however, appears feasible since, as described above, GnRH is freshly produced between two pulses, and estradiol while inhibiting transcription will diminish GnRH release since the stores are not renewed.

Endocrine Active Organs

10

Contents

10.1 Overview 270
10.2 The Endocrine System of the Brain 270
10.2.1 The Hypothalamus 270
10.2.2 The Pituitary Stalk 272
10.2.3 The Pituitary 273
10.2.4 The Pineal Gland 277
10.3 The Thyroid and Parathyroid Glands 278
10.3.1 The Thyroid Gland 278
10.3.2 The Parathyroid Glands 279
10.4 The Adrenal Glands 280
10.4.1 The Adrenal Cortex 281
10.4.2 The Adrenal Medulla 282
10.5 Endocrine Cells of the Gastrointestinal Tract 283
10.5.1 The Stomach 284
10.5.2 Duodenum and Ileum 285
10.6 Langerhans Islets of the Pancreas 285
10.7 Endocrine Cells of the Kidney 287
10.8 Gonads 288
10.8.1 Gonad Development 288
10.8.2 Male Gonads: Testes 288
10.8.3 Female Gonads: Ovaries 290
10.9 Selected Endocrine Active Organs in Invertebrates 292
10.9.1 Neurosecretion in Mollusks 292
10.9.2 Endocrine Glands in Crustaceans 294
10.9.3 Neurosecretion and Endocrine Organs in Insects 294

In this chapter, by presenting the anatomy of endocrine organs, we wish to demonstrate the differences between endocrine glands such as the adrenal glands and the thyroid gland and endocrine tissues of the gastrointestinal tract and neurosecretion in the brain.

B. Kleine, W.G. Rossmanith, *Hormones and the Endocrine System*,
DOI 10.1007/978-3-319-15060-4_10

10.1 Overview

The endocrine system appears to be hierarchically organized. Metabolism, growth, reaction to coldness, heat, or threat, and reproduction are all centrally controlled. It has not been know for long whether there is a master gland and where it is located. Many different organs, first of all the heart, were considered as controlling the inner functions of the physiology. Later, the pituitary was assumed to be the master gland since its hormones act in the thyroid gland, the adrenal glands, and the gonads. The perception of the superior function of the hypothalamus is a relatively recent development.

It is basic knowledge that all endocrine power emanates from the hypothalamus. Here the messages from the different organs are neuronally integrated. From the hypothalamus, releasing hormones are sent to the pituitary to release hormones controlling the thyroid gland, adrenal glands, and gonads. Many hormones of the gastrointestinal tract act like a type of reflex, bypassing the central control. These stimulate functional properties of other gastrointestinal tract cells directly. The organs of the gastrointestinal tract, however, are densely innervated, but the receptors of the endocrine signals from different hormone-producing cells in the gut and other organs of the digestive tract are in endocrine loops.

10.2 The Endocrine System of the Brain

At least three brain sections are part of the endocrine system: the master "gland" of the endocrine system—the hypothalamus—the pituitary or hypophysis, and the pineal gland. Whereas the hypothalamus and posterior pituitary are composed of neurons and neurosecretory cells, the anterior pituitary and the pineal gland are bona fide glands.

10.2.1 The Hypothalamus

Those brain areas at the side or in front of the third ventricle are called the hypothalamus. The third ventricle is a fluid-filled space in the diencephalon.

Like other brain areas, the hypothalamus has distinct nuclei. These are centers with many neurosecretory cells of similar function. Whereas the nerve axons which, for example, in the case of the optic nerve extend from the eye to the brain or in the case of motor nerves extend from the brain to the peripheral muscles are usually covered by a sheath of connective tissue cells which form the stable blood–brain barrier, in the hypothalamus the nerves are unsheathed. This allows multiple connections between controlling nerve cells and hormone-producing neurosecretory cells and within neurosecretory cells. Nuclei in this context are not the cellular nuclei.

Different nuclei and areas of the hypothalamus have different functions in endocrine control (Fig. 10.1, Table 10.1).

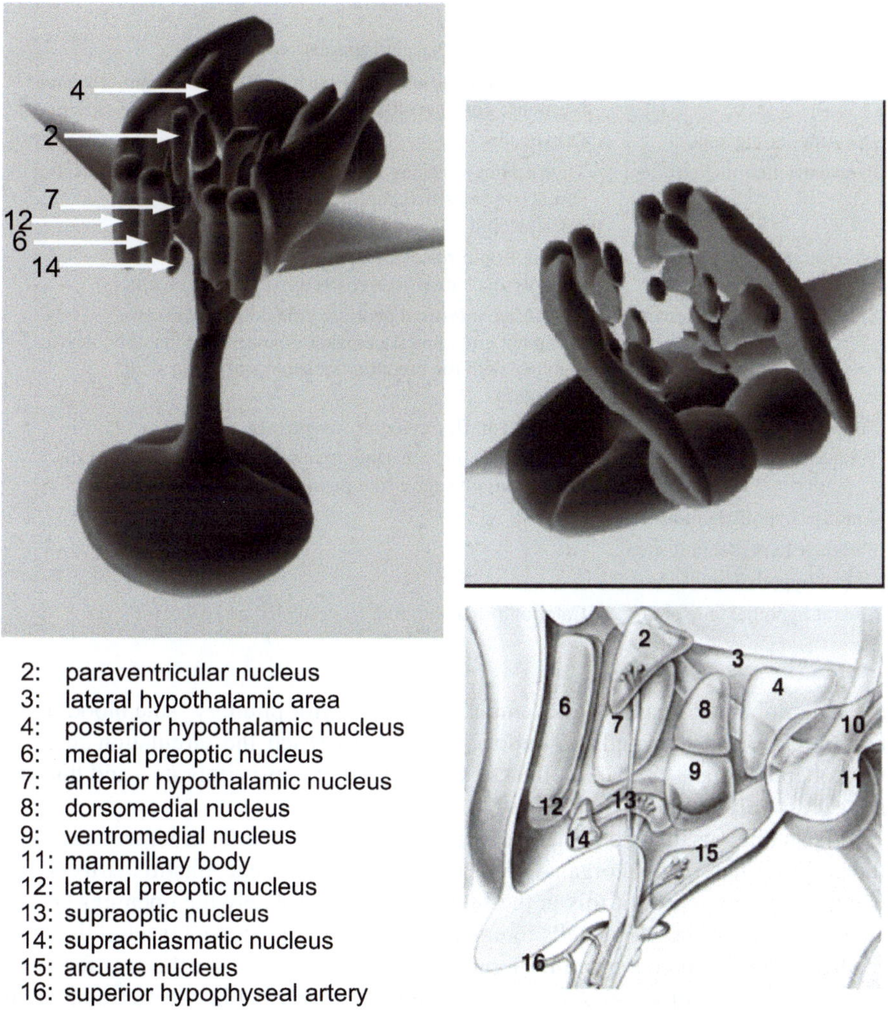

Fig. 10.1 The human hypothalamus: three-dimensional model and schematic overview (From Hirsch 1998)

Neurosecretory cells of the hypothalamus differ from other nerve cells in that they do not control other cells via synapses; rather, they release their hormones into the circulation. For this purpose, the axons of these neurosecretory cells reach into the median eminence, an area at the bottom of the hypothalamus. Therein blood capillaries of a portal system have many small openings, so-called defenestrated areas or sieve plates. Once a neurosecretory cells has released its hormones in the median eminence, these hormones reach the circulation through the little windows. Although brain arteries are usually sheathed by the blood–brain barrier, blocking, for example, lymphocyte migration into the brain, in the median eminence and

Table 10.1 Hypothalamic nuclei

Region	Hormone synthesis and actions
Arcuate nucleus	Growth-hormone-releasing hormone; gonadotropin-releasing hormone; glucagon-like peptide
Dorsomedial nucleus	Orexin
Paraventricular nucleus	Thyrotropin-releasing hormone; angiotensin II; somatostatin; vasoactive intestinal peptide; oxytocin; vasopressin; corticotropin-releasing hormone
Preoptic nucleus	Gonadotropin-releasing hormone; lipopolysaccharide-induced fever (μ-opioid receptors)
Suprachiasmatic nucleus	Vasoactive intestinal peptide; gonadotropin-releasing hormone; circadian pacemaker connected to the optic nerve, together with the preoptic nucleus involved in sleep regulation
Supraoptic nucleus	Angiotensin II; oxytocin; vasopressin
Ventromedial nucleus	Leu-enkephalin; substance P; neurotensin; glucose/insulin regulation; glucagon-like peptide; "threat, attack, flight"
Anterior hypothalamic area	
Posterior hypothalamic area	
Dorsal hypothalamic area	
Lateral hypothalamic area	Calcitonin-like peptide; 5α-testosterone reductase

other neurohemal organs—for example, in the posterior pituitary—the blood–brain barrier is nonexistent and the endocrine products reach the circulation directly. Neurohemal organs exist in mollusks as well in other non-vertebrates, and thus are not exclusively found in vertebrates.

It is an obvious question whether white blood cells, bacteria, or viruses can enter the brain via neurohemal organs—for example, tick-borne encephalitis virus. There is apparently no report of this in the literature. Most probably the openings are too small to allow white blood cells to enter. However, nasal sprays for administration of drugs which enter the brain along the olfactory nerve have been offered for a few years. The blood–brain barrier is equally permissive in the vascular organ of the lamina terminalis, in direct vicinity of the preoptic nucleus, as well as in the circumventricular organs (see Fig. 11.15); angiotensin II from the circulation, for example, can thus reach receptors on the surface of nerve cells. Via these connections, other hormones such as leptin and other neuropeptides can find their central receptors.

Neurosecretory cells of the hypothalamus are interconnected with other areas of the brain. Nerve cell axons reach certain hypothalamic nuclei and form synapses there.

10.2.2 The Pituitary Stalk

By the median eminence hormones are transported directly to the pituitary via a portal system. Special neurosecretory cells reach beyond the median eminence into the posterior pituitary, the neurohypophysis. These nerve cells and connective tissue form the pituitary stalk, the infundibulum. The blood vessels around the pituitary stalk transport the neuropeptide from the median eminence to the anterior pituitary, the pars distalis. Below the pituitary stalk there is the pituitary (see the description of calendar cells in Sect. 12.4).

10.2.3 The Pituitary

Below the hypothalamus in a hole in the sphenoid bone—the bone on which the brain lies—is the pituitary (Fig. 10.2). It is joined to the brain via the pituitary stalk and its surrounding blood vessels. The human pituitary is composed of the anterior pituitary and the posterior pituitary, aka the neurohypophysis (lobus nervosus). The human pars intermedia of the pituitary is only discernible during the fetal period and in children. In animals, the pars intermedia provides the hormones important for

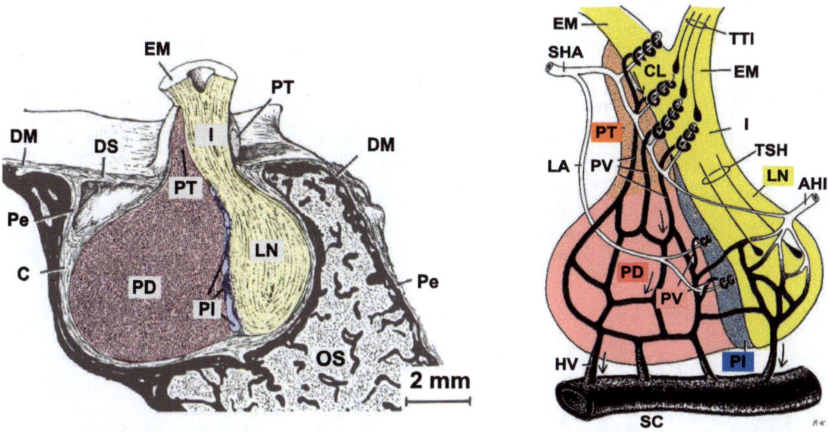

Fig. 10.2 The pituitary—an outline: general anatomy (*left*) and blood supply (*right*). In the sella turcica of the sphenoid bone (os sphenoidale, *OS*) the pituitary is placed with the adenohypophysis (anterior pituitary; pars distalis, *PD*), an intermediate region (pars intermedia, *PI*), and the neurohypophysis (posterior pituitary; lobus nervosus, *LN*). The capsule (*C*) of the pituitary is followed by the periosteum (*Pe*) of the sphenoid bone and its dura mater (*DM*). By the pituitary stalk (infundibulum, *I*), nerve axons from different hypothalamic nuclei reach the pituitary. Axons in the tuberoinfundibular tract (*TTI*) originate from neurosecretory cells of the dorsomedial nucleus, ventromedial nucleus, or arcuate nucleus ending mostly in the median eminence (eminentia mediana, *EM*); axons in the supraopticoparaventriculohypophysial tract (tractus supraopticoparaventriculohypophysialis, *TSH*) are from the preoptic nucleus or paraventricular nucleus and reach the neurohypophysis. The upper part of the anterior pituitary is called the pars tuberalis (*PT*). Using a portal system, hypothalamic neuropeptides from the median eminence get into the anterior pituitary. The superior hypophyseal artery (*SHA*) serves capillary loops (*CL*) of the median eminence and via the loral artery (*LA*) the capillary loops of the pars intermedia; the inferior hypophyseal artery (*IHA*) supplies blood to the neurohypophysis. Within the capillary loops, neuropeptides are taken up and reach the adenohypophysis via portal veins (*PV*), and anterior pituitary hormones arrive via the portal veins in the cavernous sinus (*SC*) (Modified from Krstić (1991): *left* plate 122, *right* plate 128)

coloring and color adaption to the environment. Since we humans adapt no longer by hormonal activity and coloring but by behavior to our environment (a dark red face indicates not color adaption but a strong bloodstream in the facial capillaries), this part of the pituitary could degenerate.

10.2.3.1 The Anterior Pituitary

The anterior pituitary (aka the adenohypophysis or pars distalis) contains those cells which release adrenocorticotropic hormone (ACTH), luteinizing hormone (LH), follicle-stimulating hormone (FSH), thyroid-stimulating hormone, growth hormone, and prolactin (Fig. 10.3). In the thyroid gland and in the different zones of the adrenal glands, only one hormone is made. In the anterior pituitary as well as in

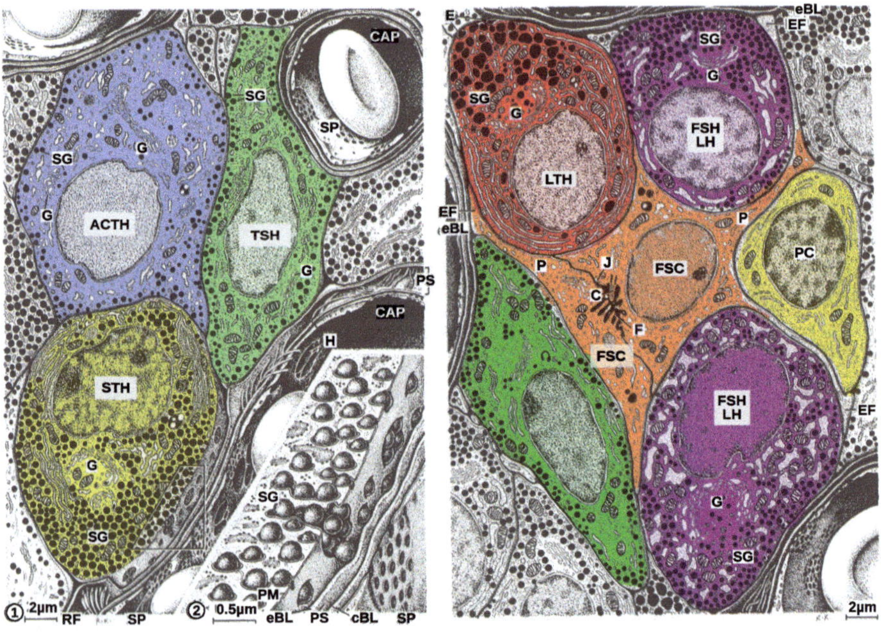

Fig. 10.3 Endocrine cells of the anterior pituitary. The following cell types can be distinguished: adrenocorticotropic cells (*ACTH*), somatotropic cells (*STH*), lactotropic cells (*LTH*), gonadotropic cells (*FSH* and *LH*), and thyrotropic cells (*TSH*). In addition there are follicular stellate cells (*FSC*) and precursor cells (*PC*) with no detectable hormone synthesis. All these hormone-synthesizing cells are characterized by a visible Golgi apparatus (*G*), numerous mitochondria, pronounced rough endoplasmic reticulum, and very strikingly, an enormous number of secretory vesicles (*SG*). On stimulation by hypothalamic releasing hormones, these vesicles fuse with the cell membrane, and release their content. Secreted hormones need only pass the short distance through through basal lamina and the perivascular space (*PS*) before reaching the capillaries via sieve plates (*SP*). Therein, capillaries are fenestrated to allow prompt uptake of hormones (*left*, magnification). Sieve plates are characteristic features of endocrine organs. *C* cilia, *CAP* capillary, *cBL* capillary basal lamina, *E* exocytosis, *eBL* external basal lamina, *EF* end feet, *H* holes, *J* junctions, *P* cytoplasmic processes, *PM* plasma membrane, *RF* reticular fibers (Modified from Krstić (1991), plates 124 and 125)

pancreatic islets, adjacent cells release different hormones. (These hormones have already been discussed in Sect. 4.4).

Adrenocorticotropic Cells

These are polygonal cells with an eccentric cellular nucleus; secretory granules are mostly found at the cell margins. These cells can be stained by neither basophilic nor acidophilic dyes.

Secretion of ACTH is controlled by corticotropin-releasing hormone.

Somatotropic Cells

These are rounded cells of medium size with a globular cellular nucleus and a conspicuous nucleolus, large electron-dense secretory granules[1] distributed all over the cytosol, large rough endoplamic reticulum cisternae, and well-developed mitochondria; these cells can be stained by acidophilic dyes.

Growth hormone secretion is controlled by growth-hormone-releasing hormone and somatostatin.

Thyrotropic Cells

These are irregularly shaped, polygonal, elongate or stellate cells with an elliptic cellular nucleus and well-developed organelles; osmiophilic[2] secretory granules are grouped in the cell periphery; granules can be stained by basophilic dyes.

Release of thyroid-stimulating hormone is under the control of hypothalamic thyrotropin-releasing hormone.

Lactotropic Cells

These are rounded cells with an oval cellular nucleus; mainly during pregnancy and lactation there are many large secretory granules. These granules are acidophilic.

Secretion of prolactin is steadily inhibited by dopamine. A functional hypothalamic prolactin-releasing hormone has not yet been discovered.

Gonadotropic Cells

These are oval cells with a rounded cellular nucleus. Osmiophilic secretory granules arise from the Golgi apparatus. The cells are basophilic.

Secretory granules contain simultaneously FSH and LH, whose release is controlled in humans by hypothalamic gonanotropin-releasing hormone type I. An individual FSH-releasing hormone has been postulated anecdotally, but the fact that both hormones are found in the same vesicles argues against a special FSH-releasing hormone.

---

[1]Electron density in electron microscopy depends on the atomic number; the higher the atomic number, the greater the density. Metals such as iron and molybdenum are heavier than hydrogen and carbon.

[2]Osmium tetroxide is used for contrast enhancement in electron microscopy.

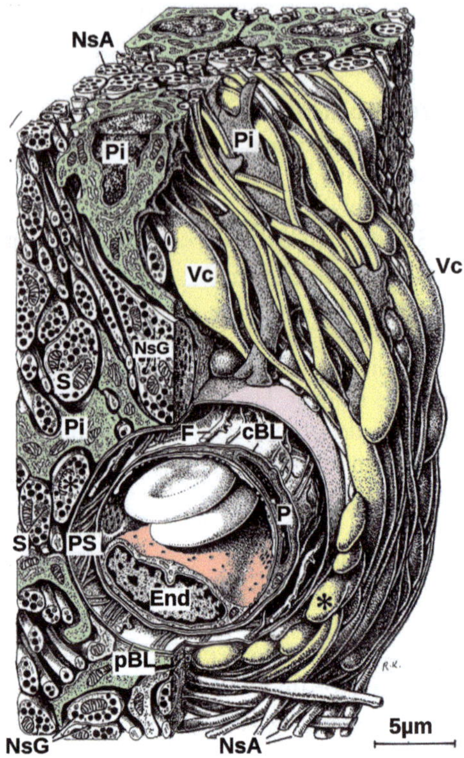

Fig. 10.4 Section of the posterior pituitary. Axons of arginine vasopressin and oxytocin neurons release vasopressin and oxytocin. Neurosecretory axons (*NsA*) end at capillaries. Neurosecretory axons are full of neurosecretory granules (*NsG*). Between the neurosecretory axons there are pituicytes (*Pi*), irregularly shaped stellate cells, whose endings are in intimate contact with neurosecretory axons. Hormones when released reach the capillaries and thus the circulation by passing first the pericapillary basal lamina (*pBL*), then the pericapillary space (*PS*), and finally the fenestrated capillary basal lamina (*cBL*). *End* endothelial cell, *F* fibrocytic process, *P* pericyte, *S* synaptic vesicle, *Vc* varix, *asterisk* enlarged axon endings (Modified from Krstić (1991), plate 127)

10.2.3.2 The Posterior Pituitary

In contrast to the anterior pituitary, in the posterior pituitary (Fig. 10.4) there is no hormone synthesis, only release. In the posterior pituitary, the axons of neurosecretory cells from the paraventricular nucleus, ventromedial nucleus, supraoptic nucleus, and arcuate nucleus end at capillaries to release their hormones into the pituitary portal system (Fig. 10.4).

The posterior pituitary is like the median eminence, a neurohemal organ where brain cells release their content into the circulation and where the blood–brain barrier is permissive, where capillaries have fenestrated sieve plates for taking up released hormones. The two type of neurosecretory cells there release two hormones with closely related structures but different functions: oxytocin and vasopressin.

10.2.3.3 The Pars Intermedia

In the intermediate region or pars intermedia located between the anterior and the posterior pituitary, proopiomelanocortin is cleaved to melanocyte-stimulating hormone (MSH) instead of ACTH as in the anterior pituitary. MSH controls coloring in amphibians, where the pars intermedia is markedly developed. MSH in humans regulates appetite and hunger, but it is not made in the pars intermedia.

10.2.4 The Pineal Gland

The pineal gland (Fig. 10.5) got its name (glandula pinealis) from its cone-like form and not because of its acervuli, crystalline calcium phosphate that looks like fir cones or raspberries. No function has ever been attributed to these acervuli; the age-dependent buildup of calcium-enriched and calcium-deficient layers points to seasonally different deposition of calcium ions. The characteristic cell of the pineal gland is the pinealocyte, which synthesizes and releases melatonin in dark—light cycles (see Sect. 7.3).

Innervation of the pineal gland occurs via nerve fibers through the pineal stalk and via septa. Adrenergic fibers of the coronary nerve connect the superior cervical ganglion (not covered by myelin sheaths) with the pineal gland. These fibers do not cross the pineal stalk. Axons reaching the pineal gland via the pineal stalk originate in the posterior commissure and the thalamus. In addition, the pineal gland

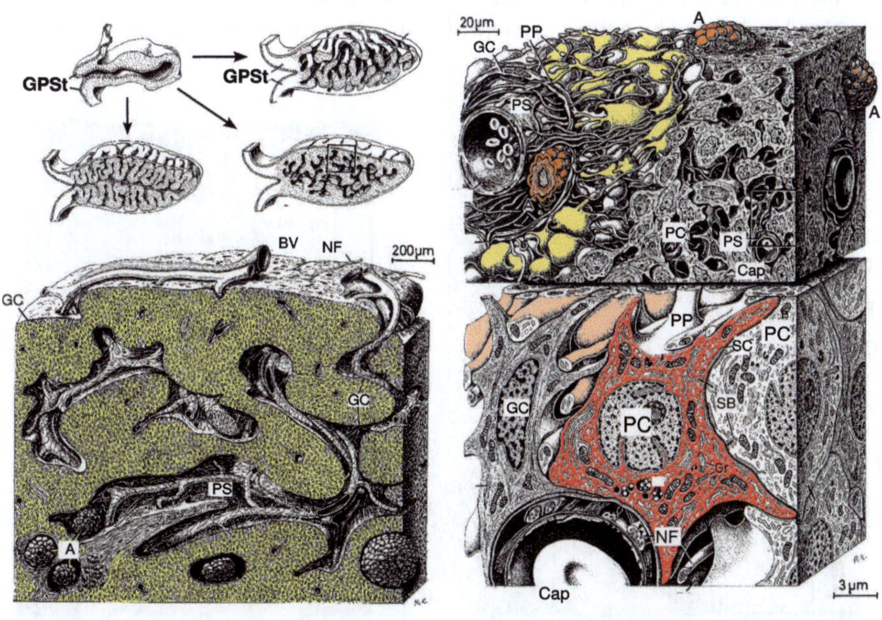

Fig. 10.5 The pineal gland. The pineal gland (glandula pinealis, epiphysis cerebri) got its name from its cone-like shape. *Upper left*: The anastomizing cords of the pineal gland develop from a diverticulum of the diencephalon by branching or crooking. Septa reach from the capsule into the inner pineal gland. Glial cells (*GC*) separate glandular tissue and septa, however in an incomplete flat layer. *Lower left*: Within spaces there are arteries, veins (*BV*), and nerves (*NF*). Within the perivascular and pericapillary spaces as well as in the parenchyma there are (visible by X-ray) acervuli (*A*) made of calcium apatite and calcium carbonate. *Upper right*: Pinealocytes secrete hormones via processes (*PP*) near capillaries. *Lower right*: The typical pineal cell is the pinealocyte (*PC*), which releases melatonin. The pinealocyte cellular nucleus is partially deeply invaginated. Smooth endoplasmic reticulum close to the membrane forms characteristic cisternae (*SC*). Synaptic bars (*SB*) occur close to the plasma membrane. *Cap* capillary, *GPSt* pineal stalk, *PS* perivascular space (Modified from Krstić 1991, plates 129 and 130)

receives signals from the retina by which the light-inhibited melatonin synthesis is controlled.

The pineal gland of amphibians reacts to light. In humans there are neuronal connections to the optic nerve, which suggests some sensing of light.

Melatonin is the characteristic hormone of the pineal gland. Its biosynthetic intermediate serotonin is a important neurotransmitter in the central nervous system. Melatonin biosynthesis is controlled by the suprachiasmatic nucleus, which triggers, together with noradrenergic nerves, in the absence of light transcription of arylalkylamine *N*-acetyltransferase (see Sect. 7.3).

10.3 The Thyroid and Parathyroid Glands

10.3.1 The Thyroid Gland

The thyroid gland (Fig. 10.6) is situated below the larynx in front of the trachea. The gland is covered by a thick capsule of connective tissue. Such connective tissue

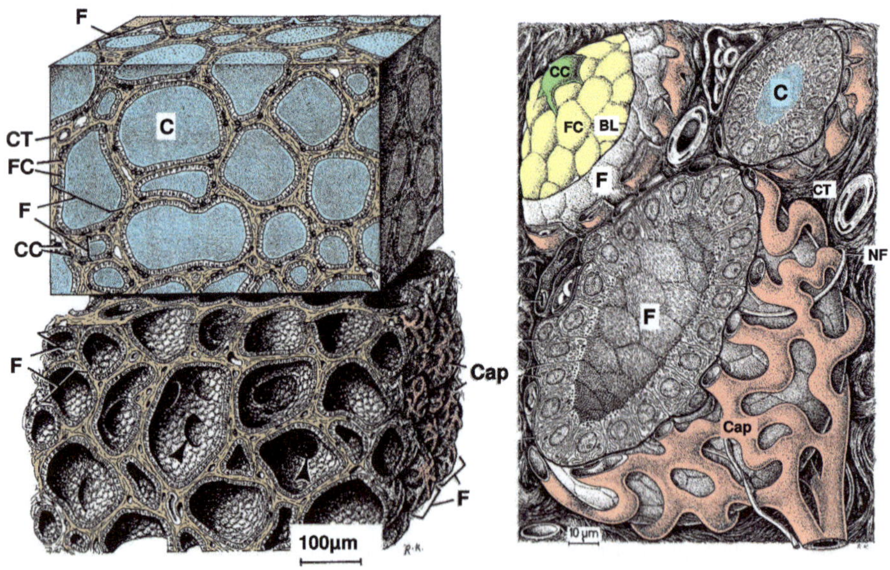

Fig. 10.6 The thyroid gland. *Upper left*: Section of the thyroid gland, follicles filled with colloid. *Lower left*: Similar section, follicles empty. *Right*: Individual follicles are magnified. Characteristic of the thyroid gland are the follicles (*F*) filled with colloid (*C*) which are lined by a unicellular follicular cell (*FC*) layer. Around the follicles there is a basal lamina (*BL*), connective tissue (CT), and a dense network of fenestrated capillaries (*Cap*). On the fenestrated capillaries there are some C cells (*CC*), which are darkened in the *upper-left image*. From the capillaries, follicular cells take up iodine, which they couple to thyroglobulin. Such iodothyroglobulin is stored in the colloid. Stimulated by thyroid-stimulating hormone, follicular cells pinocytose lipid droplets, cleave thyroxine from thyroglobulin, and release this thyroxine. C cells generate and secrete calcitonin. *NF* nerve fiber (Modified from Krstić (1991), plates 131 and 132)

also separates the characteristic follicles from each other. Follicles are formed by a unicellular layer of follicular cells surrounding an amorphic gelatinous colloid.

Follicular cells generate triiodothyronine and thyroxine—hormones involved in metabolic homeostasis. Iodide actively taken up by follicular cells is oxidized by thyroid peroxidase to iodine, which is then used to derivatize tyrosine residues of thyroglobulin. Iodinated thyroglobulin is stored in the colloid. Triggered by thyriod-stimulating hormone, iodothyroglobulin is taken up by endocytosis into vesicles called phagosomes. Phagosomes fuse with lysosomes to form phagolysosomes, where thyroxine is enzymatically removed from thyroglobulin. Thyroxine and eventually derived triiodothyronine diffuse into the capillaries surrounding the follicles.

There is another cell type in the thyroid gland, so-called C cells. These cells have osmiophilic granules where calcitonin and somatostatin are generated and stored.

10.3.2 The Parathyroid Glands

On the back of the human thyroid gland, there are four small glands, the parathyroid glands (Fig. 10.7). In reptiles, there are additional parathyroid glands found in pairs along the brain artery. Below a capsule of dense connective tissue, the glandular parenchyma is invaginated by connective-tissue-lined septa containing capillaries. From the stroma, the parenchyma is separated by a visible basal lamina.

The cross section in Fig. 10.7 shows the major cell type: the chief cell or clear cell characterized by a voluminous cellular nucleus with a clearly visible pronucleus, by the Golgi apparatus, by mitochondria and large lipofuscin granules, and by glycogen stores and lipid droplets. Secretory granules can be observed close to fenestrated capillaries.

A special feature of the parathyroid gland is oxyphilic cells conspicuously filled with mitochondria. Although the origin and function of these cells, which occur not only in thyroid carcinomas, but also in other tumors, is not fully understood, Kimura et al. (2009) have offered some explanations with respect to their generation: in mice where they stimulated thyroid cells for a long time with interferon, they found that the oxyphilic cell type arose as a consequence of some defect of a proteasome protein, LMP2, whose gene is located in major histocompatibility complex II and which might be involved in autoimmune diseases.

The main function of the parathyroid gland is synthesis and release of parathormone, which is required for functional calcium metabolism in connection with calcitonin and vitamin D_3.[3]

[3]C cells of the thyroid gland release calcitonin and somatostatin; C cells of the parathyroid gland, however, release parathormone.

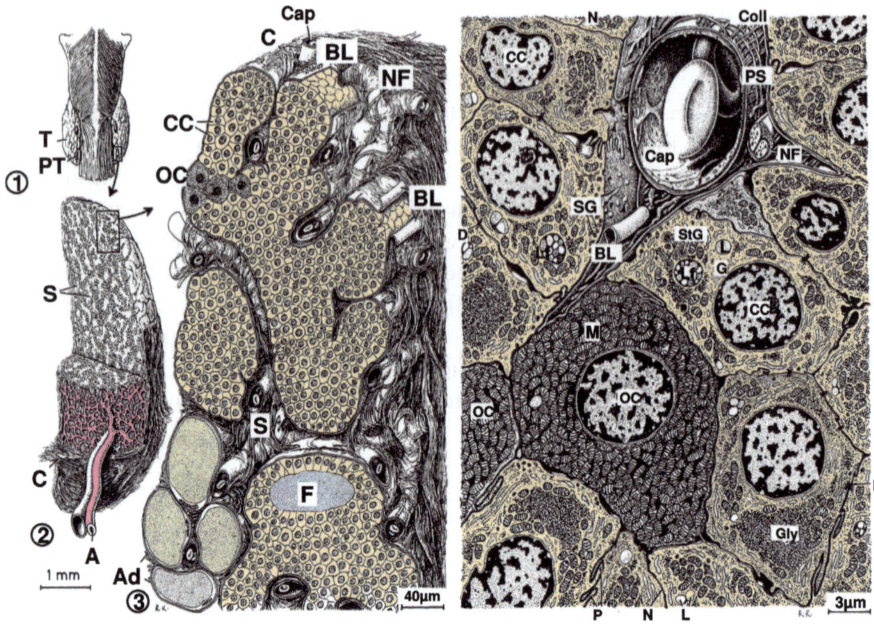

Fig. 10.7 The parathyroid gland. On the back of the thyroid gland (*T*; *1*) there are four small parathyroid glands (*PT*; *2*), a section of which is shown in *3* and on the *right* there is a magnification to show individual cells. Below a capsule (*C*) there is the stroma formed by septa (*S*) extending from the capsule into the parenchyma. Around the capillaries (*Cap*) there are collagen fibers (*Coll*). In the perivascular space (*PS*) there are nerve fibers (*NF*). A basal lamina (*BL*) separates the parenchyma cells from the perivascular space. In the parenchyma there is mostly one cell type named a chief cell or clear cell (*CC*). Less abundant are oxyphilic cells (*OC*) aka onkocytes, whose cytoplasm is completely filled by mitochondria for an unknown reason. Chief cells synthesize parathormone and release it. In contrast to pituitary or pancreatic cells, secretory granules are not a prominent feature; there are, however, larger storage granules (*StG*) containing parathormone as well. Apart from the Golgi apparatus (*G*), lipid droplets, and glycogen stores (*Gly*), there are lipofuscin vesicles (*Lf*) with remnants of indigestible degradation products. Chief cells are kept together by nexus (*N*), small desmosomes (*D*), and interdigitations formed as processes (*P*). *Ad* adipose cell, *F* follicle, *M* mitochondria, *L* lipofucsin, *SG*. secretory granula (Modified from Krstić (1991), plates 135 and 136)

10.4 The Adrenal Glands

The adrenal gland (glandula adrenalis or glandula suprarenalis) forms a cap on top of the kidney. Below a capsule there is the adrenal cortex and the adrenal medulla. Adrenal arteries enter the capsule at several sites, and nerves and lymphatic vessels enter the adrenal gland via the hilus, where the adrenal vein leaves the organ. Like all glands, the adrenal glands are covered by capsules. A yellowish cortex surrounds the white medulla.

10.4.1 The Adrenal Cortex

The adrenal cortex (Fig. 10.8) has three distinct zones, in each of which different steroids are synthesized.

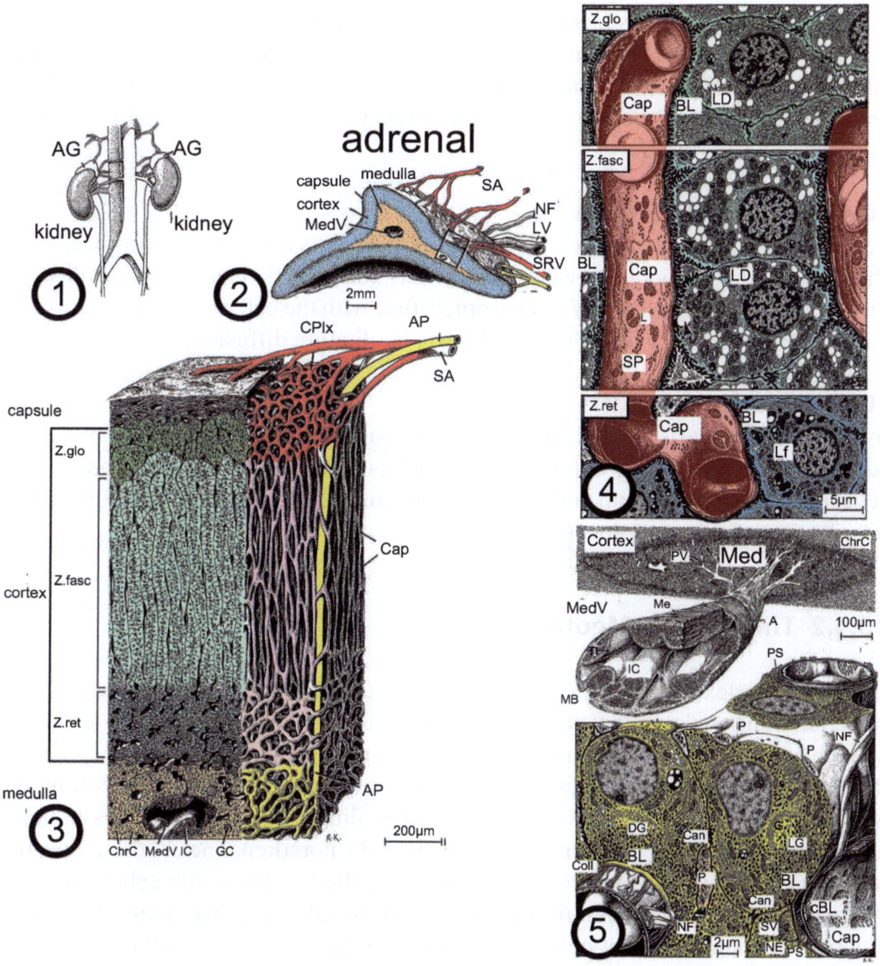

Fig. 10.8 The adrenal gland (*AG*). On the kidneys the adrenal glands are like caps consisting of cortex and medulla (*2*). The cortex is organized in three layers (*3*): the zona glomerulosa (*Z.glo*, *4 top*), the zona fasciculata (*Z.fasc*, *4 middle*), and the zona reticularis (*Z.ret*, *4 bottom*). Sections of the medulla (*Med*) and the medullary vein (*MedV*) are shown in *5*. *A* adventitia, *AP* perforating arterioles, *BL* basal lamina, *Can* canaculi, *Cap* capillary, *cBL* complete basal lamina, *ChrC* chromaffin cell, *CPlx* capsular plexus, *Coll* collagen, *DG* dark granules, *G* Golgi apparatus, *GC* ganglion cell, *IC* intimal cushion, *LD* lipid droplet, *LG* light granules, *Lf* lipofuscin, *LV* lymphatic vessels, *MB* muscle bundles, *Me* media, *MedA* medullary artery, *NE* nerve endings, *NF* nerve fibers, *P* process, *PS* perivascular space, *PV* postcapillary venules, *SA* suprarenal arteries, *SV* synaptic vesicles (Modified from Krstić (1991), plates 137, 138, and 139)

10.4.1.1 Zona Glomerulosa

In the zona glomerulosa there are round or horseshoe-shaped nests of cells, which secrete mineralocorticoids—mainly aldosterone. By renin from the juxtaglomerular cells of the kidneys, angiotensin I is generated from angiotensinogen and is further processed to angiotensin II by angiotensin-converting enzyme. In the adrenal gland, angiotensin II stimulates aldosterone synthesis. The cell surface is enlarged by microvilli, where the cells face the capillaries, thus allowing more steroids to diffuse through the membrane. Within the cells there are lipid droplets, mitochondria, rough endoplasmic reticulum, and much smooth endoplasmic reticulum. Since steroids are not stored, but are released by diffusion, steroidogenic cells do not contain secretory granules.

10.4.1.2 Zona Fasciculata

Large parallel strands of cells with many lipid droplets and spherical mitochondria shape the zona fasciculata, where cortisol is made and released, controlled by pituitary ACTH. Occasionally lipid droplets fuse with the cell membrane. Microvilli enlarge the cell surface along the capillaries, facilitating diffusion.

10.4.1.3 Zona Reticularis

In the zona reticularis the cells have larger lipofuscin granules than in the two other adrenal cortex zones. The extent of microvilli is less than described before. These cells have extensive smooth endoplasmic reticulum. Here androgens, predominantly dihydroepiandrosterone, are synthesized.

10.4.2 The Adrenal Medulla

Below the cortex, there is the medulla with neurosecretory cells. In the medulla, catecholamines, adrenaline and noradrenaline, are synthesized, stored in granules, and on cholinergic stimulation released. Again we find the secretory granules that are a characteristic feature of other hormone-producing cells. According to Krstić (1991), dense granules indicate noradrenaline and lighter granules indicate adrenaline. In animals, different types of cells make noradrenaline or adrenaline, but in humans there is only one cell type, which is called a chromaffin cell because of its easy staining with chromium salts. Chromaffin cells have processes which form a network around adjacent cells. In the medulla there are also ganglion cells, and nerve fibers forming synapses to the chromaffin cells with acetylcholine-containing storage granules. The basal lamina is uninterrupted, and perivascular spaces separate the basal lamina from the capillary basal lamina. Sieve plates facilitate transfer of hormones into the circulation.

Blood arrives in the medulla through the capsule by the suprarenal artery. Special perforating arterioles reach through the cortex directly into the medulla. The blood is collected in postcapillary venules ending in the medullary vein. This is characterized by a tunica intima and its spiral-shaped muscle fibers pumping hormone-enriched blood from the adrenal gland into the circulation.

The medulla cells are of neuronal origin. As in other neurosecretory cells, hormones are stored in vesicles and on demand are released by calcium-triggered vesicle–membrane fusion.

10.5 Endocrine Cells of the Gastrointestinal Tract

The different organs of the gastrointestinal tract (stomach, duodenum, ileum, colon, pancreas) are not primary organs of the endocrine system; however, important hormones are produced therein. In the pancreas, different endocrine cell types are collected in Langerhans islets surrounded by exocrine acinar cells. In the gastric gland and crypts of Lieberkühn in the duodenum and the ileum, there are only singular endocrine cells with their secretory granules oriented away from the gut lumen and close to the basal lamina, and these can be classified immunohistologically as hormone positive. Figure 10.9 provides an overview of the different hormones of the stomach, duodenum, pancreas, and ileum. These are indicated by arrows away from the lumen of the stomach or gut. Only a few cell types are indicated. There are, however, up to 16 different cell types characterized by different combinations of the hormones synthesized.

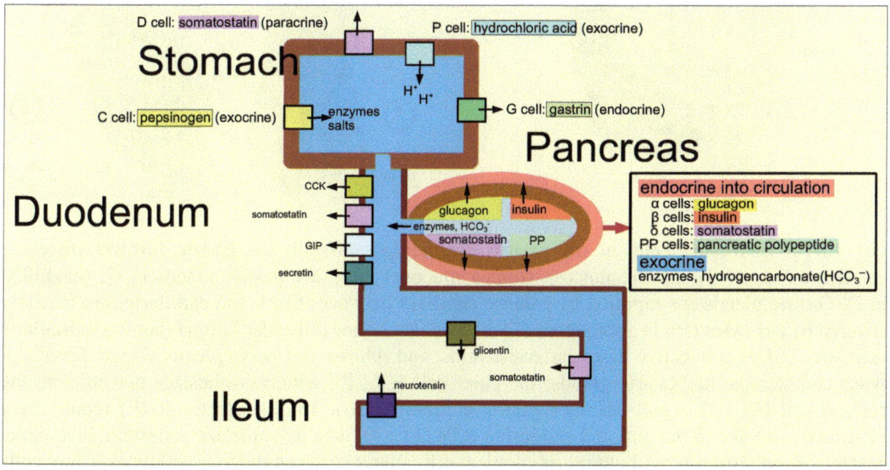

Fig. 10.9 Endocrine and exocrine secretion in the gastrointestinal tract. *Arrows into the gut lumen* signal exocrine production of acid, salts, and digestion enzymes; *arrows away from the lumen* signal endocrine/paracrine secretion of hormones. *CCK* cholecystokinin, *GIP* gastroinhibitory peptide, *PP* pancreatic polypeptide

10.5.1 The Stomach

The gastric mucosa has many furrows. At each end of such a furrow there are one or more gastric glands (Fig. 10.10). These are columns of cells with different functions: the chief cells produce pepsinogen, the precursor of the gut digestion enzyme pepsin; parietal cells secrete 0.1 N hydrochloric acid; and at the lower end of the gland, there are a few endocrine cells secreting serotonin (5-hydroxytryptamine; see Fig. 7.3), histamine, gastrin, or somatostatin.

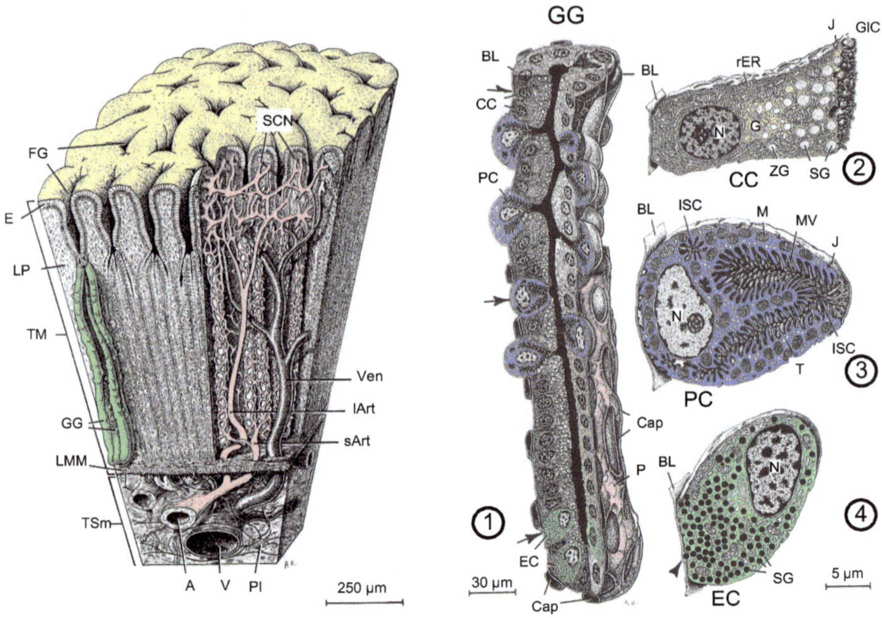

Fig. 10.10 Endocrine cells in the gastric glands. *Left*: Within the gastric furrows (faveolae gastricae; *FG*) of the gastric mucosa (tunica mucosa) there are gastric glands (*GG*; magnified in *1*). Gastric glands are supplied by a dense capillary network (*Cap*), and capillaries are partially covered by pericytes (*P*). In submucosa (*TSm*) there are below muscular lamina (lamina muscularis mucosae; *LMM*) connective tissue, arteries, veins, and submucosal nerve plexus. *Right*: Three cell types characterize the gastric glands: the chief cell (*CC*; *2*), which synthesizes pepsinogen; the parietal cell (*PC*; *3*), which secretes gastric acid and gastric intrinsic factor (*GIF*) required for vitamin B_{12} uptake in the gut; and endocrine cells (*EC*; *4*), which synthesize serotonin, histamine, gastrin, or somatostatin, without morphological differences between the four cell types. Chief cells have an extended rough endoplasmic reticulum (*rER*) and zymogen granules (*ZG*) maturing into apically placed secretory granules (*SG*). Parietal cells contain intracellular secretory channels with distinct microvilli (*MV*) and many mitochondria (*M*). Endocrine cells exhibit on their basal side numerous secretory granules (*SG*), which are already known from other endocrine active cells. *A* artery, *BL* basal lamina, *Cap* capillary, *E* epithelium, *GIC* glycocalyx, *ISC* intracellular secretory canaliculus, *J* junction, *lArt* long arteriole, *LP* lamina propria, *N* cell nucleus, *Pl* submucosal plexus, *sArt* short arteriole, *SCN* superficial capillary network, *T* tubulovesicular profile, *TM* tunica mucosa, *V* vein, *Ven* venule (Modified from Krstić (1991), plates 92 and 94)

10.5.2 Duodenum and Ileum

The endocrine cells of the duodenum, aka amine precursor uptake and decarboxylation (APUD) cells,[4] as well as those of the following ileum are found at the bases of intestinal villi in the crypts of Lieberkühn. These crypts harbor stem cells for the renewal of mucosa, Paneth cells to release enzymes and antibodies as mediators of the immune system in the gut lumen, and APUD cells, which secrete hormones on the basal cell side.

Several cell types have been distinguished. In the duodenum four different types have been differentiated:

1. *I cells.* In these cells cholecystokinin is synthesized. It promotes bile bladder contractions and thereby activation of hormone release from the pancreas.
2. *S cells.* The secretin released from these cells triggers in the pancreas secretion of hydrogencarbonate, enzymes, and hormones.
3. *K cells.* The product of K cells, gastric inhibitory polypeptide, inhibits hydrochloric acid release from parietal cells of the gastric glands. At the same time, pancreatic insulin release is stimulated.
4. *D cells.* The somatostatin made by and released from these cells inhibits release by and activity of other cells.

In the ileum, APUD cells mainly synthesize glicentin, neurotensins, and somatostatin.

10.6 Langerhans Islets of the Pancreas

The pancreas (Fig. 10.11) is situated in a loop of the duodenum. Most pancreatic cells are exocrine cells secreting enzyme precursors such as trypsinogen, chymotrypsinogen, carboxypeptidase, amylase, lipases, phospholipases, or deoxynucleases. These enzymes are stored in vesicles and on demand are released into the pancreatic ducts to enter finally the duodenum. Once there, endopeptidases generate active enzymes from the precursors.

In the exocrine parenchyma, endocrine cells are encased in so-called pancreatic or Langerhans islets (Fig. 10.11). These are covered by a thick network of fenestrated capillaries. Four different cell types have been found in the islets:

1. *α cells.* These cells synthesize glucagon (see Sect. 4.7), which antagonizes insulin actions. Glucagon activates gluconeogenesis in the liver. The cellular nucleus is deeply invaginated and shows a conspicuous nucleolus. Electron-dense secretory granules and a preferential staining by silver ions (agyrophilic) are characteristic of these cells.

[4]These cells are derived from the neural crest.

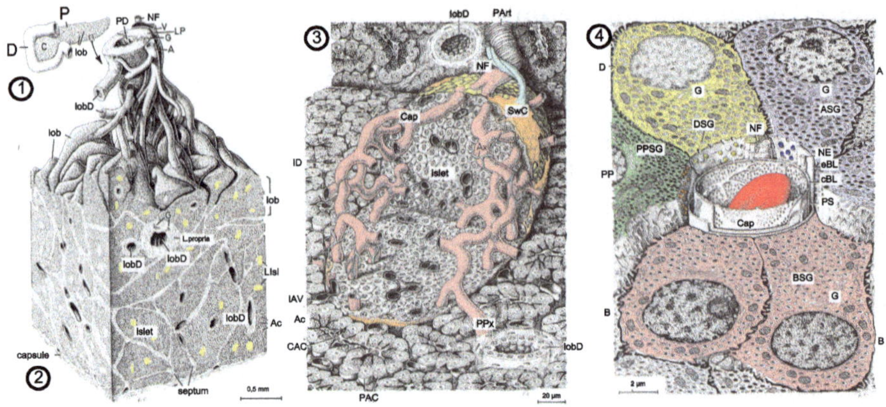

Fig. 10.11 Langerhans islets. The pancreas (*P*), made of lobes, is an exocrine and endocrine active organ directly connected to the duodenum (*D*) (*1*). In the dark tissue (*2*) islets can be observed as areas of lighter color. The pancreas generates enzymes for digestion of chyme. These enzymes are taken up by lobular channels and end in the pancreatic duct (*PD*), from where they are released into the duodenum. The *middle plate* shows the dense capillary network around and within the islets and how nerve fibers (*NF*) reach the islet cells. Islets are partially covered by Swan cells (*SwC*). Hormones released from islet cells are taken up by capillaries and are distributed via islet acinar portal veins (*IAV*) or via the periductal plexus (*PPx*). Four cell types (*right plate*)—α cells (*A*; glucagon-forming cells), β cells (*B*; insulin-releasing cells), δ cells (*D*; somatostatin-secreting cells), and PP cells (*PP*; pancreatic-polypeptide-generating cells)—can be distinguished by the shape of the cellular nucleus, the abundance and density of secretory granules, the size of the Golgi apparatus, and the number of mitochondria. Common to all types is the secretion into the perivascular space (*PS*), from where hormones reach the capillaries via sieve plates. *Ac* acini, *ASG* A cell secretory granule, *BSG* B cell secretory granule, *C* caput (head), *CAC* centroacinar cell, *Cap* capillary, *cBL* capillary basal lamina, *DSG* D cell granules, *eBL* external basal lamina, *G* gland-like outpocketing (*left*) and Golgi apparatus (*right*), *ID* intercalated duct, *LIsl* islet of Langerhans, *lob* lobule, *lobD* intralobular duct, *LP* lamina propria, *NE* nerve ending, *PAC* pancreatic acinar cell, *PArt* periductal arteriole, *PPSG* pancreatic polypeptide granules, *PS* perivascular space, *V* vein (Modified from Krstić (1991), plates 117, 120, and 121)

2. *β cells*. These synthesize insulin, a major regulator of metabolism. The cellular nucleus is oval and frequently invaginated, and the nucleolus is voluminous. Secretory granules exhibit an osmiophilic center.
3. *δ cells*. Somatostatin, the general inhibitor in the endocrine system, is made by these cells. Secretory granules contain a fine granular material, which is moderately osmiophilic.
4. *PP cells*. PP cells (aka F cells) secrete pancreatic polypeptide, which blocks exocrine pancreatic secretion (see Sect. 4.10.6). These cells are found in islets close to the duodenum, but also in association with acinar cells, and are characterized by a comparatively (to the other islet cells) large number of osmium-dense secretory granules and elliptical mitochondria.

10.7 Endocrine Cells of the Kidney

The kidney (Fig. 10.12) filters the blood. In the glomeruli, water, salts, low molecular weight proteins, and other substances can leave the arterioles, and they constitute the primary urine. In the loop of Henle, they can be resorbed by active transport

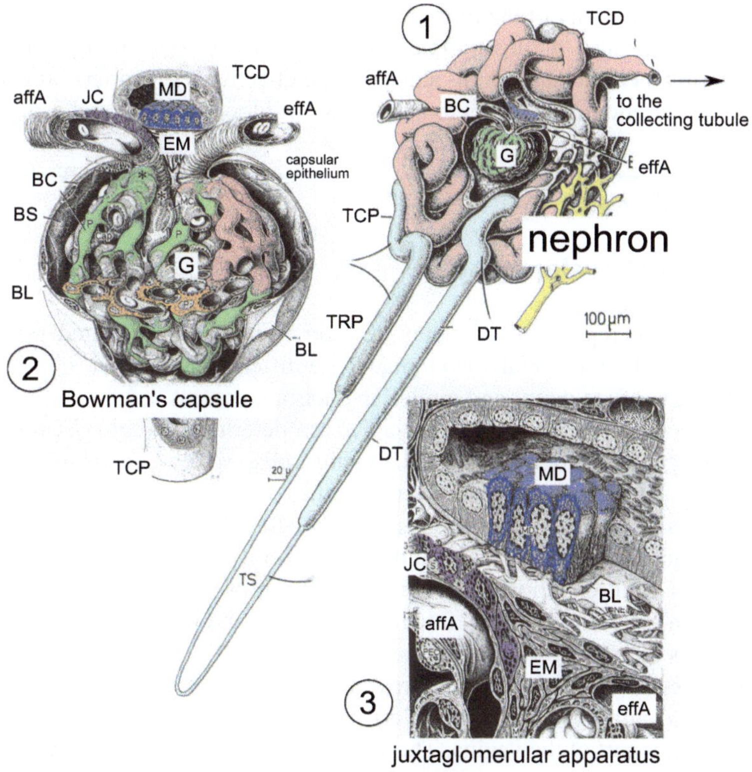

Fig. 10.12 The nephron and the juxtaglomerular apparatus. The nephron (*1*) is the functional renal subunit. By afferent arterioles (*affA*) blood enters Bowman's capsule (BC; *2*), where it is filtered within the glomerulus (*G*); the filtrate (i.e., primary urine) reaching Bowman's space (*BS*) flows from there to the proximal tubule (*PT*). This proximal tubule can be subdivided into the proximal convoluted tubule (*TCP*) and the proximal straight tubule (*P*; aka the loop of Henle). In the latter, water and electrolytes are resorbed. The primary urine reenters Bowman's capsule in the distal convoluted tubule (*DT*), which contacts Bowman's capsule at its vascular pole, forming the columnar macula densa (*MD*). Urine is finally collected via connecting pieces by the collecting duct. The cells of the afferent arterioles which are in contact with the macula densa are the juxtaglomerular cells (*JC*) constituting the juxtaglomerular apparatus. These cells have two functions: they measure the blood pressure and they release the enzyme renin, which processes angiotensinogen. The nearby cells of the macula densa are osmosensors and react to changes in osmolarity by release of sodium ions, which induce release of renin from juxtaglomerular cells. The loop of Henle is a target of aldosterone, which binds to nuclear receptors and induces expression of Na^+/K^+-ATPase, which can reabsorb sodium (as well as chloride and water) and secrete potassium. *BL* basal lamina, *effA* efferent arteriole, *EM* extraglomerular mesangium (Modified from Krstić (1991), plates 143, 145, and 152)

controlled by different hormones dependent on the actual osmolarity and blood pressure. Complex circuits (see Sect. 11.8) control the presence of transporters and receptors in the loop of the nephrons. The amount of urine filtered per day is largely dependent on these circuits.

In addition, the kidney itself, not only the adrenal gland close by, is endocrine active. The juxtaglomerular apparatus is an integral part of homeostasis by secreting renin, which is involved in osmoregulation. The cells of the macula densa are sensors of the osmolarity of the filtrate and can induce renin secretion.

The kidney is also a target organ of the endocrine system: by controlling resorption of water and salts using aquaporins and active transporters, aldosterone and atrial natriuretic peptide react to imbalances of blood pressure and osmolarity.

10.8 The Gonads

10.8.1 Gonad Development

Male and female reproductive organs develop from undifferentiated primordial gonads. At week 6 of fetal development, undifferentiated gonads, primordial urethral buds, wolffian ducts, and müllerian ducts are present.

In the presence of Y-chromosomally coded sex-determining region Y, supporting cells of the gonads differentiate into Leydig cells, which secrete testosterone. This testosterone secretion determines the further sexual development. Initiated by testosterone, androgen-specific responses are mediated by the androgen receptor: synthesis of antimüllerian hormone and the degeneration of the müllerian ducts in the male fetus, differentiation of the primordial nephros into the efferent ducts, and differentiation of the wolffian duct into the epididymis and vas deferens. The descent of the testes is triggered by androgens as well.

In the absence of sex-determining region Y (SRY) protein—that is, in females who lack the Y chromosome—the müllerian ducts develop into the uterus, fallopian tubes, and upper part of the vagina. The wollfian ducts degenerate (see Sect. 6.7).

10.8.2 Male Gonads: Testes

Within lobules separated by septa, the testis contains about 500–1,000 convoluted tubes, the seminiferous tubules. Within these, sperm are generated with the help of Sertoli cells.

A basal lamina encloses each seminiferous tubule. Layers of fibrocytes and myofibroblasts are found on top of the basal lamina. Mostly Leydig cells fill the spaces between the different seminiferous tubules. These Leydig cell are grouped around capillaries. Leydig cells secrete the male sex hormone testosterone, which in turn maintains sperm development in the nearby seminiferous tubules. Testosterone acts on other systems in skin, bone, or muscle. Sertoli cells convert testosterone into estradiol, which as an endocrine regulator inhibits gonadotropin release or initiates calcium bone deposition. Like estradiol, inhibin is made by Sertoli cells, and controls gonadotropin release in the pituitary.

10.8.2.1 Sertoli Cells

Sperm development occurs in Sertoli cells (Fig. 10.13, left) (briefly described here). The spermatogonium divides by mitosis into the primary spermatocyte. Primary spermatocytes replicate their chromosomal DNA and start the first meiotic division. Occasionally, crossing over is visible. The resulting secondary spermatocytes do not replicate their DNA, but divide directly into spermatids. These spermatids transform without any further cell division into spermatozoa with a strongly condensed chromosome (for additional information, see Krstić 1991).

Sertoli cells are large cells with a deeply invaginated cellular nucleus and a prominent voluminous nucleolus, cisternae of rough endoplasmic reticulum, and much smooth endoplasmic reticulum around the spermatids. The cytoplasm also encloses lysosomes, crystals, and residual bodies.

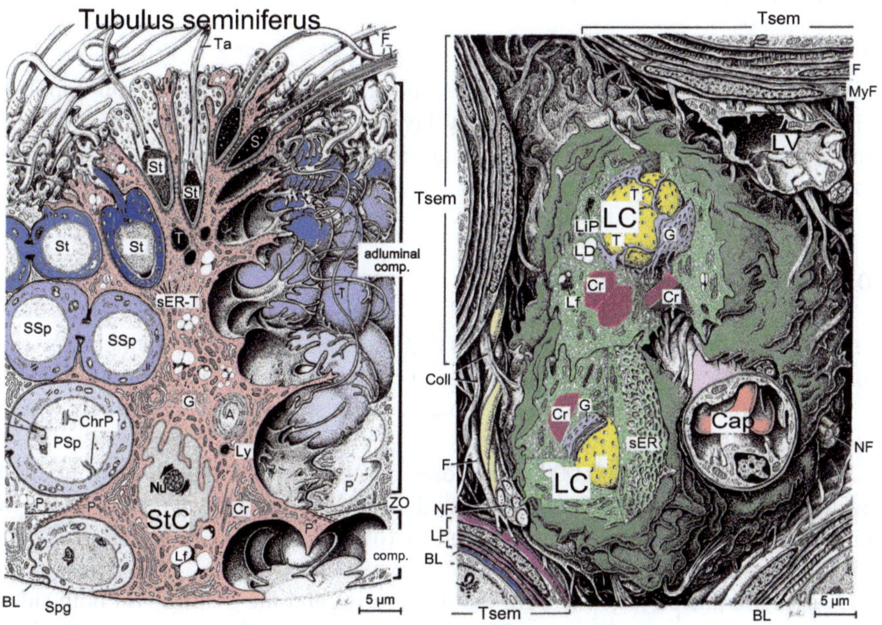

Fig. 10.13 Sertoli and Leydig cells in the testis. *Left*: Sertoli cells (*StC*) with different developmental stages of sperm. *Right*: Leydig cells (*LC*) in the interstitium. *A* annulate lamellae, *BL* basal lamina, *Cap* capillary, *ChrP* paired chromosomes, *Coll* collagen, *comp.* compartment, *Cr* crystal, *F* fibrocyte, *G* Golgi apparatus, *LD* lipid droplet, *Lf* lipofuscin, *LiP* lipochrome pigment, *LV* lymphatic vesicle, *Ly* lysosome, *MyF* myofibroblast, *NF* nerve fiber, *Nu* nucleus, *P* process, *PSp* primary spermatocyte, *S* sperm, *sER* smooth endoplasmic reticulum, *sER-T* smooth endoplasmic reticulum tubule, *Spg* spermatogonia, *SSp* secondary spermatocyte, *St* spermatid, *T* Golgi tubules, *TA* tail, *Tsem* seminiferous tubule, *ZO* zonula occludens (tight junction) (Modified from Krstić (1991), plates 164 and 169)

Sertoli cells serve spermatocytes by providing mechanical support and nutrients. Maldeveloped spermatocytes are phagocytosed by Sertoli cells and digested. Sertoli cells are endocrine active since they release androgen-binding globulin and express CYP19, which converts testosterone from the Leydig cells into estradiol.

10.8.2.2 Leydig Cells

In the spaces between the seminiferous tubules (interstitium) there are collagen fibers, nerve fibers, lymphatic vessels, and capillaries. Leydig cells, aka interstitial cells, are also found here (Fig. 10.13, right). Stimulated by luteinizing hormone, these cell synthesize testosterone. Characteristic of Leydig cells is the network of smooth endoplasmic reticulum which occupies the cytosol almost completely. Several Golgi apparatus are connected by tubules. Crystals of Reinke are landmarks of Leydig cells and of androgen-releasing tumor cells in men and women (tumors of the adrenal cortex, theca cell tumors). The chemical nature of these crystals is still unknown, although Janko and Sandberg (1970) determined the proteinaceous origin of the crystal. Lipofuscin vesicles, lipochromic pigments, and some lipid droplets complete the description of these cells.

10.8.2.3 Blood–Testis Barrier

This barrier blocks access of immune cells, for example, to the seminiferous tubules. It is formed by the basal lamina and by layers of fibrocytes and myofibroblasts. Contacts between different Sertoli cells are specially sealed in the tight junctions (zonulae occludentes).

10.8.3 Female Gonads: Ovaries

Female gonads, or ovaries, are placed on both sides of the uterus, to which they are connected via the oviducts. The medulla and cortex of ovaries are distinguishable.

The medulla contains blood and lymphatic vessels as well as nerve fibers which enter or leave the ovary at its hilus.

In the cortex about 200,000 primordial follicles are generated early in fetal development of girls. From these primordial follicles, several develop at one time in monthly episodes into primary and secondary follicles and either arrive at ovulation or become atretic.

Ovaries are placed together with the mesovaries and the oviducts on both sides of the uterus. The cortex is covered by a basal lamina and the germinal epithelium (epithelium germinalis), which should not be confounded with germ cells. The cortex consists on its outside of the tunica albuginea and below the tunica albuginea of fibroblasts and fibrocytes in characteristic whorls into which the follicles are inserted (Fig. 10.14).

A primordial follicle consists of an oocyte, a basal lamina, and a flat layer of follicular cells. During maturation to a primary follicle, the oocyte becomes larger and the follicular cells divide and form a layer of cuboidal columnar cells. The zona pellucida develops with holes for the apical poles of the follicular cells.

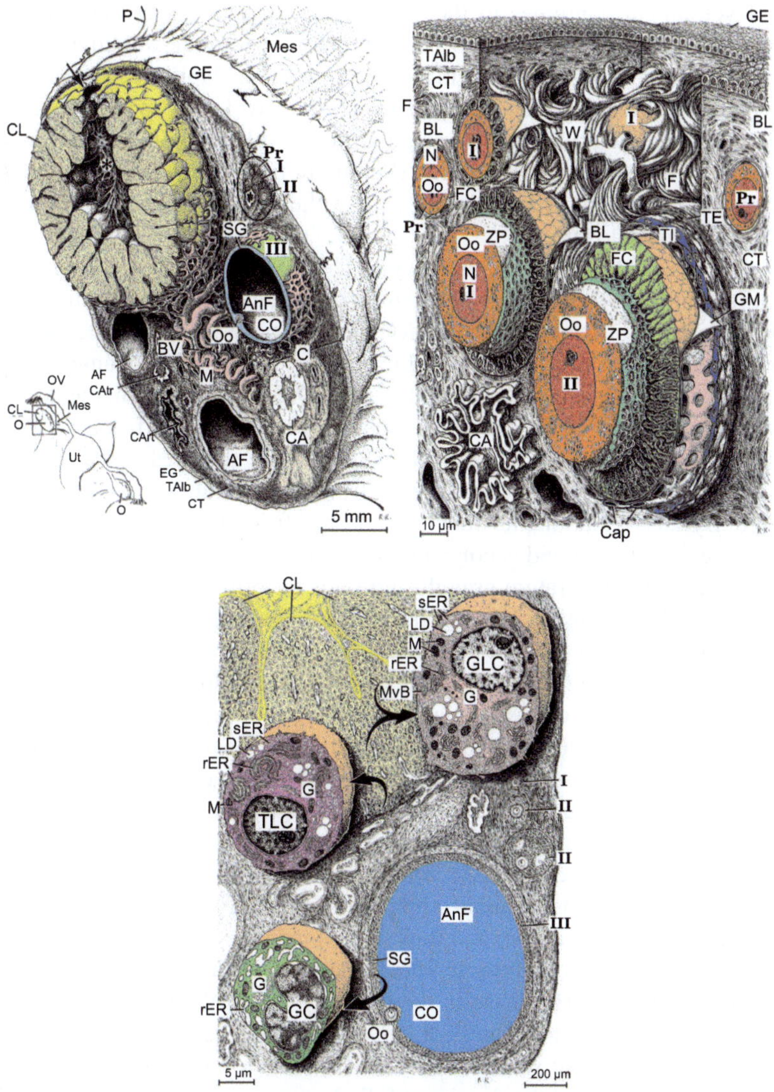

Fig. 10.14 Endocrine cells in follicular development and in the ovary. *Upper left*: A section across an ovary. *Upper right*: Primary and secondary follicles. *Bottom*: An antral follicle (*AnF*; fluid filled) and, magnified, a granulosa cell (*GC*) as well as a theca luteal cell (*TLC*) and a granulosa luteal cell (*GLC*). *AF* atretic follicle, *BL* basal lamina, *BV* blood vessels, *C* cortex, *CA* corpus albicans, *Cap* capillary, *CAtr* corpora atretica, *CL* corpus luteum, *CO* cumulus oophorus, *CT* connective tissue, *F* fibroblast and fibrocyte, *FC* follicular cell, *G* Golgi apparatus, *GE* germinal epithelium, *GM* glassy membrane, *I* primary follicle, *II* secondary follicle, *III* tertiary follicle, *LD* lipid droplet, *M* medulla (*upper left*) and mitochondrion (*Bottom*), *Mes* mesovarium, *MvB* multivesicular bodies, *N* nucleus, *O* ovary, *Oo* oocytes, *OV* oviduct, *P* peritoneum, *Pr* primordial follicle, *rER* rough endoplasmic reticulum, *sER* smooth endoplasmic reticulum, *SG* stratum granulosum (granular layer), *Ut* uterus, *TAlb* tunica albicans, *TE* theca externa, *TI* theca interna, *W* whorl, *ZP* zona pellucida (Modified from Krstić (1991), plates 186, 187, and 191)

When development continues to the secondary follicular stage, the thickness of the zona pellucida increases further, and to the one-cell-thick follicular cell layer, a second layer of cells is added. In addition, a dense capillary network around the secondary follicle develops, around which the two layers of the theca interna and theca externa are organized. During ongoing development to the tertiary stage, fluid spaces become obvious, and they eventually fuse in the antral follicle to the follicular antrum. In this stage, the layer of follicular cells is called the granular layer (stratum granulosum).

In the center of the follicle is the oocyte. This oocyte develops during follicular maturation to the largest cell of the organism. During maturation from primordial follicle to primary follicle to secondary follicle to tertiary follicle, the cell undergoes the first meiotic division, being tetraploid. Sister chromatid exchanges (crossing over) also occur at this stage. At the end of the follicular maturation, the oocyte and its associated cells called the cumulus oophorus becomes loose from the follicular wall and starts to swim in the follicular fluid. When the pressure in the follicle increases, the cumulus oophorus protrudes from the ovarian wall and eventually is disrupted. By this ovulation, the oocyte is released from the follicle and drifts in the oviduct. The follicular liquid is concomitantly released into the oviduct.

After ovulation, the stratum granulosum folds to form the corpus luteum. When stimulation of the corpus luteum by choriogonadotropin fails to occur because the egg does not nidate, the corpus luteum degenerates to the corpus albicans. Many follicles do not reach ovulation, but are arrested in development owing to lack of FSH. These follicles become atretic and atrophy to the corpora atretica.

Follicular cells, theca interna cells, and granulosa cells are endocrine active. Theca interna cells release testosterone, which the nearby granulosa cells aromatize to estradiol. After ovulation, theca luteal cells and granulosa luteal cells synthesize progesterone. In addition, granulosa luteal cells release relaxin, which blocks uterus contractions. Follicular cells secrete inhibin.

Follicular maturation is stimulated by pituitary FSH. In cows, which usually give birth to only one calf at a time, and thus like women are among the unipara, it has been estimated that the one follicle which matures first releases enough estradiol and inhibin to block FSH release in the pituitary. This follicle alone can continue development since it has become FSH independent, whereas all other follicles still depend on FSH. With the withdrawal of FSH, these follicles become atretic. The surviving follicle is called the dominant follicle. Such a mechanism most probably acts also in humans.

10.9 Selected Endocrine Active Organs in Invertebrates

10.9.1 Neurosecretion in Mollusks

More than 50 different neuropeptides are synthesized in mollusks. However, neurohemal organs have not yet developed in snails. Neurosecretory cells, instead, secrete peptides at the surface of different ganglia or at efferent nerves (pink

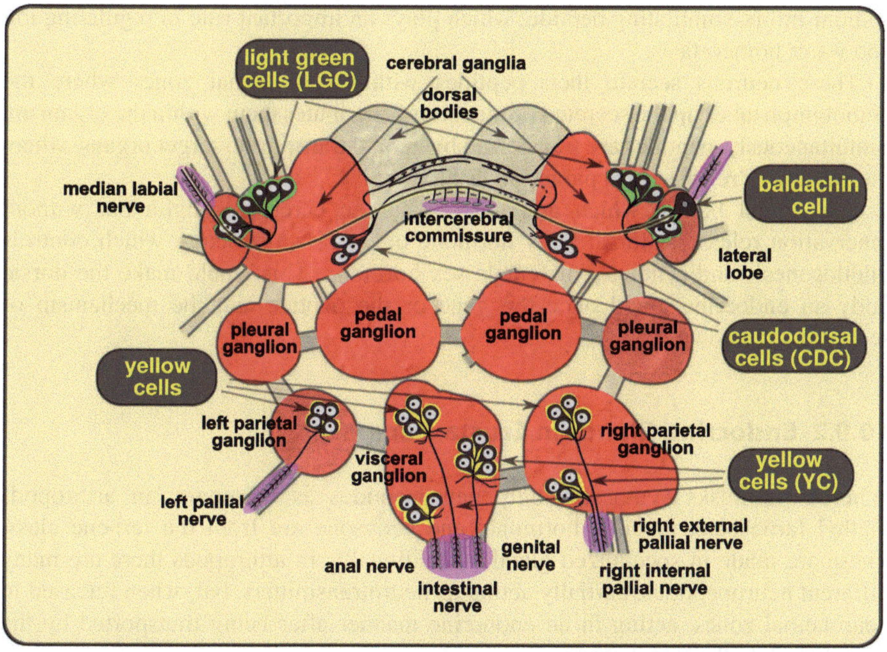

Fig. 10.15 The neuroendocrine system of mollusks (After Penzlin and Ramm 2008, p. 523)

in Fig. 10.15). An enhanced density of neurosecretion has been observed in the intercerebral commissure. There, canopy cells secrete an as yet[5] unknown peptide (perhaps ILPs). In addition, caudodorsal cells (CDCs) of the great pond snail (*Lymnaea stagnalis*) secrete egg-laying hormone (ELH), several CDC peptides, and calfluxin.

The ELH of *Aplysia californica* is released not by CDCs, but by about 400 bag cells located at the connections of pleural and abdominal ganglia. The precursor polypeptide is processed to five peptides: α-bag cell peptide (BCP), β-BCP, γ-BCP, an acid peptide, and the ELH (for additional details, see genbank/Swiss-Prot entry P01362.2). After posttranslational processing, the BCP and the ELH are sorted into different vesicles: BCPs are released in the abdominal ganglion, whereas ELHs are set free in the zone between pleural and abdominal ganglia (Fisher et al. 1988). This observation has not been corroborated in other invertebrate or vertebrate organisms, and analysis of the unorthodox mechanism is still required (Sossin et al. 1990).

The light green cells[6] of the cerebral ganglia release the insulin-like peptide of mollusks. Within the parietal and visceral ganglia there are yellow cells releasing

[5]Last checked in November 2011

[6]The light green or yellow color arises by staining sections with alcian yellow or alcian blue (Dudel et al. 2001).

sodium-influx-stimulating peptide, which plays an important rule in regulating ion and water homeostasis.

These neurons secrete their peptides within neurohemal zones where the hemolymph takes up the secreted products and distributes them within the organism. Simultaneously, peptides are transferred by axonal transport to target organs, where they act when released in a paracrine fashion.

The dorsal bodies which are attached to the cerebral ganglia but without innervation release a dorsal body hormone of unknown structure which controls vitellogenesis and synthesis of female sex hormone. This would make the dorsal body an endocrine gland once the gene or the peptide and the mechanism of secretion are known.

10.9.2 Endocrine Glands in Crustaceans

Whereas mollusks synthesize only neuropeptides as hormones, in arthropods methyl farnesoate, juvenile hormone, and ecdysone are from the terpene class. These are made in specialized glands. Additionally, in arthropods there are many different neuropeptides, partially acting as neurotransmitters, but, when released in neurohemal zones, acting in an endocrine manner after being transported by the hemolymph to the target organ.

In crustaceans (Fig. 10.16) there three major neurohemal zones: the sinus gland, where neuropeptides made in the X organ are released; the postcommissural organ, with peptides from the brain; and the pericardial organ, where neuropeptides of subesophageal origin are released.

Development and maturation of gonads are controlled by the mandibular organ (M organ) and are triggered by the methyl farnesoate (Fig. 6.3, **11**) made therein.

The molting hormone ecdysone is made in the Y organ. Its synthesis is inhibited by the molt-inhibiting hormone from the sinus gland. The conversion of ecdysone and 25-deoxyecdysone, which are both released in the Y organ, into 20-hydroxyecdysone and ponasterone A, respectively, occurs in peripheral tissue such as the testes and the midgut gland.

10.9.3 Neurosecretion and Endocrine Organs in Insects

In insects, most peptide hormones are made and released by neurosecretory cells. Chapman (1998) distinguishes three different activities: as a neurotransmitter, as a neuromodulator, or as a neurohormone. Neurohormones are secreted within neurohemal zones into the hemolymph, whereas neurotransmitters act in synapses and neuromodulators act in a paracrine way on other nerve cells. This formal separation is obsolete when certain neuropeptides act as neurotransmitters and as neurohormones and are recognized away from the releasing organ by cognate receptors. A functional distinction between the nervous system and the endocrine system is less possible in insects than in vertebrates.

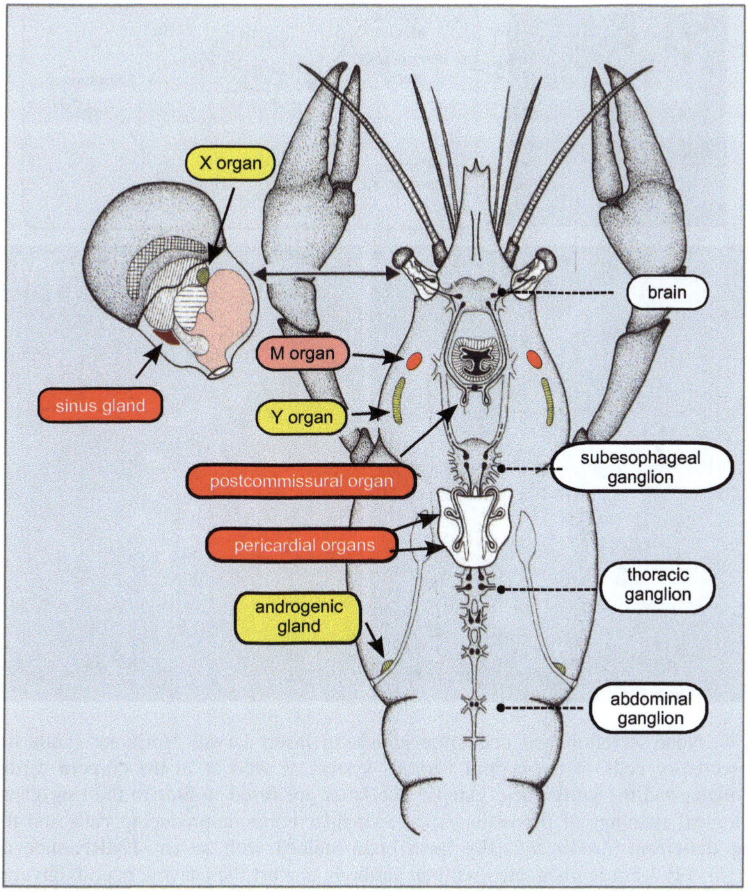

Fig. 10.16 Endocrine glands in crustaceans (After Penzlin and Ramm 2008, p. 526)

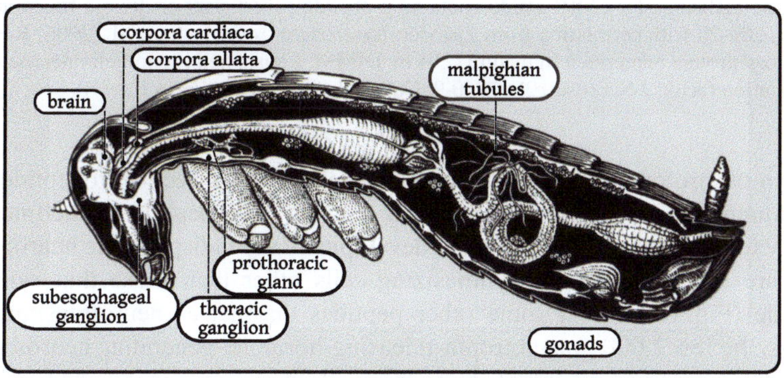

Fig. 10.17 Endocrine cells and organs in adult insects (According to Penzlin and Ramm 2008, p. 529)

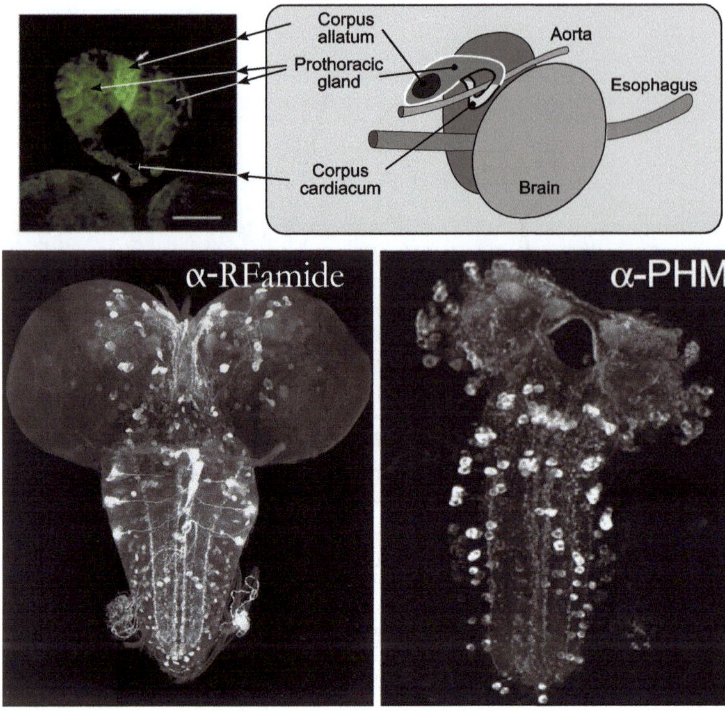

Fig. 10.18 Neurosecretion and endocrine glands in insect larvae. Hormone synthesis occurs in neurosecretory cells of the central nervous system as well as in the corpora cardiaca, the corpora allata, and the prothoracic glands. The latter are fused in flies to the ring gland (*Top*). Immunological stainings of the brain indicate singular hormone-producing cells and the axons spreading therefrom. *Lower left*: Fly larva brain stained with an anti-FMRFamide antibody. *Lower right*: Fly larva brain stained with an antibody against the enzyme peptidylglycine alpha-hydroxylating monooxygenase (*αPHM*), which converts C-terminal glycine into an amide. The transcription factor Dimm, controlling, for example, α-PHM expression was overexpressed here (*Upper left* from Chiang et al. (2002); Copyright (2002) National Academy of Sciences, U.S.A; *upper right* redrawn according to Shiga (2003); *lower left* from Merte and Nichols (2002); Drosophila melanogaster FMRFamide-containing peptides: redundant or diverse functions? Peptides, 23, 209-20 with permission from Elsevier; *lower right* from Hewes et al. (2006) Regulation of secretory protein expression in mature cells by DIMM, a basic helix-loop-helix neuroendocrine differentiation factor. J Neurosci, 26: 7860-9, Copyright Society for Endocrinoly)

The most profound difference between vertebrate and insect neuropeptides is the small number of insect neurons releasing a given neuropeptide. Sometimes there are only four or eight cells releasing a developmentally indispensable neuropeptide. There are few neuropeptide-synthesizing cells other than those that synthesize RFamide (Fig. 10.18) and some other peptides that act as neurotransmitters. In humans, the few 2,000 gonanotropin-releasing-hormone-generating neurosecretory cells are two orders of magnitude more frequent. Insects are, on the one hand, much smaller than vertebrates, and avidity and specificity of receptors has to be very high to correctly detect the low concentrations of hormones in the hemolymph.

Postembryonic insect development is finished when the last molt had occurred. In hemimetabolous and holometabolous species, metamorphosis is linked to this last molt. The molting hormone ecdysone synthesized in the prothoracic gland is not needed as before, and thus the prothoracic gland degenerates. However, ecdysone is made in adult insects in gonads and in the fat body.

Corpora cardiaca are of neuronal origin (Siegmund and Korge 2001). Corpora allata are glands proper. The characteristic hormone of the corpora allata is juvenile hormone, a sesquiterpene. It is related to the methyl farnesoate of crustaceans since methyl farnesoate is an intermediate during the synthesis of juvenile hormone. In most insects, the corpora cardiaca and corpora allata exist in pairs along the aorta (Fig. 10.17).

In the ring gland of dipteran larvae, the prothoracic glands are distinguished by their large polyploid cells and are on either side of the fused corpora allata with much smaller diploid cells. The corpora cardiaca are at either end of the ring gland next to the prothoracic glands and consist of neurosecretory cell nerve endings plus intrinsic glandular cells that secrete adipokinetic hormone. Figure 10.18 demonstrates, again, that these organs have countable numbers of cells.

Regulatory Patterns

11

Contents

11.1 Feedback Control 300
11.2 Regulator Circuits 301
11.2.1 Pressure and Stress 301
11.2.2 Calcium Metabolism 304
11.3 Regulation of Reproduction 305
11.3.1 Regulation of GnRH 305
11.3.2 Regulation of the Menstrual Cycle 307
11.3.3 Puberty 309
11.3.4 Menopause 310
11.4 Glucose Metabolism 311
11.4.1 The Origin of Blood Glucose 311
11.4.2 Regulators and Control Variables 311
11.4.3 Glucose-Dependent Gene Expression in the Liver 311
11.4.4 Glucose-Dependent Insulin Secretion in the Pancreas 312
11.4.5 Insulin-Dependent Procedures 313
11.4.6 Glucagon and Blood Glucose Level Increase 313
11.5 Appetite and Hunger 313
11.5.1 Central Nervous System 314
11.5.2 Parasympathetic Fibers of the Vagus Nerve 315
11.5.3 Sympathetic Fibers of the Splanchnic Nerve 315
11.5.4 Enteric Nervous System 315
11.5.5 Endocrine Mediators and Neuropeptides 315
11.5.6 Hormone Receptors on Nerve Cells 317
11.5.7 Mechanoreceptors 317
11.5.8 Feeding Circuits in the Brain 318
11.5.9 Hunger and Food Intake in *Drosophila melanogaster* 319
11.6 Growth 321
11.6.1 Epiphyseal Cartilages 321
11.6.2 Zonal Organization of the Epiphysis 322
11.6.3 Regulation by Hormones 322
11.7 Growth and Molt in Ecdysozoans 324
11.7.1 Regulation of Growth in Insects 326
11.7.2 Hormones and Postembryonic Development 326
11.7.3 Linkage of Growth and Metamorphosis 328

B. Kleine, W.G. Rossmanith, *Hormones and the Endocrine System*,
DOI 10.1007/978-3-319-15060-4_11

11.7.4 Regulation of Ecdysis 330
11.7.5 Postembryonic Development in Holometabolous Species 332
11.8 Regulation of Blood Pressure, Osmolarity and Blood Volume 334
11.8.1 Integration of Several Control Circuits 334
11.8.2 Osmoreceptors at the Blood–Brain Barrier 335
11.8.3 Angiotensin II Receptors at the Blood–Brain Barrier 336
11.8.4 AVP Release in the Posterior Pituitary 337
11.8.5 The Role of Oxytocin 337
11.8.6 Thirst and the Endocrine System of the Brain 337
11.8.7 Biochemistry of Water and Sodium Resorption 338

In this chapter we will present examples of endocrine regulation. A single hormone in the endocrine system is something like an individual in a family tree. Different sources exist which determine the creation and development of a single person, and his or her progeny are dependent not on him or her alone, but also on many other people.

That is like the release of a hormone. From everywhere, messengers touch the receptor of the hormone-producing cell, some force release, others retention, some new synthesis, others arrest. An endocrine cell has to integrate all these different influences and decide what to do. This cannot be achieved by intelligence, but by an interplay of control elements in the related signal transduction pathways of the different receptors or by complex interaction while activating genes, or both.

If a cell actually releases an hormone, the stimulating forces prevailed. These are, however, often blocked by high concentrations of the very same hormone, or by other, downstream hormones. For example, gonadotropin-releasing hormone (GnRH) triggers in the pituitary release of luteinizing hormone (LH) and follicle-stimulating hormone (FSH); these trigger in the gonads estradiol and testosterone synthesis. The latter are, in turn, strong inhibitors of GnRH release and LH/FSH secretion.

11.1 Feedback Control

Figure 11.1 demonstrates a simple feedback control pathway. In the thyroid, on stimulation by thyroid-stimulating hormone (TSH), thyroxine (T_4) is synthesized and released (1 in Fig. 11.1). By T_4 and by triiodothyronine (T_3) derived therefrom through the action of deiodinase, however, TSH release in the pituitary is blocked (2 in Fig. 11.1). TSH release is stimulated by thyrotropin-releasing hormone (TRH; 3 in Fig. 11.1). This TRH release (4 in Fig. 11.1) is equally blocked by T_4 (5 in Fig. 11.1), whereby T_4 has to cross the blood–brain barrier.

T_4 synthesis is additionally controlled, for example, by circadian rhythms not shown here. TRH release in the brain is multiply influenced by other neurons and hormones: by serotonin (6 in Fig. 11.1) and by noradrenaline (7 in Fig. 11.1). TSH release is negatively influenced by (hypothalamic) dopamine (8 in Fig. 11.1) and somatostatin (9 in Fig. 11.1). Estrogens in the circulation enhance TSH release

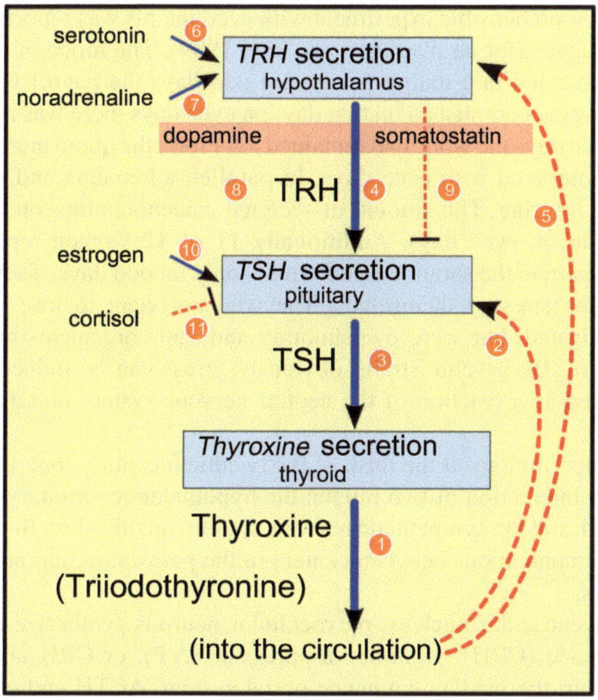

Fig. 11.1 Synthesis of thyroxine and its control. *Blue, continuous lines* indicate stimulation, *red, broken lines* indicate inhibition; *numbers* are explained in the text. *TRH* thyrotropin-releasing hormone, *TSH* thyroid-stimulating hormone

(10 in Fig. 11.1), and the glucocorticoid cortisol suppresses TSH formation (11 in Fig. 11.1).

In very similar ways, other hormones are regulated by different additional hormones, neuronal interactions, and other soluble messengers. In the following pages, we will provide several examples.

11.2 Regulator Circuits

11.2.1 Pressure and Stress

Stress was defined by Selye (1936, 1950) as "the nonspecific response of the body to any demand upon it." Hunger, injury, coldness, fear, cardiac arrest, exhaustion, capture, and arrest as well as social stress result, according to Selye, in the same unspecific reactions of the organism manifesting themselves in a strong elevation of the levels of glucocorticoids and adrenaline in the circulation. There might be specific reactions to any to these threats, called stressors; however, according to Selye, the unspecific aspect of the reaction should be called stress.

An easily comprehensible experiment with accountants was reported by Lennart Levi and colleagues (for an overview, see Levi 1989). The mode of payoff for the pay slip was changed in a daily manner. On odd days the payoff was due to the number of entries accomplished on that day; on even days there was no such check. Whereas on even days the workload remained as before, the quota increased by 14 % on odd days compared with even days. In parallel, adrenaline and noradrenaline were analyzed in urine. The amount of secreted catecholamines on odd days was 40 % above that on even days. Additionally 11 of 12 women reported fatigue, backache, and pain in the shoulders and arms—only on odd days. They felt hurried.

These women stressed themselves. They did not want to lose income, were especially motivated, but were overshooting and their organisms reacted to this psychic pressure. By psychic strain, obviously, stress can be induced. Stress can thus be regarded as a reaction of the central nervous system manifested in other organs.

Modern endocrinology at the dusk of the twentieth century took the physiology of stress as the interaction of two pillars: the hypothalamic–pituitary–adrenal axis, on the one hand, and the sympathetic nervous system, on the other. Both are coupled by mutual communications between centers in the paraventricular nucleus and the locus coeruleus.

In the paraventricular nucleus, parvocellular neurons synthesize corticotropin-releasing hormone (CRH), arginine vasopressin (AVP), or CRH and AVP. After CRH release into the median eminence portal system, ACTH and endorphins are released in the pituitary. This ACTH stimulates in the adrenal glands adrenaline production and release (see Sect. 7.1) and steroid hormone synthesis, above all synthesis of glucocorticoids.

In the locus coeruleus, noradrenergic neurons particularly stimulate sympathetic nerve cells. The locus coeruleus nerves project, for example, to the parvocellular neurons of the paraventricular nucleus, and these latter project to the noradrenergic neurons of the locus coeruleus in a mutual way; in this interaction CRH and AVP act as neurotransmitters. The entire system is further controlled by unknown zeitgebers—in any case CRH, AVP, and ACTH are released in one to three pulses per hour and in a daily rhythm where evening/night pulses are greater than those during the day (Fig. 11.2). Furthermore, many other neurons from brain areas and peripheral organs act via sympathetic and other nerves on the paraventricular nucleus and locus coeruleus neurons. The limbic system is tightly coupled to the stress response. Other hormones such as neuropeptide Y (NPY) and substance P are also involved: NPY stimulates CRH release and inhibits locus coeruleus neurons, whereas substance P acts in the opposite way (Strakis and Chrousos 1997).

The central role of the mutual interaction of the paraventricular nucleus and locus coeruleus has been questioned in recent years. In the last 10 years, several authors have reported on specific reactions to stressors which cannot be reconciled with the definition of Selye.

One aspect of stress regulation is the action of glucocorticoids in the brain. The patterns is most complex. Corticosterone and cortisol were observed in brain; released from the adrenal gland, they are transported by corticosteroid-binding

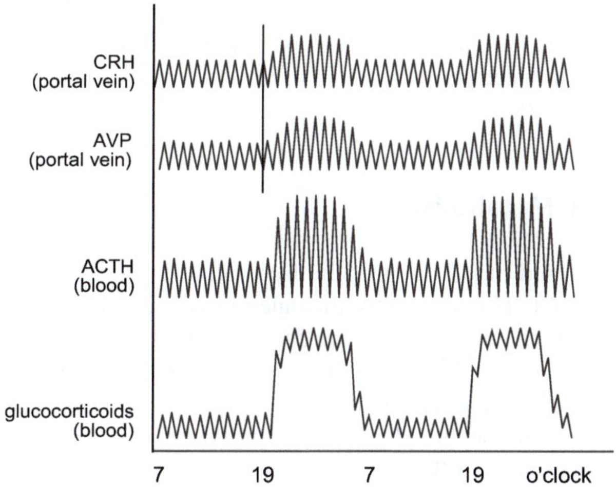

Fig. 11.2 Pulsatile and circadian activity of the hypothalamic–pituitary–adrenal axis. The pulse shift from corticotropin-releasing hormone (*CRH*) to arginine vasopressin (AVP) is indicated by the *vertical bar*. The amplitude changes in a circadian manner. Following the pulsatile CRH/AVP release, pituitary adrenocorticotropic hormone (*ACTH*) is released in pulses and stimulates adrenal synthesis and release of glucocorticoids. Elevated AVP/CRH pulses in the evening and during the night result in markedly enhanced glucocorticoid levels still oscillating with a frequency of 20 min (Redrawn from Chrousos 1998)

globulin or unbound (may be attached to albumin). Whether corticosteroid-binding globulin can cross the blood–brain barrier has not been analyzed; steroids, as far as we know, are able to do so. In the periventricular tissue around the blood capillaries, 11-β-hydroxysteroid dehydrogenase (11β-HSD) type 2 is strongly expressed, converting cortisol into the inactive 11-deoxycortisol (see Fig. 6.21).

In many animal species, corticosterone (Fig. 6.21) instead of cortisol is the active glucocorticoid. This is not converted by 11β-HSD type 2. Corticosterone, too, binds to corticosteroid-binding globulin, but with a somewhat lower affinity. Furthermore, cortisol is removed from cells of the blood–brain barrier by the multidrug resistance P-glycoprotein; this multidrug resistance protein serves, in general, to move toxic substances. Obviously, cortisol is regarded as something foreign, whereas corticosterone is not. Thus, there are two mechanisms which favor the use of corticosterone in the brain.

Corticosterone binds preferentially to mineralocorticoid receptor (MR) and with reduced activity to glucocorticoid receptor (GR); both nuclear receptors are transcription factors. To make the situation even more complex, hippocampus cells express another 11β-HSD, of type 1, which converts corticosteroid to cortisol, which, again, prefers to bind to GR (Kloet 2003). Whether 11β-HSD type 1 is widely distributed and whether its expression is required for glucocorticoid activity in the central nervous system (CNS) has not been reported.

These findings still leave open the question whether cortisol is active in the brain. Corticosterone might bind to MR and GR. Maintaining homeostasis is achieved via MR, whereas GR-induced activities seem to facilitate recovery after disturbances of homeostasis (Kloet 2003).

11.2.2 Calcium Metabolism

Since there are closed blood circuits, the free calcium concentration in blood is a fairly exact 1 mmol/l. This is exactly the calcium concentration in saltwater, from which it is inferred that all organisms are derived from saltwater organisms.

In humans calcium is taken up with the food (1 in Fig. 11.3), brought into the blood, and stored there complexed with albumin (2 in Fig. 11.3); albumin–calcium serves as a buffer for the blood calcium concentration. Calcium from blood is incorporated into bone (3 in Fig. 11.3), or filtered from blood by the kidney (4 in Fig. 11.3) and if required resorbed from the primary filtrate (5 in Fig. 11.3).

The calcium level is controlled by calcium concentration sensing receptors in the surface of parathyroid cells (6 in Fig. 11.3) or kidney cells (7 in Fig. 11.3). The sensor is a heptahelical G-protein-coupled membrane receptor. When the sensor is stimulated by elevated calcium concentrations, it activates phospholipase to enhance the level of inositol trisphosphate (IP_3); by this messenger, resorption in the kidney is inhibited and filtered calcium is secreted in urine (8 in Fig. 11.3). In the parathyroid gland, IP_3 inhibits the synthesis of parathormone (9 in Fig. 11.3).

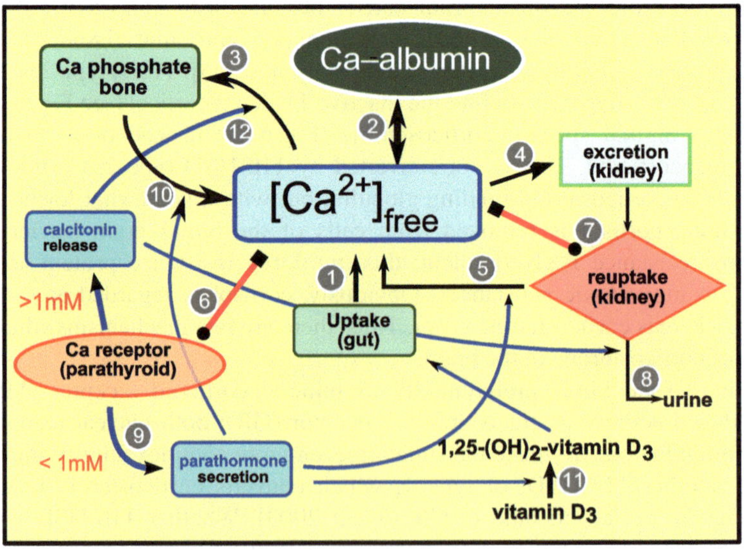

Fig. 11.3 Major components of calcium metabolism

If the calcium secretion leads to a deprivation of calcium, IP_3 release is blocked and resorption in the kidney and parathormone synthesis in the parathyroid gland are induced.

By parathormone, the level of available calcium is increased by desorption from bone (10 in Fig. 11.3). Simultaneously, the conversion of 25-hydroxyvitamin D_3 to 1,25-dihydroxyvitamin D_3 (calcitriol) (11 in Fig. 11.3) is triggered, which then allows uptake of calcium in the gut and its transport through the gut wall.

The antagonist of parathormone is calcitonin; its formation is triggered by elevated (greater than 1 mM) calcium levels. This hormone blocks resorption in the kidney and facilitates calcium incorporation into bone (12 in Fig. 11.3).

11.3 Regulation of Reproduction

Progeny and reproduction are among the most important characteristics of life. Since the debut of science, the study of these phenomena has occupied investigators. The differences between women and men have been obvious since primeval times, and children learn the functions of sexual organs at the latest in their first sex education at school.

The influence of hormones on reproduction is common knowledge to any woman taking or having taken the pill. The role of hormonal regulation during the menstrual cycle is less well known. The hormonal control of male reproductive capacities is almost unknown. The male sex hormone testosterone receives more attention when its abnormal levels are used to help defend a person accused of rape in court.

11.3.1 Regulation of GnRH

Reproductive rhythms are controlled via GnRH. Sexual activity in animals is allowed when sufficient food is available and the social fabric allows it; in many animals, a seasonal rhythm maintains control of sexual activity. Additionally, only pulsatile GnRH release induces pulsatile release of gonadotropins, which in turn regulate gonadal activity. These observations are reflected in the regulation of GnRH; neurotransmitters such as NPY, endorphins, CRH, and galanin are messengers of these environmental conditions which by controlling GnRH formation and release allow reproduction.

The perikarya of the GnRH neurons are present in the hypothalamus, mainly in the preoptic area and in the median eminence, and in addition in the lamina terminalis and septal area (Fig. 11.4). The former have been shown to be innervated strongly by other neurons, while the latter are only sparsely connected with other neurons.

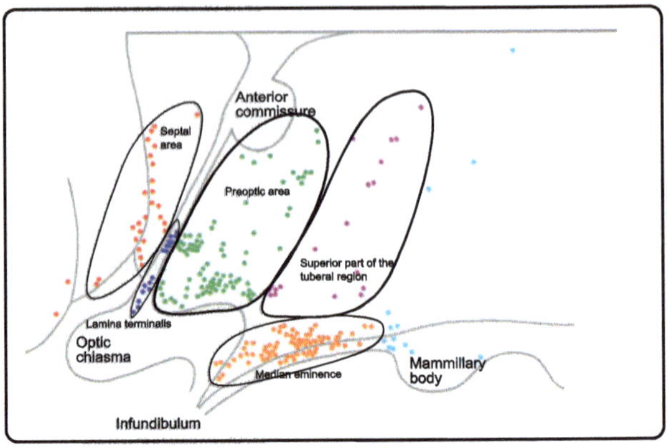

Fig. 11.4 Gonadotropin-releasing hormone (GnRH)-secreting neurons in the human hypothalamus. The perikarya of GnRH neurons are in the *labeled areas*. GnRH neurons are strongly innervated by other neurons, with the exception of those in the septal area (From Dudas and Merchenthaler 2006)

The following neurotransmitters influence GnRH neurons (Fig. 11.5): γ-aminobutyric acid (GABA), NPY, substance P, endogenous opiates such as endorphins and leu-enkephalin, CRH, galanin, catecholamines (dopamine and noradrenaline), and neurotensins (Dudas and Merchenthaler 2006) (Fig. 11.5).

Tuberoinfundibular dopaminergic (TIDA) neurons, which release dopamine, are important control elements for GnRH release. In those areas where TIDA neurons are found, there are also many NPY, galanin, endorphin, and substance P neurons, whereas CRH neurons are mainly located in the infundibulum and the preoptic area (Dudas and Merchenthaler 2006). TIDA neurons, furthermore, express estrogen receptor, allowing gonadal feedback inhibition of GnRH release (Mitchell et al. 2003). The latest addition to the list of GnRH regulators is kisspeptin (Messager et al. 2005).

Suppression of reproductive activity as a consequence of stress has been established in mammals. The central and peripheral stress systems are thought to play the prominent role: CRH directly suppresses GnRH release via synaptic contact of CRH axons to dendrites of GnRH neurons in the medial preoptic nucleus. There are differences between rodents and primates. Endogenous opiates (from proopiomelanocortin) can mediate several CRH effects, in part with respect to the menstrual cycle, to the species, or to the sex. Cytokines from the CNS are equally involved in GnRH regulation: IL-1 inhibits GnRH neuron activity and reduces GnRH synthesis and release. These IL-1 effects are mediated in part by endogenous opiates (endorphins and enkephalins) and by central prostaglandins.

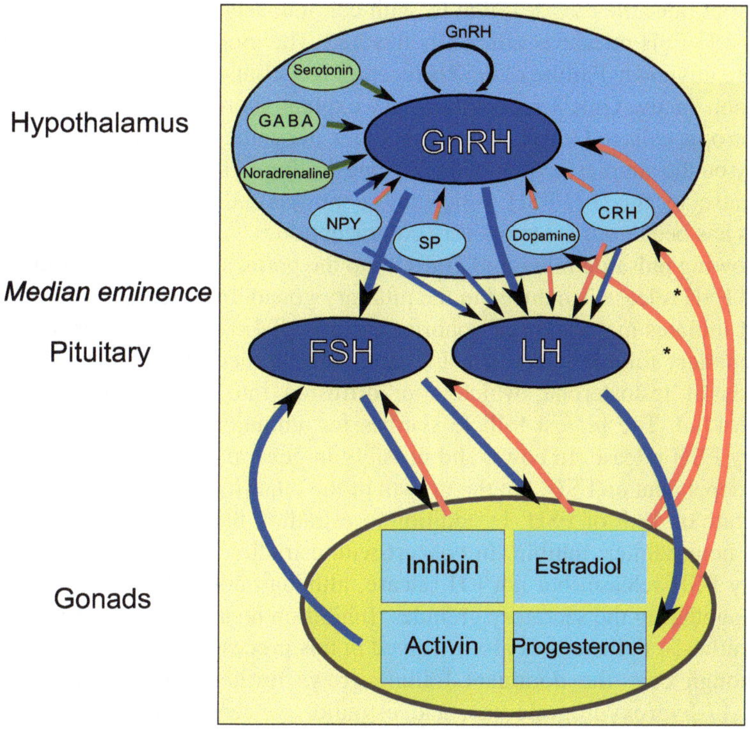

Fig. 11.5 Hormonal interplay for successful reproduction. *Blue arrows* amplifying endocrine secretion, *red arrows* mitigation of endocrine secretion, *green arrows* neurotransmitter activity [tuberoinfundibular dopaminergic neurons have been found to express estrogen receptors and may thus exert feedback control on gonadotropin-releasing hormone (*GnRH*) release], *asterisks* indicate that other neurons are potentially estrogen receptor positive and may participate in this feedback control, *CRH* corticotropin-releasing hormone, *FSH* follicle-stimulating hormone, *GABA* γ-aminobutyric acid, *LH* luteinizing hormone, *NPY* neuropeptide Y, *SP* substance P (From Dudas and Merchenthaler 2006)

11.3.2 Regulation of the Menstrual Cycle

Regulation of the female menstrual cycle occurs in four organs: in the hypothalamus, in the pituitary, in the ovaries, and in the uterus. We are talking about the hypothalamic–pituitary–gonadal axis.

The first essential variable indispensable to reproduction is *pulsatile release of GnRH* in the hypothalamus. There are only some thousand GnRH-secreting neurons in the mediobasal hypothalamus and in the preoptic nucleus. Triggered by γ-aminobutyric acid (GABA) and owing to mutual axonal connections of GnRH neurons, a coordinated secretion in the median eminence occurs. Whether galanin coexpressed in many GnRH neurons plays a role in GnRH secretion is an open question. GnRH secretion is reduced or entirely blocked by stress and its mediator CRH.

Without pulsatile release—that is, with continuously elevated or missing GnRH levels—LH/FSH release is efficiently blocked. The molecular mechanisms for this effect are unknown. Failure of GnRH receptor recycling and thus lack of cell surface expression of the GnRH receptors might explain short-term unresponsiveness of gonadotropic cells to GnRH. A possible block of synthesis has not yet been proven, but may be the simplest explanation for long-lasting LH/FSH suppression.

Because of this LH/FSH deficiency, continuous dosing of GnRH or GnRH analogs has been used for temporary contraception.[1]

Following pulsatile release of GnRH into the portal system of the infundibulum, LH and FSH release is induced in the pituitary gonadotropic cells.

FSH initiates maturation of cohorts of primordial ovarian follicles into primary and secondary follicles (meiosis). LH stimulates in theca cells around the follicles synthesis of testosterone, which after diffusion into the follicles is converted into estradiol. The largest follicle synthesizes amounts of estradiol which inhibit pituitary FSH release. In cows, and possibly in general, this follicle's growth is no longer dependent on FSH, but the growth of the other follicles still is, and thus they degenerate on lack of FSH. In addition to estradiol, the follicular fluid contains a protein heterodimer: inhibin. In concert with estradiol, inhibin blocks specifically pituitary FSH release, but not LH release, although both LH and FSH have been found together in the secretory granules. Inhibition is obviously due to blocking of FSH synthesis. Follistatin is also involved in this process.

Although only the dominant follicle grows further, androstenedione, testosterone, 17-hydroxyprogesterone, and estradiol concentrations in blood increase steadily. Shortly before ovulation there is a LH surge. Finally, the follicle bursts and liberates together with the oocyte its hormonal content. The estradiol stimulates growth of the uterine endometrium. This allows nidation of the fertilized egg within a restricted time window and which may grow into a new individual.

The corpus luteum develops from the remnants of the burst follicle. The corpus luteum provides progesterone, which inhibits degeneration and rejection of the uterine endometrium. Progesterone together with estradiol from the former follicle decreases hypothalamic GnRH release. This progesterone synthesis is maintained for only a few days and finishes if fertilization does not occur (see the next paragraph) after 7–8 days. As a consequence, blood supply to the endometrium is blocked, and the latter degenerates until it is rejected during the following menstruation. Without inhibition by estradiol and progesterone, hypothalamic GnRH release is reinitiated, FSH is made in the pituitary, and the next wave of follicles start to grow. The cycle restarts.

In the case of fertilization, the fertilized egg starts to divide while still in the ovarian duct. After the 32-cell stage, cells fuse to the morula, which gives rise to the blastocyst, with an outer layer of trophoblastic cells and the inner cell mass. The fetus develops from the inner cell mass, and the placenta develops from

[1]Permanent contraception with these analogs is not desirable owing to permanently repressed estradiol levels and the risk of osteoporosis.

the trophoblastic layer. The trophoblastic cells perform two roles: they mediate nidation into the uterine endometrium and they synthesize the pregnancy hormone choriogonadotropin. With this choriogonadotropin, progesterone synthesis in the corpus luteum is maintained, progesterone levels remain elevated, and degeneration of the endometrium is inhibited. As long as trophoblastic cells of the blastocyst and later in the placenta secrete choriogonadotropin, progesterone synthesis in the corpus luteum is ensured. In addition, progesterone blocks pulsatile GnRH release from the median eminence and LH/FSH release from the pituitary.

11.3.3 Puberty

Men and women are not fertile immediately postpartum. The reproductional competence is acquired later in life, in most individuals after the 11th to 13th birthday. Pituitary hormonal release and gonadal hormonal release in the newborn drop off within a few weeks. Menarche, the time of first ejaculation or first menstruation, is in the middle of a long period of sexual maturation starting with enhanced adrenal activity at the age of about 8 years, called adrenarche. First DHEA and DHEA sulfate secretion is enhanced, then 1–2 years later androstenedione secretion is enhanced, but the reasons for this change are still unknown. The formation of these androgens starts the growth of primary and secondary sexual organs and the development of pubic hair.

Up to puberty no pulsatile LH or FSH is measurable. The earliest measurable parameters are nightly FSH pulses strongly dependent on GnRH pulses. These GnRH pulses are induced centrally without any gonadal costimulation since puberty occurs in individuals without functional gonads. About 1 year after the first occurrence of FSH pulses, LH pulses can be measured. These FSH or LH pulses are not yet regularly timed and of constant amplitude as in adults, but are relatively rare and with largely varying frequency. As a reaction to these gonadotropin pulses and because of the development of Leydig cells in boys and the first growing ovarian follicles in girls, testosterone and estrogen levels in blood increase and they become measurable. Mood changes are explained by these irregular gonadotropin pulses and thus changing steroid levels. The first FSH pulses induce only incomplete follicular development, and ovulation does not yet occur. In the course of puberty, pulses are more regular, follicle development reaches later stages, and finally the first ovulation occurs. Menarche is reached.

In boys, the first ejaculation is not at the end of puberty, since regular GnRH pulses and coupled LH and FSH pulses are not as frequent and regular as in men, where about 18 pulses per day is normal. Enhanced FSH release forces the formation of seminiferous tubules. Under the influence of LH, precursor cells develop into Leydig cells, which mainly release testosterone.

The enhanced testosterone formation coinciding with enhanced growth hormone (GH) levels induces strong mood changes: the teenager shows—compared with childhood—an enhanced aggressive behavior although the environment, obviously, did not change: the "only" change is the maturation of the endocrine system.

Apart from the hormones of reproduction, other hormones, especially GH, insulin, and insulin-like growth factor (IGF) 1, are produced in greater amounts. With GH levels reduced, puberty is retarded.

11.3.4 Menopause

After the arrival of sexual competence, men and women are fertile for quite some time, until in women at the age of about 50 years menstruation fails to occurs. When pregnancy can be excluded—for example, in the absence of choriogonadotropin—then menopause has been reached. Since many follicles mature during the follicular stage of the menstrual cycle, with one or perhaps two arriving at ovulation, one assumes that the storage of egg cells/follicles capable of growing has been emptied. In 35 years there are about 13 cycles per year; in total 455 cycles. With 20 growing follicles per cycle, about 9,000 follicles have matured. Why the rest of the about 100,000 egg cells counted at birth do not mature is not known. About 220 follicles should mature per cycle in order to expend 100,000 primordial follicles within 455 cycles.

In men, sperms are not formed antepartum, but are first made during puberty. As long as the hormonal endowment in a man of 70 years—an age at which women are normally no longer fertile–allows there to be sufficient LH, FSH, and testosterone production, then the man is still fertile. For men andropause is presumed to be equivalent to menopause in women, but a fixed time point cannot be estimated.

Gamete development differs remarkably in men and women:

- By two meiotic divisions, four sperms arise, but a single haploid egg arises. Those chromosomes not utilized in women, which are called polar bodies, are removed from the egg.
- The first meiotic division happens in women already in utero; the second one happens after ovulation.
- In men both meiotic divisions occur during sperm development in the adult testis.

The frequency of genetically arising malformation in newborns rises with the age of both parents. There is no evidence that there is a sex bias in gamete errors. Obviously, several primary factors are balanced:

- Advantages in men: a single sperm of many fertilizes the egg, errors are diluted.
- Disadvantages in men: errors might be acquired in sperm stem cells during the long fertilization period.
- Advantages in women: after the first division an enrichment of reduplication mismatches does not occur; three quarters of the chromosomes will be removed.
- Disadvantages in women: a single egg is not selected from many other egg cells; once an error is there, it will be propagated.

In the past only the mother's age was reflected on when arguing for genetic counseling, but today the father's age is also taken into account.

11.4 Glucose Metabolism

11.4.1 The Origin of Blood Glucose

The organism derives glucose from three different sources: from food, from glucose stores, and from glucose synthesis:

1. *Glucose from food.* Glucose is taken up from the gut by a transporter protein which simultaneously takes up two sodium ions per glucose molecule (sodium–glucose cotransporter 1; Lee et al. 1994). The same transporter takes up other sugars as well.
2. *Glucose storage.* When in excess, glucose is stored as glycogen mainly in the liver. When glucose is needed, these stores can be emptied and glycogen can be converted into free glucose.
3. *Gluconeogenesis.* In an inverted process of glucose oxidation to CO_2, glucose can be newly generated from intermediates using energy. Gluconeogenesis mostly occurs in the liver.

11.4.2 Regulators and Control Variables

Insulin is the characteristic hormone of glucose regulation (see Sect. 4.7). Its antagonist is glucagon. Both hormones are released from specialized cells of the pancreas (see Sect. 10.6).

The control variable of glucose metabolism (Fig. 11.6) is the glucose concentration in blood. Dependent on the glucose concentration, there are two stages: *fasting* and *sated*. The threshold is 5 mM glucose. Below 5 mM there is fasting, above 5 mM, there is satiation.

11.4.3 Glucose-Dependent Gene Expression in the Liver

The liver switches between the two states depending on the glucose concentration: in the sated state, glucose is removed—by catabolization or by storage as glycogen; in the "fasting" state, glucose is synthesized or is provided by emptying glycogen stores. The regulating enzyme is the intracellular glucokinase, which converts glucose to glucose 6-phosphate (Glc6P), which is then used further. Intracellular regulation is possible owing to the high availability of glucose transporter 2 on the liver cell membrane (1 in Fig. 11.6).

On other cells there are glucose sensors from the family of glucose transporters: sodium–glucose cotransporter 3 (Diez-Sampedro et al. 2003). Owing to the activity of the sensors, there is a gene expression switch from glucose utilization to glucose production and vice versa (2 in Fig. 11.6). In the liver, at high glucose concentration, glucokinase expression is enhanced and thus Glc6P production is triggered, which leads either to catabolism or to glucose storage (3 in Fig. 11.6). When the glucose

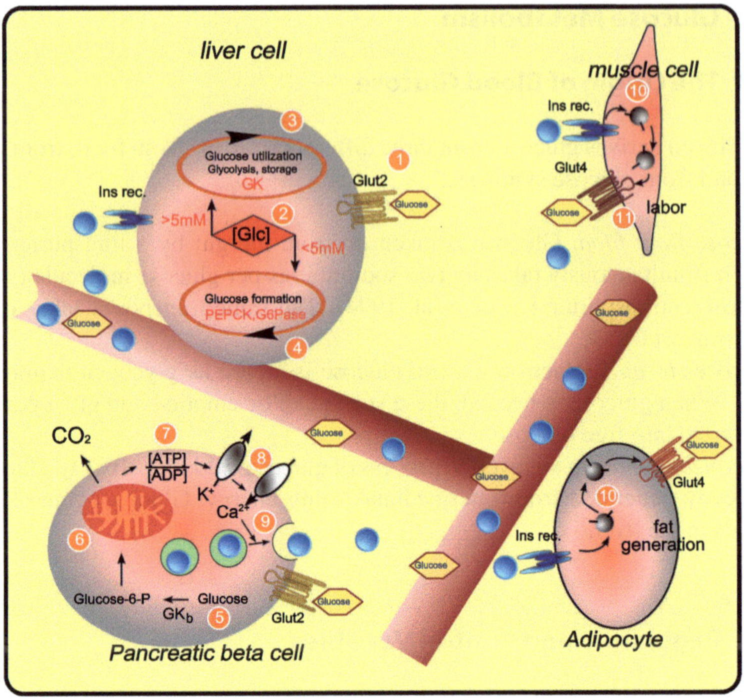

Fig. 11.6 Elements of glucose metabolism. *Blue spheres* insulin, *GK* glucokinase, GK_b glucokinase content of β cell, *Glc* glucose, *glucose-6-P* glucose 6-phosphate, *Glut* glucose transporter, *G6Pase* glucose 6-phosphatase, *Ins rec.* insulin receptor, *PEPCK* phosphoenolpyruvate carboxykinase

concentration is low, Glc6P is made available from stores or by gluconeogenesis, and the glucose produced following its cleavage is secreted into the circulation (4 in Fig. 11.6).

The rate-determining enzyme of glucose anabolism is phosphoenolpyruvate carboxykinase, whose expression is stimulated at low glucose concentration. For the conversion of Glc6P into glucose, glucose 6-phosphatase is required, and this also has a glucose-concentration-dependent expression.

Insulin has a role in glucose storage in the form of glycogen.

11.4.4 Glucose-Dependent Insulin Secretion in the Pancreas

Insulin is released in the pancreas at high glucose concentration:

1. Glucose is gated into pancreatic β cells by glucose transporter 2.
2. After phosphorylation of glucose to Glc6P (5 in Fig. 11.6), this Glc6P is catabolized in the mitochondria into CO_2 (6 in Fig. 11.6).

3. Since ATP is gained in the catabolism, the cytosolic ATP-to-ADP ratio is increased (7 in Fig. 11.6).
4. At an elevated ATP-to-ADP ratio, potassium ions are transported from the cell via a potassium channel (8 in Fig. 11.6).
5. This generates a voltage change at the membrane and induces the opening of calcium channels and an influx of calcium ions.
6. These in turn induce fusion of secretory insulin vesicles with the cell membrane (9 in Fig. 11.6), and the vesicle content is released. The released insulin acts as a hormone in the regulation of blood glucose.

11.4.5 Insulin-Dependent Procedures

In the liver, insulin stimulates Glc6P storage as glycogen. Additionally, insulin induces release of amino acids from liver cells to be used in muscle cells (insulin triggered) for protein synthesis.

In muscle cells, insulin triggers, mediated by insulin receptor (10 in Fig. 11.6), gating of preformed glucose transporter 4 from intracellular vesicles to the membrane, thus stimulating glucose uptake (11 in Fig. 11.6). The glucose taken up is used to support muscle activity.

Similarly, insulin triggers uptake of glucose into adipose cells mediated by glucose transporter 4, where glucose is used for fatty acid synthesis.

11.4.6 Glucagon and Blood Glucose Level Increase

Whereas at elevated glucose concentration insulin is secreted from β cells, at low glucose concentration pancreatic α cells secrete glucagon (see Sects. 4.7 and 10.6). In liver cells, glucagon stimulates protein kinase A, the key enzyme in gluconeogenesis and glycogen catabolism, both reactions enhancing the level of available glucose.

11.5 Appetite and Hunger

Human ingestion is controlled consciously and unconsciously (Fig. 11.7). We are able to control our eating behavior only in part: it is common experience that bursts of ravenous appetite send our good intentions packing. An autonomous regulation separated from conscious control (again in part) is active. The purpose of this control is a long-lasting constancy of body weight and therefore safeguarding of the individual's survival. To achieve this, food and energy uptake and energy consumption have to be balanced.

In this control system comprising the brain, adipose tissues, and gastrointestinal (GI) tract, hormones from the GI tract, from adipocytes, and from neurons are involved. The presentation of these circuits is limited to an overview owing to space restrictions; in addition, many details have not been fully and sufficiently analyzed.

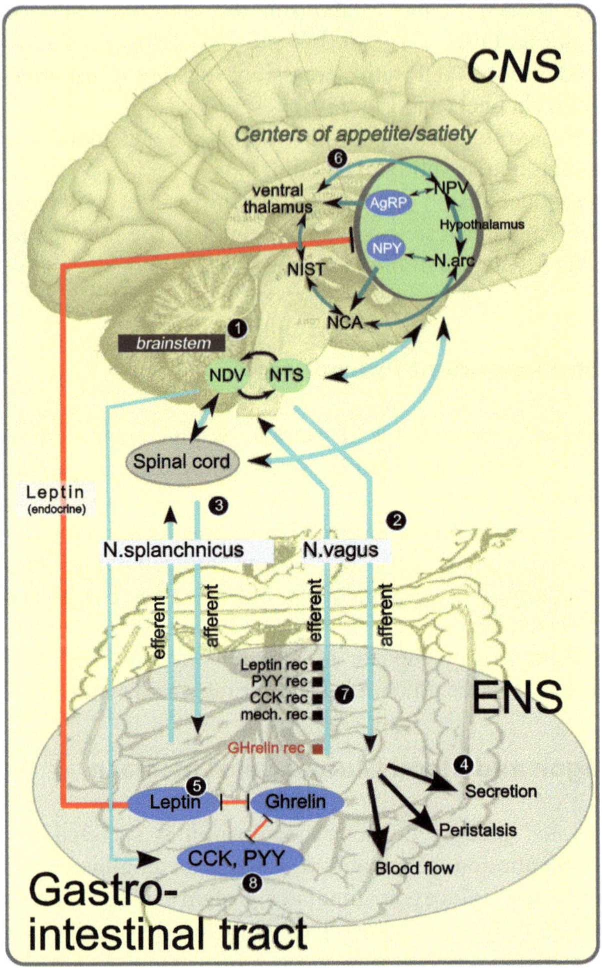

Fig. 11.7 Brain–belly interplay: regulation of appetite and hunger. *Red lines* inhibitory endocrine actions, *blue-green lines* control by nerves and synapses, *AgRP* agouti-related peptide, *CCK* cholecystokinin, *CNS* central nervous system, *ENS* enteric nervous system, *mech. rec* mechanoreceptor, *N.arc* arcuate nucleus, *NDV* dorsal motor nucleus of the vagus nerve *NCA* central nucleus of the amygdala, *NIST* bed nucleus of the stria terminalis, *NPV* paraventricular nucleus, *NPY* neuropeptide Y, *NTS* nucleus of the solitary tract, *PYY* peptide tyrosine tyrosine, *rec* receptor (From Konturek et al. 2004; Gray 1918, Fig. 715)

Konturek et al. (2004) reviewed the regulation of appetite and body weight.

11.5.1 Central Nervous System

In the CNS (Fig. 11.7) the nucleus of the solitary tract regulates gut secretion, gut movement, and blood supply. Satiety and appetite are controlled in different centers, mainly in the arcuate nucleus and the paraventricular nucleus, as well as in the ventromedial and lateral hypothalamus. Also involved are the central nucleus of the amygdala, the interstitial nucleus, and the ventral thalamus.

11.5.2 Parasympathetic Fibers of the Vagus Nerve

Starting from the brainstem (1 in Fig. 11.7), the vagus nerve interconnects multiple organs of the GI tract. About 20,000 nerves connect to different cells of the GI tract.

11.5.3 Sympathetic Fibers of the Splanchnic Nerve

Synapses of the splanchnic nerve connect the GI tract with the brain (3 in Fig. 11.7).

11.5.4 Enteric Nervous System

A specialized nervous system is generated by about 100 million nerve cells of the enteric nervous system. These nerves stimulate and inhibit the entire GI events in an intrinsic way, whereas the vagus nerve and the splanchnic nerve interconnect the GI system with the brain. Secretion of digestion enzymes, salts, or hydrochloric acid, peristalsis, and the circulation are controlled by these nerves (4 in Fig. 11.7).

11.5.5 Endocrine Mediators and Neuropeptides

The following hormones are among those involved in the balance of hunger and satiety:

- *Leptin*. Leptin (see Sect. 4.8.1) is made in adipose cells, and the amount released is dependent on their number. Using leptin, adipose cells signal the nutritional status and contribute to the regulation of the body weight in the long run. Simultaneously, leptin is involved in short-term food intake since its release is elevated postprandially. Leptin acts not only in the brain (6 in Fig. 11.7), but also via a paracrine loop in adipose tissue (Jequier and Tappy 1999) as well as in the GI tract (Meier and Gressner 2004) (5 in Fig. 11.7). In the GI tract leptin antagonizes the actions of ghrelin, whereas in the hypothalamus, especially in the arcuate nucleus, the synthesis and release of NPY and agouti-related peptide (AgRP) are blocked. This leptin action is mediated by the long splice variant (OB-Rb) of the leptin receptor.
- *Cholecystokinin* (CCK). CCK (see Sect. 4.10) is made in the I cells of the duodenum. By binding of CCK to its receptors on the ends of the vagus nerve (7 in Fig. 11.7), satiety is signaled to the brain. Antagonistic receptor blockade results in prolonged feeding.

- *Peptide tyrosine tyrosine* (PYY). PYY (see Sect. 4.10) is most probably released under central control from endocrine gut cells. PYY acts antagonistic to ghrelin (8 in Fig. 11.7).
- *NPY*. NPY is one of the most intensively studied endocrine molecules of the third millennium: more than 5,600 articles published between 2000 and 2014 had a major focus on NPY. Central NPY is a major player in food intake, together with AgRP (see Sect. 4.3.4). NPY/AgRP knockout mice are fully viable and fertile, and do not exhibit anorexia; however, when NPY is administered intracerebrally, these knockout mice demonstrate all the symptoms related to foraging.

  NPY belongs to a family of proteins characterized by the pancreatic polypeptide fold (see Sects. 4.3 and 4.10 and Fig. 4.41). It is synthesized in hypothalamic neurons, mainly in the arcuate nucleus, but also in brainstem neurons projecting into the hypothalamus and in endocrine cells of the GI tract. In the brain it acts mostly as a neurotransmitter via synapses between neurons; furthermore, NPY is released by neurosecretory cells in the median eminence and acts in the pituitary.

  Apart from its role in food intake, NPY is involved in the hypothalamic–pituitary–adrenal axis by the control of CRH, it coregulates heart frequency via NPY-immunopositive neurons, and its acts in angiogenesis and wound healing.

  A signal sequence mutation of human NPY (Leu←Pro; see the framed **L** on the dark background in Fig. 4.40) results not in adiposity, but in elevated cholesterol levels and an enhanced risk of cardiovascular diseases. In Finland, about 14 % of the population is affected. In another study, not only NPY levels were decreased, but so were noradrenaline and insulin levels, and the blood glucose level was elevated. The heart frequency was enhanced.
- *AgRP*. AgRP binds, for example, to melanocortin receptors; in appetite regulation, melanocortin 4 receptor (MC4-R) has a special role. Formed in neurosecretory cells of the basal hypothalamus (in the arcuate nucleus), AgRP binds to MC4-R mainly in centers of satiety and appetite in the paraventricular nucleus and other hypothalamic nuclei as well as in the central nucleus of the amygdala, the bed nucleus of the stria terminalis, and the ventral thalamus. By these ligand–receptor interactions, food intake is blocked.

  MC4-R has different ligands: α-melanocyte-stimulating hormone (α-MSH) and adrenocorticotropic hormone (ACTH) inhibit appetite, and AgRP and agouti protein enhance appetite. Melanocortin 3 receptors might also be involved in appetite regulation.
- *Ghrelin*. Ghrelin is a recently identified ligand for the long-known GH secretagogue receptor (GHS-R1), which when bound by small synthetic peptides and nonpeptide molecules triggers release of GH. GHS-R1a is a heptahelical membrane receptor; however, GHS-R1b, which is generated after alternative RNA splicing with retention of an intron with an early stop codon, is a heptahelical membrane protein with unknown function.

11.5.6 Hormone Receptors on Nerve Cells

Hormone and neuropeptide receptors involved in the control of the balance between hunger and satiety are present on a number of neurons in the ventromedial and arcuate hypothalamus (NPY receptor, MC4-R, leptin receptor) as well on the vagus nerve, signaling from the GI tract to the brainstem and to additional centers: PYY receptor, leptin receptor, CCK receptor, and GH secretagogue receptor.

11.5.7 Mechanoreceptors

The same vagus nerve signals extension and contraction of the stomach and gut wall to the brain with the help of mechanoreceptors.

Figure 11.7 illustrates how the regulation between the CNS and the enteric nervous systems occurs: hormones act on receptors present on nerve cells or on other cells with effector functions (acid secretion, enzyme production, muscle contraction).

The scheme sketches only a complex cross talk of endocrine and neuronal elements: in the GI tract peripheral signals, such as hunger-triggering peptides ghrelin and PYY, and satiety-inducing CCK or mechanoreceptors measuring stomach wall extension elicit in concert hunger or satiety, which ends food intake. Additional signals—for example, insulin secretion from the pancreas triggered by a carbohydrate diet—also induce satiation.

Signaling of satiety or hunger happens in the CNS. When the above-mentioned hormones enter the brain (leptin or insulin) or by neuronal connections act in the hypothalamus (PYY, CCK, ghrelin), neurotransmitter signaling is either stimulated or blocked. One example is AgRP, which by acting on MC4-R reduces appetite. Increased release of NPY, however, owing to elevated ghrelin levels, generates hunger. Since in order to maintain a constant body weight not only food intake but also energy consumption is relevant, the latter's parallel regulation of ingestion is expedient and necessary: sating neuropeptides with an action on energy consumption increase this consumption, whereas hunger-generating neuropeptides diminish it.

Apart from the interactions outlined, many other neuropeptides are involved in central regulation of appetite: for example, α-MSH, CRH, and orexins; monoamines such as serotonin also have a role.

The hormone leptin, which is released from white adipose tissue, is, in contrast to the other GI tract and pancreatic hormones/neuropeptides, an element of a long-term regulatory circuit. Its synthesis and release are dependent on the amount of fat, and thus provide a feedback between food intake and energy stores. To act in the hypothalamus (mainly in the arcuate nucleus), leptin has to enter the brain. Because it is a peptide of 16 kDa, passive diffusion across the blood–brain barrier is not possible. It has to be specifically gated across the blood–brain barrier. It could well be that the short splice variant of the leptin receptor (OB-Ra) fulfills this transport function.

During evolution organisms have been confronted with lack of food rather than with plenty of it. Adaptations have therefore evolved which react sensitively to lack of food and only insensitively to a surplus of food—that is, exceeding the required amount for homeostasis. The transport capacity for leptin is exhausted in humans and rodents of normal weight, so enhanced leptin levels in obese situations cannot be transmitted and do not find an adequate reaction in the hypothalamus. This effect is aggravated owing to a decrease of leptin transport capacity in obesity and consequently a further reduced signal transduction via OB-Rb. The result is the so-called leptin resistance.

Hunger is triggered by ghrelin in the GI tract and by ghrelin, NPY, and AgRP in the CNS. Ghrelin from the stomach acts via the vagus nerve on the CNS, where these ghrelin signals lead to NPY and AgRP release in the ventrobasal hypothalamus (arcuate nucleus). Ghrelin secretion from neurosecretory hypothalamic cells also contributes to NPY and AgRP release.

The functional antagonist of ghrelin is leptin: leptin blocks efficiently synthesis and release of ghrelin in the brain and in the GI tract and thus generates satiety. Ghrelin, in turn, blocks leptin. Leptin might penetrate the blood–brain barrier in order to reach in neurohemal organs, for example, the median eminence axons of ghrelin neurons and block its release there. In rats, it was shown that intranasally administered leptin reaches the CNS and can act there (Fliedner et al. 2006; Schulz et al. 2004).

In contrast to leptin, the other ghrelin antagonists act via receptors on the vagus nerve and not by direct action in the CNS. PYY is an effective blocker of hunger.

11.5.8 Feeding Circuits in the Brain

For a couple of years it has been well known that the lateral hypothalamic area (LHA) is crucial for the regulation of appetite and hunger:

- Numerous neurons in the LHA express melanin-concentrating hormone (MCH) and orexins (Jobst et al. 2004). These MCH/orexin neurons are blocked by synaptic contacts of proopiomelanocortin/cocaine- and amphetamine-regulated transcript (CART) neurons from the arcuate nucleus; the latter are stimulated by leptin. Leptin as a product of adipocytes thus signals into the LHA via the arcuate nucleus to the brain excess of food and blocks feeding by inhibiting the release of MCH and orexins from the LHA.
- The proopiomelanocortin/CART neurons, on the other hand, trigger the release of TRH and CRH from the paraventricular nucleus.
- In the arcuate nucleus there are some NPY/AgRP neurons expressing the leptin receptor which react to leptin. The fact that only the transcription factor SOCS-3, but not Fos, is stimulated by leptin in these cells is interpreted to mean that these neurons are inhibited by leptin. In the proopiomelanocortin/CART neurons, however, both SOCS-3 and Fos are found; therefore these cells are thought to be stimulated.

- AgRP and α-MSH compete for the same MC-R4, α-MSH being the agonist, AgRP the antagonist. In concert with NPY, AgRP stimulates hunger, whereas α-MSH blocks feeding, as leptin does. Ghrelin acts differently: it stimulates the AgRP/NPY neurons in the arcuate nucleus. NPY antagonists block the impact of ghrelin on the hypothalamus.

The role of the nucleus of the solitary tract on the regulation of food has long been underestimated:

- Being close to the area postrema where the blood–brain barrier is permissive, access of hormones to the receptors in the nucleus of the solitary trac is feasible.
- Gastrointestinal peptides such as CCK and pancreatic polypeptide as well as leptin bind to nucleus of the solitary tract receptors.
- In addition, pancreatic polypeptide, glucagon-like peptide 1, and leptin act via their receptors on the vagus nerve into the nucleus of the solitary tract.
- The melanocortin receptor is equally found in the nucleus of the solitary tract. When α-MSH or an agonists is injected intracebrally into the fourth ventricle, which is close to the nucleus of the solitary tract, feeding is blocked, whereas the injection of an antagonist induces feeding.

The nucleus of the solitary tract signals into the nuclear region of the nucleus accumbens, which is an important reward center. This action is mediated mainly by dopamine signals. When in the nucleus accumbens tyroxine hydrolase is missing, mice stop feeding. When dopamine is injected into the nucleus accumbens in these mice, they feed and favor tasty over unsavory food.

Endogenous opoids when injected in the nucleus accumbens stimulate the preference for sugar and lipid-rich food. Mutual GABAergic connections between the LHA and the nucleus accumbens possibly a have role in the choice of food. Since the nucleus accumbens bears MCH receptors, it might be influenced by synaptic contacts to MCH neurons of the LHA or by released MCH in an endocrine way.

Important players of the central food regulation are thus the arcuate nucleus, the paraventricular nucleus, the LHA, the nucleus of the solitary tract in the brainstem and finally the two regions of the nucleus accumbens in the basal forebrain. The hormones stimulating feeding are NPY, AgRP, ghrelin, MCH, and orexins; those blocking food uptake are mainly α-MSH and leptin.

11.5.9 Hunger and Food Intake in *Drosophila melanogaster*

NPY's role in food intake in vertebrates has an invertebrate analog: Neuropeptide F (NPF). Directly after molt, *Drosophila melanogaster* larvae are fixed on food intake, and some hours later they stop feeding and start wandering. Both stages, feeding and wandering, differ considerably in NPF expression: only feeding larvae express NPF in the four brain neurons and the pairwise neurons of the ventral cord; in wandering larvae these neurons are devoid of NPF. Using molecular techniques, Wu et al.

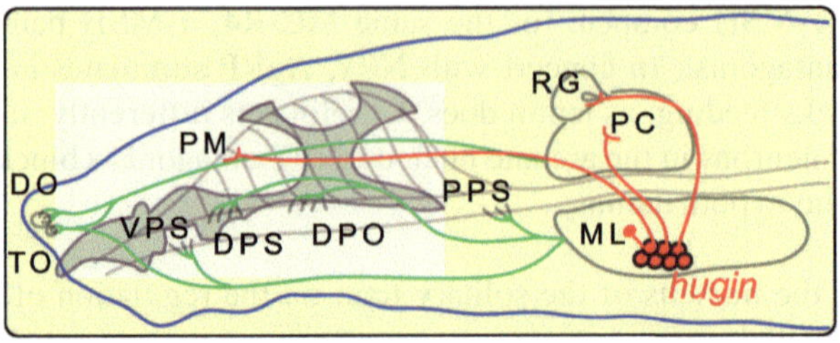

Fig. 11.8 Network of hugin neurons in the subesophageal ganglion. Hugin neurons, their expression restricted to the subesophageal ganglion, are in contact with taste nerves (*gray lines*): dorsal organ (*DO*), terminal organ (*TO*), dorsal pharyngeal sense organ (*DPS*), ventral pharyngeal sense organ (*VPS*), posterior pharyngeal sense organ (*PPS*), dorsal pharyngeal organ (*DPO*). Hugin neurons project to the maxillary lobe (*ML*), to the pharyngeal muscle (*PM*), to the ring gland (*RG*), and into the protocerebrum (*PC*) (*red lines*) (From Melcher and Pankratz 2005)

(2003b) demonstrated that overexpression of NPF in larvae prolongs the feeding phase, whereas elimination of NPF leads to avoidance of food. Absence of NPF induced aversion to a glucose-containing diet. In the laboratory, wandering larvae try to dig into agar. This behavior has also been observed in NPF-deficient larvae.

In their search for genes involved in the food intake of fly larva, Melcher et al. (2007) found two genes: *klumpfuss* (*klu*) and *pumpless* (*ppl*). Flies with defects in these genes are unable to bring food from the pharynx into the esophagus (Melcher et al. 2007; Melcher and Pankratz 2005). These flies die early from malnutrition. The gene product of *ppl* belongs to the glycine cleavage system; *klu* encodes a zinc finger protein, mainly expressed in developing nerves. In *klu* mutants, it was shown that the expression of several neuropeptides was increased: corazonin, NPF, hugin, adipokinetic hormone, pigment-dispersing factor, and cardioacceleratory peptide (CAP).

Hugin (the *D. melanogaster*[2] pyrokinin) is exclusively expressed in perikarya of the subesophageal ganglion. These neurons project into the central brain, to the ring gland (corpus cardiacum, corpus allatum), and to the pharyngeal muscle. The hugin neurons in fly larvae receive signals from several sensory organs (Fig. 11.8).

Sex peptide also stimulates food intake, at least in inseminated female flies. Sex peptide is generated in the testes, and during copulation is transferred with the sperms into the female. Activation of food intake facilitates vitellogenesis and egg deposition. When sex peptide is mutated in males (by molecular techniques), the females after copulation with these males do not lay eggs. And they are sought for copulation, as described before (see Sect. 5.4.3), by other males (Carvalho et al. 2006).

It is not yet understood how the different signals contribute to the complete situation. The relation among food intake, growth, and molting is discussed in Sect. 11.7.

---

[2]Drosophila melanogaster.

11.6 Growth

Embryonic growth requires an exact time and space control of gene expression, activator and inhibitor release, and secretion of hormones and other proteins, and a sufficient supply of maternal nutrients. During embryonic growth, the entire organism is to be built—bones and all other organs and body parts. Postpartum human growth, in the sense we discuss below, is mostly restricted to bone growth. Three stages are distinguished: the first one with rapid growth until the third year of life, then until puberty a stage with decelerated but steady growth, and during puberty again a stage with very rapid growth until the final body height is achieved.

11.6.1 Epiphyseal Cartilages

Bone growth occurs in the epiphyseal cartilages (Fig. 11.9). Here chondrocytes proliferate and thus enhance cartilage mass. Since this chondrocyte growth happens perpendicular to the bone in one bone slice, a new volume element is created as a

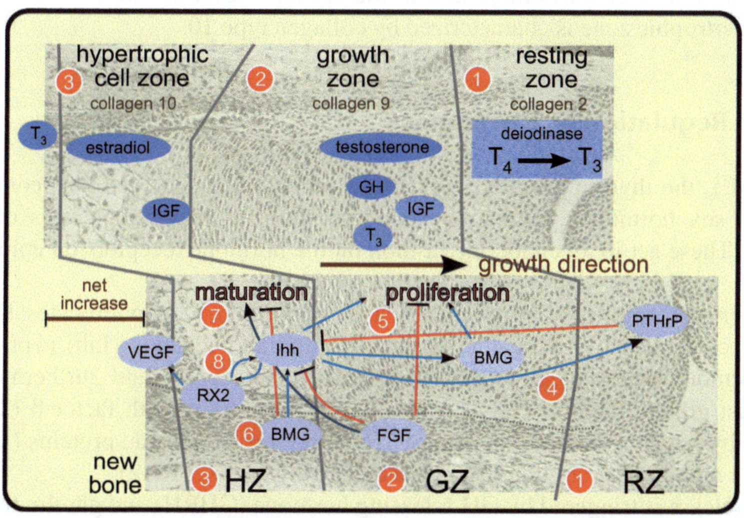

Fig. 11.9 Endocrine and paracrine control in the epiphysis. *In the upper part* the epiphyseal anatomy is depicted with the hormones active in the separate zones; *in the lower part* paracrine control by locally formed factors is depicted. For linear growth, the epiphysis has to move away from the bone's center. By the addition of new bone at the epiphyseal boundary (*left*), bone length increases. Thus, the epiphysis migrates. The epiphysis is marked by three zones: the rest zone (*RZ*), the growth zone (*GZ*), and the hypertrophic zone (*HZ*). Endocrine growth regulation is achieved by growth hormone (*GH*), insulin-like growth factor (*IGF*), thyroid hormone, corticoids, and sex hormone. A paracrine control is exerted by Indian hedgehog (*Ihh*), parathormone-related peptide (*PTHrP*), bone morphogenetic protein (*BMG*), and fibroblast growth factors (*FGF*). For the generation of new bone, vascular endothelial growth factor (*VEGF*) is required. *RX2* runx2, T_3 triiodothyronine, T_4 thyroxine (Adapted from Eerden et al. 2003; background image from Darl R. Swartz, Lafayette Center For Medical Education, Indianapolis, IN, USA)

disk in the bone and the bone's ends migrate away from each other. When you look at the X-rays of a child's hand, you will notice the epiphyses as dark cross sections: there appear more phalanges than are obvious, since in epiphyseal cartilages only cartilage is created, and this is permeable with regard to X-rays. X-ray-impermeable calcium phosphate is incorporated only at later stages.

11.6.2 Zonal Organization of the Epiphysis

The first zone, the *rest zone* (1 in Fig. 11.9), with the chondrocyte stem cells recognizable in small assemblies in sections is characterized by collagen type 2.

Close to the rest zone there is the *growth zone* (2 in Fig. 11.9). Therein chondrocytes divide and create the new volume. Since the newly formed cells cannot move to one side—owing to the rapid creation of extracellular matrix—the cells form columns. The growth zone cell are characterized by collagen type 9.

When chondrocytes have been dividing for some time, their volume strongly increases in the *hypertrophic zone* (3 in Fig. 11.9) before they die. Stimulated by vascular endothelial growth factor, blood capillaries extend into the space generated and osteoblasts migrate into the space. The latter form the new bone. The hypertrophic zone is characterized by collagen type 10.

11.6.3 Regulation by Hormones

GH, IGF1, the thyroid hormones thyroxine and triiodothyronine, glucocorticoids, and the sex hormones testosterone and estradiol are major regulators of bone growth. These act in an endocrine fashion on the hormone receptors on epiphyseal cells.

Within the epiphyses, proliferation and maturation of chondrocytes is additionally controlled by two paracrine factors—Indian hedgehog (Ihh) peptide and parathormone-related peptide (PTHrP), with its receptor—and furthermore by fibroblast growth factors (FGFs) from the transforming growth factor β (TGF-β) family, by heparan sulfate biosynthesis, by bone morphogenetic proteins (BMPs), and by a member of runt transcription factor family, runx2 (RX2).

GH release is triggered by GH-releasing hormone (GHRH) and ghrelin from somatotropic cells of the anterior pituitary and is inhibited by somatostatin. Its actions are partially direct, and in part are mediated by IGF1. Both the GH receptor and the IGF receptor need to be expressed in chondrocytes for regular bone growth. Both receptors enhance cellular division of stem cells and chondrocytes by facilitating mitosis and blocking arrest and maturation into hypertrophic chondrocytes.

There are two IGFs—IGF1 and IGF2, the latter mainly expressed before birth. They exert their functions in several control circuits:

- *Proliferation.* In fibroblasts, muscle, skin, epithelial, and bone cells, cells of male and female gonads, and several tumor cells, cell division is triggered by IGF1.
- *Cell death.* In hematopoietic cells and in some tumor cells, IGF1 blocks apoptosis.
- *Cellular differentiation.* Myoblasts, osteoclasts, osteoblasts, chondrocytes, neuronal cells, and adipocytes start to differentiate on being triggered by IGF1.
- *Cellular functions.* IGF1 actions have been described in endocrine and immune cells: IGFs stimulate hormone synthesis and release in theca and granulosa cells (see Sect. 10.8). They stimulate thymulin release from thymus epithelium. In the adrenal cortex, IGF1 enhances the number of membrane ACTH receptors and thus increases ACTH actions.

The actions of IGF1 are controlled by IGF-binding proteins (IGFBPs). IGFBPs lower the availability of IGF. In the epiphysis—in concert with IGF—they mediate mitosis and proteoglucan synthesis. These IGFBPs are in turn controlled by IGF, insulin, and TGF-β. The IGF receptor has been found on chondrocytes of the growth zone and the hypertrophic zone, but not the rest zone.

Thyroxine (T_4) is an important growth regulator. It is converted by deiodinase into the active triiodothyronine (T_3) (4 in Fig. 11.9). Deiodinase is expressed in the epiphysis and enhances by local conversion availability of T_3. The intracellular T_3 receptor is expressed in chondrocytes of the rest zone and growth zone and in osteoblasts of newly formed bone. T_3 stimulates chondrocyte maturation.

Glucocorticoids such as cortisol induce bone resorption and block bone-forming osteoblasts, which delays growth in children treated with glucocorticoids. Under normal conditions, glucocorticoids are required to block chondrocyte mitosis for T_3-induced chondrocyte conversion into hypertrophic cells. In the epiphysis, cortisol can be inactivated by a 11β-HSD.

The role of steroids appears obvious given that male sex hormones induce larger body sizes than do female ones.

The different control elements influencing the body form are far from well understood, whereas some major steps of sex-specific regulation have become apparent (Gatford et al. 1998).

The pulse frequency of GH release is effected by GHRH and somatostatin. There is evidence that in women only GHRH secretion and in men release of both GHRH and somatostatin are sex specifically controlled. In growing men compared with young women, this leads to more frequent pulses with higher amplitudes and lower nadirs. Likewise, the number of IGF and GH receptors differs with the sex as do the IGF1 levels and the concentrations of IGFBP. The synthesis of soluble GH-binding proteins (by cleavage of the membrane receptor) appears equally to differ with the sex.

In the epiphyses, estrogens and androgens have opposite functions: androgens stimulate chondrocyte growth (androgen receptor is expressed in the epiphysis), whereas estrogens stimulate chondrocyte maturation (Eerden et al. 2003) A young man with an estrogen receptor defect showed extreme growth owing to a failure to close the epiphyses. Two other individuals with aromatase defects showed similar growth. Whereas the latter two profited from estradiol substitution, the former was estradiol resistant. Sex-specific differences in the androgen–estrogen balance appear to be at the origin of different body sizes in men and women by directly influencing epiphyseal activity.

In addition to these systemic regulators, the paracrine roles of the transcription factor Ihh and PTHrP have been analyzed by Vortkamp et al. (1996) and others. The role of BMPs from the TGF-β family in heparan sulfate synthesis was also found recently.

Hedgehog proteins are so-called morphogens with indispensable function during embryonic morphogenesis. By binding to the Patched receptor, they release the protein Smoothened from Patched–Smoothened complexes, which then induces cellular reactions. Without hedgehog, Smoothened cannot be released.

The Ihh variant is expressed in chondrocytes becoming hypertrophic. Ihh diffuses in the direction of the bordering perichondrium at the epiphyseal margin and stimulates there via TGF-β PTHrP release from the rest zone (4 in Fig. 11.9). PTHrP in turn diffuses into the space in between the growth zone and the hypertrophic zone and inhibits Ihh expression (5 in Fig. 11.9). BMPs stimulate and FGFs inhibit this Ihh expression. Thus, BMPs block maturation of and stimulate chondrocyte proliferation by an Ihh-mediated BMP release in the growth zone (6 in Fig. 11.9). FGFs in turn inhibit proliferation and induce chondrocyte maturation (7 in Fig. 11.9). Furthermore, in an enhancer circuit, Ihh induces RX2 release and RX2 stimulates Ihh release (8 in Fig. 11.9).

Without control of Ihh release, dwarfism occurs. Ihh apparently controls the rate of chondrocyte maturation.

PTHrP had been related to abnormal calcification in tumors (Suva et al. 1987). Vortkamp et al. (1996) identified its role in the epiphysis.

The balance between Ihh and PTHrP has a decisive role in the control of growth, Ihh expression, and chondrocyte maturation. Mutants of Ihh, Patched, PTHrP, and their receptors impair normal growth.

11.7 Growth and Molt in Ecdysozoans

The body form of insects and crustaceans is determined not by an internal skeleton, but by an exoskeleton. This does not allow steady growth, and growth in these animals means rejection of the exoskeleton (molt) and formation of a new one. During this process, the new skeleton is formed first; thereafter, the animal leaves the old skin, and the new skeleton needs to harden, which is a lengthy process and may last several days in lobsters or several hours in dragonflies, during which time the animals are immobile and rather defenseless.

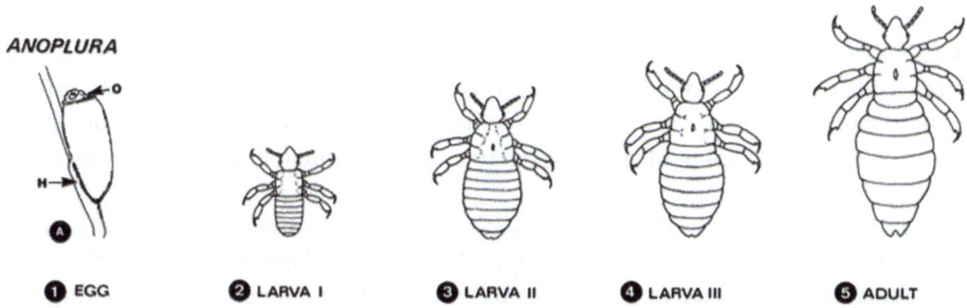

Fig. 11.10 Hemimetabolous development in lice (Anoplura). The egg (*O*) fixed on some hair (*H*) develops into a nymph, which molts twice into another, larger nymph. The third molt leads to the adult animal (From Mehlhorn 2001)

Crustaceans may molt without a limit to the number of molts. In insects several patterns can be distinguished:

- *Ametabolous insects.* The forms of the freshly molted insects resemble those of the adults. During molt it is only the size which increases, no additional organs develop, and gonads mature gradually.
- *Hemimetabolous insects.* These insects perform a partial metamorphosis. After hatching, the larvae, called nymphs, increase in size after each molt. The adult insect acquires by the last molting sexual maturation and, for example, wings. Hemimetabolous insects do not undergo pupation (Fig. 11.10).
- *Holometabolous insects.* Holometabolous insects undergo the characteristic pupation during their development. Larvae molt several times. The different stages are called instar; the fifth instar is the larva after the fifth molt. At the end of development, the larva pupates. The adult insect, the imago, ecloses from the pupa

In addition to insects and crustaceans, there are other invertebrates which molt. According to molecular genetics, Kinorhyncha, Loricifera, Priapulida, Nematoda (e.g., *Caenorhabditis elegans*), Nematomorpha, Lobopodia, Onychophora, Tardigrada, and Arthropoda share a common ancestor. All these have in common shedding of the exoskeleton. The superphylum is called Ecdysozoa.

The molting processes, and those processes during pupation, are controlled by hormones. Neuropeptides from the brain, juvenile hormone from the corpora allata, cholesterol-derived ecdysteroid from the prothoracic gland, and additional peptide hormones from the periphery are woven into a network of complex interactions.

Whether, how often, and when a larva/nymph undergoes molt depends on several environmental and intrinsic factors: Environmental factors include the temperature, the length of days or nights, and the richness of food supply and its composition. The intrinsic factors are, for example, synthesis of different hormones, such as insulin-like peptide (ILP) 2 and other ILPs in *D. melanogaster*, and synthesis of prothoracicotropic hormone (PTTH), the hormone that triggers ecdysone formation in the prothoracic gland or conversion of ecdysone into 20-hydroxyecdysone in the periphery. The process of molting is also controlled in an endocrine manner:

Generation of the new exoskeleton depends on ecdysteroids. Stripping of the old skin is also initiated by the decline in the amount of ecdysteroids. The movement whereby the animal removes its old skin is a specific behavior that is determined by hormones. In nematodes, for example, is has been shown that loss of one molt hormone blocks stripping of the exoskeleton, which ultimately leads to the animal's death.

Finally, the outcome of a molt is determined by hormones: whether the molt of a nymph again leads to the next nymph stage or whether it will be the final molt (eclosion) and lead to a sexually mature adult which acquires, for example, wings is dependent on hormones. In holometabolous species, pupation occurs before eclosion. There are, however, some hemimetabolous species with a final molt resembling pupation: the two last nymph stages of thrips (Thysanoptera) do not take up food and form a kind of cocoon. Some coccid nymph stages were also observed not to feed any further (Chapman 1998, p. 368).

11.7.1 Regulation of Growth in Insects

The life cycle of insect larvae usually has three phases: food uptake, wandering, and molting. The larvae cycle through these severalfold. The transitions are most interesting.

With use of molecular techniques it has been observed in *D. melanogaster* that the body length which a individual will reach is directly related to the expression of hormones, Especially, ILP, its receptors, and the insulin receptor substrate Chico are required for lipid utilization in flies. Without ILP, ILP receptor, or insulin receptor substrate there are developmental disturbances where larval molts are delayed and the final size is diminished. Overexpression of ILP, in contrast, results in an enhanced length. It is interesting to note that flies where ILP expression was reduced—for example, by destruction of certain ILP neurons in the brain—live longer than comparable wild-type flies. It is also noteworthy that such a phenotype of animals with defects of development and enhanced life span can equally be generated by feeding of yeast-deficient nutrition.

Short NPF controls ILP expression (see Sect. 11.5.9). Defect mutants with low or no short NPF expression are similarly development deficient and longer living than ILP- or ILP-receptor-defective animals. Overexpression of the transcription factor Foxo, which is downregulated by ILP in adipose cells, likewise enhances the life span of flies.

11.7.2 Hormones and Postembryonic Development

The hormone for development of ecdysozoans is ecdysone (Fig. 6.23, **55**), together with its derivatives. In insects, it is synthesized in the prothoracic gland, and in crayfish it is synthesized in the Y organ. Insects cannot generate cholesterol, the precursor of ecdysone; they have to acquire it together with their nutrition. Transition from one larval stage to the next is preceded and triggered by

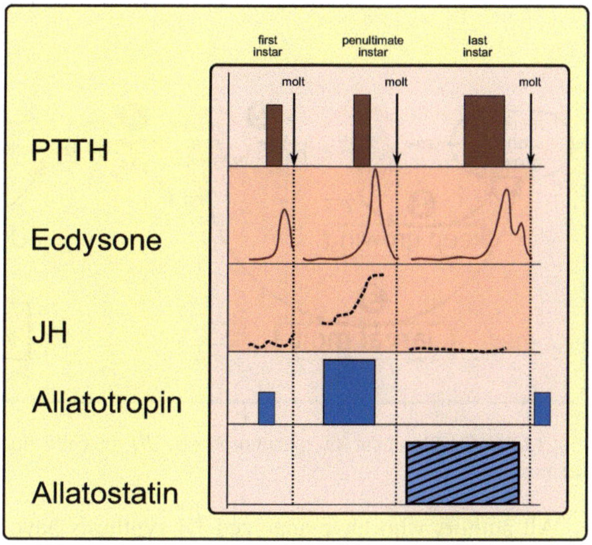

Fig. 11.11 Interactions of juvenile hormone (*JH*) and ecdysone: prothoracicotropic hormone (*PTTH*)

ecdysone bursts. Ecdysone-like compounds such as 3-dehydroecdysone, which is made in *Manduca* in parallel to ecdysone, are converted in the hemolymph into ecdysone. In peripheral tissue, mainly in the midgut, ecdysone is oxidized to 20-hydroxyecdysone (Fig. 6.23, **56**). This is the active hormone which binds to the ecdysone receptor, a nuclear receptor active as a transcription factor. Ecdysone synthesis is stimulated by PTTH.

Juvenile hormone (JH) is measurable during postembryonic development until the last larval or nymphal stage has been reached, then JH synthesis is downregulated. JH synthesis in the corpora allata is stimulated by allatotropin and is blocked by allatostatins (Fig. 11.11).

The synthesis of both terpenes—JH and ecdysone—is stimulated and repressed by neuropeptides. Allatotropin and PTTH facilitate and allatostatins and prothoracicostatic hormone suppress JH or ecdysone synthesis. PTTH induces prothoracic ecdysone synthesis by acting through its receptor Torso possibly via the extracellular-signal-regulated kinase pathway and also stimulates Ca^{2+}/calmodulin and cyclic AMP, whereas allatotropins act by hydrolysis of phosphatidylinositol. Both neuropeptides are released from a few brain neurons. By which stimuli allatotropin release is triggered is largely unknown. In adult flies (*Phornia regina*) it could be reproducibly observed that allatotropin is released 8 h after a protein meal. In larvae, however, such studies have not been reported.

The release of PTTH is coupled to the activity of the circadian clock since the zeitgeber neurons in *Rhodnius prolixus* (vector of trypanosomiasis; Chagas disease) were found in the direct neighborhood of PTTH neurons. That way, there is a circadian rhythm for PTTH release in these bugs, and for ecdysone synthesis and release. This rhythm is, in addition, dependent on daylight (Steel and Vafopoulou 2006; Vafopoulou et al. 2007). Whether the ecdysone levels in the hemolymph and peripheral tissues depend on this rhythm and whether other stimuli control the amount of ecdysone released is an open question.

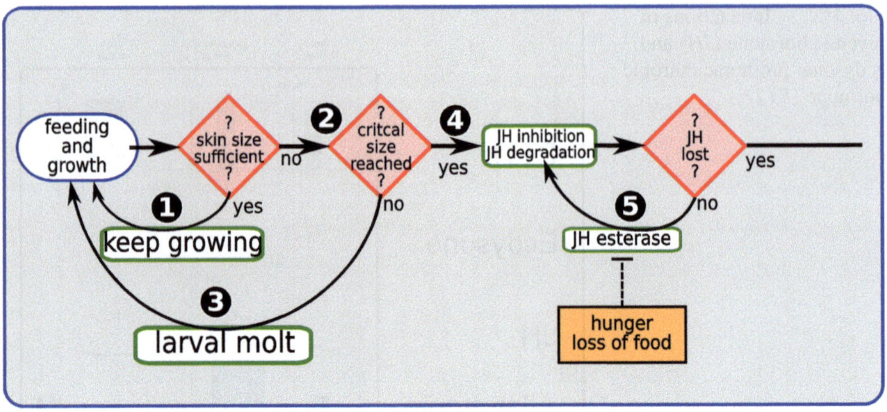

Fig. 11.12 Development of metamorphosis. *JH* juvenile hormone, *PTTH* prothoracicotropic hormone

All authors who have analyzed JH synthesis have stated that in the last larval stage of hemimetabolous insects, no JH could be found in the hemolymph. The corpora allata are still present, since in the adult animal JH is again synthesized and important for gonad maturation. There is thus a temporary JH release blockage possibly effected by allatostatins. The detailed analysis of these finding remains to be done.

11.7.3 Linkage of Growth and Metamorphosis

The question of when a holometabolous larva will pupate and when it will only undergo a larval molt is an unsolved problem of entomological endocrinology. Some years ago, it was observed that there is a critical size after which the insect can begin metaphorphosis and undergo a pupal molt. No biochemical basis for the determination of the critical size has been identified, except in hemipterans (*Dipetalogaster maximus*). There, Nijhout (1984) identified stretch receptors—that is, nerve cells that react to stretching. Similar mechanoreceptors are supposed to exist in the human atrium and to be involved in blood volume regulation via release of atrial natriuretic peptide (ANP). Stretch receptor neurons in the *D. maximus* abdomen signal to the brain. This release triggers an ecdysone pulse. If saline is injected into the abdomen of *Oncopeltus fasciatus* larvae, the PTTH–ecdysone axis can be triggered and large milkweed bugs developing after this treatment are miniaturized. In bloodsucking *Rhodnius prolixus*, a single blood meal can trigger abdominal stretch receptors to induce molting (or metamorphosis).

In the dung beetle *Onthophagus taurus*, PTTH and ecdysone are induced when dung has been taken up; depriving the larvae of the nutrition and thus bringing forth an early end of food intake can induce PTTH and later metamorphosis.

In the tabacco hornworm (*Manduca sexta*), PTTH release and ecdysone synthesis are negatively controlled by JH. After removal of the JH-forming corpora allata,

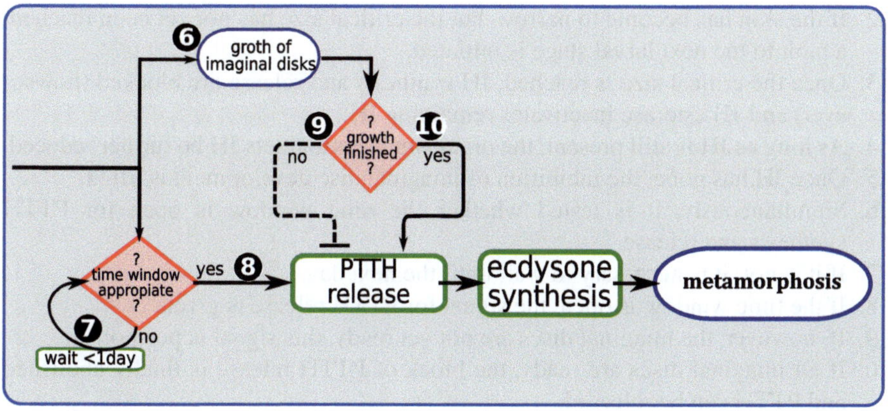

Fig. 11.12 (continued)

precocious metamorphosis occurs. On the other hand, JH administration delays metamorphosis. Once the critical size has been reached, JH generation is blocked by unknown mechanisms. In *M. sexta*, the cessation of JH synthesis occurs after the onset of feeding in the final larval instar and is thought to be due to both nervous inhibition and the loss of the methyltransferease such that the glands secrete only JH acid. Simultaneously, there is an increase in the activity of JH-degrading JH esterase, which converts JH I or JH II (Fig. 6.5, **24**) into the inactive acid (see Fig. 6.8). Expression of this enzyme is directly coupled to sufficient nutrition: when the larva fasts, JH esterase expression is blocked; with nutrients in surplus, JH is degraded and metamorphosis is initiated.

PTTH release in *M. sexta*[3] is permitted in a narrow time window; once this has passed, PTTH can be released only during the same period of the next day. Because PTTH neurons are in close contact with neurons of the circadian clock, a direct synaptic linkage is assumed.

During metamorphosis new organs develop from the imaginal disc. These do not develop as long as JH is present. After reduction of the JH levels with the help of JH esterase, the imaginal discs may mature. During their own development they signal with an unknown messenger that PTTH release should be suppressed. After reaching their final size, the imaginal discs stop this inhibitory signal for PTTH, and its release can occur.

Figure 11.12 summarizes the different control elements and developmental stages:

1. During the larval/nymph stage permanent control tests whether growth—that is, cellular division—is still possible. The "how" of this control is unknown; if growth is still allowed, feeding occurs.

[3]Manduca sexta.

2. If the skin has become to narrow, but the critical size has not yet been reached, a molt to the next larval stage is initiated.
3. Once the critical size is reached, JH synthesis and release are blocked (howsoever) and JH esterase inactivates remaining JH.
4. As long as JH is still present, the organism waits and lets JH be further reduced.
5. Once JH has gone, the inhibition of imaginal disc development is lifted.
6. Simultaneously, it is tested whether the time window is open for PTTH synthesis and release.
7. If it is not, it is necessary to wait until the next day.
8. If the time window is open, the signal for PTTH release is given.
9. If, however, the imaginal discs are not yet ready, this signal is postponed.
10. If all imaginal discs are ready, the block of PTTH release is finally abolished and PTTH can be released.

Stimulated by PTTH, ecdysone is synthesized and metamorphosis is initiated.

11.7.4 Regulation of Ecdysis

Initiated by ecdysone, a new epidermis is generated below the old skin. To trigger this, ecdysone is peripherally converted into 20-hydroxyecdysone. 20-Hydroxyecdysone interacts with a nuclear receptor dimer (ecdysone receptor/ultraspiracle[4]); the ligand–dimer complex migrates into the nucleus and acts as a transcription factor on ecdysone-responsive elements.

In the cases studied, PTTH release is brief at a particular time or in response to a particular stimulus such as stretch receptors in *R. prolixus* or *D. maximus*. It comes as a pulse and initiates ecdysone release from the prothoracic gland. Then it is thought, but it has not been proven, that ecdysone (or 20-hydroxyecdysone) acts on the prothroacic gland in a positive-feedback manner to increase the amount released. At high levels of ecdysone and 20-hydroxyecdysone, the feedback on the glands becomes negative and perhaps prothoracicostatic hormone and/or other peptides act to turn off the glands.

It has been observed that although the level of endogenous ecdysone decreases in the hemolymph, administration of exogenous ecdysone or 20-hydroxyecdysone delays ecdysis; this way, the process is regulated, but it has not been analyzed further.

When the ecdysone concentration in the hemolymph has dropped under a threshold, ecdysis is prepared. Corazonin is the first hormone known to be involved. It is expressed in lateral neurosecretory cells of the brain with axons into the corpora cardiaca and the corpora allata as well as onto neurons of the ventral nerve cord. There are no ecdysone-responsive elements in the corazonin gene. Therefore, a direct control of corazonin expression by ecdysone receptor is unlikely.

[4] An insect analog of retinoid X receptor

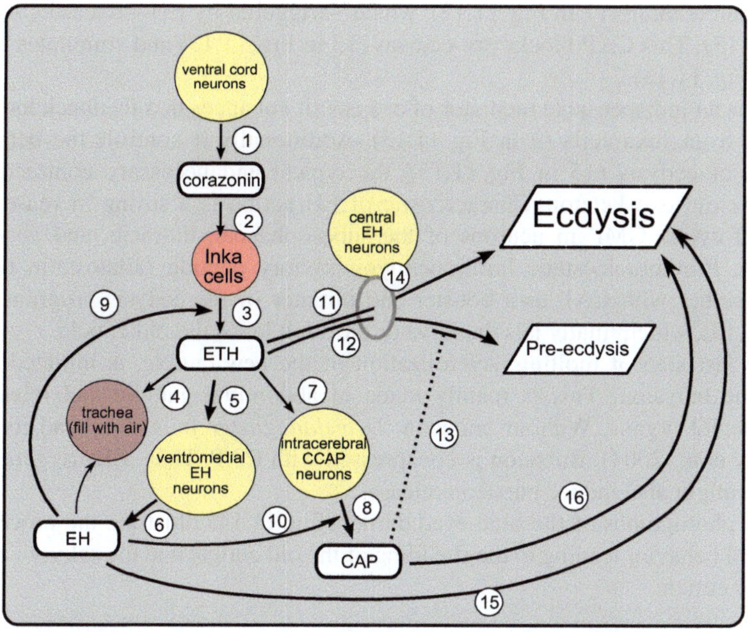

Fig. 11.13 Regulation of ecdysis. Neurons (*large circles*) or neuroendocrine cells (*small circles*) secrete hormones (*rounded rectangles*). These induce release of additional hormones (*solid lines*) or block their actions (*dotted lines*). *CAP* cardioacceleratory peptide, *CCAP* crustacean cardioacceleratory peptide, *EH* eclosion hormone, *ETH* ecdysis-triggering hormone (According to Clark et al. 2004 with additions)

To initiate ecdysis, a few neurons of the brain and the ventral cord secrete corazonin (1 in Fig. 11.13) into the hemolymph. In the Inka cells (2 in Fig. 11.13) in the vicinity of the trachea, corazonin binds to its receptors and stimulates release of ecdysis-triggering hormone (ETH).

ETH is formed in Inka cells stimulated by ecdysteroids, but is stored in vesicles. Release is triggered by corazonin (Zitnan et al. 2007; Zitnan and Adams 2005). This acts locally on the trachea (4 in Fig. 11.13), which—influenced by ETH and eclosion hormone (EH)—is inflated. Endocrine-active ETH acts on ventromedial EH-releasing neurons (5 in Fig. 11.13). This EH in turn empties the ETH stores of Inka cells to stop further ETH release before ecdysis occurs. Pre-ecdysis comprises the generation of the new exoskeleton and resorption of the old one, leaving only a marginal old skin. Further pre-ecdysis behavior is characterized in the tobacco hornworm (*M. sexta*), for example, by rhythmic contractions about 1 h before ecdysis by which the old skin is successively lost. For these contractions, pre-ecdysis-triggering hormone and ETH have to act directly on abdominal ganglia in the tobacco hornworm.

Once the level of ETH had been reduced, ecdysis can be initiated, which means the stripping of the old exuviae. ETH still acts on CAP neurons in the pars intercerebralis (7 in Fig. 11.13), which—triggered by EH—release CAP (8 in Fig. 11.13). This CAP blocks pre-ecdysis (13 in Fig. 11.13) and stimulates ecdysis (14 in Fig. 11.13).

EH is an indispensable mediator of ecdysis: It enhances in a feedback loop ETH release from Inka cells (9 in Fig. 11.13). Additionally it controls the behavioral pattern of ecdysis (15 in Fig. 11.13), the typical and necessary contractions to strip the old exoskeleton. Characteristic of EH action is a strong increase of the level of cyclic GMP in neurons of the subesophageal, thoracic, and abdominal ganglia. Prothoracicostatic hormone/myoinhibitory peptide (allatostatin type B) acts together with CAP as a booster and initiator of the ecdysis program (16 in Fig. 11.13), which finally lets the larva or the adult leave the old cuticle.

The last step of molting, sclerotization of the new cuticle, is initiated by the hormone bursicon. This is mainly made in abdominal ganglia and released in neurohemal organs. Without bursicon, *D. melanogaster* fails to spread its wings (Dewey et al. 2004). Bursicon is coexpressed with CAP. Thus, EH-triggered CAP release might also induce bursicon release.

The photographs of the blue-eyed darner (Fig. 11.14) illustrate the process: the ecdyseal behavior leading to the shedding of the old cuticle and the sclerotization of the new cuticle.

11.7.5 Postembryonic Development in Holometabolous Species

In holometabolous insects, additional hormones act during metamorphosis.

As long as JH is synthesized and reaches its targets, development whereby new organs are formed from imaginal discs is inhibited. With the JH level too low to block gene expression, the molt program occurs. With the protein Methoprene-tolerant, a basic helix–loop–helix (bHLH) protein, identified as the JH receptor, the roles of Methoprene-tolerant, germ-cell-expressed bHLH-PAS, and ultraspiricle might become clearer.

Development from a larva to an imago includes destruction of old tissues and generation of new organs: whereas the butterfly larvae have stemmata (simple eyes), the imagines possess complex compound eyes. Compound eye development from an imaginal disc is inhibited by JH. This JH action is mediated by Methoprene-tolerant (Parthasarathy et al. 2008).

Wing formation also originates from imaginal discs. Gonad development is achieved at the transition from the larva to the imago. In some insects there are gills at larval stages which are lacking in the adult. Whereas larvae take in food by grating, the butterflies use a relatively long proboscis for taking up nectar. The development of these diverse organs has not been analyzed in such detail to exclude additional endocrine actions.

The nervous system also undergoes changes during metamorphosis. For example, corazonin neurons of *D. melanogaster* become apoptotic about 6 h after initiation of metamorphosis (Choi et al. 2006). Other neurons also undergo apoptosis, with the result of a visible shrinking of the ventral nerve cord.

Fig. 11.14 Blue-eyed darner (*Aeshna cyanea*) before, during, and after ecdysis

11.8 Regulation of Blood Pressure, Osmolarity and Blood Volume

Blood pressure and osmolarity are regulated by nerve cells as well as endocrine circuits. Osmolarity regulation (with individual thresholds of 280–295 mOsm) is mainly done by hormones. Cells close to the blood–brain barrier measure osmolarity via sodium channels.

11.8.1 Integration of Several Control Circuits

For hemostasis regulation, several circuits are entangled: pressure determination and pressure control, blood volume regulation, and control of osmolarity. Since pressure is a function of volume and osmotic pressure, such an interplay can be understood: with three circuits involved, it is not easy to provide an overview.

Blood pressure is determined in the aorta wall, in sinuses of the carotid arteries, and in the periphery. The receptors and the biochemical mechanism have not yet been identified. Many nerves transfer the information into the nucleus of the solitary tract, which in turn has nerve cells controlling contraction or dilatation of vessel wall muscles (the neuronal circuitry of baroreceptor and peripheral adjustment is called baroreflex); other neurons inform the locus coeruleus and the diagonal band of Broca, from where neurons project into the paraventricular nucleus and the supraoptic nucleus, inducing AVP release in the posterior pituitary.

Osmolarity is detected by osmoreceptors (sodium channels) at the blood–brain barrier. Neuronal interplay of the subfornical organ and the vascular organ of the lamina terminalis with the paraventricular nucleus and supraoptic nucleus increases or blunts AVP and oxytocin release. AVP stimulates renal expression of aquaporin 2 (Aq2) proteins, which resorb water from the primary urine. Furthermore, resorption of sodium from the urine is reduced, with an overall effect of reduced osmolarity.

Blood volume is measured in the atria. Stretch receptors have not been characterized. By the atria, ANP is released; ANP by reduction of water resorption and sodium resorption in the kidneys and by direct effects on blood vessels contributes to volume regulation.

Following the increased AVP effect on retention of water in urine, the urine salt concentration is enhanced. An increased sodium concentration in urine is measured in cells of the macula densa and leads to renin secretion from the nearby juxtaglomerular cells. From the angiotensinogen precursor in blood, renin cleaves angiotensin I, which is converted by angiotensin-converting enzyme into angiotensin II. Angiotensin II then stimulates adrenal aldosterone synthesis and release. Aldosterone, in turn, induces fusion of sodium channels to the apical membranes of renal tubuli with the effect of sodium back-resorption. Enhanced sodium levels in urine thus lead via renin, angiotensin II, and aldosterone to an enforced sodium resorption.

On the other hand, angiotensin II acts at the blood–brain barrier and triggers the neurons in the subfornical organ and the vascular organ of the lamina terminalis, which then stimulate hypothalamic paraventricular nucleus and supraoptic nucleus magnocellular neurons to release AVP.

11.8.2 Osmoreceptors at the Blood–Brain Barrier

Nerve cells are separated from the circulation by myelin sheaths and by capillaries with thickened walls (mainly protoplasmic astrocytes surround the capillaries). There are only a few places in the brain where this blood–brain barrier is permeable. In Fig. 11.15 these areas are labeled by black circles with orange letters: the subfornical organ and the vascular organ of the lamina terminalis together with the area postrema and the choroid plexus are called circumventricular organs and are

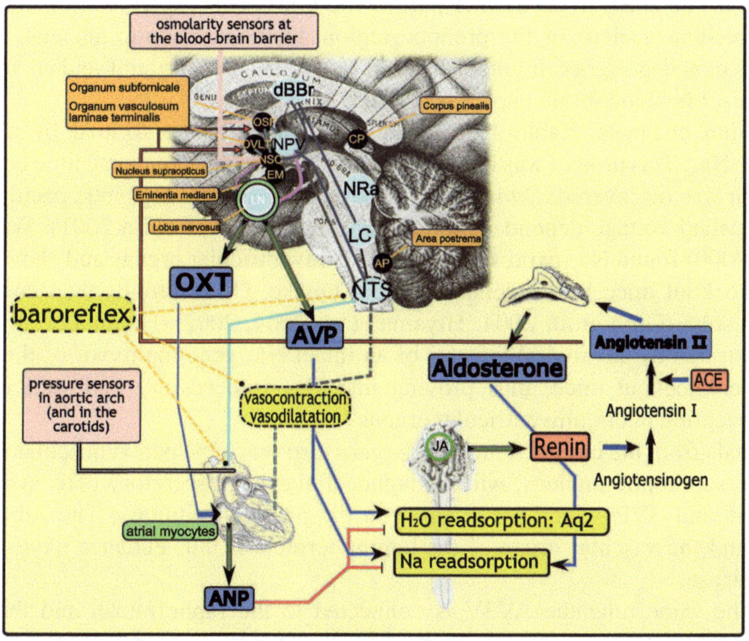

Fig. 11.15 Central and endocrine control of blood pressure and osmolarity. Hemostasis control occurs by action in concert of the brain, the brainstem, the heart plus the vessel system, the adrenal glands, and the kidneys. Several circuits are entangled: osmolarity, pressure (baro) reflex, and water balance. The major hormones are arginine vasopressin (*AVP*), aldosterone, angiotensin II, and atrial natriuretic peptide (*ANP*). Those areas where the blood–brain barrier is permeable and estimation of osmolarity via osmoreceptors or the transfer of molecules from blood to the brain is possible are labeled by *black circles* and *orange letters*. Major centers of hemostasis regulation are marked by *light blue circles*. Angiotensin II synthesis in the brain is not shown for reasons of clarity. Centers from where thirst originates are not yet known. *Blue arrows* endocrine stimulation, *red arrows* endocrine inhibition, *green arrows* regulation on axons and synapses. *ACE* angiotensin-converting enzyme, *Aq2* aquaporin 2, *dBBR* diagonal band of Broca, *JA* juxtaglomerular apparatus, *LC* locus coeruleus, *NPV* paraventricular nucleus, *NRa* raphe nucleus, *NTS* nucleus of the solitary tract, *OXT* oxytocin (Modified from Gray 1918; Krstić 1991; image by H.J. Heikenwälder)

Table 11.1 The atypical Na_x sodium channel

Name	Species	Tissue/cell type	Size (amino acids)	Gene	Chromosome locus	GenBank no.
Nav2.1	Human	Heart, uterus, muscle	1,682	*SCN7A*	2q21-23[a]	M91556
Na-G	Rat	Astrocytes	Partial sequence	*SCN7a*	3q21	M96578
SCL11	Rat	PNS (DRG)	1,702			Y09164
Nav2.3	Mouse	Heart, uterus, muscle	1,681			L36179

From Goldin (2001)
DRG dorsal root ganglion, *PNS* peripheral nervous system
[a]George et al. (1994)

located in the anterior ventricular part of the third ventricle (AV3V). Not labeled is the median nucleus of the preoptic region. In the supraoptic nucleus, median eminence and posterior pituitary as well as in the pineal gland and in the area postrema, the blood–brain barrier is opened.

Sodium channels (Table 11.1) detecting osmolarity are formed by a single protein: Na_x. This protein was known in humans and rodents for some time before its function was discovered. Owing to some structural and phylogenetic peculiarities, it was called voltage-dependent sodium channel type 2 (Goldin 2001). Watanabe et al. (2000) found Na_x expressed in the circumventricular organs and showed that Na_x knockout mice had a deficient liquid uptake. Other groups then confirmed these results (Grob et al. 2004; Hiyama et al. 2004, 2002). Hiyama et al. (2004) demonstrated by adenoviral transfer of an intact Na_x gene the repair of the defect in these knockout mice, thus proving that Na_x is necessary and sufficient for osmoreception in circumventricular organs.

Signals from the circumventricular organs also reach the paraventricular nucleus and the supraoptic nucleus, where magnocellular neurosecretory cells synthesize oxytocin and AVP, which are released in the posterior pituitary. The subfornical organ and the vascular organ of the lamina terminalis thus enhance oxytocin and AVP release.

At the same time the AV3V is connected to the raphe nuclei and the locus coeruleus by adrenergic nerves. From there the kidney is controlled by adrenergic nerves: circulation, resorption, and renin production, which, as described later, are also controlled in an endocrine manner.

11.8.3 Angiotensin II Receptors at the Blood–Brain Barrier

Angiotensin II is a key hormone of circulation control. It is derived by the endopeptidase actions of the renal enzyme renin on the angiotensinogen precursor and of angiotensin-converting enzyme on angiotensin I. Apart from stimulating adrenal aldosterone synthesis, angiotensin II plays an important role in control of AVP (and oxytocin) release, as demonstrated by the expression of angiotensin receptor in neurons close to the blood–brain barrier and by the action of serum angiotensin II levels on the brain.

At the same time—but not depicted in Fig. 11.15—angiotensin II is expressed and posttranslationally modified in neurons of the AV3V. These angiotensin II neurons were identified in the subfornical organ, vascular organ of the lamina terminalis, and median nucleus of the preoptic region, and the axons reach the paraventricular nucleus and supraoptic nucleus. With angiotensin II administered intracerebrally, treated animals stop all other activities and search for water. Intracerebral angiotensin II is thus regarded as a thirst signal. On the other hand, AVP release and oxytocin release are initiated, which facilitates water resorption in the kidney, thereby reducing water loss.

11.8.4 AVP Release in the Posterior Pituitary

AVP release and oxytocin release from the pituitary are under neuronal control: Via synapses, membrane potentials are generated which are relayed to the axon ends in the posterior pituitary and which induce fusion of secretory vesicles with the plasma membrane, whereby the vesicle content is released into the pericapillary space and via sieve plates into the capillaries.

In addition from noradrenaline (sympathetic system), angiotensin II (from angiotensinergic nerves of the subfornical organ and vascular organ of the lamina terminalis) and GABA are active as neurotransmitters. GABAergic nerves from the diagonal band of Broca route signals from the locus coeruleus and the nucleus of the solitary tract. These signals have been generated by receptors estimating blood pressure in the aortic arch or in the carotid sinuses. The baroreflex includes AVP release mediated by the nucleus of the solitary tract and the locus coeruleus and neuronal feedback on the vessel musculature as well: by vascular constriction, blood pressure is promptly enhanced, or is decreased by vasodilatation as promptly.

11.8.5 The Role of Oxytocin

Oxytocin is released from the posterior pituitary in reaction to volume increases signaled by the nucleus of the solitary tract or locus coeruleus. Oxytocin can bind oxytocin receptors in the atrium. This leads to ANP release, which triggers renal secretion of sodium and potassium, leading to volume reduction. Oxytocin receptors have also been found in the kidney, which suggests a direct oxytocin effect on renal regulation of hemostasis. Not shown in Fig. 11.15 is the observation that ANP receptors are present on oxytocin neurons in the AV3V, which creates an enhancer loop against a volume increase.

11.8.6 Thirst and the Endocrine System of the Brain

After intracerebral administration of angiotensin II, treated animals stop all other activity and search for water (see Sect. 11.8.3). Since neurons in the subfornical organ and the vascular organ of the lamina terminalis form angiotensin II themselves, we may assume that angiotensin II is the crucial signal in the CNS which signals the need for water. Those centers transforming the angiotensin II signal into conscious or unconscious activity have not yet been identified. In humans, enhanced activity has been observed in the cingulate cortex—that is, in the limbic system—by positron emission tomography with volunteers; in animals, the lateral parabrachial nucleus and the central nucleus of the amygdala are involved (McKinley and Johnson 2004) The lateral parabrachial nucleus in turn receives signals from the area postrema (where the blood–brain barrier is permeable) and from the nucleus of the solitary tract, where the signals of peripheral baroreceptors are integrated.

Since thirst originates prima facie from lack of water which cannot be compensated for by internal recruitment of water from interstitial spaces and by resorption, it is understandable that all mechanisms dealing with blood volume or osmolarity finally lead to the demand to add water from external sources.

11.8.7 Biochemistry of Water and Sodium Resorption

Blood is filtered in the renal glomeruli. The primary urine from the capsule of Bowman is transferred in the loop of Henle. Here and in the adjacent distal tubules, hormone-controlled resorption of water and ions occurs.

The expression and membrane positioning of the water transporter aquaporin 2 (Aq2) is positively controlled by AVP. This molecule is preformed in intracellular vesicles. On binding of AVP to its vasopressin receptor, vesicle fusion with the membrane occurs and Aq2 molecules allow water import into nephrocytes. On the basal side of nephrocytes, constitutively expressed aquaporin 3 and aquaporin 4 are present to transport the water back into the circulation. Without AVP, the Aq2 receptors are again internalized and may be transported back to the membrane if required.

Sodium resorption uses on the apical side the endothelial sodium channel (ENaC) consisting of three similar subunits (α, β, and γ). The usual sodium gradient between urine and cytosol is sufficient to transport sodium ions into the cell. On the basal side, sodium is pumped from the cell to the circulation by Na^+/K^+-ATPase, which pumps potassium into the cell. ENaC expression and Na^+/K^+-ATPase expression are stimulated by aldosterone. In addition to the amiloride-sensitive ENaC, there is a second sodium channel—the bumetanide-sensitive sodium cotransporter—which allows resorption of sodium ions, and also sugars, amino acids, phosphate, and sulfate. The expression of this molecule is also dependent on aldosterone.

ENaC receptors are further controlled by the binding of an ubiquitin ligase (Nedd4). Phosphorylation of ENaC blocks this binding and allows a longer presence of the channels in the apical membrane. The differential regulation of ENaC thus contains several hormone-controlled steps: transcription and translation triggered by aldosterone, membrane transport and complex formation controlled by AVP and ANP, and finally internalization and degradation by Nedd4.

Endocrine Rock 'n Roll: Rhythms and Secretion

12

Contents

12.1 A Universal Pacemaker 340
12.2 Circadian Rhythms (Pulse Frequency 24 ± 2 h) 342
12.3 Ultradian Rhythms (Pulse Frequencies Below 22 h) 343
12.4 Seasonal (Annual) Rhythms 345

Although rhythms during hormonal release have fascinated endocrinologists for a long time, many questions—especially regarding their physiological purpose—have not been answered. The origin and relevance of secretory episodes (pulses) from the hypothalamus and pituitary in their 1–3-h intervals are not known. The mechanisms, for example, of regular gonadotropin-releasing hormone (GnRH) release during the fertile years in men and women as well as in many animals defy analysis. Many elements, however, have been defined which determine in isolated neurons or larger tissue aggregates the pulse frequency and pulse amplitude.

Secretory maxima can be observed from active organs. Peak levels, with their rhythmic repetitions, are analyzed in serum. Concentration maxima (peaks or pulses) alternate periodically with minimal secretion (nadirs). Short intervals are defined as peaks with spans of 2–3 h or shorter. Serum-level oscillations during a whole day (circadian pulses), for example, are exemplified by cortisol, with a peak in the morning and a nadir in the evening. Furthermore, serum concentrations of growth hormone or melatonin increase during the night and are dependent on as well as independent of sleep. The annual or seasonal animal fertility cycles and hibernation phases have ultralong rhythms. Complex networks of regulations are at the origin of these periodic changes, and hormones are major parameters. The analysis of the causal zeitgeber of these rhythms is beyond the grasp of endocrinology.

Circadian light–dark cycles and annual temperature oscillation are periodic parameters which act extrinsically on any organism. There is, however, in plants and

B. Kleine, W.G. Rossmanith, *Hormones and the Endocrine System*,
DOI 10.1007/978-3-319-15060-4_12

animals a conserved pacemaking mechanism with similar principles in protozoans and metazoans.

12.1 A Universal Pacemaker

The control of the generation and degradation of seven proteins is sufficient to generate a rhythm of about 24 h (circadian). The major functions of these proteins are described in Table 12.1. Decisive is the feedback inhibition exerted by the protein period (PER) on the transcription of PER RNA and thus on its own synthesis. Further important aspects of this autonomous oscillation are as follows:

- Phosphorylation of PER by casein kinase 1 epsilon (CK1ϵ), which is required for PER to get into the cellular nucleus
- Binding of phosphorylated PER to cryptochrome (CRY), allowing transport of the PER–CRY dimer into the nucleus
- Constitutive transcriptional activation of *PER* by promoter binding of the dimer formed between brain and muscle aryl hydrocarbon receptor nuclear translocator (ARNT)-like protein 1 (BMAL1) and circadian locomotor output cycles kaput (CLOCK)
- Blocking of any PER transcription by binding of PER–CRY to promoter-bound BMAL1–CLOCK
- Ubiquitin-induced protein degradation—for example, by modification with ubiquitin and subsequent protease digestion in proteasomes

Translation of proteins other than PER—CRY, BMAL1, and CLOCK also occurs periodically, potentially blocked by CRY. Forger and Peskin (2003) developed a model from biochemical data and some assumptions which simulates a periodic almost 24-h autonomous oscillation. Their model takes into account the seven proteins, the transcriptional activity, nuclear–cytosol transport, translation, phosphorylation, cytosol–nuclear transport, and ubiquitin-mediated degradation.

More recently, the effect of PER–CRY on *PER* transcription has been found to result from circadian changes of histone acetylations of *PER* and *CRY*. Rev-ERBa transcribed from the antisense strand of thyroid hormone receptor controls circadian expression of BMAL1 and is itself controlled in a circadian manner, as is the protein ROR (an orphan nuclear receptor similar to retinoic acid receptor) binding to the same DNA site.

The major components of the circadian clock have thus been found: proteins with basic helix–loop–helix (bHLH) structures (Fig. 12.1) and PER–ARNT–single-minded homolog (SIM) (PAS) domains (see Table 12.1), whose transcription, translation, degradation, phosphorylation, and alternate cellular localization in the nucleus and cytosol oscillate in a circadian rhythm. The exact regulation of the cellular rhythm is not yet understood. We recognize the how, understand marginally the biochemistry, and ignore the purpose for which these rhythms are generated.

Table 12.1 Proteins of the zeitgeber

Name/OMIM entry[a]	Function/comment
BMAL1/602550	BMAL1 dimerizes to CLOCK. The dimer interacts with E-boxes: CACGTG DNA motifs with five E-boxes in the PER promoter and one in the CRY promoter. Dimer binding to the E-box activates transcription of *PER* and *CRY*. BMAL1–CLOCK levels are enough to permanently cover all E-boxes. By binding of PER–CRY dimer to a single E-box, PER and CRY transcription is inhibited. ARNT is the nuclear transport mediator for the aryl hydrocarbon receptor (with which dioxin or DDT interacts) and which stimulates *CYP1A1* and activation of other CYP enzymes in response to foreign substances. BMAL1 is very similar to ARNT. BMAL1 has bHLH and PAS domains
CLOCK/601851	CLOCK dimerizes to BMAL1 and binds as a dimer to E-boxes (see above). In 1997, 25 years after the discovery that the supraoptic nucleus plays a role in circadian oscillations, the mouse CLOCK protein was described by two groups as required for maintenance of circadian rhythms. CLOCK has, like PER and BMAL1, bHLH and PAS domains
PER1/602260	Transcription of *PER*[b] and the presence of PER obey a 24-h rhythm. *PER* is transcribed when no PER protein is present. After the cytosolic synthesis of PER, it is phosphorylated by CK1ϵ and CK2ϵ. Phosphorylated PER is co-transported with CRY into the nucleus and blocks the transcription of its own gene: classic feedback inhibition. PER proteins consist of bHLH and PAS domains, the former for DNA interaction (see Fig. 12.1), the latter for dimerization
CRY1/601933	CRY binds cytosolically phosphorylated PER. Thereafter, the complex enters the nucleus. Therein by it binding to BMAL1–CLOCK dimers, *PER* transcription is blocked. Binding by CRY makes the protein TIM (for "timeless") ready for degradation in proteasomes. Photolyases help to eliminate UV defects from DNA. They contain flavin adenine dinucleotide and pterin groups. CRY proteins are such photolyases; their role, however, appears to be to act as blue light photoreceptors. For their involvement in the circadian rhythm, the photolyase function is not important
CK1ϵ/600863	CK1ϵ cytosolically associates with PER and phosphorylates PER several times. This phosphorylated PER then couples to CRY and enters the nucleus. Casein kinases are widely distributed. They phosphorylate serine or threonine residues. CK1ϵ mutations change the circadian rhythm of flies and hamsters. The hamster mutant *tau* inhibits autophosphorylation of CK1ϵ and reduces binding to PER. Thus, nuclear availability of PER is diminished and the circadian rhythm is changed. The same applies for CK2ϵ

ARNT aryl hydrocarbon receptor nuclear translocator, *bHLH* basic helix–loop–helix, *BMAL1* brain and muscle aryl hydrocarbon receptor nuclear translocator like protein 1, *CK1ϵ* casein kinase 1 epsilon, *CK2ϵ* casein kinase 2 epsilon, *CLOCK* circadian locomotor output cycles kaput, *CRY* cryptochrome, *CYP* cytochrome P450, *PAS* period–aryl hydrocarbon receptor nuclear translocator–single-minded homolog, *PER* period

[a] Very detailed descriptions of the identification, cloning, and history of proteins can be found on the Internet: http://omim.org/entry/*number in column 1*

[b] For simplicity PER1 and PER2 are treated as one entity like CRY1 and CRY2. "PER" thus means PER1 and/or PER2

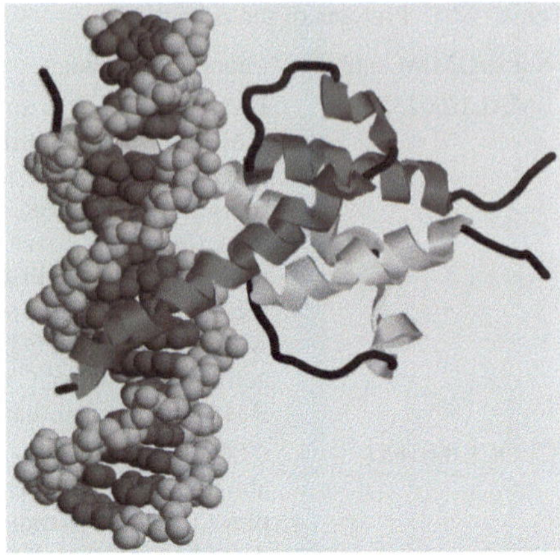

Fig. 12.1 Binding of two basic helix–loop–helix (bHLH) proteins to DNA. The long helix of the first bHLH (*dark gray*) interacts with the major DNA groove, and its short helix interacts with the second bHLH, which sinks its large helix into the next major groove (Produced with RasMol using Protein Data Bank entry 1A0A)

Neurons of the suprachiasmatic nucleus (nucleus suprachiasmaticus) are capable of maintaining, without any optical stimulation from an external light–dark cycle, a circadian rhythm for weeks. In that regard they differ from other cells where a circadian rhythm can be observed which, however, is stopped once supraoptic nucleus (nucleus supraopticus) input has ended. Thus, neurons of the supraoptic nucleus constitute the self-sufficient, superior, light-independent *zeitgeber*.

Light-independent circadian rhythms determined when animals are kept in permanent darkness differ individually in length and are about 22.5–23.5 h long. In a usual daily course with light–dark phases from the sun, the circadian rhythm is set by specialized optical nerves at sunset (reset!). Most probably all other endocrine secretory rhythms are generated with input from the supraoptic nucleus and thus accept the externally controlled day–night rhythm. A shift of the so-called biorhythm from the external day–night rhythm thus, although often stated, cannot be confirmed by endocrine analysis. Arctic winters and summers where the day–night rhythms are interrupted and permanent night shifts are exceptional conditions for circadian endocrine secretion. The endocrinology of subpolar life with very long days in summer and very short days in winter might provide valuable research themes.

12.2 Circadian Rhythms (Pulse Frequency 24 ± 2 h)

In Fig. 12.2, two different pulse rates can be distinguished: a short one with intervals of about 1 h (circhoral or ultradian) and a longer one with a 24-h oscillation (circadian). A circadian oscillation can be overlaid by an ultradian rhythm. In the release of corticotropin-releasing hormone (CRH) and arginine vasopressin (AVP), we notice circadian variability from the shift of pulse amplitudes. For glucocorticoid hormone release, a circadian rhythm is apparent with peak secretion during the night and a nadir during the day.

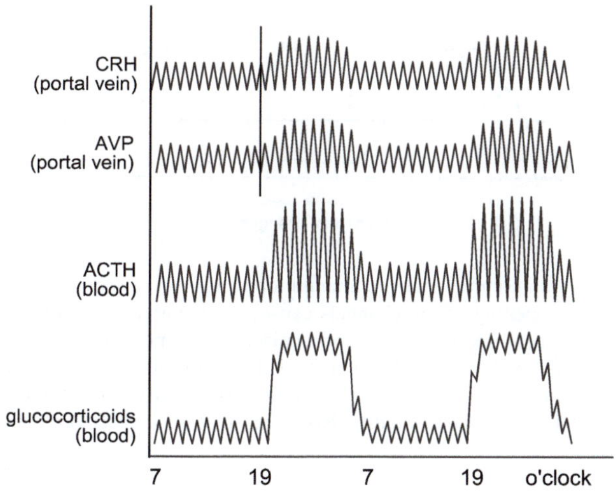

Fig. 12.2 Circadian rhythm of corticotropin-releasing hormone (*CRH*)/arginine vasopressin (*AVP*)/adrenocorticotropic hormone (*ACTH*) pulse amplitude and cortisol release. In daily rhythm, CRH and AVP pulse amplitudes oscillate under control of the suprachiasmatic nucleus. This influences, in turn, ACTH pulses, where nadirs still occur in short intervals. Adrenal glucocorticoids show these short oscillations; however, the baseline is shifted after onset of darkness (Redrawn from Chrousos 1998)

From the scheme one can conclude that circadian glucocorticoid pulses appear as a consequence of enhanced CRH amplitudes. The circadian rhythm of glucocorticoid secretion has its origin in the brain: neurons of the suprachiasmatic nucleus project to neurons of the paraventricular nucleus (nucleus paraventricularis) and trigger CRH release.

12.3 Ultradian Rhythms (Pulse Frequencies Below 22 h)

CRH is secreted as shown in Fig. 12.2 in a hourly rhythm, GnRH is secreted at 60–90-min intervals, and insulin secretion peaks every 4 min. For understanding the regularities in the generation of secretory episodes, we should regard neurosecretory cells not as hormone-producing ones, but as neurons. Nunemaker et al. (2003b) demonstrated that GnRH neurons possess several kinds of action potentials—that is, activities of ion channels and coupled calcium oscillations are at the origin of pulsatile hormone release. Most probably a coordinated influx of calcium into cells occurs, and these are discharged and therefore secrete hormone simultaneously (Fig. 12.3). With regard to insulin secretion from β cells, similar electric pulses and calcium oscillations precede the insulin secretions. Enhancing the intracellular calcium will trigger fusion of secretory granules to the cell membrane, where the granular content is freed (Sect. 11.4, Fig. 11.6).

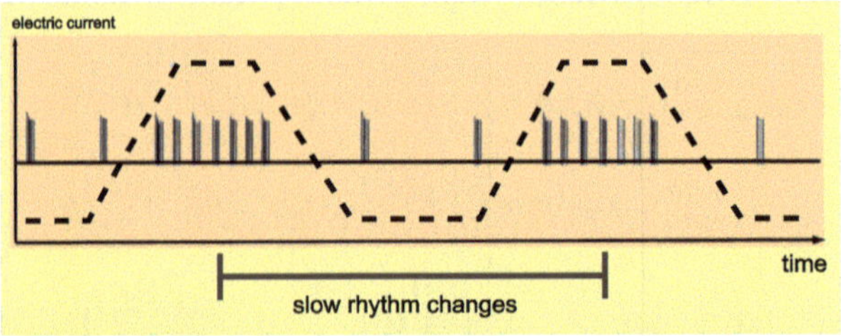

Fig. 12.3 Rhythm of autonomic episodic channel openings in gonadotropin-releasing hormone (GnRH) neurons. The opening of ion channels causes the membrane potential to change. If the membrane potential is measured for long periods, episodes become apparent where rapid openings and closures occur (so-called bursts). The frequency of bursts follows a slower rhythm depicted by the *dashed line*. In the same rhythm, secretory granules fuse to the cell membrane and thus release hormones, here GnRH, in a periodic manner (Redrawn from Nunemaker et al. 2003b)

With use of GnRH tumor lines (GT1-7) it was shown that sodium and calcium channels are mainly responsible for transmembrane ion currents; these effects could also, however, be generated by coordinated potassium ion influx. Different calcium channel types were involved (Nunemaker et al. 2003a).

Insulin-producing β cells as well as GnRH neurons occur together with many other cells of similar type. When pulses of insulin or GnRH are generated, secretion has to be coordinated. For this, the slower rhythm—that is, the burst frequency (Fig. 12.3)—has to be kept in time. A mechanism for this has yet to be found. In GnRH neurons, axonal interplay using GABAergic and noradrenergic neurons can be assumed, whereas in β cells, no mechanism has yet been proposed. Since the pancreas and the islets are intensively innervated, a noradrenergic control might perhaps coordinate insulin release.

The pulsatile GnRH release during the menstrual cycle is modulated by estradiol. Again, the slow rhythmic frequency of burst episodes is changed, not their intensity or the variation of rare versus frequent bursts. It appears that estradiol interacts where the coordination is controlled.

Current knowledge suggests that GnRH neurons generate their burst rhythm autonomously. Single cells or cell lines from GnRH neurons demonstrate rhythmic channel openings and secret GnRH in an episodic manner. What is missing is information about the biochemical origin of these episodes. For other hormone-producing cells, any proteins or genes have been found generating regular rhythms in second or minute intervals.

We do know, however, that permanent application of GnRH or its synthetic derivatives blunts any episodic GnRH release. The GnRH receptor is found not only on luteinizing hormone/follicle-stimulating-hormone-producing gonadotropic cells of the pituitary, but also in the hypothalamus. This makes GnRH when it is secreted hypothalamically in a paracrine way or as neurotransmitter a candidate for the signal that coordinates release from GnRH neurons in the median eminence (eminentia mediana). A permanently elevated GnRH level would, in turn, result in hypothalamic GnRH receptors being occupied, induce their internalization and degradation, and thus mediate a potential GnRH receptor blockage of GnRH synthesis. If we assume a GnRH–GnRH receptor interaction results in periodic stimulation of GnRH synthesis, permanently elevated levels of GnRH (or GnRH agonists as well) would disturb this rhythm. For the explanation of the effects of a GnRH receptor antagonist, receptor blockage suffices.

12.4 Seasonal (Annual) Rhythms

The length of the light–dark phases of a day changes with the season. Simultaneously, the median temperature varies. Existential functions such as collecting food, hibernation, and reproduction are coupled to these rhythms. All wild animals are subject to these rhythms, whereas domestic animals are no longer fully dependent on warm–cold phases or summer–winter rhythms.

Endocrine adaption of sexual activity with the seasonal rhythm has been analyzed in only a few examples (Fig. 12.4). These examples exhibit so many species characteristics that we restrict ourselves to basic determination of seasonal rhythm.

In recent years a new role of the upper part of the anterior pituitary, the pars tuberalis, was found. Pars tuberalis cells play a role in seasonal endocrine control.

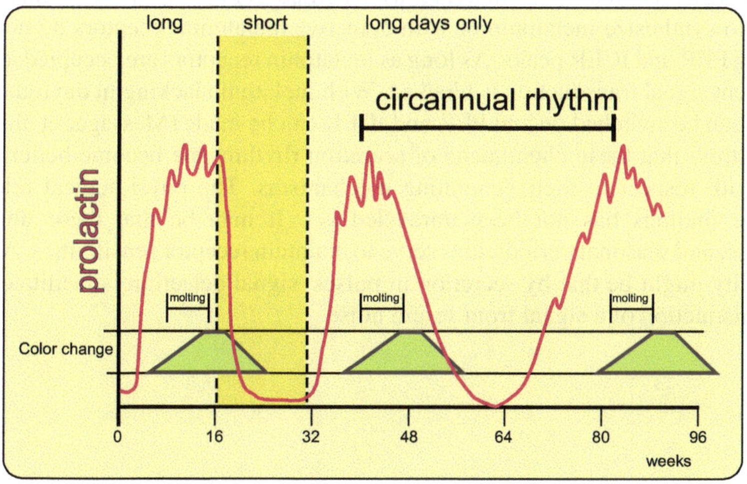

Fig. 12.4 Annual rhythm of prolactin release and coat color changes in the Soay sheep. Under the influence of clock genes and the seasonly determined melatonin, pituitary prolactin release is triggered in the light season. This, in turn, controls color coat changes and the start of the molt. The basic rhythm is prolonged when the day length (experimentally maintained) is not shortened (*right*). The level of prolactin in blood still decreases, and a coat color change occurs. Yet, a new rise in prolactin level followed by another coat color change and molt occurs without changes in the light–dark rhythms, but with longer intervals than in daylight-controlled phases (Redrawn from Lincoln et al. 2003)

These cells, called calendar cells, express the clock genes (see earlier) *PER*, *CRY*, *BMAL1*, and *CLOCK*. In the pars tuberalis, however, their expression, in contrast to that in the suprachiasmatic nucleus, appears to be controlled by melatonin. One premise for this is melatonin receptor expression in these cells. Melatonin serum levels are strongly dependent on the length of darkness (see Sect. 7.3), and thus levels are low in summer and high in winter. Calendar cells integrate the amount of melatonin present and differentiate short (6–10-h) from long (12–16-h) melatonin receptor interaction episodes.

As a consequence, in seasons with short nights, the calendar cells secrete a prolactin-stimulating hormone, tuberalin, whose identity has not yet been established. As described in Sect. 4.5.2, a prolactin-stimulating hormone in sensu stricto has not been found in the hypothalamus of animals and humans; in contrast, prolactin release is strongly blocked by dopamine. Such a tuberalin might be a de facto prolactin-releasing hormone or an endogenous dopamine antagonist, binding to but not activating the dopamine receptor.

In contrast to the suprachiasmatic nucleus, the amplitude of PER expression and that of inducible cyclic adenosine monophosphate (cAMP) early repressor (ICER) are seasonally controlled: PER and ICER are synthesized mainly on long days; on short days very little if any PER or ICER is synthesized. By injection of melatonin, PER and ICER expression is delayed. PER and ICER expression starts when melatonin is removed. With high levels of melatonin removed (dark season), more PER and ICER are made than in light seasons. Mice lacking the pineal gland or those unable to synthesize melatonin or with defective melatonin receptors do not show morning PER and ICER peaks. As long as melatonin receptors are occupied, cAMP-dependent signal transduction is blocked. With melatonin lacking in daylight, these signals can be switched on and PER and ICER can be made (Messager et al. 1999).

We stress that these phenomena of secretion rhythms are become better understood with respect to their generating mechanisms. The physiological relevance of these findings has not been unraveled yet. It may be that these ultradian, circadian, and seasonal periodicities serve to maintain receptor sensitivities. Another possibility might be that by secretion in pulses, signal detection is facilitated with easier distinction of a signal from serum noise.

Evolution of the Endocrine System 13

Contents

13.1 Division of Labor 349
13.2 Evolution of Neuropeptide Hormones 349
13.3 Evolution of Glycoprotein Hormones 350
13.4 Insulin and Insulin-Like Proteins 351
13.5 Evolution of CYP Enzymes and Steroid Hormones 351
13.6 Catecholamine Evolution 353
13.7 Conclusion 353

During the evolution of life, singled-celled organisms developed into multicellular ones and these multicellular organisms became plants or animals. When animals got organs with diverse functions, they had to invent communication between these to establish homeostasis. The devices of this communication are nerves and hormones.

As shown in previous chapters, the biochemical roots of the endocrine systems are beyond the kingdom Metazoa. Potentially steroidogenic enzymes have been found in bacteria, as well as in plants. The roots of the "tree of life" are buried below thick sediments. For Fig. 13.1 we have used the poster of Westheide and Rieger (2007) as a reference. The sketch accompanying the fact sheet on invertebrate neuropeptides is based on information in this poster. Figure 13.1 contains two numbers which indicate the periods where separation of protostomes from deuterostomes and chordates from hemichordates or echinoderms occurred. Scientific names are listed together with common names in the Appendix .

For the text that follows, we assume that the presence of a homologous gene in two species means that the common ancestor already had this gene. If fish and humans have a gene with the same function, then this gene was present when amphibians separated from the fish lineage. If the fruit fly expresses the same cytochrome P450 (CYP) enzyme, then this gene was already present when the lineages giving rise to arthropods and chordatas separated. Such an assumption is based on the sequence analysis of many proteins, especially that of CYP. If a

B. Kleine, W.G. Rossmanith, *Hormones and the Endocrine System*,
DOI 10.1007/978-3-319-15060-4_13

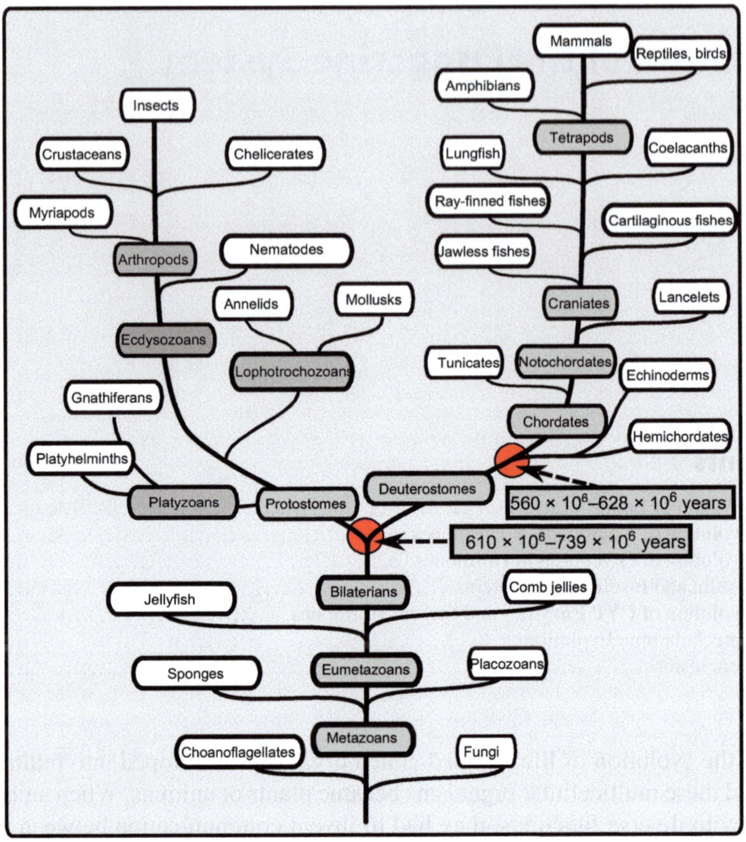

Fig. 13.1 Metazoan evolution (Modified from Wenger and Hoogewijs 2010)

function in two species is due to a structurally similar protein and a gene with a related sequence, then there are better reasons to accept segregation from a common ancestor than a multiple recreation with stunning similarity.

The evolution of the endocrine system as described in humans dates to the beginning of life: cholesterol was already made by protozoans. The last intermediate of cholesterol biosynthesis, 7-dehydrocholesterol (Fig. 6.4, **18**), is isomerized to vitamin D_3 by UV light. This reaction is as old as 7-dehydrocholesterol biosynthesis. Thus, we suspect that vitamin D_3 has been made for billions of years.

In contrast to these very old molecular origins, the control of calcium homeostasis by vitamin D_3 is much more recent. As long as metazoans lived in the ocean, the calcium concentration in the body fluid was determined by the calcium concentration in the ocean: about 1 mM.

The presence of a molecule in a species at the beginning of evolution does not argue for an ancient function and action of this hormone as they are found today in humans. Hanke (1970) argues that compounds did exist before they acquired an endocrine function.

Most interestingly, freshwater fish species as well as all land-living animals from amphibians on regulate their serum calcium levels to the calcium concentration of saltwater, as do humans, although the ocean has not been their natural environment for the last 100 million years. For the maintenance of the saltwater calcium levels in blood, freshwater fish and land-living animals have developed parathormone. In fish, parathormone-related peptides are made in corpuscles of Stannius, and in humans they are made in the parathyroid gland. It is possible that parathormone is the last hormone appearing in the evolution of the endocrine system.

In tunicates, iodine is fixed using mucopolysaccharides. Starch, however, can fix iodine without coupling iodine covalently. The development of thyroid peroxidase, which oxidizes iodide to elementary iodine, which is then transferred to the tyrosine residues of thyroglobulin and where one ring is transferred to another iodinated tyrosine residue, is not in any way related to fixing iodine with mucopolysaccharides and mucus. Phagocytosis of the iodine-containing colloid and the enzymatic digestion which gives rise to thyroxine might have been possible in ancestors of fish. The thyroid peroxidase and thyroxine release, however, first occurred in fish, including saltwater fish.

13.1 Division of Labor

Certain eukaryotic single-celled organisms are able to adapt their movement to an external signal and thus to swim/drift in the direction of a food source. The nature of such a signal inducing a directed movement is not known. From the time when single-celled organisms aggregated to multicellular aggregates with assigned roles, intraorganismal communication to balance offer and demand was established. Nerves and neurosecretory cells served for these purposes. The simplest metazoans known, cnidarians belonging to coelenterates, express precursor proteins giving rise to neuropeptides. Nerves in sensu stricto or ganglia have been found only in later animals, molluscs, or annelids. The development of nerves and neuropeptides—among other—distinguishes metazoans (multicellular animals) from plants and fungi.

13.2 Evolution of Neuropeptide Hormones

Since earliest animals expressed neuropeptides, this characteristic has been common throughout the animal kingdom. Neuropeptides normally[1] arise as translational products of specialized neurosecretory cells where precursor proteins are

[1] In placozoans no neurons exist; however, neuropeptides have recently been found in their genome (Nikitin 2014) and in specialized neurosecretory cells (Smith et al. 2014).

successively processed by cell-type-specific enzymes, prohormone convertases, endopeptidase, peptidylglycine alpha-amidating monooxygenase, and if an N-terminal glutaminyl residue is present, glutaminyl cyclase, where mature neuropeptides are stored in granules and where these granules are fused to the cell membrane on interaction of a ligand with a membrane receptor, inducing a rise of intracellular calcium concentration which triggers vesicle–membrane fusion.

The catalog of so-called **FMRF**amide peptides starts in cnidarians (see Fig. 13.1). In insects—for example, in cockroaches—there is a leucosulfakinin (**EQFEDYGHMRF**amide) or a leucomyosuppressin (p**EDVDHVFLRF**amide) (first identified in *Leucophaea maderae*), and in crustaceans, there is **SQRNFLRF**amide/**TQRNFLRF**amide. The FMRF motif is still present in an extended version in met-enkephalin **YGGFMRF**. The fact that from a precursor protein several identical copies of thyrotropin-releasing hormone or different peptides such as in proopiomelanocortin can be synthesized is equally transferred from the most primitive animals.

Some functions which are controlled in humans by hormones other than neuropeptides are regulated in early animals by neuropeptides. The calcium level maintained in vertebrates by vitamin D_3 and its derivatives is regulated—for example, in earthworms (*Eisenia fetida*)—by neuropeptides. Although vitamin D_3 could perhaps be used in earthworms, as described earlier, it is not used for calcium homeostasis. One could argue that earthworms do not get enough UV light to isomerize dehydrocholesterol in sufficient quantities (see Sect. 6.11).

13.3 Evolution of Glycoprotein Hormones

With the identification of the GPa2–GPb5 heterodimer in flies (Sudo et al. 2005), it has become obvious that glycoprotein hormones have an ancient origin. Thyrotropin-releasing-hormone-like proteins have been found in different invertebrates: for example, in cnidarians, mollusks, and nematodes. A leucine-rich receptor is also present throughout the metazoan kingdom. There are, however, some important differences. The fly and the human GPb5 lack the sixth pair of cysteines which stabilizes the seat belt of the β chain around the α chain (see Fig. 4.22). For this reason the association of GPa2 and GPb5 might not be as strong as that in thyroid-stimulating hormone, follicle-stimulating hormone, or luteinizing hormone. The separation of GPa1 and GPa2 from GPb5 from the β chains of thyroid-stimulating hormone, luteinizing hormone, follicle-stimulating hormone, choriogonadotropin, and their receptors occurred at the origin of vertebrates. GenBank, however, apart from arthropods and vertebrates, contains only a β-chain homolog from nematodes, and no α-chain homolog, although several species linking flies and vertebrates to the ancestor have been fully sequenced. It is worth mentioning that arthropods have another cysteine-knot hormone, bursicon (Sect. 5.5.5). Sudo et al. (2005) have demonstrated that the GPa2–GPb5 heterodimer and the bursicon heterodimer act on specific G-protein-coupled receptors (DLGR1 and DLGR2, respectively) without any cross-reaction.

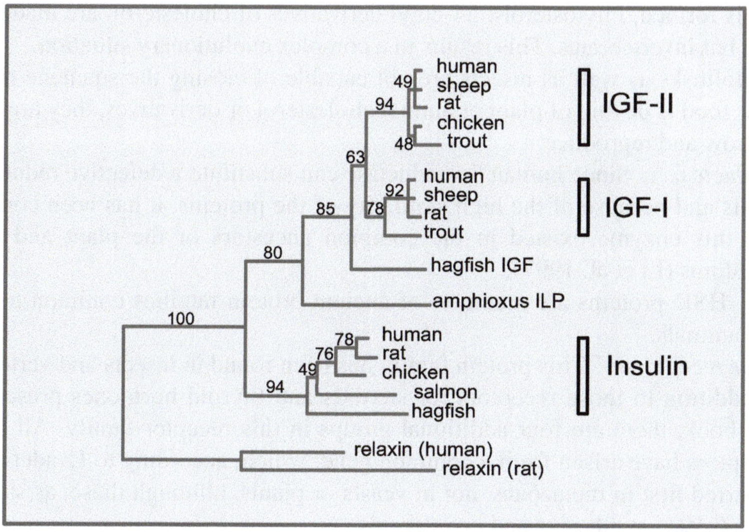

Fig. 13.2 Tree of vertebrate insulin evolution. Separation of insulin and insulin-like growth factor (*IGF*) predates vertebrate evolution and was present already in hagfish. *ILP* insulin-like peptide (Redrawn from Ellsworth et al. 1994)

13.4 Insulin and Insulin-Like Proteins

The vertebrate species analyzed all possess insulin, insulin-like growth factor (IGF) 1, and IGF2 (see Fig. 13.2). In agnathans, insulin and only one IGF-like protein have been identified. Insulin homologs with the three segments B chain, C peptide, and A chain on one precursor, disulfide bridges, and the C peptide removed in posttranslational processing have been found with differing numbers of genes in arthropods and placozoans (see Sect. 5.1.2; see also Nikitin 2014), and thus probably occur in all protostomes.

13.5 Evolution of CYP Enzymes and Steroid Hormones

To generate a steroid and for the steroid to become active, different CYP enzymes, one or two hydroxysteroid dehydrogenases (HSDs), and 5α-reductase have to convert cholesterol, and finally a nuclear receptor (or a membrane receptor) has to interact with the hormone and induce signal transduction or gene activation:

Cholesterol biosynthesis. For the cyclization of squalene to the gonane-ring system, oxygen is required, as for further conversion to cholesterol. Aerobic bacteria, cyanobacteria, plants, and protozoa as well as invertebrates and vertebrates should be able to perform this cyclization. The intermediate in most taxa is cycloartenol (Fig. 6.25, **61**), whereas in fungi and vertebrates lanosterol (Fig. 6.4,

16) is formed. Phytosterols, 24-ethyl derivatives of cholesterol, are made by all taxa but invertebrates. This results in a complex evolutionary situation.

Mollusks as well as insects are not capable of closing the squalene rings. If their food is devoid of plant or animal cholesterol or derivatives, they are unable to grow and reproduce.

5α-Reductase. Since human 5α-reductase can substitute a defective reductase in plants and because of the high similarity of the proteins, it has been concluded that this enzyme existed in the common ancestors of the plant and animal kingdoms (Li et al. 1997).

HSD. HSD proteins are members of ancient protein families common to plants and animals.

Nuclear receptors. This protein family has been found in insects and vertebrates. In addition to those receptors for steroids and thyroid hormones presented in this book, there are four additional groups in this receptor family. All nuclear receptors have arisen from a common gene, which, according to Laudet (1997), occurred first in metazoans, not in yeasts or plants, although these, as shown in Sect. 6.10.2, synthesize and use steroids.

Steroid hormone receptors arose from a receptor common to protostomes and deuterostomes. The molecule in sea slugs (*Aplysia*) is homologous to vertebrate proteins. In the evolution of insects, this gene has been lost. At the beginning of vertebrate evolution, the gene was duplicated twice. Thereafter, the five receptors for estrogens, androgens, glucocorticoids, mineralocorticoids, and gestagens developed separately.

In agnathous lampreys, Thornton (2001) cloned nuclear receptors and suggested that the estrogen receptor is the primordial steroid receptor. This suggests that all enzymes required for estrogen synthesis were already expressed and functionally active in these early vertebrates. Baker (2002), however, claims that several saturated compounds could bind to the estrogen receptor and initiate gene activation.

Dehydroepiandrosterone, for example, can be converted by 17β-HSD to 5α-androstane-3β,17β-diol, which has been shown to activate the estrogen receptor in spite of the lack of the aromatic ring.

CYP. CYP enzymes are found in bacteria. The enzyme CYP51, which oxidizes the 14-methyl group of lanosterol, has been found in all eukaroytes analyzed. CYP11A1, the side-chain-cleavage enzyme, appears to be restricted to vertebrates. Insects, for example, lack this enzyme. CYP19 (aromatase) is equally restricted to vertebrates and their immediate precursors.

More surprising are findings that mussels and mollusks contain estrogens. Thirty-five years ago, progesterone was found in mussels for the first time (Saliot and Barbier 1971). 17β-HSD converting, for example, androstenedione into testosterone has been found (Varaksina and Varaksin 1988). In octopuses and snails, enzymatic activity was observed which converted androstenedione to other androgens or estrogens. The observation of estradiol in scallops supports these findings.

Whether these findings reflect pollution of the environment or estrogen biosynthesis has not been answered since the original observations. The necessary enzymes, particularly CYP19, have not been found in 30 years. With respect to the fact that these animals are unable to take up cholesterol (see above) and most probably lack any receptor for estrogens or any other steroid hormone, such findings should be treated with caution. Thornton, who discovered a primordial nuclear receptor in the sea slug (*Aplysia*), himself argued that this receptor was nonfunctional with respect to steroid binding. One study (Pazos et al. 2003) did not identify either pregnenolone or progesterone.[2]

The traditional assumption that steroid hormones and steroids can be found only in vertebrates has been refuted, since plants as well as fungi and insects can make and use steroids (see Sect. 6.10).

Until proven otherwise, we can state that vertebrates are distinguished from other metazoans, fungi, plants, and monocellular organisms by CYP11A1, the side-chain-cleavage enzyme, which gives rise to pregnenolone, which is modified to corticoids and sex steroids. Those plant or arthropod enzymes found are not able to cleave the side chain of cholesterol.

Steroid hormone receptors, as far as we know, are functional only in vertebrates, whereas thyroid hormone receptors, vitamin D_3 receptors, or retinoic acid receptors were already functional in earlier metazoans.

13.6 Catecholamine Evolution

Catecholamines or rather the enzymes of catecholamine biosynthesis were probably present early in evolution. Hydroxylase activity, substituting the aromatic ring of tyrosine by a hydroxyl group, occurs in protozoans. The decarboxylase which makes dopamine from DOPA is not specifically expressed in endocrine active cells. The neurotransmitter noradrenaline and the sympathetic nervous system appear old as well. Adrenaline biosynthesis appears to be more recent since the gene is conserved in zebrafish and mammals.

13.7 Conclusion

The human endocrine system is constituted from very old elements (catecholamines, CYP, neuropeptides, insulin), more recent ones (iodine fixation, glycoprotein hormones, steroids), and primate- or human-specific ones (Y-chromosome-dependent expression, some CYP enzymes).

[2]A pubmed search in 2015 still confirmed that there are reliably called CYP19 proteins only in lancelets and vertebrates.

Part III

Hormones and Medicine

Diseases of the Endocrine System 14

Contents

14.1 Defects in the CNS/Hypothalamus........ 359
14.1.1 Kallmann Syndrome........ 359
14.1.2 Craniopharyngioma........ 359
14.2 Pituitary Defects........ 359
14.2.1 Genetic Deficits........ 359
14.2.2 Pituitary Tumors........ 360
14.2.3 Disturbances of Water Homeostasis........ 362
14.3 Thyroid Diseases........ 363
14.3.1 Lack of Thyroxine Synthesis: Hypothyroidism........ 363
14.3.2 Thyroxine Excess: Thyrotoxicosis........ 363
14.4 Dysfunctions of the Endocrine Pancreas........ 364
14.4.1 Pancreatic Tumors........ 364
14.4.2 Diabetes Mellitus........ 364
14.5 Adrenal Dysfunctions........ 365
14.5.1 Congenital Adrenal Hyperplasia........ 365
14.5.2 Hypercorticism........ 366
14.5.3 Catecholamine-Releasing Tumors........ 367
14.5.4 Autoimmune Adrenalitis (Addison Disease)........ 367
14.5.5 Aldosterone Deficiencies........ 368
14.6 Multiple Endocrine Neoplasia........ 368
14.7 Fertility Dysfunctions and Malfunctions of Reproductive Organs........ 369
14.7.1 Genetic Defects Affecting the Formation of Sex Organs........ 371
14.7.2 Gene Defects Influencing the Hypothalamus and Pituitary........ 372
14.7.3 Gene Defects of GnRHR, Gonadotropin Synthesis, and Recognition........ 373
14.7.4 Gene Defects Affecting Steroidogenesis........ 374

The endocrine system is so intimately tied to elementary body functions that malfunctions within the endocrine system are pathogenic and potentially life threatening. To understand human endocrinology, the elucidation of defects is even more required because from the defects the healthy, normal situation might be inferred. If, for example, an enzyme defect inhibits steroid hormone biosynthesis

B. Kleine, W.G. Rossmanith, *Hormones and the Endocrine System*,
DOI 10.1007/978-3-319-15060-4_14

and results in major, fatal imbalances of water homeostasis, we recognize the central role of steroidogenic enzymes for the control of renal functions. Since aromatase mutations not only cause sex reversal, but also lead to growth failures, we perceive the action of estrogen for chondrocyte maturation (see Sect. 11.6.3). Experiments in humans, for example, to knock out target genes by molecular biological means which in mice have given stunning and astonishing results are generally forbidden; therefore, endocrinologists depend on the analysis of genetic variations, even though this may be very difficult in certain situations. Information on genes defects in endocrine diseases will complete the discussion of the endocrine system and its elements, especially with respect to an extended description of genes and their products.

We describe different forms of malfunctions: In the endocrine system, release of hormones might be too high or too low. An untimely release of a hormone might occur and disturb homeostasis. The most frequent faults are oversecretion or undersecretion or tumors of endocrine active cells. Other failures involve autoimmune diseases where the immune systems rejects endocrine cells as "foreign" and destroys them, with the consequence of major endocrine dysfunctions. And there are genetic defects: either inherited or individually acquired. These might result in disturbances of the endocrine circuits presented so far. Some genetic defects give rise to infertility and are not handed down.

Many defects involve the fetus in utero. If elementary functions are blocked, fetal death occurs. In these instances, the analysis of the causes is most often impossible. If maternal hormones, however, can substitute for the lacking fetal ones, the child may survive until parturition. Once the umbilical cord has been cut, the newborn is in mortal danger in congenital adrenal hyperplasia, for example, since no cortisol is provided by the adrenal glands. The lack of aldosterone biosynthesis is as dangerous for the newborn. If identified early, such diseases can be controlled.

Other malfunctions become apparent after several weeks or much later. A growth hormone (GH) deficit is not life threatening but may lead to much reduced growth. When the defect is related to the biosynthesis of GH, recombinant GH can be used as a substitute, and its use may lead to relatively normal growth. The individual is, however, resistant to GH therapy if the GH receptor is defective either by mutations or by deletions of the receptor gene. Here the therapy is inefficient. There are, for example, two murine strains with inherited obesity: In the *ob/ob* mouse, the leptin gene is mutated. Here obesity is controlled by administration of leptin, and adipose tissue already acquired is removed. In the *db/db* mouse, however, the leptin receptor is mutated and administered leptin does not change adiposity. In human society adiposity, however, is rarely caused genetically. Not enough exercise or activity and an extremely lipid-rich diet are causal for most of the adiposeness.

Since many genetic defects of the endocrine systems involve the CNS, they might be reflected in mental development of a child.

14.1 Defects in the CNS/Hypothalamus

14.1.1 Kallmann Syndrome

Precursor cells of gonadotropin-releasing hormone (GnRH) neurons migrate from the olfactory anlage into the preoptic area of the hypothalamus. This migration depends on neuronal cell adhesion molecule. In Kallmann syndrome, this migration does not occur and there is no development of regular functional GnRH neurons. Thus, GnRH release fails to occur, with the consequence of maldevelopment of the gonads (hypogonadotropic hypogonadism). This developmental failure is often coupled to a subdeveloped or missing olfactory sense (anosmia).

14.1.2 Craniopharyngioma

Craniopharyngioma is the most frequent tumor in children. Some remnants of Rathke's pouch develop slowly into a tumor with preferred location at the optic chiasm and the sella turcica, thus before and under the hypothalamus. Although it is benign in origin, craniopharyngioma is listed among malignant diseases since it extends into the hypothalamic and pituitary areas and causes massive disturbances of the endocrine system.

Although there are disturbances of the endocrine system, these are mostly undetected and the tumor is recognized as such when optical malfunctions or behavioral abnormalities are apparent. Treatment involves surgical removal of the tumor, radiotherapy, long-lasting endocrine control, and hormone substitution (Grossman 1992; Gerok et al. 2000).

14.2 Pituitary Defects

14.2.1 Genetic Deficits

The anterior pituitary develops from Rathke's pouch by time- and space-controlled expression of several pituitary specific transcription factors: Pit1, Pitx1, Pitx2, HESX1, PROP1, LHX3, LHX4, GATA2, SF1, and Egr1. Pitx2 and LHX3 seem to stimulate expansion of the pituitary anlage by activating HESX1 and PROP1, which in turn cooperate to stimulate Pit1. Defects of any of these transcription factors cause pituitary maldevelopment. In empty sella syndrome, cells from the adenohypophysis are found outside the sella turcica and vasopressin/oxytocin release occurs ectopically. This syndrome, in addition, might be caused by necrosis, infarct, or radiation.

In the healthy pituitary, gonadotropic cells develop under the influence of GATA2, SF1, and Egr1, all activated by Pitx2. For the development of somatotropic and thyrotropic cells, Pit1 is activated by Pitx2. Maturation of gonadotropic, thyrotropic, and somatotropic cells and their hormone biosynthesis are under the control of Pitx1.

In cases of combined pituitary hormone deficiency (CPHD), mutations in the following genes have been reported (Achermann et al. 2002):

- *HESX1*. Mutations in this gene are causal for CNS defects and pituitary damage connected with faulty gonadotropin release and reduced GH secretion.
- *LHX3*. If this transcription factor is mutated, patients present with cryptorchism, micropenis, and delayed puberty.
- *PROP1*. CPHD in 50 different families has been linked to mutations of *PROP1*.

14.2.2 Pituitary Tumors

14.2.2.1 Prolactinoma

A prolactin-secreting tumor is the most frequent cause of prolactin overproduction (hyperprolactinemia) and the most frequent pituitary tumor. Growth of lactotropic cells because of lack of dopamine control appears causal. This lack of dopamine control may arise by defects of dopamine secretion or by disrupted dopamine transport in the portal system. Patients with hyperprolactinemia present with fertility disturbances: Estradiol and progesterone levels are reduced, and pulsatile follicle-stimulating hormone (FSH) and luteinizing hormone (LH) secretion is affected, and therefore so are spermatogenesis, the menstrual cycle, and oocyte maturation. In addition to headache, galactorrhea is observed in women.

The standard therapy uses a dopamine agonist such as bromocriptine (an ergoline derivative). Surgical treatment is performed when there is resistance to the drug therapy. The tumor is reached through the nose (transsphenoidal surgery) and electively removed. This treatment is most often well tolerated without a prolonged postoperative hospital stay.

14.2.2.2 Corticotroph Adenoma (Cushing Syndrome)

Adrenocorticotropic hormone (ACTH) biosynthesis and release are controlled by corticotropin-releasing hormone (CRH). Excessive ACTH release causes Cushing syndrome (see Sect. 14.5). This might be caused by a hypothalamic tumor or hyperplasia of corticotropic pituitary cells. In 4–24 % of transsphenoidally treated patients, the hyperplasia is caused by defective hypothalamic CRH control; more frequent, however, is primary corticotropinoma. There are additional observations where nonendocrine tumors release CRH (paraneoplasia), causing hyperplastic growth of corticotropic cells.

The surgical removal of ACTH-secreting pituitary tumors is the therapy of choice.

14.2.2.3 Somatotroph Adenoma (Acromegaly)

GH synthesis and release are stimulated by GH-releasing hormone (GHRH) and are inhibited by somatostatin (see Sect. 4.4). Acromegaly is most often caused by a

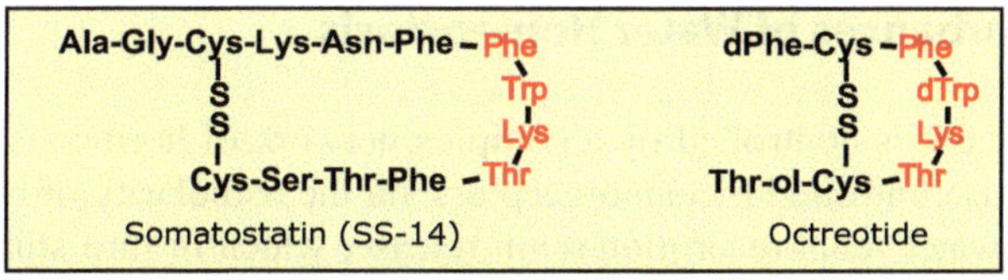

Fig. 14.1 Long-lasting somatostatin agonist: octreotide

GH-secreting tumor, and thereby through enhanced secretion of insulin-like growth factor 1. Rarely, acromegaly is initiated by a GHRH-secreting hypothalamic tumor. Indications for acromegaly are changed modified facial features—for example, brow protrusion, lower jar protrusion, or teeth gapping.

Occurring before puberty, this tumor results in giantism. As a result of this and any other sufficiently large pituitary tumor, the optic nerve might be influenced, with impaired vision or a reduced field of vision. Excessive prolactin release (see Sect. 14.2.2.1) and infertility are often linked to acromegaly.

Therapy involves transsphenoidal surgery, radiotherapy, and treatment with somatostatin analogs or a GH receptor antagonist. The long-lasting somatostatin agonist octreotide (Fig. 14.1) has been shown to control GH secretion and thus the tumor in a large number of acromegaly patients.

14.2.2.4 Gonadotroph Adenoma

Tumors of gonadotropic cells are rare, and were described first in 1974. Gonadotropic cells secrete LH and FSH controlled by periodic secretions of GnRH. The two subunits of the glycoprotein hormones (α chain and β chain) are normally found in equal numbers in serum. Patients with gonadotropinoma, however, have an increased proportion of the α chain. In male patients, the tumor presents with enlarged testes and an extension of the seminiferous tubules. In women, owing to persistent stimulation of the ovaries, the menstrual cycle is disturbed.

Surgical removal is the therapy of choice since expansion of this tumor might cause visual impairment.

14.2.2.5 Thyrotroph Adenoma

For this very rare tumor, two variants can be distinguished: primary adenoma and the so-called feedback adenoma. The former causes hyperthyroidism—excessive release of thyroid hormone (see below); the latter is caused by strongly reduced or missing levels of thyroid hormones. Feedback adenomas occur since suppression of thyrotropin-releasing hormone (TRH) or thyroid-stimulating hormone (TSH) release is not inhibited by circulating thyroxine. Thyrotropic cells thus are not blocked, synthesize TSH, and proliferate. Feedback adenomas react to thyroxine by stopping TSH synthesis, and thus are easy to treat. Primary adenomas, however, have to be treated by surgery, chemotherapy, or radiotherapy.

14.2.3 Disturbances of Water Homeostasis

Water homeostasis is controlled by a complex network of hormones: *vasopressin*, whose release is controlled by osmoreceptors via the osmolarity of blood, initiates by enhancing water renal resorption renin release, which in turn stimulates aldosterone synthesis and release. In the kidney, expression and localization of aquaporins and sodium transporters are tightly regulated. Central neuronal networks measuring and setting dilatation or constriction of arteries, and pressure sensors in carotids or in the aortic arch and atrial myocytes estimating the vessel volume or by release of atrial natriuretic peptide as a vasopressin antagonist contribute to this regulation. Centrally triggered is the thirst which makes us drink and which is triggered by changes of osmolarity.

Disturbances of water balance cause polyuria, which is identified by enhanced and salt-deprived urine excretion (*insipidus* in "diabetes insipidus" means tasteless, sallow urine).

The cause of central diabetes insipidus is insufficient or totally lacking vasopressin biosynthesis in the hypothalamic neurons. Genetic modifications might be inherited or acquired and might be linked to other diseases: diabetes mellitus (DM), visual impairment, or deafness (Wolfram syndrome). Diabetes insipidus might also occur as a consequence of other illnesses such as trauma, tumors, or infections. In nephrogenic diabetes insipidus, vasopressin binds in the kidney to the vasopressin receptor, which in turn stimulates adenylate cyclase and thus increases intracellular levels of cyclic adenosine monophosphate (cAMP). This increase of cAMP concentration is lacking in some patients; in other patients, cAMP levels increase, but the following events are inhibited. Apart from genetic defects, metabolic anomalies (increased or diminished calcium concentration), drug use (e.g., lithium salts), sickle cell anemia, and chronic nephropathies have been reported as causal.

Primary polydipsia is characterized by insatiable thirst—patients drink much more liquid than is physiologically required. As causal for this disease, psychotic conditions (e.g., schizophrenia) have been observed.

Apart from diabetes insipidus, there are patients where the serum sodium concentration is abnormally enhanced (above 140 mmol/l). Such hypernatremia might occur after taking an emetic or certain baby food, or after sodium bicarbonate infusion following cardiac arrest. Insufficient water uptake has been observed in cases of pituitary tumors, aneurysms, shock or hydrocephalus, or when movement is limited because of apoplexy or after vomiting, diarrhea, or severe burns, all combined with enhanced liquid loss.

14.3 Thyroid Diseases

14.3.1 Lack of Thyroxine Synthesis: Hypothyroidism

This disease manifests itself by slowing down all metabolism without any apparent involvement of the thyroid gland. A rapid diagnosis, however, is of utmost importance before other failures occur. This is particularly relevant in the diagnosis of inborn hypothyroidism, where genetic deficits block thyroxine synthesis. When the malfunction is diagnosed early, if thyroxine is administered, the imminent consequences such as cretinisms, growth retardation, and mental retardation can be blocked.

In adults, hypothyroidism manifests itself as dry skin, feeling cold more frequently, adding weight, or reduced activity. In the attempt to compensate for the thyroid hormone deficit, a goiter develops: Since thyroxine (and deiodothyronine) does not inhibit TRH and TSH release from the hypothalamus and the pituitary, respectively, the thyroid gland is persistently stimulated. Such stimulation leads to cellular proliferation. A similar phenomenon occurs in the case of iodine deficiency: all enzymes are functional, but cannot act since the iodine uptake does not suffice for the necessary thyroxine amounts.

Today, iodine deficits can be estimated from urine. By use of iodinated salt, iodine deficiency can be avoided.

Apart from iodine deficits, hypothyroidism might be caused by enzyme defects—of thyroid peroxidase or deiodinase—by mutations of TSH or the TSH receptor, and also by inherited errors of the nuclear thyroid hormone receptor.

Finally, the disease might be caused by an autoimmune reaction where antibodies against thyroid peroxidase block thyroxine synthesis and cause tissue destruction.

14.3.2 Thyroxine Excess: Thyrotoxicosis

If the serum thyroxine content is much enhanced compared with that of healthy individuals, the disease is called thyrotoxicosis (Basedow disease, Graves disease, Flajani disease). The cause is an enhanced stimulation of thyroid cells via the thyrotropin receptor. This stimulation may be due to pathologically increased levels of TSH, which in turn can be caused by a thyrotroph adenoma. Thyroiditis may also transiently cause thyrotoxicosis. Adenoma of thyroid cells is an additional potential cause, as are antibodies against the TSH receptor which trigger cells to enhance thyroxine synthesis and cellular proliferation.

14.4 Dysfunctions of the Endocrine Pancreas

14.4.1 Pancreatic Tumors

All four islet cell types which release insulin, glucagon, pancreatic polypeptide (PNP), or somatostatin might undergo neoplasia. PNP-secreting tumors are the slowest growing tumors, exhibiting latency periods of 25 years or more, almost without clinical symptoms (Wynick and Bloom 1992, Chap. 38 and footnote on p. 208). In addition, most tumors secreting vasoactive intestinal peptide (see Sect. 4.10.4) stem from the pancreas. The tumors of any of these pancreatic cells are slow growing; however, when they are discovered, most often liver metastases have already occurred.

The therapy tries to remove the primary tumor and the metastases. With liver metastases, one tries in many cases to stop tumor blood supply by blocking the tumor-serving arteries.

14.4.2 Diabetes Mellitus

Diabetes mellitus (DM) is found as three different types: type 1, which manifests itself most often in children and young adults; type 2, which occurs during aging; and gestational diabetes, which occurs as a complication of pregnancy. There are rare cases of inherited DM with genetic defects of the insulin secretion mechanisms, cystic-fibrosis-related diabetes, and steroid diabetes due to elevated levels of corticoids.

DM type 1 is an autoimmune disease against insulin-secreting β cells causing an insulin deficiency. The disposition for DM type 1 is inherited together with certain MHC alleles: HLA-DR3, HLA-DR4, or HLA-DQ(nonAsp57). Additionally, viral infections are thought to allow the disease to manifest itself. Coxsackie virus, mumps virus, rubella virus, and Epstein–Barr virus have been found to be involved.

In cases of DM type 1, the blood glucose level increases since the insulin-stimulated storage of glucose as glycogen fails, since protein biosynthesis, which is initiated by insulin, does not occur and proteins are degraded instead, and since fat is metabolized and not—triggered by insulin—produced. These three reactions instead of using glucose produce glucose and thus pathologically increase the blood glucose level.

DM type 2 is characterized not by a similar insulin deficiency (in contrast, sufficient insulin is produced), but by a failure to utilize insulin called insulin resistance and therefore a failure to react to insulin by reducing blood glucose levels. Mainly in adipose patients, this insulin resistance occurs where at first the maintenance of normal blood glucose levels requires a certain degree of insulin hypersecretion, leading to hyperglycemia and thereafter to hyperinsulinemia. Later this disease is characterized by a hyperglycemia with concomitant hyperinsulinemia, where the levels of insulin receptors are reduced on otherwise reactive cells. The demand for persistent insulin production finally exhausts the β cells. With an inherited disposition, DM type 2 finally manifests itself. A diet low in carbohydrates and weight reduction may delay DM type 2 or prevent it. The mechanisms of DM type 2 are currently being intensively investigated.

As a consequence of other metabolic dysfunctions, DM may occur, for example, in patients with Cushing syndrome or in patients who are receiving corticoid therapy. In the latter case, the blood glucose level is normal, but the kidney does not retain glucose as much as under normal conditions, and thus the urine contains glucose.

DM type 1 requires continuous blood glucose control and insulin medication. DM type 2 needs the correct diet, weight reduction, sporting activity, and medication such as insulin and drugs such as biguanide, α-glucosidase inhibitors, sulfonyl urea, and repaglinide.

14.5 Adrenal Dysfunctions

The adrenal gland is most important for steroidogenesis. In the adrenal cortex the corticoids cortisol and aldosterone as well as several androgens are synthesized. In addition, in the adrenal medulla the catecholamines adrenaline and noradrenaline are released. For the biosynthesis of these hormones, a set of enzymes are required as described in Chaps. 6 and 7. Furthermore, the adrenal gland is controlled by different external signals. Irregularities of these control variables and lack of individual enzymes are at the origin of major pathological situations and may be fatal. In addition, adrenal cells might degenerate to tumors, with the consequence of endocrine disturbances.

14.5.1 Congenital Adrenal Hyperplasia

Steroidogenesis in the adrenal cortex is controlled by hypophyseal ACTH. Enlargement of the adrenal cortex occurs when a hormone such as cortisol is missing which would usually inhibit ACTH release by negative feedback. A persistent uncontrolled increase in ACTH concentration—as, for example, in the case of 21-hydroxylase (CYP21) deficiency—initiates division of adrenal cells, which are still unable to release cortisol. This defect manifesting itself as hyperplasia might be inborn (congenital) or might be acquired (late onset).

A defect of any of the enzymes listed in Fig. 14.2 might cause major developmental deficiencies and life-threatening metabolic diseases. Any of these enzymes might selectively be lacking. Since the placenta and the maternal organism can substitute for a deficiency of fetal hormones, such malfunctions might not be lethal in utero.

Several of these defects are recognized postpartum as deformations of sex organs. Complete lack of 3β-hydroxysteroid dehydrogenase, CYP11A1, or CYP17, or androgen receptor inactivity presents as pseudohermaphroditism with a male genome but female genitalia (see Sect. 10.8.1). Since defects of 3β-hydroxysteroid

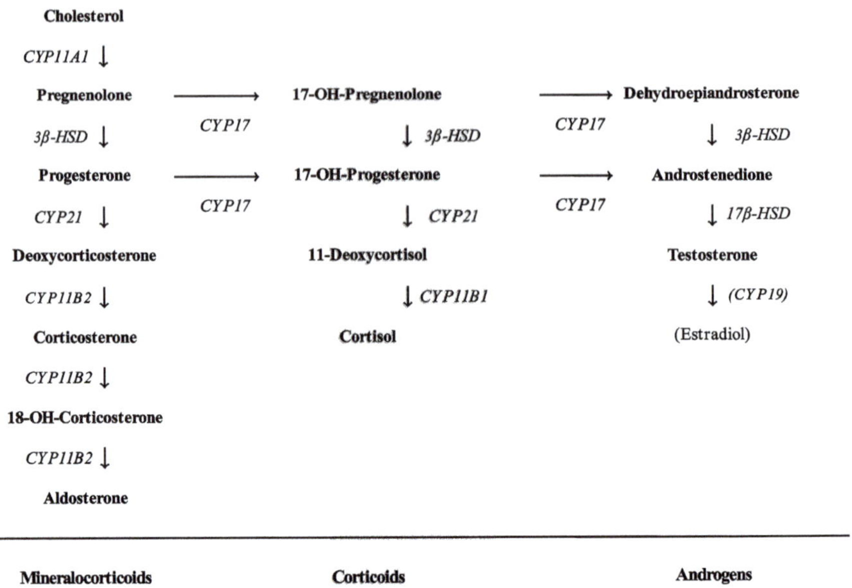

Fig. 14.2 Steroid biosynthesis in the adrenal gland. The enzymes (in *italics*) are described in Sect. 6.5. *HSD* hydroxysteroid dehydrogenase

dehydrogenase and CYP11A1 affect mineralocorticoid biosynthesis, a salt loss crisis can occur, which without treatment is fatal.

For diagnosis and control of therapy, it is necessary to estimate the concentrations of cortisol, ACTH, testosterone, 17α-progesterone, and aldosterone. Owing to the lack of certain enzymes, corticoids and sex hormones are present in defect-specific patterns. Owing to lack of, for example, CYP17, aldosterone levels increase since aldosterone synthesis is augmented. When cortisol cannot be made owing to a CYP11B1 defect, the concentration of 11-deoxycortisol is increased . These defects can be treated only with lifelong substitution by glucocorticoids and mineralocorticoids.

14.5.2 Hypercorticism

This severe disease was first described by Harvey W. Cushing and is named after him: Cushing syndrome. Elevated cortisol levels occur, on the one hand, as described in Sect. 14.2.2 by increased ACTH release—for example, owing to hyperplasia of CRH- or ACTH-secreting cells. On the other hand, such processes are triggered by primary cortisol-releasing adrenal tumors. Causal, in addition, is ectopic ACTH release from tumors not related to the hypothalamic–pituitary–adrenal axis—for example, small cell lung carcinoma, pancreatic carcinoma, thymoma, ovarian carcinoma, and prostate carcinoma. Ectopic ACTH production and the thereby initiated hypercorticism are examples of so-called paraneoplasia.

Particular features of Cushing syndrome are rapid gain of weight of the face and trunk, proximal muscle weakness, disturbed glucose tolerance, virilization (hirsutism), hypertension, and often mental problems.

The favored therapy is first surgical removal of the adenoma, and if this fails, bilateral adrenalectomy and subsequent persistent hormone substitution.

14.5.3 Catecholamine-Releasing Tumors

Pheochromocytoma is a rare catecholamine-secreting tumor, which grows separately and slowly, but which may, however, reach a mass of more than 1 kg. Chromaffin granules containing adrenaline and noradrenaline together with chromogranin are characteristic. Malignancy can be identified by determining karyotype aberrations: most tumors with abnormal—that is, aneuploid—karyotype are malignant and tend to metastasize.

Like pheochromocytoma, there are extra-adrenal chromaffin tumors releasing catecholamines called paraganglioma. Furthermore, there are ganglioma of neural origin which also release catecholamines.

The diagnosis of these diseases has to discriminate them from other causes of hypertension. In addition to catecholamine-releasing tumors there are other neoplasias causing hypertension: renin-secreting, ACTH-secreting, vasopressin-secreting, and parathormone-secreting tumors. Estimation of metadrenaline (metanephrine), an adrenaline metabolite, in urine is a fairly sensitive measure of pheochromocytomas. Moreover, plasma noradrenaline/adrenaline levels are estimated in the 24-h urine. With blocking of the thyroid gland (to stop uptake of radioactive iodine), computer tomography and radionuclide scanning are performed after uptake of *meta*-[^{123}I]iodobenzylguanidine.

The surgical removal of a pheochromocytoma often leads to complete remission of the disorder. Malignant degenerations have been treated in more than 100 patients by *meta*-[^{123}I]iodobenzylguanidine, with remission rates of 50 % and a similar reduction of symptoms.

14.5.4 Autoimmune Adrenalitis (Addison Disease)

Autoimmune adrenal hyperplasia (Addison disease) is characterized by antibodies against the steroidogenic cells of the adrenal cortex. As one target, the ACTH receptor has been identified. Antibodies against other intracellular targets—for example, CYP21—have also been observed. No particular cause of this autoimmune reaction has been found. In mice, Husebye et al. (2006) have shown that sections of CYP21 are immunogenic and stimulate T-lymphocyte responses. Antibodies against CYP21 have been found in patients, but anti-CYP21 antibodies might accompany

Addison disease, and whether anti-CYP21 T lymphocytes are causal remains to be shown. A genetic disposition has also been found.

Substitution using orally administered glucocorticoid and mineralocorticoid drugs is required.

14.5.5 Aldosterone Deficiencies

Aldosterone regulates sodium and potassium levels in blood. Aldosterone biosynthesis is controlled by angiotensin II, by elevated potassium levels, and by ACTH. The enzymes involved in the adrenal cortex are CYP21 and CYP11B2. Any defect or failure of these enzymes causes aldosterone deficiency and subsequently failure of sodium resorption in the kidney, a type of diabetes insipidus. Deficiencies of renin, angiotenin-converting enzyme, and angiotensin and most forms of congenital adrenal hyperplasia (see Sect. 14.5.1) result in pathologically reduced blood aldosterone levels.

Enhanced aldosterone concentrations are symptoms of primary aldosterone-releasing tumors. Some diuretics result in strongly reduced potassium levels. In hypertension patients undergoing diuretic therapy and with very low plasma potassium levels, one patient in 20 had a unilateral primary aldosterone-secreting tumor or bilateral adrenal hyperplasias.

In addition to surgical excision of the tumor, the aldosterone-secreting adenoma might be treated with spironolactone. Aldosterone can be substituted.

14.6 Multiple Endocrine Neoplasia

Multiple endocrine neoplasia (MEN) is characterized by tumors occurring simultaneously in different endocrine organs (Table 14.1). The two types of the disease are caused by mutations of two genes. In MEN type 1, *MEN1*, a tumor suppressor

Table 14.1 Types of multiple endocrine neoplasia (*MEN*)

MEN subtype	Tumor	Frequency (%)
1	Parathyroid gland tumors[a]	95
	Gastrinoma (pancreas)	40
	Insulinoma (pancreas)	
	Adenohypophyseal tumors[b]	30
2	Medullary thyroid carcinoma[c]	90
	Pheochromocytoma[d]	50
	Parathyroid adenoma	20

[a]Parathyroid-hormone-secreting C-cell tumor
[b]Prolactinoma › somatotropinoma › corticotropinoma
[c]Tumor of calcitonin-secreting C cells, situated on top of thyroid follicular cells
[d]Tumor of adrenaline-secreting chromaffin cells

gene expressed in many cells involved in cell division control and DNA repair, is mutated. The *MEN1* mutation is recessively inherited or individually acquired. MEN type 2, in contrast, is caused by persistent activation of the *RET* proto-oncogene due to mutation. Such a variant is dominantly inherited. The *RET* gene product is membrane tyrosine kinase, which on dimerization triggered by other proteins (such as persephin and neurturin; Hansford and Mulligan 2000) stimulates cellular division.

In both types, the particular role of endocrine cells has not yet been elucidated.

14.7 Fertility Dysfunctions and Malfunctions of Reproductive Organs

The biological component of human reproduction depends on successful maternal and paternal meiosis, correct fusion of gametes, regular assembly of sex organs, timely onset of puberty, and last but not least functional steroidogenesis and its correct hormonal regulation by hypothalamic centers, pituitary glands, and adrenal glands.

If meiosis, which generates haploid egg and sperm cells, does not proceed correctly, there are numerical malformations such as Klinefelter syndrome (44 autosomes and two X chromosomes plus one Y chromosome) and Turner syndrome (44 autosomes and only one X chromosome). Meiotic errors might occur in either the maternal or the paternal genome. One of the most cited parameters for these aberrations is parental age: The first meiotic division including crossing over occurs in the female fetus in utero. The follicular ovarian egg cells are arrested in development. The distribution of chromosomes to the egg cells and to polar bodies in the second meiotic division, however, might be affected by environmental and age-dependent factors and might thus give rise to incorrect chromosomal arrangements during cell division in the egg cell. Sperms, in contrast, are found only in adolescent and adult males, which makes them susceptible to age and environmental malformations.

For the egg and sperm to fuse, sperms have to be sufficiently motile. The mechanism for penetration of a sperm into the oocyte has to be intact.

The anlage of sex organs does not depend on hormones; however, sex organ development and the formation of secondary sex organs is hormone controlled (see later). Additional gene products required, for example, for testis formation are of nonendocrine origin: the cystic fibrosis gene (*CFTR*) encodes a sodium channel which when defective causes cystic fibrosis. The same defect in males causes severe gonadal dysfunctions—the seminiferous tubules are not properly formed. Such missing or insufficient development of sex organs is termed hypogonadism.

Via pulsatile release of GnRH, in adults the hypothalamus controls release of LH and FSH. These in turn stimulate maturation of gametes in males and females. LH and FSH themselves are feedback controlled during the menstrual cycle by the ovaries. The maturing follicle controls by estradiol and inhibin release pituitary LH and FSH release and via estradiol the uterine readiness for nidation. The corpus luteum arising from the follicular cells after ovulation synthesizes progesterone, which is indispensable for maintenance of pregnancy since progesterone maintains blood supply to the uterus with the nidated egg. For persistent progesterone synthesis by the corpus luteum, the embryo itself synthesizes choriogonadotropin (CG) in the trophoblastic cells and later in the syncytiotrophoblastic layer of the placenta.

A regular hormone synthesis is thus indispensable for reproductive success. Dysfunctions in establishing GnRH neurons (in Kallmann syndrome and a defective *KAL* gene) and defects in the assembly of the pituitary (and the adrenal gland) block sexual maturation.

Mutations of the gonadotropins LH, FSH, and CG cause infertility. Mutated receptors for these hormones and a defective GnRH receptor (GnRHR) are also causes of sterility.

In the following section, we summarize those genes and their products known to influence reproduction. Figure 14.3 provides an overview.

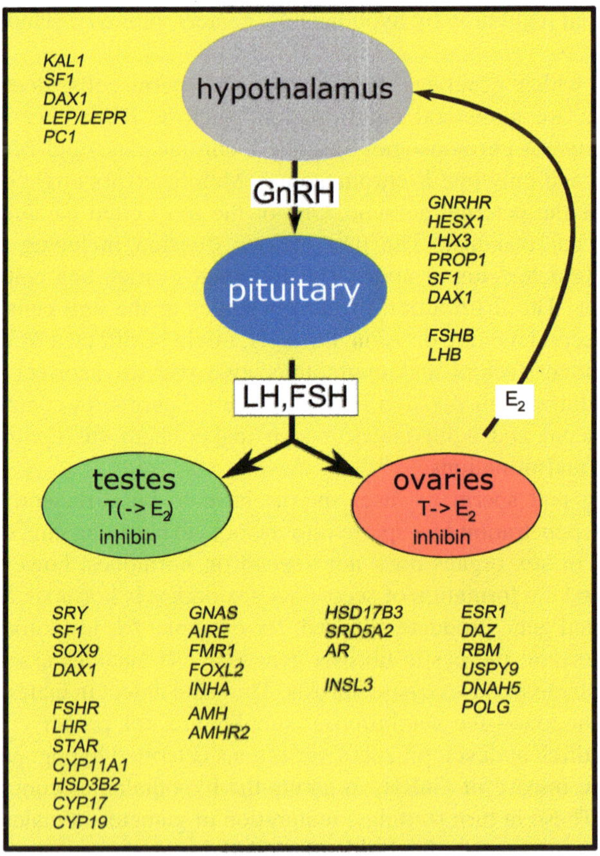

Fig. 14.3 Genetic defects causing infertility (see the comments in the text). E_2 estradiol, *ER* estrogen receptor, *FSH* follicle-stimulating hormone, *GnRH* gonadotropin-releasing hormone, *GnRHR* gonadotropin-releasing hormone receptor, *LH* luteinizing hormone, *T* testosterone

14.7.1 Genetic Defects Affecting the Formation of Sex Organs

Detects in the following genes affect the formation of sex organs (the information provided here is from Achermann et al. 2002; Shah et al. 2003):

- *SRY* (chromosome Y).

Description: Sex-determining gene; after activation of SRY by transcription factors, steroidogenic factor 1 (SF1) is transcribed; afterward, all the following steps such as Leydig and Sertoli cell generation, synthesis of antimüllerian hormone (AMH), and thus formation of the male phenotype occur.
Pathology: Defects lead (in spite of a male genotype) to a female phenotype (pseudohermaphroditism; male sex reversal).

- *SF1* (chromosome 9).

Segregation: Autosomal dominant/autosomal recessive.
Description: Steroidogenesis and AMH are stimulated.
Pathology: Primary adrenocortical deficiency.

- *NROB1* (DAX1, chromosome X).

Segregation: X-chromosomally inherited.
Description: Nuclear receptor antagonizing SRY and SF1.
Pathology: Primary adrenocortical deficiency, hypogonadism, pseudohermaphroditism, XY sex reversal.

- *SOX9* (chromosome 17).

Segregation: Autosomal dominant.
Description: Transcription factor, activated by SRY.
Pathology: *SOX9* mutations block development of the male sexual organ, XY sex reversal.

- *LEP* (leptin, chromosome 7).

Segregation: Autosomal recessive.
Description: See Sect. 4.8.1.
Pathology: Defect mutations induce obsessive gluttony and adiposity with precocious bone aging and hypogonadal hypogonadism manifesting itself by lack of puberty.

- *LEPR* (leptin receptor, chromosome 1).

Segregation: Autosomal recessive.
Description: See Sect. 8.5; the long-active signal transducing OB-Rb variant is expressed particularly in the hypothalamus, mediating leptin actions, and also on GnRH neurons.
Pathology: Defect mutations lead to hypogonadism; see *LEP* above.

- *AMH* (chromosome 19).

Segregation: Autosomal recessive.
Description: AMH (related to inhibin and activin) is a product of fetal Sertoli cells and induces the müllerian duct to degenerate and block female sex organ development. From animal experiments, it is known that AMH blocks premature follicle maturation.
Pathology: See *AMHR2*.

- *AMHR2* (chromosome 12).

Segregation: Autosomal recessive.
Description: Receptor for AMH related to transforming growth factor β receptor.
Pathology: Defects in AMH and AMH receptor result in spite of a male genotype owing to lack of müllerian duct repression in partially female sex organs. The afunctional uterus and vagina are prone to neoplastic degeneration.

14.7.2 Gene Defects Influencing the Hypothalamus and Pituitary

Defects in the following genes influence the hypothalamus and pituitary:

- *KAL1* (chromosome X).

Segregation: X-chromosomally inherited.
Description: KAL1/anosmin 1 controls the migration of GnRH neurons during embryogenesis (see Sect. 14.1).
Pathology: Hypogonadism and anosmia.

- *HESX1* (chromosome 3).

Segregation: Autosomal recessive.
Description: See Sect. 14.2.1.
Pathology: Septo-optic hypoplasia (de Morsier syndrome): optic nerve defects and lack of septum pellucidum, CPHD (see Sect. 14.2.1).
And alternatively,
Segregation: Autosomal dominant.
Description: See Sect. 14.2.1.
Pathology: Isolated GH deficiency.

- *LHX3* (chromosome 9).

Segregation: Autosomal recessive.
Description: See Sect. 14.2.1
Pathology: CPHD (does not apply to corticotropic cells).

- *PROP1* (chromosome 5).

Segregation: Autosomal recessive.
Description: See Sect. 14.2.1.

Pathology: CPHD (does not apply to corticotropic cells).

- *SF1*

  See Sect. 14.7.1.

14.7.3 Gene Defects of GnRHR, Gonadotropin Synthesis, and Recognition

The following genes are of relevance here:

- *GNRHR* (chromosome 4).

Segregation: Autosomal recessive.
Description: GnRHR is a heptahelical G-protein-coupled receptor (see Sect. 8.2) with a very short intracellular C-terminus. It is expressed in the pituitary and the placenta. In the pituitary, it mediates LH and FSH release, and in the placenta it triggers CG release.
Pathology: Idiopathic hypogonadism (not inherited) in about 20 % of patients has been related to mutations of GnRHR. Often there are two defective alleles, both with reduced GnRH binding or G-protein activation. These mutations might lead to mild types of GnRH resistance (administered GnRH is ineffective), but result in severe dysfunctions such as micropenis, hidden or maldescended testes, lack of puberty, and missing GnRH pulses.

- *LHB*/*LHR* (chromosome 19/2).

Segregation: Autosomal recessive/autosomal recessive.
Description: See Sect. 4.4/heptahelical G-protein-coupled membrane receptor, see Sect. 8.2.1.
Pathology: One patient with LH deficiency where the receptor binding was inhibited presented with delayed puberty, low testosterone level, and blocked spermatogenesis. Male sex organ development in utero was due to placental CG binding to the LH receptor as well. Prolonged CG administration led to testis enlargement, virilization (pubic hair), and increase of sperm numbers, but reproductive success could not be achieved.
In the case of *LHR* mutations, in male fetuses sexual development is impaired owing to lack of or reduction of the levels of testosterone and dihydrotestosterone (DHT): in order of decreasing severity, there is lack of any virilization, hypospadias, micropenis, and lack of puberty. In spite of normal puberty, in women with homozygous *LHR* mutations, menstrual cycle abnormalities up to amenorrhoea have been identified. *LHB* mutations with different mutated alleles result in female infertility.

- *FSHB*/*FSHR* (chromosome 11/2).

Segregation: Autosomal recessive/autosomal recessive.

Description: See Sect. 4.4/heptahelical G-protein-coupled membrane receptor.

Pathology: Mice without FSH or FSH recepror are fertile in spite of reduced testicular volume and partially blocked spermatogenesis. Human males with *FSHR* mutations have various defects of spermatogenesis. Spermatogenesis is blocked with FSH defect mutations. In women, analogous defects cause delayed puberty, amastia, and missing menarche. Mutations may be homozygous or heterozygous with different defects on both alleles. The cysteine knot is particularly involved, thus reducing the stability of the protein fold and blocking the formation of dimers with the α chain. In individual cases, substituting FSH helps to achieve normal development, ovulation, and childbirth.

14.7.4 Gene Defects Affecting Steroidogenesis

Errors in androgen and estrogen synthesis also inhibit reproduction. Since these hormones are required for sex organ formation, inborn steroidodysgenesis is often apparent at birth. There are, however, cases of pseudohermaphroditism—for example, with a deficiency of α-reductase, which converts testosterone into DHT. DHT is required for penis development; without DHT, male sex organs are present, but the penis is lacking and a female phenotype occurs.

Defects in the following genes affect steroidogenesis:

- *STAR* (chromosome 8).

Segregation: Autosomal recessive.
Description: See Sect. 6.4.

- *CYP11A1* (chromosome 15).

Segregation: Autosomal dominant.
Description: See Sect. 6.5.1.

- *HSD3B2* (chromosome 1).

Segregation: Autosomal recessive.
Description: See Sect. 6.5.2.

- *CYP17* (chromosome 10).

Segregation: Autosomal recessive.
Description: See Sect. 6.5.3.

- *HSD17B3* (chromosome 9).

Segregation: Autosomal recessive.
Description: See Sect. 6.5.4.
Pathology: Defects in these five genes lead to primary adrenal hypoplasia, as well as absence of puberty. During puberty, girls develop male characteristics, and boys lack spermatogenesis.

- *CYP19* (chromosome 15).

Segregation: Autosomal recessive.
Description: See Sect. 6.5.9.
Pathology: CYP19 (aromatase) defects have been observed only recently: A pregnant woman developed virilization (hirsutism, lowering of voice). The child was a female hermaphrodite. Such a maternal virilization could occur since the placental CYP19 protecting the mother from the fetal testosterone was afunctional, which demonstrates the placental CYP19 is of fetal origin. Since estrogens cannot be made by the child, androgenic intermediates and thus testosterone are active. If testosterone is lacking owing to SRY deficits and when the timed sequence of first SRY expression, second Sertoli and Leydig cell formation, third testosterone and AMH synthesis, and fourth degeneration of müllerian ducts fails, both wolffian and müllerian ducts remain and develop, giving rise to both male and female sex organs, and thus to the hermaphrodite phenotype. Whether other estrogen-dependent functions such as sexual central imprinting, growth, and puberty were affected was not reported.
Mutations leading to elevated CYP19 activity give rise in male patients to mastopathy before puberty and hypogonadism due to strongly elevated estrogen levels. Estradiol levels can be normalized by aromatase inhibitors. Since the publication cited a father and son, fertility was not compromised (Shozu et al. 2003).

- *SRD5A2* (chromosome 15).

Segregation: Autosomal recessive.
Description: See Sect. 6.5.5.
Pathology: 5α-Reductase defects cause, owing to lack of DHT, insufficient virilization during puberty; in male fetuses, sex characteristics are incompletely developed since testosterone cannot substitute for DHT.

- *AR* (chromosome X).

Segregation: X-chromosomally inherited.
Description: The androgen receptor is a nuclear receptor (see Sect. 8.1).
Pathology: Androgen resistance is the result of defect mutations of the androgen receptor blocking any testosterone or DHT effects. Owing to the location on the X chromosome, a complete failure to react to testosterone/DHT occurs, and thus sex reversal occurs. Many different intermediate forms of incomplete male sex organ development have been observed (Quigley et al. 1995).

- *ESR1* (chromosome 6).

Segregation: Autosomal recessive.
Description: The estrogen receptor α is the nuclear receptor for estradiol and related estrogens in reproduction.
Pathology: Until 2002 a single *ESR1* mutation had been reported in a man (Smith et al. 1994). Only in 2013 a mutation was reported in a women (Quaynor et al.

2013). Since estradiol has a major role in testicular development and function (estradiol being made by Leydig cells), *ESR1* mutations influence testicular development and impair or influence fertility (O'Donnell et al. 2001). Given the fact that there are different genes for estrogen receptor α (*ESR1*) and estrogen receptor β (*ESR2*), obviously these cannot substitute for each other.

Boosting Performance: Legally or Not 15

Contents

15.1 Anabolic Steroids 378
15.2 β_2-Agonists 379
15.3 Peptide Hormones 379
 15.3.1 Gonadotropins 379
 15.3.2 Corticotropin 380
 15.3.3 GH and Insulin-Like Growth Factor 380
 15.3.4 Hypothalamic Releasing Hormones 380
15.4 Beta Blockers 381
15.5 Erythropoietin 381

The (ab)use of analeptics in daily life, above all as performance-enhancing drugs in endurance sports, has been made possible by new insights into endocrine regulation of basic metabolic circuits. National and international antidoping rules and regulations list the following compound classes as forbidden: analeptics, anesthetics, anabolics, diuretics, peptide hormones, beta blockers, and substances masking forbidden compounds.

Anabolic agents are almost exclusively preparations containing testosterone or its chemically synthesized derivatives. There are also agonists of noradrenaline/adrenaline (β_2-agonists), pituitary peptide hormones as well as their hypothalamic releasing hormones—among those already discussed are adrenocorticotropic hormone (ACTH), choriogonadotropin (CG), growth hormone (GH), insulin, and luteinizing hormone (LH)—and aromatase inhibitors. Antagonists of noradrenaline/adrenaline, so-called beta blockers, are also listed as forbidden drugs. Also to be mentioned is erythropoietin, *the* drug of the last decade in endurance sports.

B. Kleine, W.G. Rossmanith, *Hormones and the Endocrine System*,
DOI 10.1007/978-3-319-15060-4_15

15.1 Anabolic Steroids

Synthetic testosterone analogs are considered as anabolic steroids (Fig. 15.1). By addition of a chlorine (clostebol) or the introduction of an additional unsaturated bond (methenolone or methandrostenolone) or a new methyl group at position 17 (methandrostenolone) or position 1 (methenolone), substances are generated whose actions as a sex hormone are reduced compared with the action of testosterone but whose metabolic activity is enhanced. These anabolic drugs stimulate muscle formation and reduce fat deposits.

Tetrahydrogestrinone is a functional, but not a structural testosterone analog, and has a C-18 rather than a C-19 backbone, lacking the methyl group which is removed by aromatase (see Fig. 6.19) as in nandrolone or 19-norandrostenedione (Fig. 15.1). It was developed as a doping agent and never has never been used for medical treatment. The drug was found only after a syringe was anonymously handed to a US laboratory working for the US Anti-Doping Agency (Knight 2003).

Applied in females, these drugs induce virilization with typical male hair growth (hirsutism), lowering of the vocal range, menstrual irregularities, and enlargement of the clitoris.

In men, there is mastopathy, reduction of the testes, and lower sperm numbers after taking anabolics.

The myocardial enlargement without simultaneous generation of new blood vessels is life threatening. Moreover, lipid metabolism is impaired, and the HDL

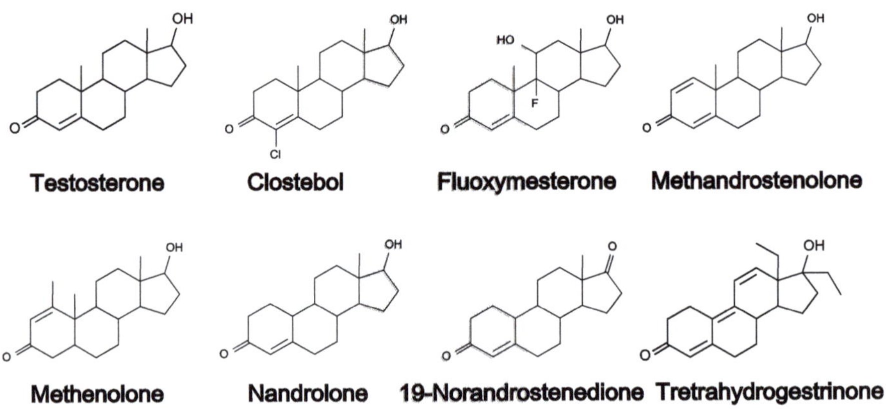

Fig. 15.1 Anabolic steroids derived from testosterone. By addition of a chlorine or fluorine (clostebol, fluoxymesterone), by the introduction of an additional unsaturated bond (methenolone and methandrostenolone), by removal of the C-19 methyl group (nandrolone and 19-norandrostenedione), or by addition of an additional methyl group at position 1 (methenolone) or position 17 (fluoxymesterone and methandrostenolone), testosterone is changed in a way that aromatization to estradiol is completely blocked (see Fig. 6.19). Subsequently, enhanced muscle formation by long-living testosterone or its derivatives occurs (see Fig. 6.10 for the enumeration)

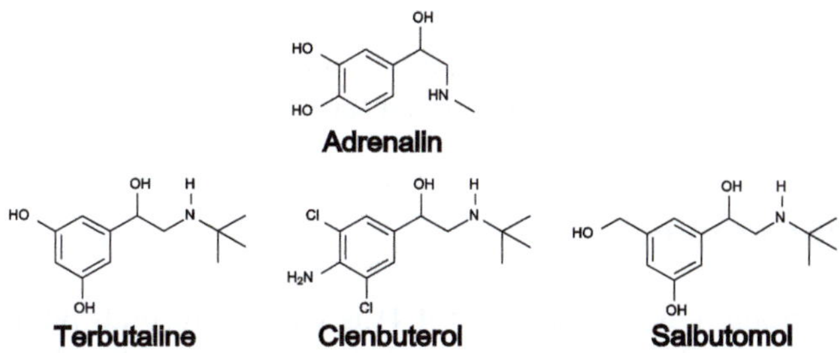

Fig. 15.2 (Nor)adrenaline agonists/β_2-agonists

level is reduced in favor of LDL, which might lead to calcific sclerosis. Liver dysfunctions up to liver carcinomas might be caused by prolonged abuse of anabolics. Mental actions of anabolics have also been observed: euphoria and enhanced aggression. There is the risk of addiction.

15.2 β_2-Agonists

Drugs such as noradrenaline and adrenaline triggering the adrenergic receptors are named agonists. Those exclusively stimulating β-adrenergic type 2 receptors are called β_2-agonists. These substances are listed as anabolics because due their action protein synthesis is stimulated in muscle cells and new muscle mass can be generated. Since clenbuterol or salbutamol (Fig. 15.2) are needed by asthmatics for better respiration, the application of these drugs by inhalation has been allowed when prescribed by a physician.

15.3 Peptide Hormones

The anabolic actions of peptide and glycoprotein hormones and those of their releasing hormones are not immediately evident. Because these occur naturally, drug abuse is difficult to identify, and it is suspected that these substances have squeezed out the use of anabolic steroids.

15.3.1 Gonadotropins

CG is actually the pregnancy hormone (see Sect. 4.4). Owing to its large homology to LH (Fig. 4.23) it binds to the LH receptor on testicular Leydig cells and triggers testosterone synthesis. For this reason, CG is put on the same level as anabolic testosterone analogs and its use is forbidden in men.

If CG is identified in men, this is 100 % indicative (if doping has not occurred) of a germ cell tumor. In women, CG identified outside pregnancy is a tumor marker for choriocarcinoma, which when diagnosed early has a good prognosis.

15.3.2 Corticotropin

Taking corticotropin (ACTH; see Sect. 4.4) is equated to oral, intramuscular, or intravenous administration of corticosteroids since these are stimulated by ACTH. Corticosteroids are forbidden because of their anti-inflammatory properties and especially their pain-threshold-increasing actions.

Tetracosactrin (Synacthen) is an entirely synthetic preparation of the first 24 amino acids of corticotrophin (ACTH1-24) which has also been used as a doping agent to stimulate corticoids. Its detection is possible under optimal conditions (Thomas et al. 2009).

15.3.3 GH and Insulin-Like Growth Factor

GH and its insulin-like growth factor (IGF) mediators are major regulators of anabolism. They make liver glucose available for energy and protein metabolism in muscles, and they stimulate bone growth (see Sect. 11.6) and muscle formation. Therefore, intake of GH and IGF is considered as doping and is forbidden. When anabolic steroids are taken, adverse events occur fast and most often reversibly, but this is not the case when GH or IGF is taken: adverse events occur slowly and are not reversed after intake has stopped. In addition, taking IGF might induce a severe hypoglycemia.

15.3.4 Hypothalamic Releasing Hormones

Use synthetic releasing hormone analogs is put on the same level as taking peptide or glycoprotein hormones. Since we have shown in this book that biosynthesis and release of these hormones are controlled by a multilayered complex endocrine and neuronal control, the drug (ab)use of these releasing hormones appears as a major assault against essential physiological functions. Regulation of GH is influenced by episodic pulsatile and circadian factors. To apply further releasing hormones appears counterproductive. Enhanced GH pulse amplitudes (higher peaks, lower nadirs) and, on average, reduced GH levels in men compared with women are related to greater muscle mass and lower male body fat (Gatford et al. 1998).

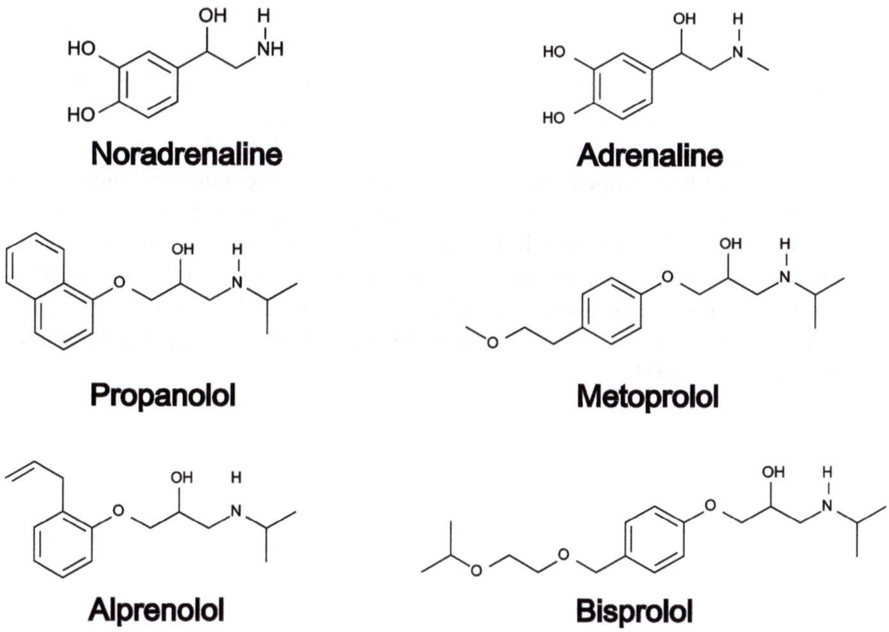

Fig. 15.3 Noradrenaline antagonists: beta blockers

15.4 Beta Blockers

Those drugs called beta blockers occupy β-adrenergic receptors on nerve endings of the sympathetic nervous system and thus inhibit noradrenaline and adrenaline actions. The known beta blockers are synthetic analogs of catecholamines. Examples are shown in Fig. 15.3. Beta blockers improve symptoms such as hypertension, circulatory disorders of the heart, myocardial infarction, and arrhythmia.

They are listed among doping drugs since they help to overcome nervousness and stress in sports where concentration and silence are required.

15.5 Erythropoietin

Erythropoietin has become the "drug of choice" in endurance sports. Released from the kidney, erythropoietin stimulates proliferation and maturation of reticulocytes into erythrocytes in bone marrow; thus, the oxygen capacity is increased.

Erythropoietin is necessary for dialysis patients, where because of renal insufficiencies erythropoietin synthesis is reduced, which in turn reduces erythrocyte formation, and oxygen capacity with insufficient peripheral oxygen supply.

In healthy people erythropoietin increases the red blood cell count. If the blood density is too high and circulation is no longer provided, thrombosis might occur.

In the brain, lungs, and myocardial vessels, such thrombosis is immediately life threatening.

Since aged erythrocytes are normally eliminated in the liver, an athlete with a too high blood density (hematocrit) is banned from competitions.

Since erythropoietin is continuously produced by the kidney, an intake of erythropoietin is almost impossible to detect if the drug is the intact and unmodified human erythropoietin. For this reason there is only a temporary ban if an athlete's hematocrit is too high. Commercially available erythropoietin preparations, however, contain modified erythropoietins. Such a difference can well be identified by, for example, antibodies or chemical analysis, a fact which allowed the detection of doping by the cross-country skier Johann Mühlegg (among others) at the winter Olympic Games in 2002.

Appendix

16

Contents

16.1 List of Species 383
16.2 Glossary 385
16.2.1 Cell Components 385
16.2.2 Cell Structure 386
16.2.3 Intracellular Vesicles and Their Transport 387
16.2.4 Additional Definitions 388
16.2.5 Amino Acids 390
16.3 PyMOL Scripts 392
16.3.1 Glycoprotein Hormone α Chain 392
16.3.2 Growth Hormone/Prolactin Using PyMOL 393
16.3.3 CYP51 394
16.3.4 CYP19 395

16.1 List of Species

Table 16.1 Species

Scientific name	Vernacular name	Superior taxon	Class/phylum
Acheta domesticus	House cricket	Gryllidae	Insecta
Achlya ambisexualis		Peronosporomycetes	Fungi
Acyrthosiphon pisum	Pea aphid	Aphidoidea	Insecta
Aedes aegyptii	Yellow fever mosquito	Culicidae	Insecta
Aeshna cyanea	Blue-eyed darner	Odonata	Insecta
Agelenopsis aperta	Desert grass spider	Agelenidae	Arachnida
Agrotis ipsilon	Dark sword-grass	Noctuidae	Insecta
Anopheles gambiae		Culicidae	Insecta
Apis melifera	Honeybee	Apidae	Insecta

(continued)

B. Kleine, W.G. Rossmanith, *Hormones and the Endocrine System*,
DOI 10.1007/978-3-319-15060-4_16

Table 16.1 (continued)

Scientific name	Vernacular name	Superior taxon	Class/phylum
Aplysia californica	California sea hare	Anaspidea	Mollusca
Aplysia kurodai	Kuroda's sea hare	Anaspidea	Mollusca
Armadillidium vulgare	Common pill bug	Isopoda	Crustacea
Bombyx mori	Silk moth	Lepidoptera	Insecta
Brugia malayi		Filarioidea	Nematoda
Busycon contrarium	Lightning whelk	Neogastropoda	Mollusca
Caenorhabditis elegans		Rhabditida	Nematoda
Callinectes sapidus	Atlantic blue crab	Brachyura	Malacostraca
Cancer magister	Dungeness crab	Brachyura	Malacostraca
Carausius morosus	Indian stick insect	Neoptera	Insecta
Carcinus maenas	European shore crab	Brachyura	Malacostraca
Ciona intestinalis	Vase tunicate	Cionidae	Urochordata
Dipetalogaster maxima	(Blood sucking bug)	Heteroptera	Insecta
Drosophila melanogaster	Fruit fly	Neoptera	Insecta
Eisenia fetida	Redworm	Lumbricidae	Annelida
Euphyllia anchora	Anchor coral	Scleractinia	Cnidaria
Helicoverpa armigera	Cotton bollworm	Noctuidae	Insecta
Helicoverpa assulta	Oriental tobacco budworm	Noctuidae	Insecta
Helicoverpa zea	Cotton earworm	Noctuidae	Insecta
Helix aspersa	Garden snail	Pulmonata	Gastropoda
Helix pomatia	Burgundy snail	Pulmonata	Gastropoda
Homarus americanus	American lobster	Astacidea	Malacostraca
Leucophaea maderae	Madeira cockroach	Blattodea	Insecta
Locusta migratoria	Migratory locust	Cealifera	Insecta
Lumbricus terrestris	Common earthworm	Lumbricidae	Annelida
Lymnea stagnalis	Great pond snail	Pulmonata	Gastropoda
Macrobrachium rosenbergii	Giant river prawn	Caridea	Malacostraca
Macrocallista nimbosa	Sunray Venus clam	Veneroida	Bivalvia
Manduca sexta	Tobacco hornworm	Lepidoptera	Insecta
Mytilus edulis	Blue mussel	Mytiloida	Bivalvia
Neobellieria bullata	Grey flesh fly	Neoptera	Insecta
Oncopeltus fasciatus	Mildweed bug	Hemiptera	Insecta
Orconectes immunis	Calico crayfish	Astacidea	Malacostraca
Orconectes limosus	Spiny-cheek crayfish	Astacidea	Malacostraca
Pagurus bernhardus	Common hermit crab	Pleocyemata	Malacostraca
Pandalus borealis	Great northern prawn	Caridea	Malacostraca
Pandalus jordani	Pacific/ocean shrimp	Caridea	Malacostraca
Pacifastacus leniusculus	Signal crayfish	Astacidea	Malacostraca
Penaeus aztecus	Brown shrimp	Penaeidae	Malacostraca
Penaeus japonicus	Kuruma prawn	Penaeidae	Malacostraca
Penaeus vannamei	Pacific white shrimp	Penaeidae	Malacostraca
Phormia regina	Black blow fly	Neoptera	Insecta

(continued)

Table 16.1 (continued)

Scientific name	Vernacular name	Superior taxon	Class/phylum
Procambarus clarkii	Red swamp crayfish	Astacidea	Malacostraca
Psacothea hilaris	Yellow spotted longicorn beetle	Neoptera	Insecta
Rhodnius prolixus	(Triatomid bug)	Heteroptera	Insecta
Romalea microptera	Eastern lubber grasshopper	Caelifera	Insecta
Schistocerca gregaria	Desert locust	Caelifera	Insecta
Schistocerca nitens	Gray bird grasshopper	Caelifera	Insecta
Strongylocentrotus purpuratus	Purple sea urchin	Echinoida	Echinodermata
Tenebrio molitor	Mealworm	Coeleoptera	Insecta
Tribolium castaneum	Red flour beetle	Coeleoptera	Insecta
Trichoplax adhaerens		–	Placozoa
Uca puligator	Sand fiddler crab	Brachyura	Malacostraca

16.2 Glossary

16.2.1 Cell Components

ATP: Adenosine triphosphate (ATP) is a molecule generated during intracellular oxidation of glucose and which stores energy. By enzymatic transfer of phosphate, the energy is also transferred.

Cyclic AMP: Cyclic adenosine monophosphate (cAMP) is generated from ATP by adenylate cyclase. Adenylate cyclase is, for example, stimulated by the hormone glucagon. Within the cell, cAMP has function a hormone has in the organism: it is a messenger. With hormones seen as primary messengers, cAMP is the prototype of second messengers. Other second messengers are cyclic guanosine monophosphate, diacyl glycerol, inositol trisphosphate, and nitrogen monoxide.

Enzymes: Enzymes are protein molecules performing chemical reactions.

Molecules: A cell consists of amino acids, proteins, glycerol derivatives, fatty acids, lipids, large and small sugar molecules, deoxyribonucleic acid (DNA), ribonucleic acid (RNA), and a multitude of other substances, such as vitamins, salts, trace elements, and other organic compounds. To obtain information about all these substances, please consult a biochemistry textbook.

Nucleic acids: Nucleic acids are built with a chain where a sugar and a phosphate alternate. To each sugar a so-called base is linked, the sequence of the base determining genetic information. Two different sugars are used differing from each other by a single oxygen atom. Genetic information in all living cells from bacteria to plants to animals consists of DNA using deoxyribose with four bases—adenine, cytidine, guanine, and thymine. Within the cell, information from the cellular nucleus is transcribed into RNAs built with a ribose phosphate backbone and where together with adenine, cytidine and guanine, uracil is used instead of thymine. There are functionally different RNAs: messenger RNA (mRNA), ribosomal RNA, and transfer RNA (tRNA).

In viruses, genetic information consists of DNA or RNA.

Proteins: Proteins are described in detail in Chap. 4.

Ribonucleic acid (RNA): See nucleic acids.

Ribosomes: A ribosome consists of two ribosomal RNA strands and a multitude of proteins. It has two different subunits. Ribosomes serve to translate genetic information from the mRNA in a polypeptide sequence. The amino acids are transferred by tRNAs to the ribosomes.

RNA cap: To the 5′ end of mRNA, a guanosine phosphate is linked to form a guanosine 5′-phosphate 5′-RNA cap. In addition, the guanine at the 7-position is methylated. The RNA cap is a signal for export from the cellular nucleus and enhances RNA stability by providing degradation protection.

Transfer RNA: Transfer RNAs (tRNAs) are loaded by enzymes with amino acids which will then be brought to the ribosome by these aminoacyl-tRNAs. For any of the 20 amino acids, there is at least one tRNA; for selenocysteine (used in deiodinase, which converts thyroxine into triiodothyronine in the thyroid gland), another tRNA is necessary.

16.2.2 Cell Structure

Cell membrane: The animal cell is enclosed by a cell membrane separating the interior from the environment and keeping the cell contents together. Cell membranes have distinguishable inner and outer faces. They are built from lipids, proteins, and sugar-modified lipids.

Cell nucleus: In eukaryotic cells, genetic information is enclosed by a membrane forming the cellular nucleus. The membrane is called a nuclear membrane. The cellular nucleus consists of DNA strands, protein such as histones which pack DNA closely, and enzymes replicating or transcribing DNA. By the binding of further proteins to gene segments, the activity of transcribing enzymes for the respective segment can be controlled. Thus, these DNA-binding proteins regulate the activity of genes—that is, a DNA stretch with genetic information for a certain protein.

Cellular interior, the cytosol or cytoplasm: The cytosol is the intracellular space filled with protein, salts, and small and large membrane-enclosed droplets within the cellular membrane. Signals from the cellular membrane to the cellular nucleus have to be mediated in the cytosol. This cytosol is by no means only a thick solution; a dense fiber network called the cytoskeleton links different components into functional units.

Cytoskeleton: The cytoskeleton consists of protein fibers. It fixes the position of the cellular nucleus within the cell, as well as the other intracellular, membrane-enclosed compartments; that is, mitochondria or vesicles such as the Golgi apparatus, the endoplasmic reticulum (ER), and secretory granules (with, e.g., hormones to be released on triggering) are maintained in place by the cytoskeleton. Even cell-to-cell connections are stabilized by the cytoskeleton.

Endoplasmic reticulum: The endoplasmic reticulum (ER) is a special intracellular compartment enclosed by an individual membrane. Its outer face is related to the inner face of the cell membrane, and its inner face is related to the outer side of the cell membrane. The ER is the place where those metabolic steps happen which determine whether a freshly made protein is targeted to the cell membrane or even further for secretion. In addition to the smooth ER, there is the rough ER.

Golgi apparatus: The Golgi apparatus looks like a pile of folded pizza getting larger with greater distance from the cellular nucleus. In the Golgi apparatus, proteins are modified with sugar residues. Such sugars serve as target signals and are used to sort the proteins into the different compartments. These compartments form by invagination of the Golgi apparatus and separation from it and are transported with the help of the cytoskeleton and helper proteins.

Mitochondria: In mitochondria glucose is used for ATP formation. This ATP is used in almost all metabolic steps. For this ATP formation, glucose, oxygen, and a set of enzymes are required, the latter located at the inner mitochondrial membrane.

Nucleolus: With use of electron microscopy, round structures in the cellular nucleus are visible, and represent ribosome-forming sites. The ribosomal RNAs are transcribed from the chromosome and loaded with the different proteins imported into the cellular nucleus.

Rough endoplasmic reticulum: The rough ER is the place of protein formation. The growing proteins in the rough ER are, however, transferred further into the rough ER through a pore. The many ribosomes on the outer (cytosolic) surface of the rough ER have a rough appearance when viewed by electron microscopy.

Secretory granules: These intracellular structures are hormone stores. Almost all peptide hormones, catecholamines, and melatonin are stored in vesicles. These arise by invagination of the Golgi apparatus or the ER. By intracellular transport, these granules are stored close to the cell membrane.

16.2.3 Intracellular Vesicles and Their Transport

Endocytosis and exocytosis: Endocytosis is a mechanism where particles, bacteria, or other solid material outside the cell become enclosed by the cellular membrane to become a membrane-enclosed vesicle within the cell. Most important in the process are clathrin molecules organizing a frame around the vesicle together with the cytoskeleton and which are visible under the electron microscope (coated pits).

The process in the opposite direction, from the cell to the exterior, is called exocytosis. By means of exocytosis, peptides, proteins, and other amino acid derived hormones are released. Most significant is the act of vesicle membrane fusion to the plasma membrane facilitated by soluble *N*-ethylmaleimide sensitive factor attachment receptors (SNAREs) and induced by calcium ions.

Lysosome: A lysosome is one of the Golgi apparatus derived vesicles packed with enzymes to digest proteins and lipids from, for example, bacteria. Such a lysosome fuses to a phagosome, by which process the enzymes can reach the phagosomal content and digest it. Such a process happens in the thyroid gland, where thyroglobulin is stored in the follicular lumen and phagocytosed on demand, and thyroxine is generated by digestion of thyroglobulin within phagolysosomes (see Chap. 7). Bacteria are often phagocytosed and digested in phagolysosomes.

Phagosome: A phagosome is an intracellular vesicle formed by particle endocytosis. Such particles might be a bacterium, a virus, a crystal, or another foreign particle.

Pinocytosis: Liquid droplets can also be taken up by cells and enclosed within a membrane. Such a process is called pinocytosis; the intracellular vesicle is also called a phagosome.

16.2.4 Additional Definitions

14-3-3 proteins: 14-3-3-proteins are ubiquitous in eukaryotic cells. They preferentially bind to phosphorylated serines and thus control protein functions in many cells

Acidophilic/basophilic: Cells or the cellular compartment that are stained by the acidic eosin are called acidophilic. Cells that are stained by a basic dye are named basophilic.

Affinity: Reversible chemical reactions at equilibrium are characterized by reaction constants indicating whether the reactions occur autonomously or, for example, only because of catalysis. Antibody–antigen reactions are equally reversible reactions under the law of mass action, with the equilibrium the more on the side of the antibody–antigen complex the higher the antibody affinity for a given antigen. Affinity is thus the attracting force of the antibody for the antigen.

Avidity: In an immunoreaction where many different, even structurally diverse antibodies participate with individual affinities, the force of the reaction cannot be measured. When compared with other immunoreactions with the same antibody but a different antigen, the reactions can be compared and given a scale. Avidity is the term by which these different reactions are compared.

Crossing over: During meiosis, sister chromosomes associate lengthwise. Thereafter, genetic exchange happens, with entire chromosomal sections exchanged.

C-terminus, N-terminus: Amino acids have an amino group and a carboxy group. For a peptide bond, the carboxy group of one amino acid reacts with the amino group of the next. One amino group and one carboxy group are retained. These are called the N-terminal group (for NH_2, i.e., amino) and the C-terminal group (for COOH, i.e., carboxy).

Genetic code: The bases of trinucleotides form the code of amino acids. The four bases are uracil (U), cytidine (C), adenine (A), and guanine (G). The following table shows the translation code, the *genetic code*, from RNA to amino acid occurring in the ribosomes. The trinucleotide ACA would be translated to threonine (Thr), and GAA would be translated to glutamine (Gln). Some trinucleotides cause strand termination, such as UAA and UAG.

1st position	2nd position				3rd position
	U	C	A	G	
U	Phe	Ser	Tyr	Cys	U
	Phe	Ser	Tyr	Cys	C
	Leu	Ser	Stop	Stop	A
	Leu	Ser	Stop	Trp	G
C	Leu	Pro	His	Arg	U
	Leu	Pro	His	Arg	C
	Leu	Pro	Gln	Arg	A
	Leu	Pro	Gln	Arg	G
A	Ile	Thr	Asn	Ser	U
	Ile	Thr	Asn	Ser	C
	Ile	Thr	Lys	Arg	A
	Met	Thr	Lys	Arg	G
G	Val	Ala	Asp	Gly	U
	Val	Ala	Asp	Gly	C
	Val	Ala	Glu	Gly	A
	Val	Ala	Glu	Gly	G

HLA: HLA molecules are protein products of the major histocompatibility complex gene locus. HLA proteins have pockets for peptide fragments derived by digestion of cellular peptides in the proteasome. Such HLA–peptide complexes on the surface of a cell signal to T lymphocytes that such a cell belongs to the respective individual. Virus-infected cells incorporate viral peptides into the HLA–peptide pockets. Since T lymphocytes have not learned of these virus—HLA complexes when primed for their host, such complexes are recognized as foreign and the cells are attacked and destroyed by T lymphocytes.

Imprinting Imprinting denotes the fact that certain alleles are inactivated because of their paternal or maternal origin. In about 50 human genes, genes on one allele cannot be activated since they are strongly inactivated by DNA methylation and association with inactivator proteins. Since this inactivation has already occurred in the fertilized egg, it must be of paternal or maternal origin. The inactivation is maintained in the progeny after cell division. If imprinting is defective—for example, in inactivator protein or DNA methyltransferase mutants—severe developmental complications arise. *IGF2* is an example of an imprinted gene.

Karyotype: A karyotype is the description of the number and the appearance of chromosomes in the cell analyzed. Diploid karyotype means that with the exception of the sex chromosomes all chromosomes appear as pairs (in normal cells). In a haploid karyotype, chromosomes appear once, and the sex chromosomes are X *or* Y (in the egg or sperm). In chromosomal aberrations, individual chromosomes are lacking or exist in triplicate. Sometimes chromosomes exist in four copies (tetraploid). In plants, there are eightfold and further multiplied karyotypes.

Signal peptide: Membrane and secretory proteins are translated not in the cytosol, but on the surface of the ER and are stored therein. Their synthesis is directed to the ER by the so-called signal peptide. When peptide synthesis starts, the first 20–35 amino acids, the signal peptide, are translated. Signal recognition particles bind to these and target the complex of RNA and ribosome plus nascent chain to the ER membrane. There a pore is formed through which the growing chain is directly transfered into the ER. In the ER the signal peptide is cleaved by the signal peptidase. Protein folding does not happen until the chain has reached the ER.

Synapse: Two neurons form a synapse with their axons at the contact site, where neurotransmitters are secreted, and diffuse through the narrow cleft to postsynaptic receptors.

Tumor: A tumor is a space-demanding proliferation of cells, synonymous with neoplasia. There are benign and malignant tumors. Depending on the kind of tumor and the type of the tissue pushed away, pathological conditions up to death may occur. Malignant tumors are those where secondary tumors (metastases) due to tumor cell dissemination are formed.

16.2.5 Amino Acids

Table 16.2 Amino acids, three-letter and one-letter codes, and structures

Amino acid	3-letter code	1-letter code	Structure
Alanine	Ala	A	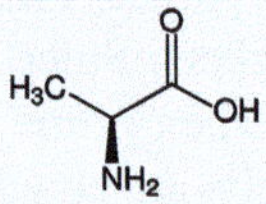
Arginine	Arg	R	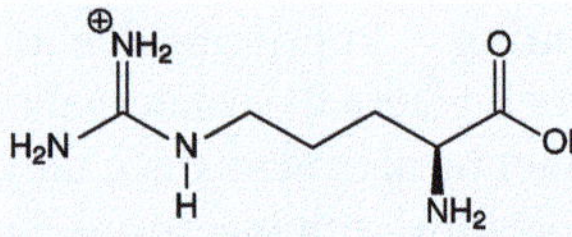
Asparagine	Asn	N	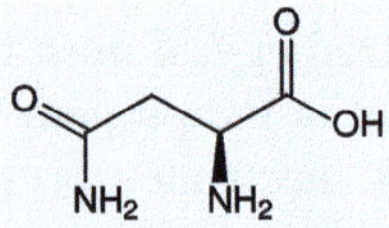
Aspartic acid	Asp	D	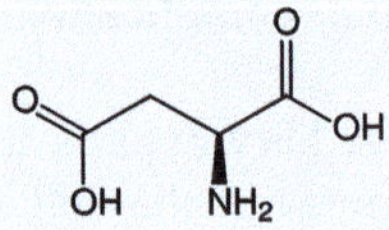

(continued)

Table 16.2 (continued)

Amino acid	3-letter code	1-letter code	Structure
Cysteine	Cys	C	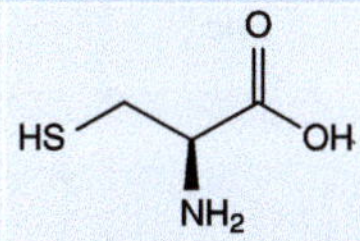
Glutamic acid	Glu	E	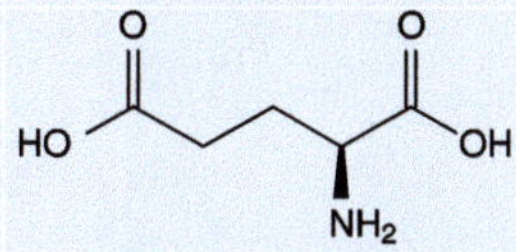
Glutamine	Gln	Q	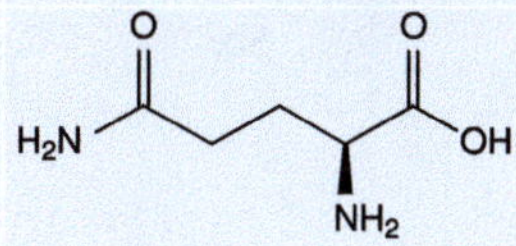
Glycine	Gly	G	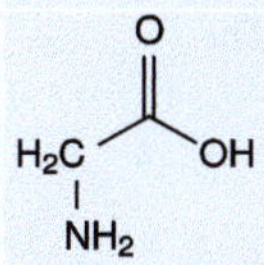
Histidine	His	H	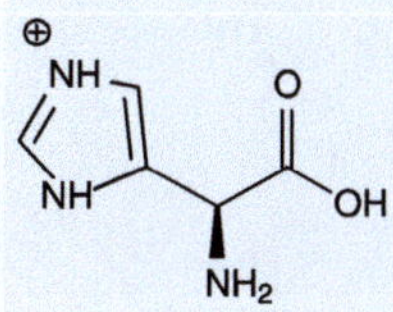
Isoleucine	Ile	I	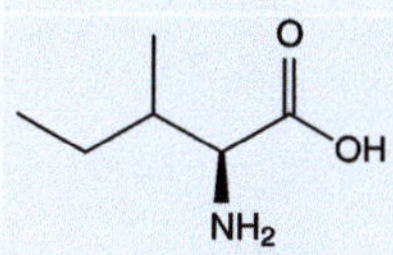
Leucine	Leu	L	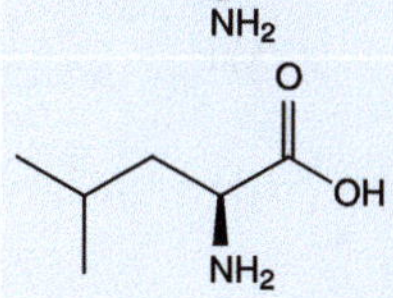
Lysine	Lys	K	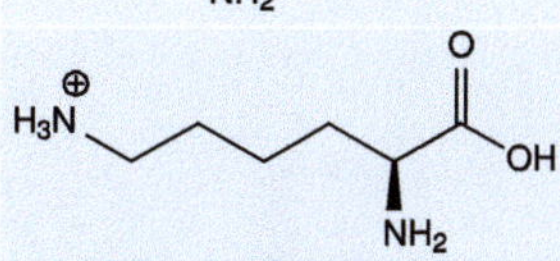
Methionine	Met	M	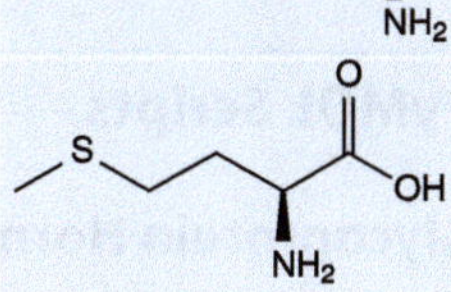

(continued)

Table 16.2 (continued)

Amino acid	3-letter code	1-letter code	Structure
Phenylalanine	Phe	F	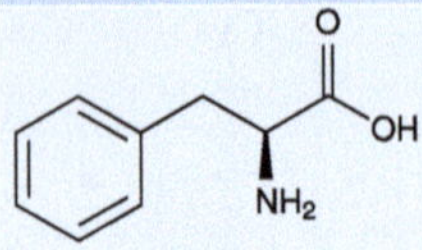
Proline	Pro	P	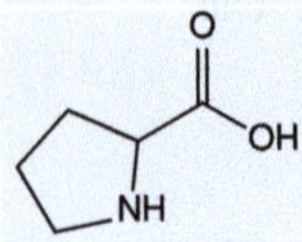
Serine	Ser	S	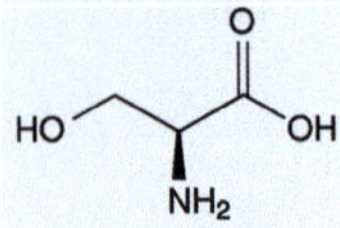
Threonine	Thr	T	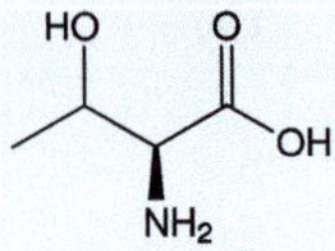
Tryptophan	Try	W	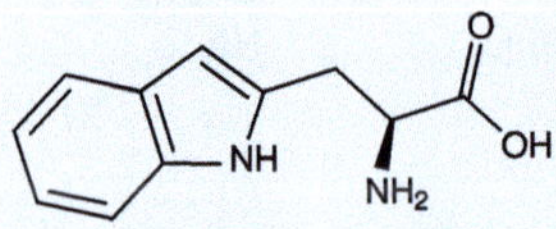
Tyrosine	Tyr	Y	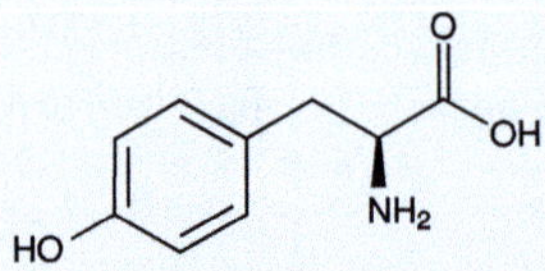
Valine	Val	V	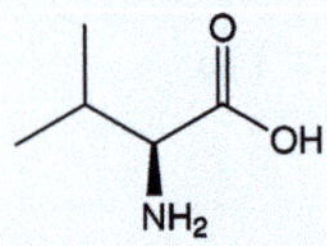

16.3 PyMOL Scripts

16.3.1 Glycoprotein Hormone α Chain

```
    #load ~/Science/pymol/TSHalpha.pml

load ~/Science/strukturen/gonadotropin/1HRP.pdb
cmd.hide("everything","1HRP")
select alphachain, chain a
select betachain, chain b
select helices, ss h
select helicesalpha, alphachain and helices
```

```
select betas, ss s
select betasA, betas and alphachain
select schwefel, name SG
select schwalpha, schwefel and alphachain
select cysteins, resn cys
select cysalpha, cysteins and alphachains
select cysalpha, cysteins and alphachain
select mychainalpha1, (resi 28–32) and alphachain
select mychainalpha2, (resi 82–84) and alphachain
select myringalpha, mychainalpha1 or mychainalpha2
select mypierce, (resi 10+60) and alphachain
select piercesulfur, (name SG) and mypierce
cmd.hide("everything","schwefel")
set sphere_scale, 0.4, cysteins
set sphere_scale, 0.4, mypierce
set sphere_scale, 0.3, myringalpha
set sphere_scale, 0.5, piercesulfur
cmd.color(13,"piercesulfur")
set_view (\\
    –0.721869946,   –0.651766121,    0.232594654,\\
    –0.195120096,   –0.130773291,   –0.972020864,\\
     0.663950145,   –0.747057140,   –0.032772087,\\
     0.000103876,    0.000121470, –174.335067749,\\
    15.360612869,   34.682048798,    3.615572453,\\
   135.404708862,  213.288848877,  –20.000000000 )

cmd.color("grey70" ,"alphachain")
cmd.show("cartoon" ,"alphachain")
util.cba(6,"myringalpha",_self=cmd)
cmd.show("sticks"    ,"mypierce")
cmd.show("sticks"    ,"myringalpha")
cmd.color(13,"schwalpha")
cmd.show("cartoon"   ,"betasA")
cmd.show("spheres"   ,"cysteins")
util.cba(154,"mypierce",_self=cmd)
cmd.show("sticks"    ,"cysalpha")
cmd.show("spheres"   ,"schwalpha")
cmd.hide("everything","betachain")

set sphere_scale, 0.5, piercesulfur
cmd.color(13,"piercesulfur")

stereo swap
stereo crosseye
ray 1000,1000
png ~/Science/pymol/TSHalpha.png
save TSHalpha.png
```

16.3.2 Growth Hormone/Prolactin Using PyMOL

```
 load ~/Science/strukturen/Prolaktin/1N9D_Prolaktin.pdb
cmd.hide("everything","1N9D_Prolaktin")
cmd.show("cartoon"   ,"1N9D_Prolaktin")
```

```
cmd.spectrum("count",selection="(1N9D_Prolaktin)&*/ca")
set_view (\
     0.080638364,   -0.958228469,   -0.274401009,\
     0.513196349,    0.275917083,   -0.812711596,\
     0.854475081,   -0.075285666,    0.514008284,\
     0.000000000,    0.000000000, -196.476440430,\
    -0.272903442,    0.056584358,    0.227214813,\
   154.903518677,  238.049346924,  -20.000000000 )
stereo crosseye
stereo swap
ray 1000,1000
png ~/Science/pymol/PRL.png
```

(The growth hormone drawing using 1HGU.pdb was done with an almost identical script.)

16.3.3 CYP51

```
stereo off
load ~/Science/strukturen/CYP/CYP51/1EA1.pdb
hide everything
sele chainA, chain a
cmd.color(5278,"chainA")
sele haem, resi 460
sele helices, (ss h) and chainA
sele sheets, (ss s) and chainA
show cartoon, chainA
util.cba(13,"haem",_self=cmd)
set sphere_scale, 0.7, haem
show spheres, haem
sele hemcys, resi 394
cmd.color(5259,"hemcys")
cmd.show("spheres"   ,"hemcys")
sele CysS, hemcys and name "SG"
cmd.color(13,"CysS")
cmd.show("sticks", "hemcys")
set sphere_scale, 0.4, hemcys
set sphere_scale, 0.7, CysS
cmd.spectrum("count",selection="helices",byres=1)
cmd.color(8,"sheets")
util.cba(144,"haem",_self=cmd)
util.cba(6,"hemcys",_self=cmd)
set\_view (\
     0.368408203,   -0.457030654,   -0.809582591,\
    -0.865943551,    0.148186371,   -0.477703989,\
     0.338291317,    0.877034962,   -0.341159761,\
    -0.000179388,    0.000013747, -398.444000244,\
   -16.277500153,   -4.500016689,   64.140602112,\
   368.594787598,  428.240142822,  -20.000000000 )

ray 1000,1000
png ~/Science/strukturen/CYP/CYP51/CYP51_mono.png
stereo swap
```

```
stereo crosseye
stereo on
ray 1100,1000

png ~/Science/strukturen/CYP/CYP51/CYP51_stereo.png
stereo off
```

16.3.4 CYP19

```
load ~/Science/strukturen/CYP/CYP19/3EQM.pdb
cmd.hide("everything","3EQM")
#select the heme residue
select heme, resi 600
# sphere diameter
set sphere_scale, 0.4, heme
#color scheme for heme:
util.cba(5,"heme",_self=cmd)
#show sticks with larger spheres
cmd.show("spheres"   ,"heme")
cmd.show("sticks"    ,"heme")

# select androstendion
select androst, resi 601
set sphere_scale, 0.6, androst
util.cba(144,"androst",_self=cmd)
cmd.show("sticks"    ,"androst")
cmd.show("spheres"   ,"androst")

select mysheets, ss s
color white, mysheets
show cartoon, mysheets

select myhelices, ss h
cmd.spectrum("count",selection="myhelices",byres=1)
set cartoon_oval_length, 0.5, myhelices
set cartoon_oval_width, 0.13, myhelices
set cartoon_fancy_helices,1

show cartoon, myhelices

select cpychain, resi 45-496
select loops, ss l+"" and cpychain
cmd.color(6,"loops")
show cartoon, loops

select cnineteen, name C19
select ceighteen, name C18


select ringone, name C1+C2+C3+C4+C5+C10
cmd.color(34,"ringone")

cmd.color(8,"cnineteen")
```

```
cmd.color(11,"ceighteen")

_ set_view (\
_     0.314305723,    0.076195419,   -0.946257472,\
_     0.248245627,    0.955489814,    0.159396768,\
_     0.916289091,   -0.285004675,    0.281401485,\
_     0.000000000,    0.000000000, -208.213180542,\
_    83.395401001,   50.151500702,   46.364837646,\
_   112.259994507,  304.166656494,  -20.000000000 )


ray 1000, 1000
png ~/Science/pymol/CYP19.png
```

Erratum to: Hormones and the Endocrine System

Textbook of Endocrinology

Bernhard Kleine and Winfried G. Rossmanith
Authors

ISBN 978-3-319-15059-8

Erratum to:
B. Kleine, W.G. Rossmanith, *Hormones and the Endocrine System*,
DOI 10.1007/978-3-319-15060-4

DOI 10.1007/978-3-319-15060-4_17

The parts/sections listed below have been replaced by the following corrections:

Chapter 4:

Page 41, Line 18
β2-Agonists, stimuli of the β2-catecholamine receptor, induce GH release presumably by stopping SST secretion.

Page 48, Footnote 9
SST secretion is inhibited by the pancreatic polypeptide (PP) (section 4.10.6; Kim W, et al., FEBS Letters, 588:3233–3239)

Page 89, Footnote 28
Vasotocin regulates oocytes maturation and ovulation in fish.
(Joy KP, Chaube R (2015)
Vasotocin – A new player in the control of oocyte maturation and ovulation in fish. Gen Comp
Endo 221:54–63)

The updated original online version for this chapter can be found at DOI 10.1007/978-3-319-15060-4_4

B. Kleine, W.G. Rossmanith, *Hormones and the Endocrine System*,
DOI 10.1007/978-3-319-15060-4_17

Chapter 6:

Fact sheet 6.6

Fact sheet 6.6: 5α-Reductase

Structure:	no crystallization reported, member of the isoprenyl-cysteine carboxyl methyl-transferase (ICMT) family
Gene:	SRD5A1: Chromosome 5 locus p15.3, 5 exons SRD5A2: Chromosome 2 locus p23, 5 exons
Topology:	ER membrane protein; SRD5A1: widely distributed; SRD5A2: preferentially in androgen target tissue
Function:	reduces testosterone to dihydrotestosterone (type 2), progesterone to dihydroprogesterone and androstenedione to androstanedione

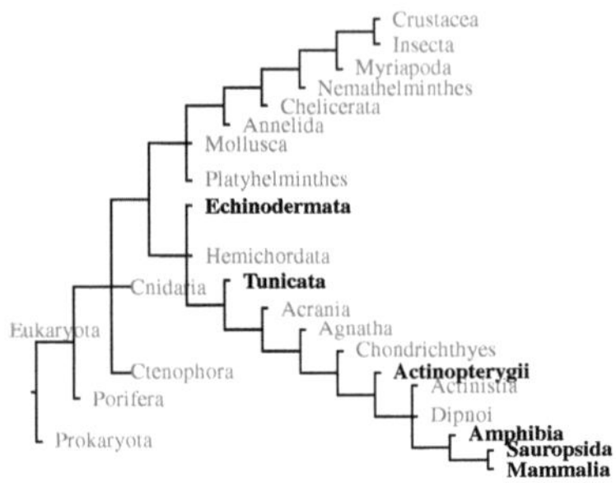

The updated original online version for this chapter can be found at DOI 10.1007/978-3-319-15060-4_6

Chapter 7:

Page 244, Figure 7.3

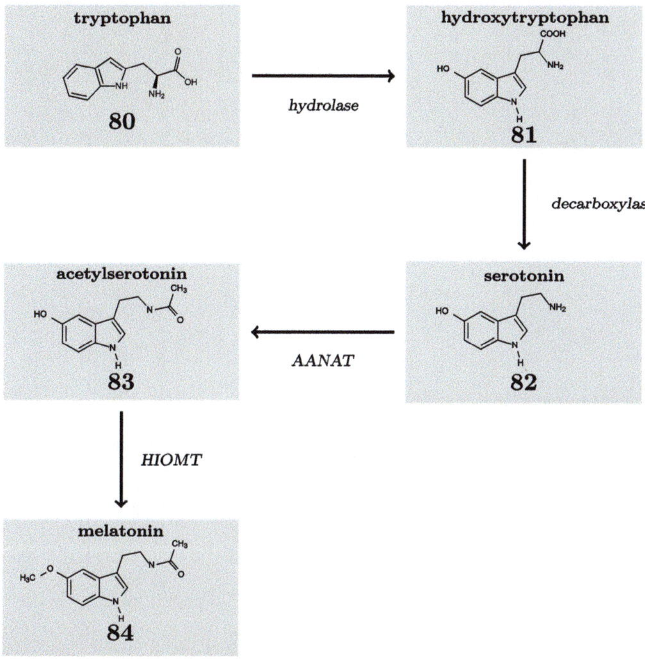

Fig. 7.3 Melatonin synthesis from L-tryptophan. *AANAT* arylalkylamine *N*-acetyltransferase, *HIOMT* hydroxyindole *O*-methyltransferase

The updated original online version for this chapter can be found at DOI 10.1007/978-3-319-15060-4_7

References

Achermann JC et al (2002) Genetic causes of human reproductive disease. J Clin Endocrinol Metab 87:2447–'2454

Adams BA et al (2002) Three forms of gonadotropin-releasing hormone, including a novel form, in a basal salmonid, *Coregonus clupeaformis*. Biol Reprod 67:232–239

Adams BA et al (2003) Six novel gonadotropin-releasing hormones are encoded as triplets on each of two genes in the protochordate, *Ciona intestinalis*. Endocrinology 144:1907–1919

Adams ME (2003) Juvenile hormones. In: Resh VH, Cardé RT (eds) Encyclopedia of insects, chap 143. University of California, Riverside: Academic Press, San Diego/London/Burlington, pp 611–618

Ager EI et al (2008a) Evolution of the CDKN1C-KCNQ1 imprinted domain. BMC Evol Biol 8:163

Ager EI et al (2008b) Expression and protein localisation of IGF2 in the marsupial placenta. BMC Dev Biol 8:17

Alsop D, Vijayan M (2009) The zebrafish stress axis: molecular fallout from the teleost-specific genome duplication event. Gen Comp Endocrinol 161:62–66

Alvarez E, Cahoreau C, Combarnous Y (2009) Comparative structure analyses of cystine knot-containing molecules with eight aminoacyl ring including glycoprotein hormones (GPH) alpha and beta subunits and GPH-related a2 (GPA2) and b5 (GPB5) molecules. Reprod Biol Endocrinol RB&E 7:90

An S et al (2012) Insect Neuropeptide Bursicon Homodimers Induce Innate Immune and Stress Genes during Molting by Activating the NF-kB Transcription Factor Relish. PLoS ONE 7: e34510

Anastasi A, Erspamer V, Endean R (1967) Isolation and structure of caerulein, an active decapeptide from the skin of *Hyla caerulea*. Experientia 23:699–700

Anastasi A et al (1969) Structure and pharmacological actions of phyllocaerulein, a caerulein-like nonapeptide: its occurrence in extracts of the skin of *Phyllomedusa sauvagei* and related *Phyllomedusa* species. Br J Pharmacol 37:198–206

Anastasi A et al (1970) Presence of caerulein in extracts of the skin of *Leptodactylus pentadactylus labyrinthicus* and of *Xenopus laevis*. Br J Pharmacol 38:221–228

Andersson S et al (1996) Molecular genetics and pathophysiology of 17 beta-hydroxysteroid dehydrogenase 3 deficiency. J Clin Endocrinol Metab 81:130–136

Asakawa A et al (2005) Stomach regulates energy balance via acylated ghrelin and desacyl ghrelin. Gut 54:18–24

Attia GR et al (2000) Progesterone receptor isoform a but not b is expressed in endometriosis. J Clin Endocrinol Metab 85:2897–2902

Audsley N, Weaver RJ (2006) Analysis of peptides in the brain and corpora cardiaca-corpora allata of the honey bee, *Apis mellifera* using MALDI-TOF mass spectrometry. Peptides 27:512–520

Audsley N, Goldsworthy GJ, Coast GM (1997) Quantification of Locusta diuretic hormone in the central nervous system and corpora cardiaca: influence of age and feeding status, and mechanism of release. Regul Pept 69:25–32

B. Kleine, W.G. Rossmanith, *Hormones and the Endocrine System*,
DOI 10.1007/978-3-319-15060-4

Audsley N et al (2008) Allatoregulatory peptides in Lepidoptera, structures, distribution and functions. J Insect Physiol 54:969–980

Baggerman G et al (2002) Peptidomics of the larval *Drosophila melanogaster* central nervous system. J Biol Chem 277:40368–40374

Baker ME (2002) Recent insights into the origins of adrenal and sex steroid receptors. J Mol Endocrinol 28:149–152

Bartke A (2004) Prolactin in the male: 25 years later. J Androl 25:661–666

Bauer JW et al (2008) Galanin family of peptides in skin function. Cell Mol Life Sci CMLS 65:1820–1825

Baxter RC (2000) Insulin-like growth factor (IGF)-binding proteins: interactions with IGFs and intrinsic bioactivities. Am J Physiol Endocrinol Metab 278:E967–E976

Bechtold DA, Luckman SM (2007) The role of RFamide peptides in feeding. J Endocrinol 192:3–15

Bentley GE et al (2008) Gonadotropin-inhibitory hormone and its receptor in the avian reproductive system. Gen Comp Endocrinol 156:34–43

Berglund MM, Hipskind PA, Gehlert DR (2003) Recent developments in our understanding of the physiological role of PP-fold peptide receptor subtypes. Exp Biol Med (Maywood) 228:217–244

Bertucci F, Birnbaum D (2009) Distant metastasis: not out of reach any more. J Biol 8:28

Bettio A, Dinger MC, Beck-Sickinger AG (2002) The neuropeptide Y monomer in solution is not folded in the pancreatic-polypeptide fold. Protein Sci 11:1834–1844

Bilban M et al (2004) Kisspeptin-10, a KiSS-1/metastin-derived decapeptide, is a physiological invasion inhibitor of primary human trophoblasts. J Cell Sci 117:1319–1328

Birch NP et al (1991) Differential glycosylation of N-POMC1–77 regulates the production of gamma 3-MSH by purified pro-opiomelanocortin converting enzyme. A possible mechanism for tissue-specific processing. FEBS Lett 290:191–194

Bird IM, Mason JI, Rainey WE (1998) Protein kinase A, protein kinase C, and ca(2+)-regulated expression of 21-hydroxylase cytochrom P450 in H295R human adrenocortical cells. J Clin Endocrinol Metab 83:1592–1597

Bishop CA, O'Shea M, Miller RJ (1981) Neuropeptide proctolin (H-Arg-Tyr-Leu-Pro-Thr-OH): immunological detection and neuronal localization in insect central nervous system. Proc Natl Acad Sci U S A 78:5899–5902

Boler J et al (1969) The identity of chemical and hormonal properties of the thyrotropin releasing hormone and pyroglutamyl-histidyl-proline amide. Biochem Biophys Res Commun 37:705–710

Boonyaratanakornkit V et al (2001) Progesterone receptor contains a proline-rich motif that directly interacts with SH3 domains and activates c-src family tyrosine kinases. Mol Cell 8:269–280

Borovsky D (2003) Trypsin-modulating oostatic factor: a potential new larvicide for mosquito control. J Exp Biol 206:3869–3875

Borovsky D et al (1990) Mosquito oostatic factor: a novel decapeptide modulating trypsin-like enzyme biosynthesis in the midgut. FASEB J 4:3015–3020

Böttner B, Denner K, Bernhardt R (1998) Conferring aldosterone synthesis to human CYP11B1 by replacing key amino acid residues with CYP11B2-specific ones. Eur J Biochem 252:458–466

Boushey RP, Drucker DJ (2003) Gastrointestinal hormones and gut endocrine tumors. In: Larsen PR, Kronenberg HM, Melmed S, Polonsky K (eds) Williams textbook of endocrinology, chap 38, 10th edn. Saunders, Philadelphia, pp 1777–1796

Breves JP et al (2010) Gene expression of growth hormone family and glucocorticoid receptors, osmosensors, and ion transporters in the gill during seawater acclimation of mozambique tilapia, *Oreochromis mossambicus*. J Exp Zool Part A Ecol Genet Physiol 313:432–441

Broad PM et al (1989) Structure and methylation of the human calcitonin/alpha-CGRP gene. Nucleic Acids Res 17:6999–7011

Brogiolo W et al (2001) An evolutionarily conserved function of the Drosophila insulin receptor and insulin-like peptides in growth control. Curr Biol 11:213–221

Bundgaard JR, Rehfeld JF (2008) Distinct linkage between post-translational processing and differential secretion of progastrin derivatives in endocrine cells. J Biol Chem 283:4014–4021

Bures EJ et al (1998) Determination of disulfide structure in agouti-related protein (AGRP) by stepwise reduction and alkylation. Biochemistry 37:12172–12177

Burgus R et al (1969) Structure moleculaire du facteur hypothalamique hypophysiotrope TRF d'origine ovine: mise en evidence par spectrometrie de masse de la sequence PCA-His-Pro-NH2. C R Acad Sci Hebd Seances Acad Sci D 269:1870–1873

Burke RD et al (2006) A genomic view of the sea urchin nervous system. Dev Biol 300:434–460

Bylemans D et al (1994) Sequencing and characterization of trypsin modulating oostatic factor (TMOF) from the ovaries of the grey fleshfly. *Neobellieria (Sarcophaga) bullata*. Regul Pept 50:61–72

Cameo P et al (2004) Chorionic gonadotropin and uterine dialogue in the primate. Reprod Biol Endocrinol 2:50

Carolsfeld J et al (2000) Primary structure and function of three gonadotropinreleasing hormones, including a novel form, from an ancient teleost, herring. Endocrinology 141:505–512

Cartier EA et al (2010) A biochemical and functional protein complex involving dopamine synthesis and transport into synaptic vesicles. J Biol Chem 285:1957–1966

Carvalho GB et al (2006) Allocrine modulation of feeding behavior by the Sex Peptide of Drosophila. Curr Biol 16:692–696

Castro LF, Santos MM, Reis-Henriques MA (2005) The genomic environment around the Aromatase gene: evolutionary insights. BMC Evol Biol 5:43

Catania A (2007) The melanocortin system in leukocyte biology. J Leukoc Biol 81:383–392

Cazzamali G, Saxild N, Grimmelikhuijzen C (2002) Molecular cloning and functional expression of a Drosophila corazonin receptor. Biochem Biophys Res Commun 298:31–36

Cazzamali G et al (2005) The Drosophila gene CG9918 codes for a pyrokinin-1 receptor. Biochem Biophys Res Commun 335:14–19

Chang J-C et al (2009) Receptor guanylyl cyclases in Inka cells targeted by eclosion hormone. Proc Natl Acad Sci U S A 106:13371–13376

Chantalat L et al (1995) The crystal-structure of wild-type growth-hormone at 2.5 Ångstrom resolution. Prot Pept Lett 2:333–340

Chapman RF (1998) The insects: structure and function, 4th edn. Cambridge University Press, Cambridge/New York/Melbourne, p 819

Chemelli RM et al (1999) Narcolepsy in orexin knockout mice: molecular genetics of sleep regulation. Cell 98:437–451

Chen EY et al (1989) The human growth hormone locus: nucleotide sequence, biology, and evolution. Genomics 4:479–497

Chiang AS et al (2002) Insect NMDA receptorsmediate juvenile hormone biosynthesis. Proc Natl Acad Sci U S A 99:37–42

Choi MY et al (2003) Identification of a G protein-coupled receptor for pheromone biosynthesis activating neuropeptide from pheromone glands of the moth *Helicoverpa zea*. Proc Natl Acad Sci U S A 100:9721–9726

Choi YJ, Lee G, Park JH (2006) Programmed cell death mechanisms of identifiable peptidergic neurons in *Drosophila melanogaster*. Development 133:2223–2232

Christie AE et al (2007) Midgut epithelial endocrine cells are a rich source of the neuropeptides APSGFLGMRamide (*Cancer borealis* tachykinin-related peptide Ia) and GYRKPPFNGSI-Famide (Gly1-SIFamide) in the crabs *Cancer borealis*, *Cancer magister* and *Cancer productus*. J Exp Biol 210:699–714

Chrousos GP (1998) Ultradian, circadian, and stress-related hypothalamic-pituitary-adrenal axis activity-a dynamic digital-to-analog modulation. Endocrinology 139:437–440

Chua SC et al (1987) Cloning of cDNA encoding steroid 11 beta-hydroxylase (P450c11). Proc Natl Acad Sci U S A 84:7193–7197

Chung JS, Webster SG (2004) Expression and release patterns of neuropeptides during embryonic development and hatching of the green shore crab, *Carcinus maenas*. Development 131:4751–4761

Chung JS, Webster SG (2005) Dynamics of in vivo release of molt-inhibiting hormone and crustacean hyperglycemic hormone in the shore crab, *Carcinus maenas*. Endocrinology 146:5545–5551

Chusho H et al (2001) Dwarfism and early death in mice lacking C-type natriuretic peptide. Proc Natl Acad Sci U S A 98:4016–4021

Chuva de Sousa Lopes SM et al (2006) Patterning the heart, a template for human cardiomyocyte development. Dev Dyn 235:1994–2002

Clark AC, del Campo ML, Ewer J (2004) Neuroendocrine control of larval ecdysis behavior in *Drosophila*: complex regulation by partially redundant neuropeptides. J Neurosci 24:4283–4292

Colwell CS et al (2003) Disrupted circadian rhythms in VIP- and PHI-deficient mice. Am J Physiol Regul Integr Comp Physiol 285:R939–R949

Conlon JM (2002) The origin and evolution of peptide YY (PYY) and pancreatic polypeptide (PP). Peptides 23:269–278

Conlon JM et al (1988) Insulin and other islet hormones (somatostatin, glucagon and PP) in the neuroendocrine system of some lower vertebrates and that of invertebrates–a minireview. Horm Metab Res 20:406–410

Copenhaver PF, Truman JW (1986a) Identification of the cerebral neurosecretory cells that contain eclosion hormone in the moth *Manduca sexta*. J Neurosci 6:1738–1747

Copenhaver PF, Truman JW (1986b) Metamorphosis of the cerebral neuroendocrine system in the moth Manduca sexta. J Comp Neurol 249:186–204

Cottrell GA (1997) The first peptide-gated ion channel J Exp Biol 200:2377–2386

Coulter CL, Jaffe RB (1998) Functional maturation of the primate fetal adrenal in vivo: 3. Specific zonal localization and developmental regulation of CYP21A2 (P450c21) and CYP11B1/CYP11B2 (P450c11/aldosterone synthase) lead to integrated concept of zonal and temporal steroid biosynthesis. Endocrinology 139:5144–5150

Cox KJ et al (1997) Cloning, characterization, and expression of a G-protein-coupled receptor from *Lymnaea stagnalis* and identification of a leucokinin-like peptide, PSFHSWSamide, as its endogenous ligand. J Neurosci 17:1197–1205

Crapo LM (1985) Hormones, the messengers of life. W.H. Freeman & Co, New York, p 194

Cruz-Bermudez ND, Marder E (2007) Multiple modulators act on the cardiac ganglion of the crab, *Cancer borealis*. J Exp Biol 210:2873–2884

Csoknya M et al (2005) Neurochemical characterization of nervous elements innervating the body wall of earthworms (Lumbricus, Eisenia): immunohistochemical and pharmacological studies. Cell Tissue Res 321:479–490

Dai L et al (2008) Identification, developmental expression, and functions of bursicon in the tobacco hawkmoth, *Manduca sexta*. J Comp Neurol 506:759–774

Daniel AR, Lange CA (2009) Protein kinases mediate ligand-independent derepression of sumoylated progesterone receptors in breast cancer cells. Proc Natl Acad Sci U S A 106:14287–14292

Darrouzet E et al (1997) Hydroxy juvenile hormones: new putative juvenile hormones biosynthesized by locust corpora allata in vitro. Biochem Biophys Res Commun 240:752–758

Davis MM et al (2007) A neuropeptide hormone cascade controls the precise onset of post-eclosion cuticular tanning in *Drosophila melanogaster*. Development 134:4395–4404

De Vriese C, Delporte C (2008) Ghrelin: a new peptide regulating growth hormone release and food intake. Int J Biochem Cell Biol 40:1420–1424

Dejda A, Sokolowska P, Nowak JZ (2005) Neuroprotective potential of three neuropeptides PACAP, VIP and PHI. Pharmacol Rep 57:307–320

Delgado M, Pozo D, Ganea D (2004) The significance of vasoactive intestinal peptide in immunomodulation. Pharmacol Rev 56:249–290

Denef C (2008) Paracrinicity: the story of 30 years of cellular pituitary crosstalk. J Neuroendocrinol 20:1–70

Dewey EM et al (2004) Identification of the gene encoding bursicon, an insect neuropeptide responsible for cuticle sclerotization and wing spreading. Curr Biol 14:1208–1213

Dias JA (2001) Is there any physiological role for gonadotrophin oligosaccharide heterogeneity in humans? II. A biochemical point of view. Hum Reprod 16:825–830

Dierick HA, Greenspan RJ (2007) Serotonin and neuropeptide F have opposite modulatory effects on fly aggression. Nat Genet 39:678–682

Dierks A et al (2010) Identification of adrenal genes regulated in a potassium-dependent manner. J Mol Endocrinol 45:193–206

Diez-Sampedro A et al (2003) A glucose sensor hiding in a family of transporters. Proc Natl Acad Sci U S A 100:11753–11758

Dinulescu DM, Cone RD (2000) Agouti and agouti-related protein: analogies and contrasts. J Biol Chem 275:6695–6698

Dircksen H et al (2001) Crustacean hyperglycaemic hormone (CHH)-like peptides and CHH-precursor-related peptides from pericardial organ neurosecretory cells in the shore crab, *Carcinus maenas*, are putatively spliced and modified products of multiple genes. Biochem J 356:159–170

Dockray GJ (1999) Topical review. Gastrin and gastric epithelial physiology. J Physiol 518(Pt 2):315–324

Dos Santos S et al (2009) Distinct expression patterns of glycoprotein hormone-alpha2 and -beta5 in a basal chordate suggest independent developmental functions. Endocrinology 150:3815–3822

Ducret E, Anderson GM, Herbison AE (2009) RFamide-related peptide-3, a mammalian gonadotropin-inhibitory hormone ortholog, regulates gonadotropin-releasing hormone neuron firing in the mouse. Endocrinology 150:2799–2804

Dudas B, Merchenthaler I (2006) Three-dimensional representation of the neurotransmitter systems of the human hypothalamus: inputs of the gonadotrophin hormone-releasing hormone neuronal system. J Neuroendocrinol 18:79–95

Dudel J, Menzel R, Schmidt RF (2001) Neurowissenschaft: Vom Molekül zur Kognition, 2 aktualis. Springer, Berlin/Heidelberg/New York/Tokyo

Dunlop FM et al (2003) A compound heterozygote case of type II aldosterone synthase deficiency. J Clin Endocrinol Metab 88:2518–2526

Duvaux-Miret O et al (1992) Immunosuppression in the definitive and intermediate hosts of the human parasite *Schistosoma mansoni* by release of immunoactive neuropeptides. Proc Natl Acad Sci U S A 89:778–781

Ebrahim IO et al (2003) Hypocretin (orexin) deficiency in narcolepsy and primary hypersomnia. J Neurol Neurosurg Psychiatry 74:127–130

Ebrahim IO, Peacock KW, Williams AJ (2005) Posttraumatic narcolepsy–two case reports and a mini review. J Clin Sleep Med 1:153–156

Eckert M, Agricola H, Penzlin H (1981) Immunocytochemical identification of proctolinlike immunoreactivity in the terminal ganglion and hindgut of the cockroach *Periplaneta americana* (L.). Cell Tissue Res 217:633–645

van der Eerden BC, Karperien M, Wit JM (2003) Systemic and local regulation of the growth plate. Endocr Rev 24:782–801

Ellsworth DL, Hewett-Emmett D, Li WH (1994) Evolution of base composition in the insulin and insulin-like growth factor genes. Mol Biol Evol 11:875–885

Erickson JD et al (1996) Distinct pharmacological properties and distribution in neurons and endocrine cells of two isoforms of the human vesicular monoamine transporter. Proc Natl Acad Sci U S A 93:5166–5171

Erspamer V (1971) Biogenic amines and active polypeptides of the amphibian skin. Ann Rev Pharmacol 11:327–350

Erspamer V et al (1972) Occurrence of bombesin and alytesin in extracts of the skin of three European discoglossid frogs and pharmacological actions of bombesin on extravascular smooth muscle. Br J Pharmacol 45:333–348

Evans PD (1984) Studies on the mode of action of octopamine, 5-hydroxytryptamine and proctolin on a myogenic rhythm in the locust. J Exp Biol 110:231–251

Falkmer S et al (1985) Phylogenetical aspects on islet hormone families: a minireview with particular reference to insulin as a growth factor and to the phylogeny of PYY and NPY immunoreactive cells and nerves in the endocrine and exocrine pancreas. Peptides 6(Suppl 3):315–320

Farkash Y, Timberg R, Orly J (1986) Preparation of antiserum to rat cytochrome P-450 cholesterol side chain cleavage, and its use for ultrastructural localization of the immunoreactive enzyme by protein A-gold technique. Endocrinology 118:1353–1365

Fernandez R et al (1995) The Drosophila insulin receptor homolog: a gene essential for embryonic development encodes two receptor isoforms with different signaling potential. EMBO J 14:3373–3384

Fink JW et al (2005) Seasonal changes in mesotocin and localization of its receptor in the prostate of the brushtail possum (*Trichosurus vulpecula*). Biol Reprod 72:470–478

Fiore MMD et al (2000) Mammalian and chicken I forms of gonadotropin-releasing hormone in the gonads of a protochordate, Ciona intestinalis. Proc Natl Acad Sci U S A 97:2343–2348

Fisher JM et al (1988) Multiple neuropeptides derived from a common precursor are differentially packaged and transported. Cell 54:813–822

Fitzpatrick DA, O'Halloran DM, Burnell AM (2006) Multiple lineage specific expansions within the guanylyl cyclase gene family. BMC Evol Biol 6:26

Fliedner S, Schulz C, Lehnert H (2006) Brain uptake of intranasally applied radioiodinated leptin in Wistar rats. Endocrinology 147:2088–2094

Flier JS (2004) Obesity wars: molecular progress confronts an expanding epidemic. Cell 116:337–350

Florio P et al (2004) Inhibins and activins in pregnancy. Mol Cell Endocrinol 225:93–100

Forger DB, Peskin CS (2003) A detailed predictive model of the mammalian circadian clock. Proc Natl Acad Sci U S A 100:14806–14811

Fournier D et al (2008) Design and synthesis of bisubstrate inhibitors of type 1 17beta-hydroxysteroid dehydrogenase: overview and perspectives. Eur J Med Chem 43:2298–2306

Freeman ME et al (2000) Prolactin: structure, function, and regulation of secretion. Physiol Rev 80:1523–1631

Fujisawa Y et al (1999) The *Aplysia mytilus* inhibitory peptide-related peptides: identification, cloning, processing, distribution, and action. J Neurosci 19:9618–9634

Fukamachi S et al (2009) Dual control by a single gene of secondary sexual characters and mating preferences in medaka. BMC Biology 7:64

Furukawa Y et al (2001) The enterins: a novel family of neuropeptides isolated from the enteric nervous system and CNS of *Aplysia*. J Neurosci 21:8247–8261

Gaede G (2004) Regulation of intermediary metabolism and water balance of insects by neuropeptides. Ann Rev Entomol 49:93–113

Gammie SC, Truman JW (1997) Neuropeptide hierarchies and the activation of sequential motor behaviors in the hawkmoth, *Manduca sexta*. J Neurosci 17:4389–4397

Gammie SC, Truman JW (1999) Eclosion hormone provides a link between ecdysis-triggering hormone and crustacean cardioactive peptide in the neuroendocrine cascade that controls ecdysis behavior. J Exp Biol 202:343–352

García-Villoria J et al (2010) X-inactivation of HSD17B10 revealed by cDNA analysis in two female patients with 17β-hydroxysteroid dehydrogenase 10 deficiency. Eur J Hum Genet 18:1353–1355

Gatford KL et al (1998) Sexual dimorphism of the somatotrophic axis. J Endocrinol 157:373–389

Genazzani AD et al (1996) Long-term GnRH-agonist administration revealed a GnRH-independent mechanism stimulating FSH discharge in humans. Eur J Endocrinol 134:77–83

George A Jr et al (1994) Assignment of a human voltage-dependent sodium channel alpha-subunit gene (SCN6A) to 2q21-q23. Genomics 19:395–397

Gerok W et al (2000) Endokrinologie In: Gerok W et al (eds) Die Innere Medizin, chap 12, 10th edn. Schattauer Stuttgart, New York, pp 935–1102

Ghosh D et al (2009) Structural basis for androgen specificity and oestrogen synthesis in human aromatase. Nature 457:219–223

Gilbert LI, Rybczynski R, Warren JT (2002) Control and biochemical nature of the ecdysteroidogenic pathway. Ann Rev Entomol 47:883–916
Goldin AL (2001) Resurgence of sodium channel research. Ann Rev Physiol 63:871–894
Gonzalez-Rey E et al (2006) Cortistatin, a new antiinflammatory peptide with therapeutic effect on lethal endotoxemia. J Exp Med 203:563–571
Goodman W, Granger N (2005) The juvenile hormone. In: Gilbert LI, Iatrou K, Gill SS (eds) Molecular insect science. Endocrinology, chap 7, vol 3. Elsevier, Boston, pp 320–408
Goto-Kazeto R et al (2004) Localization and expression of aromatase mRNA in adult zebrafish. Gen Comp Endocrinol 139:72–84
Gray H (1918) In: Lewis WH (ed) Anatomy of the human body. Online version Bartleby.com 2000. Lea & FEbiger
Green VL et al (1999) 17Beta-hydroxysteroid dehydrogenase type 1, 2, 3, and 4 expression and enzyme activity in human anterior pituitary adenomas. J Clin Endocrinol Metab 84:1340–1345
Grob M, Drolet G, Mouginot D (2004) Specific Na^+ sensors are functionally expressed in a neuronal population of the median preoptic nucleus of the rat. J Neurosci 24:3974–3984
Grossman A (ed) (1992) Clinical endocrinology. Blackwell Scientific Publications, Oxford/-Paris/Berlin/Wien
Guilgur LG et al (2006) Evolution of GnRH ligands and receptors in gnathostomata. Comp Biochem Physiol A Mol Integr Physiol 144:272–283
Gutierrez JA et al (2008) Ghrelin octanoylation mediated by an orphan lipid transferase. Proc Natl Acad Sci U S A 105:6320–6325
Gwee PC et al (2008) Sequence and organization of coelacanth neurohypophysial hormone genes: evolutionary history of the vertebrate neurohypophysial hormone gene locus. BMC Evol Biol 8:93
Hagan JJ et al (1999) Orexin A activates locus coeruleus cell firing and increases arousal in the rat. Proc Natl Acad Sci U S A 96:10911–10916
Han SK et al (2005) Activation of gonadotropin-releasing hormone neurons by kisspeptin as a neuroendocrine switch for the onset of puberty. J Neurosci 25:11349–11356
Hanke W (1970) Hormone. Fortschr Zool 20:318–380
Hansford JR, Mulligan LM (2000) Multiple endocrine neoplasia type 2 and RET: from neoplasia to neurogenesis. J Med Genet 37:817–827
Hartfelder K (2000) Insect juvenile hormone: from "status quo" to high society. Braz J Med Biol Res 33:157–177
Helfrich-Forster C et al (2002) The extraretinal eyelet of *Drosophila*: development, ultrastructure, and putative circadian function. J Neurosci 22:9255–9266
Henry JP et al (1994) Biochemistry and molecular biology of the vesicular monoamine transporter from chromaffin granules. J Exp Biol 196:251–262
Hewes RS, Truman JW (1991) The roles of central and peripheral eclosion hormone release in the control of ecdysis behavior in Manduca sexta. J Comp Physiol [A] 168:697–707
Hewes RS et al (2006) Regulation of secretory protein expression in mature cells by DIMM, a basic helix-loop-helix neuroendocrine differentiation factor. J Neurosci 26:7860–7869
Hirata T et al (1988) Structures and actions of *Mytilus* inhibitory peptides. Biochem Biophys Res Commun 152:1376–1382
Hirsch M (1998) interBRAIN: Topographische Anatomie des ZNS des Menschen (Software). Springer. 978-3540146438
Hirsch PF, Baruch H (2003) Is calcitonin an important physiological substance? Endocrine 21:201–208
Hiyama TY et al (2002) Na(x) channel involved in CNS sodium-level sensing. Nat Neurosci 5:511–512
Hiyama TY et al (2004) The subfornical organ is the primary locus of sodium-level sensing by Na(x) sodium channels for the control of salt-intake behavior. J Neurosci 24:9276–9281
Hnasko R et al (1997) Two distinct pituitary cell lines from mouse intermediate lobe tumors: a cell that produces prolactin-regulating factor and a melanotroph. Endocrinology 138:5589–5596

Hobson S-A et al (2008) Galanin acts as a trophic factor to the central and peripheral nervous systems. Cell Mol Life Sci CMLS 65:1806–1812

Hofer S, Homberg U (2006) Evidence for a role of orcokinin-related peptides in the circadian clock controlling locomotor activity of the cockroach *Leucophaea maderae*. J Exp Biol 209:2794–2803

Hoff AO et al (2002) Increased bone mass is an unexpected phenotype associated with deletion of the calcitonin gene. J Clin Invest 110:1849–1857

Hogben LT, Slome D (1931) The pigmentary effector system. VI. the dual character of endocrine co-ordination in amphibian colour change. Proc R Soc Lond Ser B 108:10–53

Holman G, Cook B, Nachman R (1986a) Isolation, primary structure and synthesis of two neuropeptides from *Leucophaea maderae*: members of a new family of Cephalomyotropins. Comp Biochem Physiol C 84:205–211

Holman G, Cook B, Nachman R (1986b) Primary structure and synthesis of two additional neuropeptides from *Leucophaea maderae*: members of a new family of Cephalomyotropins. Comp Biochem Physiol C 84:271–276

Honegger H-W et al (2008) Bursicon, the tanning hormone of insects: recent advances following the discovery of its molecular identity. J Comp Physiol A 194:989–1005

Houweling AC et al (2005) Comparative analysis of the natriuretic peptide precursor gene cluster in vertebrates reveals loss of ANF and retention of CNP-3 in chicken. Dev Dyn 233:1076–1082

Hsu SY, Nakabayashi K, Bhalla A (2002) Evolution of glycoprotein hormone subunit genes in bilateralmetazoa: identification of two novel human glycoprotein hormone subunit family genes, GPA2 and GPB5. Mol Endocrinol 16:1538–1551

Hua YJ et al (1999) Identification of a prothoracicostatic peptide in the larval brain of the silkworm, *Bombyx mori*. J Biol Chem 274:31169–31173

Huising MO, Flik G (2005) The remarkable conservation of corticotropin-releasing hormone (CRH)-binding protein in the honeybee (*Apis mellifera*) dates the CRH system to a common ancestor of insects and vertebrates. Endocrinology 146:2165–2170

Huising MO, Kruiswijk CP, Flik G (2006) Phylogeny and evolution of class-I helical cytokines. J Endocrinol 189:1–25

Hull K, Marler R, Harvey S (2006) Neural calcitropic peptides: immunoreactive characterization in fish and invertebrates. Neurosci Lett 404:15–19

Hummon AB et al (2006) From the genome to the proteome: uncovering peptides in the Apis brain. Science 314:647–649

Husebye ES et al (2006) The substrate-binding domain of 21-hydroxylase, the main autoantigen in autoimmune Addison's disease, is an immunodominant T cell epitope. Endocrinology 147:2411–2416

Huybrechts J, De Loof A, Schoofs L (2004) Diapausing Colorado potato beetles are devoid of short neuropeptide F I and II. Biochem Biophys Res Commun 317:909–916

Iino M et al (2007) Progesterone: its occurrence in plants and involvement in plant growth. Phytochemistry 68:1664–1673

Ikeda E (1913) Kimon Rimensen. In: Ikeda E (ed) Experimental anatomy and physiology of *Bombyx mori*. Meibundo, Tokyo, pp 242–243

Ikemoto T, Park MK (2006) Molecular and evolutionary characterization of the GnRH-II gene in the chicken: distinctive genomic organization, expression pattern, and precursor sequence. Gene 368:28–36

Ishizaki H, Suzuki A (1994) The brain secretory peptides that control moulting and metamorphosis of the silkmoth, *Bombyx mori*. Int J Dev Biol 38:301–310

Iwakoshi E et al (2002) Isolation and characterization of a GnRH-like peptide from *Octopus vulgaris*. Biochem Biophys Res Commun 291:1187–1193

Iwanov PP, Mescherskaya KA (1935) Die physiologischen Besonderheiten der geschlechtlich unreifen Insektenovarien und die zyklischen Veränderungen ihrer Eigenschaften. Zool J Physiol 55:281–348

Jagge CL, Pietrantonio PV (2008) Diuretic hormone 44 receptor in Malpighian tubules of the mosquito *Aedes aegypti*: evidence for transcriptional regulation paralleling urination. Insect Mol Biol 17:413–426

Janko AB, Sandberg EC (1970) Histochemical evidence for the protein nature of the Reinke crystalloid. Obstet Gynecol 35:493–503

Jena BP (2004) Discovery of the Porosome: revealing the molecular mechanism of secretion and membrane fusion in cells. J Cell Mol Med 8:1–21

Jequier E, Tappy L (1999) Regulation of body weight in humans. Physiol Rev 79:451–480

Jimenez-Liñan M, Rubin BS, King JC (1997) Examination of guinea pig luteinizing hormone-releasing hormone gene reveals a unique decapeptide and existence of two transcripts in the brain. Endocrinology 138:4123–4130

Jindra M, Palli SR, Riddiford LM (2012) The juvenile hormone signaling pathway in insect development. Ann Rev Entomol 58:181–204

Jobst EE, Enriori PJ, Cowley MA (2004) The electrophysiology of feeding circuits. Trends Endocrinol Metab 15:488–499

Johnsen AH (1998) Phylogeny of the cholecystokinin/gastrin family. Front Neuroendocrinol 19:73–99

Johnsen AH, Rehfeld JF (1990) Cionin: a disulfotyrosyl hybrid of cholecystokinin and gastrin from the neural ganglion of the protochordate Ciona intestinalis. J Biol Chem 265:3054–3058

Johnson EC et al (2003) Identification of Drosophila neuropeptide receptors by G protein-coupled receptors-beta-arrestin2 interactions. J Biol Chem 278:52172–52178

Jones G et al (2013) Ligand binding pocket function of Drosophila USP is necessary for metamorphosis. Gen Comp Endocrinol 182:73–82

Jones JI, Clemmons DR (1995) Insulin-like growth factors and their binding proteins: biological actions. Endocr Rev 16:3–34

Josso N et al (2005) AMH and AMH receptor defects in persistent Müllerian duct syndrome. Hum Reprod Update 11:351–356

Kaiya H et al (2003a) Amidated fish ghrelin: purification, cDNA cloning in the Japanese eel and its biological activity. J Endocrinol 176:415–423

Kaiya H et al (2003b) Identification of tilapia ghrelin and its effects on growth hormone and prolactin release in the tilapia, *Oreochromis mossambicus*. Comp Biochem Physiol B Biochem Mol Biol 135:421–429

Kakizawa S et al (1997) Possible involvement of somatolactin in the regulation of plasma bicarbonate for the compensation of acidosis in rainbow trout. J Exp Biol 200:2675–2683

Karolyi IJ et al (1999) Altered anxiety and weight gain in corticotropin-releasing hormone-binding protein-deficient mice. Proc Natl Acad Sci U S A 96:11595–11600

Kataoka H et al (1989) Isolation and identification of a diuretic hormone from the tobacco hornworm, *Manduca sexta*. Proc Natl Acad Sci U S A 86:2976–2980

Katayama H et al (2003) The solution structure of molt-inhibiting hormone from the Kuruma prawn *Marsupenaeus japonicus*. J Biol Chem 278:9620–9623

Kawada T et al (2011) Peptidomic analysis of the central nervous system of the protochordate, Ciona intestinalis: homologs and prototypes of vertebrate peptides and novel peptides. Endocrinology, 152:2416–2427

Kawakami A et al (1989) Structure and organization of four clustered genes that encode bombyxin, an insulin-related brain secretory peptide of the silkmoth *Bombyx mori*. Proc Natl Acad Sci U S A 86:6843–6847

Kawakoshi A et al (2003) A single and novel natriuretic peptide is expressed in the heart and brain of the most primitive vertebrate, the hagfish (*Eptatretus burgeri*). J Mol Endocrinol 31:209–220

Kawakoshi A et al (2006) Identification of a natriuretic peptide (NP) in cyclostomes (lamprey and hagfish): CNP-4 is the ancestral gene of the NP family. Gen Comp Endocrinol 148:41–47

Kawamoto T et al (1992) Role of steroid 11 beta-hydroxylase and steroid 18-hydroxylase in the biosynthesis of glucocorticoids and mineralocorticoids in humans. Proc Natl Acad Sci U S A 89:1458–1462

Kawauchi H et al (1983) Characterization of melanin-concentrating hormone in chum salmon pituitaries. Nature 305:321–323
Kean L et al (2002) Two nitridergic peptides are encoded by the gene capability in *Drosophila melanogaster*. Am J Physiol Regul Integr Comp Physiol 282:R1297–R1307
Keeler C, Dannies PS, Hodsdon ME (2003) The tertiary structure and backbone dynamics of human prolactin. J Mol Biol 328:1105–1121
Keenan DM, Roelfsema F, Veldhuis JD (2004) Endogenous ACTH concentration-dependent drive of pulsatile cortisol secretion in the human. Am J Physiol Endocrinol Metab 287:E652–E661
Kellogg J et al (1989) Corticotropin releasing hormone (CRH) maps to human chromosome 8 and identifies a TaqI RFLP. Cytogenet Cell Genet 51:1022–1022
Kelly SP, Peter RE (2006) Prolactin-releasing peptide, food intake, and hydromineral balance in goldfish. Am J Physiol Regul Integr Comp Physiol 291:R1474–R1481
Kim H-Y et al (2004a) Solution structure of human orexin-A: regulator of appetite and wakefulness. J Biochem Mol Biol 37:565–573
Kim YJ et al (2004b) Corazonin receptor signaling in ecdysis initiation. Proc Natl Acad Sci U S A 101:6704–6709
Kim Y et al (2015) Ghrelin is required for dietary restriction-induced enhancement of hippocampal neurogenesis: lessons from ghrelin knockout mice. Endocr J 62(3):269–275
Kimura HJ et al (2009) Immunoproteasome overexpression underlies the pathogenesis of thyroid oncocytes and primary hypothyroidism: studies in humans and mice. PLoS One 4:e7857
Kirchgessner AL, Liu M (1999) Orexin synthesis and response in the gut. Neuron 24:941–951
Kirchgessner AL (2002) Orexins in the brain-gut axis. Endocr Rev 23:1–15
Kleine B et al (2001) Expression of galanin in human placenta. Mol Hum Reprod 7:379–385
de Kloet ER (2003) Hormones, brain and stress. Endocr Regul 37:51–68
Klovins J et al (2004) The melanocortin system in fugu: determination of POMC/AgRP/MCR gene repertoire and synteny, as well as pharmacology and anatomical distribution of the MCRs. Mol Biol Evol 21:563–579
Knight J (2003) Drugs in sport: no dope. Nature 426:114–115
Kojima M et al (1999) Ghrelin is a growth-hormone-releasing acylated peptide from stomach. Nature 402:656–660
Konturek SJ et al (2004) Brain-gut axis and its role in the control of food intake. J Physiol Pharmacol 55:137–154
Kotaki T et al (2009) Structure determination of a new juvenile hormone from a heteropteran insect. Org Lett 11:5234–5237
Kotani M et al (2001) The metastasis suppressor gene KiSS-1 encodes kisspeptins, the natural ligands of the orphan G protein-coupled receptor GPR54. J Biol Chem 276:34631–34636
Kramer SJ et al (1991) Identification of an allatostatin from the tobacco hornworm *Manduca sexta*. Proc Natl Acad Sci U S A 88:9458–9462
Kreshchenko ND (2008) Functions of flatworm neuropeptides NPF, GYIRF and FMRF in course of pharyngeal regeneration of anterior body fragments of planarian, *Girardia tigrina*. Acta Biol Hung 59(Suppl.):199–207
Kreshchenko ND et al (2008) Effects of neuropeptide F on regeneration in *Girardia tigrina* (Platyhelminthes). Cell Tissue Res 331:739–750
Krstić RV (1991) Human microscopic anatomy. An atlas for students of medicine and biology. Springer, Berlin/Heidelberg/New York/Tokyo
Kubli E (2003) Sex-peptides: seminal peptides of the *Drosophila* male. Cell Mol Life Sci 60:1689–1704
Lapthorn AJ et al (1994) Crystal structure of human chorionic gonadotropin. Nature 369:455–461
Lauber M, Muller J (1989) Purification and characterization of two distinct forms of rat adrenal cytochrome P450(11) beta: functional and structural aspects. Arch Biochem Biophys 274:109–119
Laudet V (1997) Evolution of the nuclear receptor superfamily: early diversification from an ancestral orphan receptor. J Mol Endocrinol 19:207–226

de Lecea L et al (1996) A cortical neuropeptide with neuronal depressant and sleep-modulating properties. Nature 381:242–245
de Lecea L et al (1998) The hypocretins: hypothalamus-specific peptides with neuroexcitatory activity. Proc Natl Acad Sci U S A 95:322–327
Lee G, Bahn JH, Park JH (2006) Sex- and clock-controlled expression of the neuropeptide F gene in *Drosophila*. Proc Natl Acad Sci U S A 103:12580–12585
Lee JH et al (1996) KiSS-1, a novel human malignant melanoma metastasis-suppressor gene. J Natl Cancer Inst 88:1731–1737
Lee WS et al (1994) The high affinity Na^+/glucose cotransporter. Re-evaluation of function and distribution of expression. J Biol Chem 269:12032–12039
Lerch M, Mayrhofer M, Zerbe O (2004) Structural similarities of micelle-bound peptide YY (PYY) and neuropeptide Y (NPY) are related to their affinity profiles at the Y receptors. J Mol Biol 339:1153–1168
Levi L (1989) Occupational stressors, biological stress and workers' health. J UOEH 11:229–245
Levin ER (2005) Integration of the extranuclear and nuclear actions of estrogen. Mol Endocrinol 19:1951–1959
Levin ER (2008) Rapid signaling by steroid receptors. Am J Physiol Regul Integr Comp Physiol 295:R1425–R1430
Lewis UJ, Singh RN, Lewis LJ (1989) Two forms of glycosylated human prolactin have different pigeon crop sac-stimulating activities. Endocrinology 124:1558–1563
Li C, Kim K, Nelson LS (1999) FMRFamide-related neuropeptide gene family in *Caenorhabditis elegans*. Br Res 848:26–34
Li J et al (1997) Conservation of function between mammalian and plant steroid 5alpha-reductases. Proc Natl Acad Sci U S A 94:3554–3559
Li WX (2005) Functions and mechanisms of receptor tyrosine kinase Torso signaling: lessons from Drosophila embryonic terminal development. Dev Dyn 232:656–672
Lin L et al (1999) The sleep disorder canine narcolepsy is caused by a mutation in the hypocretin (orexin) receptor 2 gene. Cell 98:365–376
Lin SY et al (2003) Regulation of ovarian function by the TGF-beta superfamily and follistatin. Reproduction 126:133–148
Lincoln GA, Andersson H, Loudon A (2003) Clock genes in calendar cells as the basis of annual timekeeping in mammals–a unifying hypothesis. J Endocrinol 179:1–13
Lingueglia E et al (1995) Cloning of the amiloride-sensitive FMRFamide peptidegated sodium channel. Nature 378:730–733
Loi PK et al (2001) Identification, sequence and expression of a crustacean cardioactive peptide (CCAP) gene in the moth *Manduca sexta*. J Exp Biol 204:2803–2816
Lou H, Gagel RF (1998) Alternative RNA processing–its role in regulating expression of calcitonin/calcitonin gene-related peptide. J Endocrinol 156:401–405
Lovejoy DA et al (1992) Distinct sequence of gonadotropin-releasing hormone (GnRH) in dogfish brain provides insight into GnRH evolution. Proc Natl Acad Sci U S A 89:6373–6377
Lukacik P et al (2007) Structural and biochemical characterization of human orphan DHRS10 reveals a novel cytosolic enzyme with steroid dehydrogenase activity. Biochem J 402:419–427
Luo CW et al (2005) Bursicon, the insect cuticle-hardening hormone, is a heterodimeric cystine knot protein that activates G protein-coupled receptor LGR2. Proc Natl Acad Sci U S A 102:2820–2825
Mahendroo MS, Russell DW (1999) Male and female isoenzymes of steroid 5alpha-reductase. Rev Reprod 4:179–183
Marca AL et al (2009) Anti-Mullerian hormone (AMH): what do we still need to know? Hum Reprod 24:2264–2275
Marti T et al (1987) Microanalysis of the amino acid sequence of the eclosion hormone from the tobacco hornworm *Manduca sexta*. FEBS Lett 219:415–418
Martin CJ (1927) Life and work (of Ernest Henry Starling). Br Med J 1:900–906
Martínez A (2006) A new family of angiogenic factors. Cancer Lett 236:157–163

Martínez A et al (2005) Gastrin-releasing peptide (GRP) induces angiogenesis and the specific GRP blocker 77427 inhibits tumor growth in vitro and in vivo. Oncogene 24:4106–4113
Maruyama K et al (1992) Determination of disulfide bond arrangement in bombyxin-IV, an insulin superfamily peptide from the silkworm, *Bombyx mori*, by combination of thermolysin digestion of natural peptide and selective synthesis of disulfide bond isomers. J Protein Chem 11:13–20
Maruyama M et al (1999) Immunocytochemical localization of prolactin-releasing peptide in the rat brain. Endocrinology 140:2326–2333
Matsuo H et al (1971) Structure of the porcine LH- and FSH-releasing hormone. I. The proposed amino acid sequence. Biochem Biophys Res Commun 43:1334–1339
McDearmid JR, Brezina V, Weiss KR (2002) AMRP peptides modulate a novel K(+) current in pleural sensory neurons of *Aplysia*. J Neurophysiol 88:323–332
McKinley MJ, Johnson AK (2004) The physiological regulation of thirst and fluid intake. News Physiol Sci 19:1–6
Mehlhorn H (2001) Pediculus. In: Mehlhorn H (ed) Encyclopedic reference of parasitology. Springer, Berlin/Heidelberg, pp 467–467
Meier U, Gressner AM (2004) Endocrine regulation of energy metabolism: review of pathobiochemical and clinical chemical aspects of leptin, ghrelin, adiponectin, and resistin. Clin Chem 50:1511–1525
Melcher C, Bader R, Pankratz MJ (2007) Amino acids, taste circuits, and feeding behavior in *Drosophila*: towards understanding the psychology of feeding in flies and man. J Endocrinol 192:467–472
Melcher C, Pankratz MJ (2005) Candidate gustatory interneurons modulating feeding behavior in the Drosophila brain. PLoS Biol 3:e305
Meng X et al (2002) The *Drosophila* hugin gene codes for myostimulatory and ecdysis-modifying neuropeptides. Mech Dev 117:5–13
Merchenthaler I (1998) LHRH and sexual dimorphism. Ann N Y Acad Sci 863:175–187
Merte J, Nichols R (2002) *Drosophila melanogaster* FMRFamide-containing peptides: redundant or diverse functions? Peptides 23:209–220
Messager S et al (1999) Decoding photoperiodic time through Per1 and ICER gene amplitude. Proc Natl Acad Sci U S A 96:9938–9943
Messager S et al (2005) Kisspeptin directly stimulates gonadotropin-releasing hormone release via G protein-coupled receptor 54. Proc Natl Acad Sci U S A 102:1761–1766
Miller MA (1998) Regulation of galanin in memory pathways. Ann N Y Acad Sci 863:323–341
Mitani F et al (1999) Development of functional zonation in the rat adrenal cortex. Endocrinology 140:3342–3353
Mitchell V et al (2003) A confocal microscopic study of gonadotropin-releasing hormone (GnRH) neuron inputs to dopaminergic neurons containing estrogen receptor alpha in the arcuate nucleus of GnRH-green fluorescent protein transgenicmice. Neuroendocrinology 77:198–207
Miyamoto K et al (1983) Evidence that chicken hypothalamic luteinizing hormone-releasing hormone is [Gln8]-LH-RH. Life Sci 32:1341–1347
Miyamoto K et al (1984) Identification of the second gonadotropin-releasing hormone in chicken hypothalamus: evidence that gonadotropin secretion is probably controlled by two distinct gonadotropin-releasing hormones in avian species. Proc Natl Acad Sci U S A 81:3874–3878
Miyazaki M, Takai S (2006) Tissue angiotensin II generating system by angiotensin-converting enzyme and chymase. J Pharmacol Sci 100:391–397
Mochiduki A et al (2010) Stress response of prolactin-releasing peptide knockout mice as to glucocorticoid secretion. J Neuroendocrinol 22:576–584
Moeller G, Adamski J (2006) Multifunctionality of human 17beta-hydroxysteroid dehydrogenases. Mol Cell Endocrinol 248:47–55
Moeller G, Adamski J (2009) Integrated view on 17beta-hydroxysteroid dehydrogenases. Mol Cell Endocrinol 301:7–19
Moncaut N et al (2005) Five gonadotrophin-releasing hormone receptors in a teleost fish: isolation, tissue distribution and phylogenetic relationships. J Mol Endocrinol 34:767–779

Montero M et al (2000) Molecular evolution of the growth hormone-releasing hormone/pituitary adenylate cyclase-activating polypeptide gene family. Functional implication in the regulation of growth hormone secretion. J Mol Endocrinol 25:157–168

Moore KL (2003) The biology and enzymology of protein tyrosine O-sulfation. J Biol Chem 278:24243–24246

Moran TH, Kinzig KP (2004) Gastrointestinal satiety signals II. Cholecystokinin. Am J Physiol Gastrointest Liver Physiol 286:G183–G188

Morash B et al (1999) Leptin gene expression in the brain and pituitary gland. Endocrinology 140:5995–5998

Morishita F et al (1997) A novel D-amino-acid-containing peptide isolated from *Aplysia heart.* Biochem Biophys Res Commun 240:354–358

Morishita F et al (2003a) Distribution and function of an *Aplysia* cardioexcitatory peptide, NdWFamide, in pulmonate snails. Peptides 24:1533–1544

Morishita F et al (2003b) Distribution of the *Aplysia* cardioexcitatory peptide, NdWFamide, in the central and peripheral nervous systems of *Aplysia.* Cell Tissue Res 312:95–111

Morris DJ (1993) Reduced aldosterone metabolites in hypertension. J Steroid Biochem Mol Biol 45:19–25

Muller B et al (2001) Ubiquitous expression of the calcitonin-I gene in multiple tissues in response to sepsis. J Clin Endocrinol Metab 86:396–404

Müller EE, Locatelli V, Cocchi D (1999) Neuroendocrine control of growth hormone secretion. Physiol Rev 79:511–607

Munyiri FN, Ishikawa Y (2004) Endocrine changes associated with metamorphosis and diapause induction in the yellow-spotted longicorn beetle, *Psacothea hilaris.* J Insect Physiol 50:1075–1081

Murakami I et al (2007) Corticotropin-releasing hormone or dexamethasone regulates rat proopiomelanocortin transcription through Tpit/Pitx-responsive element in its promoter. J Endocrinol 193:279–290

Nagalakshmi VK et al (2004) The presence of *Drosophila melanogaster* sex peptidelike immunoreactivity in the accessory glands of male *Helicoverpa armigera.* J Insect Physiol 50:241–248

Nagasaki H et al (2006) Differential expression of the thyrostimulin subunits, glycoprotein alpha2 and beta5 in the rat pituitary. J Mol Endocrinol 37:39–50

Nagasawa H et al (1986) Amino acid sequence of a prothoracicotropic hormone of the silkworm *Bombyx mori.* Proc Natl Acad Sci U S A 83:5840–5843

Nahon J-L (2006) The melanocortins and melanin-concentrating hormone in the central regulation of feeding behavior and energy homeostasis. C R Biol 329:623–638

Nakabayashi K et al (2002) Thyrostimulin, a heterodimer of two new human glycoprotein hormone subunits, activates the thyroid-stimulating hormone receptor. J Clin Invest 109:1445–1452

Nässel DR (1999) Tachykinin-related peptides in invertebrates: a review. Peptides 20:141–158

Nässel DR et al (2008) A large population of diverse neurons in the *Drosophila* central nervous system expresses short neuropeptide F, suggesting multiple distributed peptide functions. BMC Neurosci 9:90

Navenot JM et al (2005) Kisspeptin-10-induced signaling of GPR54 negatively regulates chemotactic responses mediated by CXCR4: a potential mechanism for the metastasis suppressor activity of kisspeptins. Cancer Res 65:10450–10456

Ngamvongchon S et al (1992) Primary structures of two forms of gonadotropin-releasing hormone, one distinct and one conserved, from catfish brain. Mol Cell Neurosci 3:17–22

Nichols R (2003) Signaling pathways and physiological functions of *Drosophila melanogaster* FMRFamide-related peptides. Ann Rev Entomol 48:485–503

Nichols R (2007) The first nonsulfated sulfakinin activity reported suggests nsDSK acts in gut biology. Peptides 28:767–773

Nijhout H (1984) Abdominal stretch reception in *Dipetalogaster maximus* (Hemiptera: reduviidae). J Insect Physiol 30:629–633

Nikitin M (2014) Bioinformatic prediction of *Trichoplax adhaerens* regulatory peptides. Gen Comp Endocrinol 212:145–155

Nunemaker CS, DeFazio RA, Moenter SM (2003a) Calcium current subtypes in GnRH neurons. Biol Reprod 69:1914–1922

Nunemaker CS et al (2003b) Gonadotropin-releasing hormone neurons generate interacting rhythms in multiple time domains. Endocrinology 144:823–831

O'Donnell L et al (2001) Estrogen and spermatogenesis. Endocr Rev 22:289–318

Ogawa Y et al (2001) Brain natriuretic peptide appears to act locally as an antifibrotic factor in the heart. Can J Physiol Pharmacol 79:723–729

Okabe M, Graham A (2004) The origin of the parathyroid gland. Proc Natl Acad Sci U S A 101:17716–17719

Okada SL et al (2006) A glycoprotein hormone expressed in corticotrophs exhibits unique binding properties on thyroid-stimulating hormone receptor. Mol Endocrinol 20:414–425

Okubo K et al (2000) A novel form of gonadotropin-releasing hormone in the medaka, *Oryzias latipes*. Biochem Biophys Res Commun 276:298–303

Ono M et al (1990) cDNA cloning of somatolactin, a pituitary protein related to growth hormone and prolactin. Proc Natl Acad Sci U S A 87:4330–4334

Osugi T et al (2006) Evolutionary origin and divergence of PQRFamide peptides and LPXRFamide peptides in the RFamide peptide family. Insights from novel lamprey RFamide peptides. FEBS J 273:1731–1743

Oumi T et al (1994) Annetocin: an oxytocin-related peptide isolated from the earthworm, *Eisenia foetida*. Biochem Biophys Res Commun 198:393–399

Page NM et al (2003) Characterization of the endokinins: human tachykinins with cardiovascular activity. Proc Natl Acad Sci U S A 100:6245–6250

Park Y, Kim YJ, Adams ME (2002) Identification of G protein-coupled receptors for *Drosophila* PRXamide peptides, CCAP, corazonin, and AKH supports a theory of ligand-receptor coevolution. Proc Natl Acad Sci U S A 99:11423–11428

Parthasarathy R, Tan A, Palli SR (2008) bHLH-PAS family transcription factor methoprene-tolerant plays a key role in JH action in preventing the premature development of adult structures during larval-pupal metamorphosis. Mech Dev 125:601–616

Patel RT et al (2005) Activation of the lipid droplet controls the rate of lipolysis of triglycerides in the insect fat body. J Biol Chem 280:22624–22631

Patel YC (1999) Somatostatin and its receptor family. Front Neuroendocrinol 20:157–198

Pauli GF et al (2010) Occurrence of progesterone and related animal steroids in two higher plants. J Nat Prod 73:338–345

Payne AH, Hales DB (2004) Overview of steroidogenic enzymes in the pathway from cholesterol to active steroid hormones. Endocr Rev 25:947–970

Pazos AJ et al (2003) Sterol composition of gonad, muscle and digestive gland of *Pecten maximus* from Malaga (South Spain). Comp Biochem Physiol B Biochem Mol Biol 134:435–446

Pedram A et al (2007) A conserved mechanism for steroid receptor translocation to the plasma membrane. J Biol Chem 282:22278–22288

Peltoketo H et al (Aug. 1999a) 17beta-hydroxysteroid dehydrogenase (HSD)/17-ketosteroid reductase (KSR) family; nomenclature and main characteristics of the 17HSD/KSR enzymes. J Mol Endocrinol 23:1–11

Peltoketo H et al (1999b) Two 17beta-hydroxysteroid dehydrogenases (17β-HSDs) of estradiol biosynthesis: 17β-HSD type 1 and type 7. J Steroid Biochem Mol Biol 69:431–439

Penzlin H, Ramm K (2008) Lehrbuch der Tierphysiologie, 7th edn. Spektrum, Heidelberg

Perillo M, Arnone MI (2014) Characterization of insulin-like peptides (ILPs) in the sea urchin *Strongylocentrotus purpuratus*: insights on the evolution of the insulin family. Gen Comp Endocrinol 205:68–79. The 17th International Congress of Comparative Endocrinology

Pernet V, Anctil M, Grimmelikhuijzen CJ (2004) Antho-RFamide-containing neurons in the primitive nervous system of the anthozoan *Renilla koellikeri*. J Comp Neurol 472:208–220

Phlippen MK et al (2000) Ecdysis of decapod crustaceans is associated with a dramatic release of crustacean cardioactive peptide into the haemolymph. J Exp Biol 203:521–536

Pietrantonio PV et al (2005) The mosquito *Aedes aegypti* (L.) leucokinin receptor is a multiligand receptor for the three *Aedes* kinins. Insect Mol Biol 14:55–67

Pissios P, Bradley RL, Maratos-Flier E (2006) Expanding the scales: the multiple roles of MCH in regulating energy balance and other biological functions. Endocr Rev 27:606–620
Podust LM, Poulos TL, Waterman MR (2001) Crystal structure of cytochrome P450 14alpha -sterol demethylase (CYP51) from *Mycobacterium tuberculosis* in complex with azole inhibitors. Proc Natl Acad Sci U S A 98:3068–3073
Powell JF et al (1994) Three forms of gonadotropin-releasing hormone characterized from brains of one species. Proc Natl Acad Sci U S A 91:12081–12085
Powell JF et al (1996) Two new forms of gonadotropin-releasing hormone in a protochordate and the evolutionary implications. Proc Natl Acad Sci U S A 93:10461–10464
Predel R et al (2005) Tachykinin-related peptide precursors in two cockroach species. FEBS J 272:3365–3375
Price DA, Greenberg MJ (1977) Structure of a molluscan cardioexcitatory neuropeptide. Science 197:670–671
Proux JP et al (1987) Identification of an arginine vasopressin-like diuretic hormone from *Locusta migratoria*. Biochem Biophys Res Commun 149:180–186
Qu D et al (1996) A role for melanin-concentrating hormone in the central regulation of feeding behaviour. Nature 380:243–247
Quaynor SD et al (2013) Delayed puberty and estrogen resistance in a woman with estrogen receptor α variant. N Engl J Med 369:164–171
Quigley CA et al (1995) Androgen receptor defects: historical, clinical, and molecular perspectives. Endocr Rev 16:271–321
Rao K (2001) Crustacean pigmentary-effector hormones: chemistry and functions of RPCH, PDH, and related peptides. Amer Zool 41:364–379
Rehfeld JF et al (2001) The predominant cholecystokinin in human plasma and intestine is cholecystokinin-33. J Clin Endocrinol Metab 86:251–258
Rewitz KF et al (2009) The insect neuropeptide PTTH activates receptor tyrosine kinase torso to initiate metamorphosis. Science 326:1403–1405
Rizwan MZ et al (2009) Cells expressing RFamide-related peptide-1/3, the mammalian gonadotropin-inhibitory hormone orthologs, are not hypophysiotropic neuroendocrine neurons in the rat. Endocrinology 150:1413–1420
Robas N, Mead E, Fidock M (2003) MrgX2 is a high potency cortistatin receptor expressed in dorsal root ganglion. J Biol Chem 278:44400–44404
Robins T et al (2006) Molecular model of human CYP21 based on mammalian CYP2C5: structural features correlate with clinical severity of mutations causing congenital adrenal hyperplasia. Mol Endocrinol 20:2946–2964
Roch G J et al (2014) GnRH receptors and peptides: skating backward. Gen Comp Endocrinol 209:118–134
Roche J et al (1950) The thyroglobulin of dogs treated with thyrostimulin or with thyroid antagonists derived from thiouracil. Comptes Rendus Des Séances De La Société De Biologie Et De Ses Filiales 144:917–919
Rocheville M et al (2000a) Receptors for dopamine and somatostatin: formation of hetero-oligomers with enhanced functional activity. Science 288:154–157
Rowe M L et al (2014) Neuropeptides and polypeptide hormones in echinoderms: new insights from analysis of the transcriptome of the sea cucumber Apostichopus japonicus. Gen Comp Endocrinol 197:43–55
Rocheville M et al (2000b) Subtypes of the somatostatin receptor assemble as functional homo- and heterodimers. J Biol Chem 275:7862–7869
Rustay NR et al (2005) Galanin impairs performance on learning and memory tasks: findings from galanin transgenic and GAL-R1 knockout mice. Neuropeptides 39:239–243
Saavedra JM (1992) Brain and pituitary angiotensin. Endocr Rev 13:329–380
Sailer AW et al (2001) Identification and characterization of a second melanin-concentrating hormone receptor, MCH-2R. Proc Natl Acad Sci U S A 98:7564–7569
Sakamoto, H. (2011) Gastrin-releasing peptide system in the spinal cord mediates masculine sexual function. Anat Sci Int 86:19–29

Saliot A, Barbier M (1971) Isolation of progesterone and several ketosteroids of the female part of the gonads of the scallop *Pecten maximus*. Biochimie 53:265–266

Santen RJ et al (2009) History of aromatase: saga of an important biological mediator and therapeutic target. Endocr Rev 30:343–375

dos Santos MT et al (2011) Human stanniocalcin-1 interacts with nuclear and cytoplasmic proteins and acts as a SUMO E3 ligase. Mol Biosyst 7:180–193

Sasaki K, Morishita F, Furukawa Y (2004) Peptidergic innervation of the vasoconstrictor muscle of the abdominal aorta in *Aplysia kurodai*. J Exp Biol 207:4439–4450

Sato Y et al (1993) Precursor polyprotein formultiple neuropeptides secreted from the suboesophageal ganglion of the silkworm *Bombyx mori*: characterization of the cDNA encoding the diapause hormone precursor and identification of additional peptides. Proc Natl Acad Sci U S A 90:3251–3255

Sauman I, Reppert SM (1996) Molecular characterization of prothoracicotropic hormone (PTTH) from the giant silkmoth *Antheraea pernyi*: developmental appearance of PTTH-expressing cells and relationship to circadian clock cells in central brain. Dev Biol 178:418–429

Schneider O, Nau R, Michel U (2000) Comparative analysis of follistatin-, activin beta A and activin beta B-mRNA steady-state levels in diverse porcine tissues by multiplex S1 nuclease analysis. Eur J Endocrinol 142:537–544

Schooley D, Baker F (1985) Juvenile hormone biosynthesis. In: Kerkut GA, Gilbert LI (eds) Comprehensive insect physiology, biochemistry, and pharmacology, vol 7, 1st edn. Pergamon Press, New York, pp 363–389

Schuetz P, Albrich W, Mueller B (2011) Procalcitonin for diagnosis of infection and guide to antibiotic decisions: past, present and future. BMC Med 9:107

Schulz C, Paulus K, Lehnert H (2004) Central nervous and metabolic effects of intranasally applied leptin. Endocrinology 145:2696–2701

Sellami A, Agricola HJ, Veenstra J (2011) Neuroendocrine cells in *Drosophila melanogaster* producing GPA2/GPB5, a hormone with homology to LH, FSH and TSH. Gen Comp Endocrinol 170:582–588

Selye H (1936) A syndrome produced by various nocuous agents. Nature (Lond.) 148:84–85

Selye H (1950) The physiology and pathology of exposure to stress. Acta, Montreal

Shah K et al (2003) The genetic basis of infertility. Reproduction 126:13–25

Sheridan MA, Kittilson JD, Slagter BJ (2000) Structure-function relationships of the signaling system for the somatostatin peptide hormone family. Am Zool 40:269–286

Sherwood N et al (1983) Characterization of a teleost gonadotropin-releasing hormone. Proc Natl Acad Sci U S A 80:2794–2798

Sherwood NM et al (1986) Primary structure of gonadotropin-releasing hormone from lamprey brain. J Biol Chem 261:4812–4819

Sherwood NM, Krueckl SL, McRory JE (2000) The origin and function of the pituitary adenylate cyclase-activating polypeptide (PACAP)/glucagon superfamily. Endocr Rev 21:619–670

Shiga S (2003) Anatomy and functions of brain neurosecretory cells in diptera. Microsc Res Tech 62:114–131

Shimasaki S et al (1988) Primary structure of the human follistatin precursor and its genomic organization. Proc Natl Acad Sci U S A 85:4218–4222

Shiraishi E et al (2008) Mullerian inhibiting substance is required for germ cell proliferation during early gonadal differentiation in medaka (Oryzias latipes). Endocrinology 149:1813–1819

Shozu M et al (2003) Estrogen excess associated with novel gain-of-function mutations affecting the aromatase gene. N Engl J Med 348:1855–1865

Shumyatsky GP et al (2002) Identification of a signaling network in lateral nucleus of amygdala important for inhibiting memory specifically related to learned fear. Cell 111:905–918

Siebel AL, Bathgate RA, Parry LJ (2005) Differential expression of mesotocin receptors in the uterus and ovary of the pregnant tammar wallaby. Reproduction 129:639–649

Siegmund T, Korge G (2001) Innervation of the ring gland of *Drosophila melanogaster*. J Comp Neurol 431:481–491

Sikder D, Kodadek T (2007) The neurohormone orexin stimulates hypoxia-inducible factor-1 activity. Genes Dev 21:2995–3005

Simpson ER et al (1994) Aromatase cytochrome P450, the enzyme responsible for estrogen biosynthesis. Endocr Rev 15:342–355

Siwicki KK et al (1985) Proctolin in the lobster nervous system. Peptides 6(Suppl 3):393–402

Smit AB et al (1994) Structural characterization of a *Lymnaea* putative endoprotease related to human furin. FEBS Lett 343:27–31

Smith EP et al (1994) Estrogen resistance caused by a mutation in the estrogen-receptor gene in a man. N Engl J Med 331:1056–1061

Smith CL et al (2014) Novel cell types, neurosecretory cells, and body plan of the early-diverging metazoan trichoplax adhaerens. Curr Biol 24(14):1565–1572

Soller M et al (2006) Sex-peptide-regulated female sexual behavior requires a subset of ascending ventral nerve cord neurons. Curr Biol 16:1771–1782

Sossin WS, Sweet-Cordero A, Scheller RH (1990) Dale's hypothesis revisited: different neuropeptides derived from a common prohormone are targeted to different processes. Proc Natl Acad Sci U S A 87:4845–4848

Southey BR, Rodriguez-Zas SL, Sweedler JV (2006) Prediction of neuropeptide prohormone cleavages with application to RFamides. Peptides 27:1087–1098

Southey BR, Sweedler JV, Rodriguez-Zas SL (2008) Prediction of neuropeptide cleavage sites in insects. Bioinformatics 24:815–825

de Souza FS et al (2005) Subfunctionalization of expression and peptide domains following the ancient duplication of the proopiomelanocortin gene in teleost fishes. Mol Biol Evol 22:2417–2427

Sower SA et al (1993) Primary structure and biological activity of a third gonadotropin-releasing hormone from lamprey brain. Endocrinology 132:1125–1131

Spijker S et al (2004) Stimulus-dependent regulation and cellular expression of genes encoding neuropeptides, prohormone convertases, alpha-amidating enzyme and 7B2 in identified *Lymnaea* neurons. J Neurochem 90:287–296

Spittaels K et al (1996) Insect neuropeptide F (NPF)-related peptides: isolation from Colorado potato beetle (*Leptinotarsa decemlineata*) brain. Insect Biochem Mol Biol 26:375–382

Stangier J et al (1987) Unusual cardioactive peptide (CCAP) from pericardial organs of the shore crab *Carcinus maenas*. Proc Natl Acad Sci U S A 84:575–579

Stangier J et al (1992) Orcokinin: a novel myotropic peptide from the nervous system of the crayfish, *Orconectes limosus*. Peptides 13:859–864

Starratt AN, Brown BE (1975) Structure of the pentapeptide proctolin, a proposed neurotransmitter in insects. Life Sci 17:1253–1256

Stay B, Tobe SS (2007) The role of allatostatins in juvenile hormone synthesis in insects and crustaceans. Ann Rev Entomol 52:277–299

Steel CG, Vafopoulou X (2006) Circadian orchestration of developmental hormones in the insect, *Rhodnius prolixus*. Comp Biochem Physiol A Mol Integr Physiol 144:351–364

Stoffel-Wagner B (2001) Neurosteroid metabolism in the human brain. Eur J Endocrinol 145:669–679

Strakis C, Chrousos G (1997) Hypothalamic hormones. In: Conn P, Melmed S (eds) Endocrinology: basic and clinical principles, chap 13. Humana Press, Totowa, pp 185–209

Sudo S et al (2005) Heterodimeric fly glycoprotein hormone-alpha2 (GPA2) and glycoprotein hormone-beta5 (GPB5) activate fly leucine-rich repeat-containing G protein-coupled receptor-1 (DLGR1) and stimulation of human thyrotropin receptors by chimeric fly GPA2 and human GPB5. Endocrinology 146:3596–3604

Suva LJ et al (1987) A parathyroid hormone-related protein implicated in malignant hypercalcemia: cloning and expression. Science 237:893–896

Suzuki C et al (2007) The LIM domain homeobox gene ISL-1 is a positive regulator of glycoprotein alpha 2 (GPA2), a subunit of thyrostimulin. Regulatory Peptides 142:60–67

Suzuki C et al (2009) Inflammatory cytokines regulate glycoprotein subunit beta5 of thyrostimulin through nuclear factor-kappaB. Endocrinology 150:2237–2243

Takei Y (2001) Does the natriuretic peptide system exist throughout the animal and plant kingdom? Comp Biochem Physiol B Biochem Mol Biol 129:559–573

Takeishi Y et al (2007) Linkage disequilibrium analyses of natriuretic peptide precursor B locus reveal risk haplotype conferring high plasma BNP levels. Biochem Biophys Res Commun 362:480–484

Tamura N et al (2000) Cardiac fibrosis in mice lacking brain natriuretic peptide. Proc Natl Acad Sci U S A 97:4239–4244

Tando Y, Kubokawa K (2009) Expression of the gene for ancestral glycoprotein hormone beta subunit in the nerve cord of amphioxus. Gen Comp Endocrinol 162:329–339

Taniyama S et al (1999) Changes in the levels of mRNAs for GH/prolactin/somatolactin family and Pit-1/GHF-1 in the pituitaries of pre-spawning chum salmon. J Mol Endocrinol 23:189–198

Tartaglia LA (1997) The leptin receptor. J Biol Chem 272:6093–6096

Tatemoto K et al (1983) Galanin - a novel biologically active peptide from porcine intestine. FEBS Lett 164:124–128

Teixeira J, Maheswaran S, Donahoe PK (2001) Müllerian inhibiting substance: an instructive developmental hormone with diagnostic and possible therapeutic applications. Endocr Rev 22:657–674

Tessmar-Raible K et al (2007) Conserved sensory-neurosecretory cell types in annelid and fish forebrain: insights into hypothalamus evolution. Cell 129:1389–1400

Thirumalai V, Marder E (2002) Colocalized neuropeptides activate a central pattern generator by acting on different circuit targets. J Neurosci 22:1874–1882

Thomas A et al (2009) Determination of Synacthen in urine for sports drug testing by means of nano-ultra-performance liquid chromatography/tandem mass spectrometry. *Rapid Commun.* Mass Spectrom 23:2669–2674

Thomas P et al (2005) Identity of an estrogen membrane receptor coupled to a G protein in human breast cancer cells. Endocrinology 146:624–632

Thompson TB et al (2004) Beta A versus beta B: is it merely a matter of expression? Mol Cell Endocrinol 225:9–17

Thornton JW (2001) Evolution of vertebrate steroid receptors from an ancestral estrogen receptor by ligand exploitation and serial genome expansions. Proc Natl Acad Sci U S A 98:5671–5676

Thurston RJ, Korn N (2000) Spermiogenesis in commercial poultry species: anatomy and control. Poult Sci 79:1650–1668

Tillman JA et al (1999) Insect pheromones–an overview of biosynthesis and endocrine regulation. Insect Biochem Mol Biol 29:481–514

Torday JS, Rehan VK (2007) The evolutionary continuum from lung development to homeostasis and repair. Am J Physiol Lung Cell Mol Physiol 292:L608–L611

Torfs P et al (2001) Pyrokinin neuropeptides in a crustacean. Isolation and identification in the white shrimp *Penaeus vannamei*. Eur J Biochem 268:149–154

Tortorella C, Neri G, Nussdorfer GG (2007) Galanin in the regulation of the hypothalamic-pituitary-adrenal axis. Int J Mol Med 19:639–647

Tostivint H et al (2006) Comparative genomics provides evidence for close evolutionary relationships between the urotensin II and somatostatin gene families. Proc Natl Acad Sci U S A 103:2237–2242

Trube A, Audehm U, Dircksen H (1994) Crustacean cardioactive peptide-immunoreactive neurons in the ventral nervous system of crayfish. J Comp Neurol 348:80–93

Trudeau VL (2009) Really old hormones up to new tricks: glycoprotein hormone subunits may have roles in development. Endocrinology 150:3446–3447

Truman JW (1992) The eclosion hormone system of insects. Prog Brain Res 92:361–374

Truman JW (2005) Hormonal control of insect ecdysis: endocrine cascades for coordinating behavior with physiology. Vitam Horm 73:1–30

Truman JW, Copenhaver PF (1989) The larval eclosion hormone neurones in *Manduca sexta*: identification of the brain-proctodeal neurosecretory system. J Exp Biol 147:457–470

Tsutsui K (2009) A new key neurohormone controlling reproduction, gonadotropin-inhibitory hormone (GnIH): biosynthesis, mode of action and functional significance. Prog Neurobiol 88:76–88

Tsutsui K, Osugi T (2009) Evolutionary origin and divergence of GnIH and its homologous peptides. Gen Comp Endocrinol 161:30–33

Tsutsui K, Ubuka T (2014) Breakthrough in neuroendocrinology by discovering novel neuropeptides and neurosteroids: 1. Discovery of gonadotropin-inhibitory hormone (GnIH) across vertebrates. Gen Comp Endocrinol 205:4–10

Tsutsui K et al (2000) A novel avian hypothalamic peptide inhibiting gonadotropin release. Biochem Biophys Res Commun 275:661–667

Twan WH et al (2006) The presence and ancestral role of gonadotropin-releasing hormone in the reproduction of scleractinian coral, *Euphyllia ancora*. Endocrinology 147:397–406

Ubuka T et al (2005) Melatonin induces the expression of gonadotropin-inhibitory hormone in the avian brain. Proc Natl Acad Sci U S A 102:3052–3057

Ugleholdt R et al (2006) Prohormone convertase 1/3 is essential for processing of the glucose-dependent insulinotropic polypeptide precursor. J Biol Chem 281:11050–11057

Utz S et al (2008) Mas-allatotropin in the developing antennal lobe of the sphinx moth *Manduca sexta*: distribution, time course, developmental regulation, and colocalization with other neuropeptides. Dev Neurobiol 68:123–142

Vafopoulou X, Steel CG, Terry KL (2007) Neuroanatomical relations of prothoracicotropic hormone neurons with the circadian timekeeping system in the brain of larval and adult *Rhodnius prolixus* (Hemiptera). J Comp Neurol 503:511–524

Varaksina GS, Varaksin AA (1988) Localization of 17 beta-hydroxysteroid dehydrogenase in the gonads of bivalve mollusks–the sea pecten (*Patinopecten yessoensis* Jay) and Gray's mussel (*Crenomytilus grayanus* Dunker). Arkh Anat Gistol Embriol 95:79–82

Veelaert D et al (1997) Isolation and characterization of an adipokinetic hormone release-inducing factor in locusts: the crustacean cardioactive peptide. Endocrinology 138:138–142

Veenstra JA (1989) Isolation and structure of corazonin, a cardioactive peptide from the American cockroach. FEBS Lett 250:231–234

Veenstra JA (2000) Mono- and dibasic proteolytic cleavage sites in insect neuroendocrine peptide precursors. Arch Insect Biochem Physiol 43:49–63

Veenstra JA, Agricola HJ, Sellami A (2008) Regulatory peptides in fruit fly midgut. Cell Tissue Res 334:499–516

Vehovszky A et al (2005) Crustacean cardioactive peptide (CCAP)-related molluscan peptides (M-CCAPs) are potential extrinsic modulators of the buccal feeding network in the pond snail *Lymnaea stagnalis*. Neurosci Lett 373:200–205

Vilim FS et al (1999) Gene for pain modulatory neuropeptide NPFF: induction in spinal cord by noxious stimuli. Mol Pharmacol 55:804–811

Villeval JL et al (1994) Autocrine stimulation by erythropoietin (EPO) requires Epo secretion. Blood 84:2649–2662

Vortkamp A et al (1996) Regulation of rate of cartilage differentiation by Indian hedgehog and PTH-related protein. Science 273:613–622

Wall MA et al (1995) The structure of the G protein heterotrimer Gi alpha 1 beta 1 gamma 2. Cell 83:1047–1058

Wan G, Chan KM (2010) A study of somatolactin actions by ectopic expression in transgenic zebrafish larvae. J Mol Endocrinol 45:301–315

Wang XL et al (2000) Transcriptional regulation of human 11beta-hydroxylase (hCYP11B1). Endocrinology 141:3587–3594

Wang ZN, Bassett M, Rainey WE (2001) Liver receptor homologue-1 is expressed in the adrenal and can regulate transcription of 11 beta-hydroxylase. J Mol Endocrinol 27:255–258

Wasielewski O, Rosinski G (2007) Gonadoinhibitory effects of Neb-colloostatin and Neb-TMOF on ovarian development in the mealworm, *Tenebrio molitor* L. Arch Insect Biochem Physiol 64:131–141

Watanabe E et al (2000) Nav2/NaG channel is involved in control of salt-intake behavior in the CNS. J Neurosci 20:7743–7751

Watanabe S, Kaneko T (2010) Prolactin-releasing peptide receptor expressed in the pituitary in mozambique tilapia *Oreochromis mossambicus*: an aspect of prolactin regulatory mechanisms. Gen Comp Endocrinol 167:27–34

Wei Z et al (2000) Sulfakinins reduce food intake in the desert locust, *Schistocerca gregaria*. J Insect Physiol 46:1259–1265

Westheide W, Hoogewijs D (2010) Regulated oxygen sensing by protein hydroxylation in renal erythropoietin-producing cells. Am J Physiol Renal Physiol 298:F1287–F1296

Westheide W, Rieger R (2007) Systematic-Poster: Zoologie – Metazoa – Vielzellige Tiere, 2nd edn. Spektrum Akademischer Verlag, Heidelberg

Wheeler DE, Buck N, Evans JD (2006) Expression of insulin pathway genes during the period of caste determination in the honey bee, *Apis mellifera*. Insect Mol Biol 15:597–602

White BH, Ewer J (2013) Neural and hormonal control of postecdysial behaviors in insects. Ann Rev Entomol 59:363–381

White PC, Speiser PW (2000) Congenital adrenal hyperplasia due to 21-hydroxylase deficiency. Endocr Rev 21:245–291

White RB et al (1998) Second gene for gonadotropin-releasing hormone in humans. Proc Natl Acad Sci U S A 95:305–309

Whitmore TE et al (2000) Human secretin (SCT): gene structure, chromosome location, and distribution of mRNA. Cytogenet Cell Genet 90:47–52

Williams CM (1956) The juvenile hormone of insects. Nature 178:212–213

Wimalawansa SJ (1996) Calcitonin gene-related peptide and its receptors: molecular genetics, physiology, pathophysiology, and therapeutic potentials. Endocr Rev 17:533–585

Winther AM, Acebes A, Ferrus A (2006) Tachykinin-related peptides modulate odor perception and locomotor activity in *Drosophila*. Mol Cell Neurosci 31:399–406

Wichtl M (1999) Terpene. In: Rimpler H (ed) Biogene Arzneistoffe, chap 12, 2nd edn. Deutscher Apotheker Verlag, Stuttgart, pp 415–86

Wu C et al (2003a) Furin-mediated processing of Pro-C-type natriuretic peptide. J Biol Chem 278:25847–25852

Wu Q et al (2003b) Developmental control of foraging and social behavior by the *Drosophila* neuropeptide Y-like system. Neuron 39:147–161

Wu Y et al (2006) Promoter hypermethylation of progesterone receptor isoform B (PR-b) in endometriosis. Epigenetics 1:106–111

Wynick D, Bacon A (2002) Targeted disruption of galanin: new insights from knock-out studies. Neuropeptides 36:132–144

Wynick D, Bloom FE (1992) In: Grossman A (ed) Clinical endocrinology. Blackwell Scientific Publications, Oxford/Paris/Berlin/Wien

Xu M, Volkoff H (2007) Molecular characterization of prepro-orexin in Atlantic cod (*Gadus morhua*): cloning, localization, developmental profile and role in food intake regulation. Mol Cell Endocrinol 271:28–37

Yamanaka N et al (2006) Regulation of insect steroid hormone biosynthesis by innervating peptidergic neurons. Proc Natl Acad Sci U S A 103:8622–8627

Yamanaka N et al (2008) Neuropeptide receptor transcriptome reveals unidentified neuroendocrine pathways. PLoS ONE 3:e–3048

Yamanaka N et al (2010) Bombyx prothoracicostatic peptides activate the sex peptide receptor to regulate ecdysteroid biosynthesis. Proc Natl Acad Sci U S A, 107:2060–2065

Yang S-Y, He XY, Schulz H (2005) Multiple functions of type 10 17beta-hydroxysteroid dehydrogenase. Trends Endocrinol Metab 16:167–175

Yang J et al (2008) Identification of the acyltransferase that octanoylates ghrelin, an appetite-stimulating peptide hormone. Cell 132:387–396

Yapici N et al (2008) A receptor that mediates the post-mating switch in *Drosophila* reproductive behaviour. Nature 451:33–37

Yasothornsrikul S et al (2003) Cathepsin L in secretory vesicles functions as a prohormone-processing enzyme for production of the enkephalin peptide neurotransmitter. Proc Natl Acad Sci U S A 100:9590–9595

Yasuda A et al (1994) Characterization of crustacean hyperglycemic hormone from the crayfish (*Procambarus clarkii*): multiplicity of molecular forms by stereoinversion and diverse functions. Gen Comp Endocrinol 95:387–398

Yoo MS et al (2000) Molecular cloning, distribution and pharmacological characterization of a novel gonadotropin-releasing hormone ([trp8]GnRH) in frog brain. Mol Cell Endocrinol 164:197–204

van Zeijl CJJ et al (2011) Acute inflammation increases pituitary and hypothalamic glycoprotein hormone subunit b5 mRNA expression in association with decreased thyrotrophin receptorm-RNA expression inmice. Journal of Neuroendocrinology 23:310–319

Zhang JV et al (2005) Obestatin, a peptide encoded by the ghrelin gene, opposes ghrelin's effects on food intake. Science 310:996–999

Zhang L et al (2008) Molecular cloning, expression pattern, and immunocytochemical localization of a gonadotropin-releasing hormone-like molecule in the gastropod mollusk, *Aplysia californica*. Gen Comp Endocrinol 156:201–209

Zhang YH, Youm JB, Earm YE (2008) Stretch-activated non-selective cation channel: a causal link between mechanical stretch and atrial natriuretic peptide secretion. Prog Biophys Mol Biol

Zhao Z, Zera AJ (2004) The hemolymph JH titer exhibits a large-amplitude, morph-dependent, diurnal cycle in the wing-polymorphic cricket, *Gryllus firmus*. J Insect Physiol 50:93–102

Zheng J et al (2008) Studies of a receptor guanylyl cyclase cloned from Y-organs of the blue crab (*Callinectes sapidus*), and its possible functional link to ecdysteroidogenesis. Gen Comp Endocrinol 155:780–788

Zhu Y, Bond J, Thomas P (2003) Identification, classification, and partial characterization of genes in humans and other vertebrates homologous to a fish membrane progestin receptor. Proc Natl Acad Sci U S A 100:2237–2242

Zhu Y et al (2003) Cloning, expression, and characterization of a membrane progestin receptor and evidence it is an intermediary in meiotic maturation of fish oocytes. Proc Natl Acad Sci U S A 100:2231–2236

Zitnan D, Adams M (2005) Neuroendocrine regulation of insect ecdysis. In: Gilbert LI, Iatrou K, Gill SS (eds) Molecular insect science. Endocrinology, vol 3, chap 3.1. Elsevier, Boston, pp 1–60

Zitnan D et al (1996) Identification of ecdysis-triggering hormone from an epitracheal endocrine system. Science 271:88–91

Zitnan D et al (2007) Complex steroid-peptide-receptor cascade controls insect ecdysis. Gen Comp Endocrinol 153:88–96

Index

14-3-3-proteins, 245, 388
7b2 protein (SGNE1), 26

A
Acervuli, 277
Acetylcholine, 13, 34, 53, 102, 240, 250, 251, 282
Acetyl-coenzyme A, 192
Acetylserotonin-*O*-methyltransferase, 244
Acheta domesticus, 183
Achlya ambisexualis, 232
Acid secretion, 98, 317
Acromegaly, 3, 6, 41, 79, 360–361
Activin, 116, 117, 119, 257
O-Acyltransferase, 29
Acyrthosiphon pisum, 152
Addison disease, 367–368
Adenine, 21, 385, 388
Adenohypophysis, 59–81, 273, 274, 359
Adenoma
  adrenal, 215, 216
  corticotroph, 360
  feedback, 361
  gonadotroph, 361
  insulinoma, 368
  somatotroph, 360
  thyrotroph, 361
Adenylate cyclase, 44, 52, 94, 245, 254, 362, 385
Adipokinetic hormone (AKH), 139–142, 173
Adipose cells, 91, 95, 280, 313, 315, 326
Adiuretin, 86–88
Adrenal cortex, 62, 206, 221, 226, 280–282, 290, 323, 365, 367, 368
Adrenal dysfunction, 365–368
Adrenal gland, 15, 34, 42, 212, 214, 225–226, 266, 270, 274, 280–283, 288, 302, 335, 358, 365, 366, 370
Adrenal hyperplasia, 214, 367, 368
Adrenaline, 6, 7, 17, 34, 64, 227, 237–240, 263, 282, 301, 302, 353, 365, 367, 377, 379, 381
  antagonist (*see* Beta blockers)
Adrenal medulla, 16, 59, 237–239, 282–283, 365
Adrenarche, 225, 309
α_1—Adrenergic neurons, 32
Adrenergic receptors, 240, 379
Adrenocorticotropic cells, 64, 274, 275
Adrenocorticotropic hormone (ACTH), 15, 26, 30, 32, 34, 59, 61–66, 88, 98, 214, 215, 226–228, 250, 251, 274–276, 302, 303, 316, 323, 343, 360, 365–368, 377, 380
Adrenodoxin, 203, 216
Adrenodoxin reductase, 203
Aedes aegyptii, 148, 150, 155, 164, 165, 170, 184, 383
Aeshna cyanea, 332, 333, 383
Agelenopsis aperta, 144
Agouti-related protein (AgRP), 45, 46, 67, 97, 314–319
Agrotis ipsilon, 150, 383
Aldosterone, 17, 122, 123, 126, 213–216, 226, 228, 258, 259, 282, 287, 288, 334–336, 338, 358, 365, 366, 368
  corticoids, 228–229
  CYP11B2, 215–216
  deficiencies, 368
Aldosterone synthase. *See* CYP11B2
Allatostatins (AST), 135, 160, 199, 327, 328
  type A, 135, 180
  type B, 135, 160, 180, 332
Allatotropins, 135, 156, 178, 179, 181, 189, 327
Allopregnanolone, 229, 230
Alternate splicing, 22, 97, 99, 114, 119, 136, 139, 179, 219, 258

B. Kleine, W.G. Rossmanith, *Hormones and the Endocrine System*,
DOI 10.1007/978-3-319-15060-4

γ—Aminobutyric acid (GABA), 34, 51, 53, 230, 306, 307, 338
Amygdala, 40, 54, 116, 314–316, 338
central nucleus, 316, 338
Anabolic drugs, 378
Androgenization, 225
Androgen resistance, 375
Androgens, 201, 205, 212, 214, 219, 220, 223, 225, 226, 266, 282, 309, 324, 352, 365. *See also* Testosterone; Dihydrotestosterone
5α-Androstane-3α, 17β-diol, 212, 352
Angiotensin converting enzyme (ACE)
brain steroids, 229–230
renin, brain expression, 122, 123
vasoactive intestinal peptide, 105
Angiotensin I, 123, 128, 282, 334, 336
Angiotensin II, 122, 123, 228, 250, 272, 282, 334–338, 368
Angiotensinogen, 122, 123, 228, 282, 287, 334, 336
Angiotensinase 228
Anopheles gambiae, 27, 150, 185, 383
Anosmia, 359
Anterior pituitary. *See also* Adenohypophysis
ACTH, 63–64
CG, 69–76
endorphins, 64–65
FSH, 69–76
growth hormone, 76–79
LH, 69–76
melanocortins, 66–67
POMC, 59–63
prolactin, 79–81
TSH, 67–69
Antheridiol, 223, 231
Antiandrogens, 224
Antiestrogens, 225
Antimüllerian hormone (AMH), 120–122, 224, 288, 371, 372, 375
Antimüllerian hormone related peptide (AMHrP) 144, 146, 147
Antrum, 101, 292
Aorta, 144, 146, 334
Apis mellifera, 134, 139, 150, 152, 168, 169, 200
Aplysia californica, 37, 134, 143–146, 293, 384
Aplysia kurodai, 143, 144, 146, 384
Appetite, 34, 55, 66, 67, 98, 276, 313–324
Aquaporin, 105, 126, 334, 335, 338, 362
Area postrema, 319, 335, 338
Arginine vasopressin (AVP), 33, 34, 84, 86–88, 302, 303, 334–338, 342, 343. *See also* Vasopressin
Aromatase. *See* CYP19
Aromatase inhibitors, 220, 375, 377
Arylalkylamine *N*-acetyltransferase (AANAT), 244, 245, 278
Atrial natriuretic peptide (ANP), 87, 124–127, 328, 334, 335, 337, 338, 362
heart development, 126
Atrial myocytes, 124, 126, 362
Autocrine, 12, 117, 118, 132
Autoimmune adrenalitis, 367–368
Avidity, 92, 295, 388
Axonal transport, 30, 85, 86, 135, 149, 294

B
Bactrocera dorsalis, 171
Baroreflex, 123, 334, 337
Basedow disease, 363
Beta blockers, 377, 381
Blastocyst, 308, 309
Blood–brain barrier, 2, 14, 30, 106, 257, 271, 272, 300, 303, 317–319, 334–338
Blood pressure, 2, 16, 49, 88, 122, 123, 228, 287, 334–338
Blood supply, 273, 308, 314, 370
Blood-testis barrier, 290
Blood volume regulation, 328, 334–338
Body weight, 55, 56, 313–317
Bombesin, 102, 115–116, 250
Bombina bombina, 115
Bombyx mori, 137, 138, 150, 159, 160, 170, 179, 384
Bombyxin, 16, 92, 137–139
Bone mineralization, 126
Bone morphogenetic protein (BMP), 199, 322, 324
Brain and muscle aryl hydrocarbon receptor nuclear translocator like protein 1 (BMAL1), 340, 341, 346
BMAL1-CLOCK-dimer, 341
Brain natriuretic peptide (BNP), 124–126
Brain neurons, 142, 158, 162, 183, 319, 327
Brainstem, 32, 33, 41, 315–317, 319, 335
Brassinolide, 230
Brattleboro rat, 86
Bromocriptin, 81, 360
Bursicon, 172–174, 332, 350
burs, 173
pburs, 173
Busycon contrarium, 174, 184

C
Cachexia, 226
Caenorhabditis elegans, 134, 152, 154, 161, 175–177, 325, 384

Caeruleins, 104
Calcitonin (CT), 99–101, 127, 148, 232, 251, 272, 278, 279, 305, 368
  CALCA gene, 99, 127
  CALCB gene; 99
Calcitonin-gene-related peptide (CGrP), 99–101
  CALCA gene, 99, 127
Calcitonin-related peptides, 148
Calcitriol, 128, 232–233, 235, 236, 305. *See also* 1,25-Dihydroxyvitamin D_3
Calcium
  intracellular, 98, 247, 266
  metabolism, 279, 304–305
  store, 247, 264
Calcium channels, 264, 313
Calcium concentration, 100, 127, 128, 130, 135, 263, 264, 304, 348–350, 362
Calcium sensor, 127, 130
Calfluxin, 293
Callinectes sapidus, 183
Calmodulin, 216, 327
Cancer borealis, 141, 384
Capillaries. *See* Fenestrated capillaries
Capillary wall, 130
Carcinus maenas, 140, 141, 183, 384
Cardioacceleratory peptide (CAP), 141–143, 150, 169, 171, 320, 331, 332
  Cap2b gene, 150
  Molluscs CAP (M-CAP), 142, 143
Cardioexcitatory peptide, 143–144
Casein kinase 1 epsilon ($CK1_ε$), 340, 341
Castration, 6
Catecholamines, 17, 32, 34, 237–240, 263, 282, 302, 306, 353, 365, 367, 381, 387
Cells
  adrenocorticotropic, 64, 274, 275
  APUD, 285
  C-cell, 368
  chief, 128, 279, 280, 284
  chromaffin, 238, 281, 282, 367, 368
  corticotropic cells, 26, 32, 33, 59, 61, 64, 360, 372, 373
  folliculostellate pituitary, 119
  gonadotropic
    FSH, 72, 73, 118, 119, 266, 274, 275, 308, 361
    LH, 69, 73, 119, 266, 274, 275, 308, 361
  granulosa, 291
  I cell, 286
    cholecystokinin, 103, 285, 315
  Inka, 168–171, 189, 331, 332
  islet, 4, 107, 115, 286, 364 (*see also* Cells, pancreatic islet)
  juxtaglomerular, 122, 216, 287, 334
  lactotropic, 31, 79–81, 83, 118, 239, 274, 275, 360
  laterodorsal neurosecretory, 159
  Leydig cells (LC), 69, 71, 81, 219, 225, 266, 288–290, 309, 335, 376, 379
  melanotropic, 59
  muscle, 85, 91, 103, 106, 124, 139, 153, 157, 177, 258, 313, 379
  pancreatic islet
    α, 286, 313
    β, 286, 312
    δ, 103, 285, 286
    PP, 108, 110, 286
  pinealocyte, 277
  pyramidal cells of the hippocampus, 31
  Sertoli, 72, 120, 121, 219, 225, 288–290, 371
  somatotropic 47, 48, 96, 274, 275, 360
  stomach, 96–98, 101–105, 113, 283, 284, 317, 318
    G-cells, 102
    parietal, 102–104, 284
  theca, 69, 71, 219, 225, 290, 308
  thyrotropic, 67, 81, 274, 275, 359, 361
Central nervous system (CNS), 32, 34, 40, 41, 53, 62, 66, 80, 102, 103, 108, 111, 137, 163, 169, 172, 173, 184, 185, 187, 227, 229, 230, 250, 251, 278, 296, 302, 303, 306, 314, 315, 317, 318, 338, 359, 360
Cephalomyotropins, 155
Cephalatoxin, 86, 88
Cerebral cortex, 31, 48
Chambon, Pierre, 248
Chelicerates, 137, 155, 174
  leucokinin receptor, 135
CHH. *See* Crustacean hyperglycemice hormone
CHH precursor related peptide (CPrP), 136
Cholecalciferol, 233. *See also* Vitamin D_3
Cholecystokinin (CCK), 103–104, 315, 317, 319
Cholesterol, 17, 192, 194–195, 201–203, 221, 222, 229, 230, 232, 233, 239, 263, 316, 326, 351–353
Chondrocytes, 128, 321–324
  growth, 321, 324
  maturation, 323, 324, 358
Chondrodysplasia, 126, 128
Chorionic gonadotropin (CG), 23, 65, 69, 70, 75, 76, 251, 292, 309, 310, 350, 370, 377
Chorionic somatomammotropins (CHS), 76
Chromosome 1
  ANP, 124

BNP, 124
cortistain, 48
HSD3B2, 374
kisspeptin, 50
leptin receptor, 371
TSH-β, 69
Chromosome 2
CNP, 124
glucagon, 93
inhibin-β-B, 116
POMC, 51–63
PRL, 83
SRD5A2, 211
urocortin, 34
Chromosome 3
CCK, 103
ghrelin, 96
HESX1, 372
somatostatin, 46
TRH, 31
Chromosome 4
LH-R, 373
Chromosome 5
follistatin, 119
PC1, 26
PROP1, 372
SRD5A1, 211
stanniocalcin, 129
Chromosome 6
CYP21A2, 213
FSH, 72
LH/CG, 69
PRL, 79
TSH, 67
Chromosome 7
Epo, 130
GnIH, 42
leptin, 95
NPY, 108
Chromosome 8
CRH, 32
CYP11B1, 214
CYP11B2, 215
GnRH, 35
STAR, 374
Chromosome 9
HSD17B3, 374
LHX3, 372
SF1, 371
Chromosome 10
CYP17, 206, 374
Chromosome 11
CT, 99
FSH, 72
galanin, 51
GPHA2B5, 73
insulin, 89
PTH, 127
secretin, 104
Chromosome 12
AMHR2, 372
IGF1, 92
MCH, 54
Chromosome 14
GPB5, 73
Chromosome 15
CYP11A1, 203
CYP19, 217
Chromosome 16
agouti-related protein, 45
Chromosome 17
gastrin, 101
GH, 76
GIP, 107
HCRT, 56
PNP, 108
PYY, 108
SOX9, 371
Chromosome 18
gastrin-leasing peptide, 115
Chromosome 19
AMH, 120, 372
CG, 69
LH, 69
MIF, 120
Chromosome 20
AVP, 86
GHRH, 40
GnRH, 35
OXT, 84
Chromosome X
DAX1, 371
KAL, 372
AR, 375
Chromosome Y
SRY, 371
Ciona intestinalis, 37, 38, 104, 105, 123, 134, 158, 219, 384
Cionin, 104, 158
Circadian locomotor output cycles kaput (CLOCK), 340, 341
Circadian rhythm, 159, 186, 228, 300, 327, 340–343
Circumventricular organs, 272, 335, 336
Clenbuterol, 379
Clostebol, 378
CNP gene, 124–126
Coldness, 65, 240, 270, 301

Collagen
type 2, 322
type 9. 322
type 10, 322
Colloostatin, 165, 166
Combined pituitary hormone deficiency (CPHD), 360
Congenital adrenal hyperplasia (CAH), 213, 215, 365–366
Corazonin, 168–169, 189, 320, 330–332
dark color inducing hormone, 169
Corin, 125
Corpora allata, 178, 179, 181, 200, 296, 297, 325, 327, 328, 330
Corpora cardiaca, 136, 139, 142, 148, 149, 171, 178, 179, 186, 296, 297
Corpuscles of Stannius, 128–130, 349
Corpus luteum, 71, 72, 80, 120, 122, 291, 292, 308, 309, 370
Corticosteroid-binding globulin, 226, 303
Corticosteroids, 17, 34, 380
Corticosterone, 7, 213, 215, 216, 226, 229, 302–304
Corticotropin-releasing hormone (CRH), 7, 15, 32–35, 251, 275, 303, 307, 342, 343, 360
Cortisol, 7, 17, 32, 34, 64, 201, 205, 213–215, 226–228, 301–304, 323, 339, 343, 358, 365, 366
Cortistatin (CST), 48
Craniopharyngioma, 359
CREB-binding protein (CBP), 69
Cretinisms, 363
Crustacean, 16, 17, 91, 135–137, 140–143, 151, 153, 156, 158, 163–168, 172, 181–183, 188, 189, 199, 294, 295, 297, 324, 325, 350
Crustacean cardioacceleratory peptide (CCAP), 141–143, 331
Crustacean eye stalk, 163, 167
Crustacean hyperglycemic hormone (CHH), 135–137, 163, 167, 171
CT. *See* Calcitonin
Cushing, Harwey Williams, 5, 367
Cushing syndrome, 360, 365–367
Cyclic adenosine monophosphate (cAMP), 26, 44, 47, 52, 54, 64, 66, 72, 140, 142, 148, 151, 159, 215, 254, 255, 259, 346, 352, 362, 385
Cyclic guanosine monophosphate (cGMP), 126, 136, 170, 254
CYP11A1, 202–205, 215, 216, 218, 219, 223–225, 228, 230, 353, 365, 366, 374
CYP11B1, 202, 213–215, 226, 266
CYP11B2, 202, 213–216, 228, 229, 368
CYP17, 202, 205–206, 213, 218, 219, 225, 226, 266, 365, 366, 374
CYP17A1, 224
CYP19, 202, 216–221, 266, 290, 352, 353, 375, 395–396
CYP21, 202, 215, 225, 226, 228, 229, 231–214, 365, 367, 368
CYP21A2, 213–216
CYP24, 203, 204, 235, 236
CYP27, 235
CYP51, 193, 194, 202, 206, 352, 394
Cysteine knot, 23, 24, 68, 70, 74, 117, 173, 350, 374
Cystic fibrosis, 364, 369

D

D-amino acid, 143, 144
D-isomer, 144
Danio rerio, 51, 62, 63, 219
DAX-1, 217
transcription factor, 217
DDT. *See* Dichlorodiphenyltrichloroethane
7-Dehydrocholesterol, 194, 201, 233, 234, 248
Dehydrocholesterol reductase, 223
Dehydroepiandrosterone (DHEA), 205, 214, 219, 221, 225, 229, 264–266, 309, 352
Dehydroepiandrosterone sulfate (DHEA sulfate), 219, 225, 309
Dehydro-oogoniol, 223, 231
Deiodinase, 241, 243, 300, 323, 363, 386
Deoxycorticosterone, 213, 216, 229
11-Deoxycortisone, 213, 214, 306, 366
25-Deoxyecdysone, 230, 294
Deoxyribonucleic acid (DNA), 2, 21, 64, 157, 248, 249, 289, 340–342, 369, 385, 386, 389
DH. *See* Diapause hormone
DHEA. *See* Dehydroepiandrosterone
DHT. *See* Dihydrotestosterone
Diabetes insipidus, 4, 362, 368
central, 362
nephrogenic, 362
Diabetes mellitus (DM), 4, 107, 362, 364–365, 373
type 1, 364, 365
type 2, 364, 365
Diagonal band of Broca, 335, 337
Diapause, 151
Diapause hormone (DH), 147, 149, 150
Dicentrachus labrax, 112
Dichlorodiphenyltrichloroethane (DDT), 224, 225, 341
Diencephalon, 36, 270, 277
Digestive tract hormones, 101–112

Dihydrotestosterone (DHT), 40, 202, 211, 212, 220, 221, 223, 229, 249, 374, 375
L-Dihydroxyphenylalanine (L-DOPA) carboxylase, 142
1,25-Dihydroxyvitamin D_3, 220, 232–236, 305
Dimer
  BMAL1/CLOCK dimer, 341
  dimer formation, 74, 79, 80, 116, 118, 224, 247–249, 341, 369
  nuclear receptor, 330
Dipetalogaster maximus, 328, 330
Diphosphomevalonate decarboxylase, 193
Disulfide bridge, 23–24, 46, 54, 55, 57, 58, 67, 68, 70, 79, 80, 82, 86, 88, 90, 112, 117, 121, 124, 125, 135, 136, 142, 159, 162, 167, 171, 172, 180, 181, 255, 351
DiuH. *See* Diuretic hormone
Diuresis, 88, 136, 147, 148, 153
  leucokinin, 153
Diuretic hormone (DiuH), 147–148, 189
DM. *See* Diabetes mellitus
DNA. *See* Deoxyribonucleic acid
Dopamine, 13, 15, 17, 32, 34, 41, 46, 80, 81, 105, 237–239, 250, 251, 255, 275, 300, 306, 319, 346, 353, 360
  PRL, 239, 275, 360
Dopamine agonist, bromocriptine, 81, 360
Dopamine β-hydroxylase, 238, 239
Dorsal raphe nucleus, 57
Drosophila melanogaster, 16, 74, 92, 108, 128, 134, 137–140, 142, 150, 154, 156, 158, 159, 162, 163, 170, 172, 175, 177, 179, 181, 183, 184, 186–188, 197, 198, 319–320, 325, 326, 332, 384
Duodenum, 101–106, 283, 285, 286, 315
Dwarfism, 79, 230, 234, 324

E

Ecdysis, 17, 169–172, 330–333
Ecdysis-triggering hormone (ETH), 168–171, 181, 189, 331, 332
  INKA cells, 168–171, 189, 331, 332
Ecdysone, 159–161, 167–170, 178, 192, 196, 200, 230, 294, 295, 297, 325–328, 330
  balanced with JH, 178, 327
  Y organ, 230, 294, 326
Ecdysone biosynthesis
  JH synthesis, blocked by PTSH, 161
  MIH control, 167
  triggered by PTTH, 159, 161
Ecdysozoans, 136, 196, 324–326
Echinoderms, 18, 32, 85, 88, 92, 116, 134, 151, 170, 174, 347
Eclosion hormone (EH), 170–172, 189, 331–332
  defective mutants, 171
  isolation, 170
  membrane guanylate cyclase, 171
  neurons, 171, 331
Egg. *See also* Oocyte
  disposition, 163
Egghead protein, 162
Egg-laying hormone (ELH), 293
EH. *See* Eclosion hormone
Eisenia fetida, 177, 350, 384
Embryogenesis, 230, 372
Eminentia mediana. *See* Median eminence
Empty sella syndrome, 359
ENaC. *See* Endothelial sodium channel
Enchondromatosis, 128
Endocytosis, 279, 387
Endogenous opioids, 32, 112
Endokinins, 113, 114
Endometriosis, 12, 220, 222
  progesterone receptor isomers, 222
Endoplasmic reticulum (ER), 16, 22, 23, 25, 205, 211, 215, 221, 249, 370, 386, 387, 390
  smooth ER, 204–205, 221, 277, 282, 289–291, 387
Endorphin
  β—endorphin, 26, 29, 34, 41, 49, 59, 61, 64, 65, 112
  and enkephalins, 65, 112–113
Endothelial sodium channel (ENaC), 177, 228, 338
Enkephalin, 65, 112–113
  Leu-Enkephalin, 272
  Met-Enkephalin, 113, 350
Enteric nervous system (ENS), 56, 57, 106, 108, 112–116, 310, 314, 315
Enterins, 135, 144–147, 189
Enteroendocrine cells
  APUD, 285
  crypts of Lieberkühn, 283, 285
  endocrine middle gut, NPF, 186
  endocrine of the duodenum, 285
Enzymes
  digestion enzymes, 101, 165, 315
  steroidogenic, 192, 202–211, 225, 347, 358
Epinephrine. *See* Adrenaline
Epiphyseal cartilage, 321–322
Epiphysis, 277, 321–324
  cerebri (*see* Pineal gland)
  growth zone, 321–324
  hypertrophic zone, 321–324
  rest zone, 321–324

Epithelial cells, 102, 177, 323
EPO. *See* Erythropoietin
ER. *See* Endoplasmic reticulum
Ergocalciferol, 235
Ergosterol, 223
Erythropoietin (EPO), 79, 130–132, 257, 258, 377, 381–382
  doping, 132
Esophagus, 146, 320
Esterification, 97
  ghrelin, 29
Estradiol, 08, 15, 17, 40, 51, 53, 71–73, 99–101, 192, 201, 207, 210, 212, 216–221, 223–225, 229, 241, 258, 259, 266, 267, 288–290, 292, 300, 307, 322, 324, 344, 352, 360, 369, 370, 375, 376, 378
Estrogens, 4, 12, 79, 100, 201, 205, 212, 216, 217, 219–221, 223–225, 249, 258, 259, 265, 266, 300, 306, 307, 309, 324, 352, 353, 358, 370, 374–376
ETH. *See* Ecdysis-triggering hormone
Euphyllia ancora, 218
Evans, Ronald, 248
Exocrine, 12, 62, 98, 101, 108, 250, 283, 285, 286
Exocytosis, 274, 387
Exopeptidase E, 36
Exoskeleton, 324–326, 331, 332
Eye stalk, 135, 140, 163, 167, 168, 182

F
Farnesyl pyrophosphatase, 194
Farnesyl synthase, 193
Fasting, 257, 311
Fatigue, 240, 302
Feedback control, 40, 300–301, 307
Feedback inhibition, 34, 64, 225, 306, 340
Feedback loops, 34, 332
Feeding, 1, 33, 45, 46, 51, 54–57, 139, 148, 156, 157, 162, 200, 315, 318–320, 326, 329
Females, 6, 13, 14, 17, 40, 43, 52, 53, 71, 82, 87, 121, 122, 161–163, 200, 207, 212, 219, 223–225, 231, 264, 288, 290–292, 294, 307, 320, 323, 365, 369, 371–375, 378
  anabolic drugs, 378
Fenestrated capillaries, 14, 30, 240, 274, 276, 278, 279
Fetus, 72, 121, 192, 212, 219, 224, 226, 264, 288, 308, 358, 369, 373, 375
Fibroblast growth factor (FGF), 321, 322
Fluoxymesterone, 378
FMRFamide, 174–178, 296, 350
  human FMRF, 112
FMRFamide like peptides, 175–177
Follicle-stimulating hormone (FSH), 5, 7, 15, 23, 24, 30, 35, 36, 44, 67, 69–73, 75, 111, 117–120, 220, 225, 251, 252, 264–266, 274, 275, 292, 300, 307–310, 344, 350, 360, 361, 369, 370, 373, 374
  pulses, 309
Follistatin (FST), 117–120, 308
  heterodimers, 119
Food intake, 49, 55, 95–99, 103, 108, 158, 174–178, 186, 315–317, 319–320, 328
Foraging, 184–186, 200, 316
Fruit flies, 155, 161–163, 165, 170–173, 175, 198, 347, 384. *See also Drosophila melanogaster*
FSH. *See* Follicle-stimulating hormone
FSHR. *See* Receptor, Follicle stimularing hormone
FST. *See* Follistatin
Fungi, 13, 206, 223, 349, 351, 356, 383
  steroids, 230–232
Furin, 27, 28, 47, 57, 112, 115, 125, 127, 129, 157, 159, 175, 177, 178, 180, 185, 187, 188
  PC, 47, 57, 129, 175, 177, 178, 185

G
GABA. *See* γ-Aminobutyric acid
Galactorhea, 360
Galanin, 7, 41, 51–53, 111, 180, 181, 250, 251, 305–307
Galanin-related peptide, 52
Ganglia
  abdominal, 146, 148, 156, 173, 174, 188, 293, 295, 331, 332
  buccal, 144–146
  cerebral, 146, 186, 293, 294
  dorsal root, 336
  heart, 141
  head, 156,
  hypocerebral, 186
  parietal, 293
  pedal, 145, 146, 293
  pleural 146, 293
  spinal, 53
  superior cercical, 277
  stomatogastric, 140, 151, 153
  subesophageal 142, 149, 173, 174, 294, 295, 320, 332
  terminal,142
  thoracic,142, 148, 156, 172, 173, 295, 332
  visceral, 293
Gastric inhibitory peptide (GIP), 94, 101, 106–107, 115, 283, 286

Gastrin, 4, 5, 101–105, 107, 115–116, 132, 157, 158, 189, 250, 251, 284, 368
Gastrin-releasing peptide (GRP), 102, 115–116, 251
Gastrinoma, 368
Gastrointestinal (GI) tract, 47, 62, 97, 101, 105–107, 110, 113, 140, 189, 208, 227, 254, 269, 270, 283–285, 313, 315–318
GCG. *See* Glucagon
G-CSF. *See* Granulocyte colony stimulating factor
GDP. *See* Guanosine diphosphate
Geranyl pyrophosphate, 193
Gestagens, 17, 201, 352
Ghrelin, 29, 40, 48, 49, 67, 96–99, 316–319, 322
  CNS, 319
  esterification, 29, 97
  leptin antagonist, 315
  PYY antagonist, 316
GHRH. *See* Growth-hormone-releasing hormone
Giantism, 361
GIH. *See* Gonad-inhibiting hormone
GIP. *See* Gastric inhibitory peptide
GK. *See* Glucokinase
Glc6P. *See* Glucose 6-phosphate
Glc6Pase. *See* Glucose 6-phosphatase
GlcT. *See* Glucose transporter
Glial cell missing (Gcm) protein, 128
Glicentin, 283, 285
Glicentin-related peptide (GrPP), 93, 94
Glucagon (GCG), 49, 93–94, 101, 106, 107, 115, 251, 272, 285, 286, 311, 313, 319, 364, 385
Glucagon-like peptide 1 and 2 (GLP1 and GLP2), 93, 94, 108
Glucocorticoids, 15, 32, 64, 201, 213, 220, 226, 249, 301–303, 322, 323, 342, 343, 352, 366, 368
  in the brain, 302
  chondrocytes, 323
  pulses, 343
  stimulation by ACTH, 64, 302
Glucokinase (GK), 311, 312
Gluconeogenesis, 94, 110, 226, 227, 240, 285, 311–313
Glucose
  concentration in blood, 311
  glycogen metabolism, 311, 313
  metabolism, 95, 311–313
  storage, 312, 313
  uptake 57, 91, 227, 313
Glucose 6-phosphatase (Glc6Pase), 94, 312
Glucose 6-phosphate (Glc6P), 311–313
Glucose transporter (GlcT), 91
Glumitocin, 85
Glutaminyl cyclase, 36, 350
Glycogen
  catabolism, 311, 313
  degradation, 94
Glycoprotein hormones, 23, 24, 67, 69, 70, 72, 75–76, 131, 173, 252, 361, 379, 380, 392–393. *See also* Chorionic gonadotropin; Follicle-stimulating hormone; Luteinising hormone; Luteinising hormone
  common α chain, 67
  evolution, 350, 351
  GPHA2B5, 73–75
GnIH. *See* Gonadotropin-inhibiting hormone
GnRH. *See* Gonadotropin-releasing hormone
Goiter, 4, 363
Gonadal development, 17, 224–225
  fetal development, 225
Gonad-inhibiting hormone (GIH), 135, 136, 163–164, 167, 189
Gonadotropic cells
  FSH, 72, 73, 118, 119, 266, 274, 275, 308, 361
  LH, 69, 73, 119, 266, 274, 275, 308, 361
Gonadotropin-inhibiting hormone (GnIH), 42–45
  RFRP, 43
Gonadotropin-releasing hormone (GnRH), 6, 15, 29, 30, 35–39, 43, 44, 50, 51, 53, 69, 71–73, 81, 134, 225, 264–267, 272, 305–310, 343–345, 359, 369–373
  activity, dependent pulsatile release, 36
  elevated levels, 345
  GnRH-II, 36
  neurons
    catecholamines, 306
    CRH, 305, 306
    defective in Kallmann syndrome, 370
    GABA, 306
    galanin, 305, 306
    NPY, 305, 306
    SP, 306
    synthesis, 42
  pulses, 41, 380
  regulation, IL-1, 306
  release, 41, 48, 79, 97, 322, 323, 360
Gonadotropins, 16, 30, 44, 67, 69, 70, 74, 81, 119, 288, 289, 305, 309, 360, 370, 373–374, 379–380
GPcR. *See* Receptor, G-protein coupled membrane
G-proteins, 29, 54, 94, 254

Granules. *See* Secretory granules
Granulocyte colony stimulating factor (G-CSF)
family, 95
leptin, 95
Graves disease, 363
Growth, 1, 91, 93, 108, 136–139, 158, 215, 223, 225, 230, 321–333, 358, 363, 375
Growth hormone (GH), 3, 5, 15, 30, 41, 46, 76–79, 82, 95, 255, 257, 258, 274, 275, 309, 321, 322, 339, 358, 377, 380, 393
feedback control, gender specificity, 40
gene cluster, chromosome 17, 79
insulin, and insulin-like growth factor (IGF) 79, 361, 380
pulsatile release, 79
pulse frequency, sex controlled, 323
release, 41, 48, 76, 79, 97, 322, 323, 360
Growth-hormone-releasing hormone (GHRH), 15, 30, 40–42, 46–48, 53, 76, 130, 251, 322, 323, 360, 361
doping, 380
GHRH (1–37), 40
GHRH (1–40), 40
sexually dimorphic expression, 40
GRP. *See* Gastrin-releasing peptide
Gryllus bimaculatus, 157
Gryllus rubens, 200
GT1-7 cell line, 344
GTP. *See* Guanosine triphosphate
Guanine, 21, 385, 386, 388
Guanosine diphosphate (GDP), 252–254
Guanosine nucleotide exchange, 253
Guanosine triphosphate (GTP), 249, 252–254
hydrolysis, 253
Guanylate cyclase, 124, 125, 127, 135, 136, 166, 167, 254
cGMP, 254
receptor, 127
Gut, 48, 49, 52, 53, 62, 103, 105, 106, 108, 137, 138, 144, 145, 164, 165, 179, 232, 270, 283, 284, 305, 311
foregut, 186
midgut, 136, 138, 153, 155, 156, 186, 194
movement, 98, 314, 327
musculature, 112, 151, 153, 155, 157, 158

H

Hanke, Winfried, 349
Heart
contraction, 143
frequency, 141–148, 316
ganglion, 141
oxytocin, 337
Helicoverpa armigera, 188, 384
Helicoverpa assulta, 150, 384
Helicoverpa zea, 161, 162, 384
Helix aspersa, 177, 384
Hemimetabolous, 295, 325, 326, 328
Hemolymph, 16, 17, 135, 136, 140, 142, 147, 148, 153, 159, 162, 168, 178, 186, 188, 196, 199, 200, 294, 295, 327, 328, 330, 331
Heparan sulfate, 119, 322, 324
HESX1 protein, 359, 360, 372
Hibernation, 43, 339, 345
HIOMT. *See* Hydroxyindole-*O*-methyl transferase
Hippocampus, 31, 32, 35, 40, 48, 303
Hirsutism, 212, 367, 375, 378
HLA-DQ (nonAsp57), 364
HLA-DR3, 364
HLA-DR4, 364
Homarus americanus, 164, 384
Homeostasis
calcium, 128, 348
metabolic, 138, 202, 279
Homo-dimer, 74, 129, 173, 241, 255
PTTH, 159
Homomevalonate, 197
Hox genes, 111
HPAA. *See* Hypothalamic-pituitary-adrenal axis
HPGA. *See* Hypothalamic-pituitary-gonadal axis
HPTA. *See* Hypothalamic-pituitary-thyroid axis
HSD. *See* 17β-Hydroxysteroid dehydrogenases (HSD)
Hugin, 150, 320
18-Hydroxycorticosterone, 216
20-Hydroxyecdysone, 230, 325, 330
Hydroxyflutamide, 224
Hydroxyindole *O*-methyltransferase (HIOMT), 244, 245
17-Hydroxypregnenolone, 205
7α-Hydroxyprogesterone, 230
17-Hydroxyprogesterone, 308
Hydroxysteroid dehydrogenase (HSD)
HSD3A, 229
HSD3B, 205, 224, 225, 228, 229
HSD3B2, 204, 374
HSD17B, 202, 206–208, 224, 225, 263, 351
HSD17B3, 207, 208, 374
5-Hydroxytryptamine, 244. *See also* Serotonin
Hypercorticism, 366–367
Hyperinsulinemia, 26
Hypernatremia, 362
Hyperprolactinemia, 360

Hyperthyroidism, 361
Hypoglycaemia, 26, 41, 240, 380
Hypogonadism, 33, 359, 369, 371–373
Hypothalamic-pituitary-adrenal axis (HPAA), 15, 33, 302, 303, 366
Hypothalamic-pituitary-gonadal axis (HPGA), 15, 264–267, 307
Hypothalamic-pituitary-thyroid axis (HPTA), 15, 31, 74
Hypothalamus, 30–58, 270–272, 359, 372–373
Hypothyroidism, 363

I

I cells, 286
  cholecystokinin, 103, 285, 315
ICER. *See* Inducible cyclic AMP early repressor
ICMT. *See* Isoprenylcysteine carboxyl methyltransferase
IGF. *See* Insulin-like growth factor
IGFBP. *See* Insulin-like growth factor binding protein
Ihh. *See* Indian hedgehog
IkB protein, 66
ILP. *See* Insulin-like peptide
Imprinting, 91, 93, 375, 389
Indian hedgehog (Ihh), 321, 322, 324
Inducible cyclic AMP early repressor (ICER), 346
Infertility, 4, 36, 224, 264, 358, 361, 370, 373
Inhibin, 71, 116–120, 225, 257, 289, 292, 308, 369, 372
  Inhibin-β-A, 116
  Inhibin-β-B, 116
Inner cell mass, 308
Inner mitochondrial membrane, 214, 215, 226, 387
  CYP11A1, 203, 228
Inositol trisphosphate, 254, 304
Inotocin, 85, 88, 148
Insects, 5, 14–17, 34, 35, 53, 75, 85, 101, 111, 119, 123, 134, 136–143, 147, 150–153, 155, 156, 158, 160, 161, 163, 165, 166, 168–170, 174, 175, 179, 181–183, 186–189, 196, 200, 230, 231, 294–297, 324–326, 328, 332, 350, 352, 353
Instars, 196
Insulin, 5, 49, 51, 89–92, 94, 105, 107, 108, 115, 255, 272, 283, 285, 311–313, 316, 343, 344, 351, 364
  evolution, 351
Insulin-like growth factor (IGF)
  doping, 380
  IGF1, 54, 76, 79, 92, 93, 322, 323
  IGF2, 90–93, 323, 351, 389
Insulin-like growth factor binding protein (IGFBP), 92, 93, 323
Insulin-like peptide (ILP), 16, 137–139, 189, 293, 325, 326, 351
  dILPs, 138
  ILP1 for royal jelly, 139, 162
Insulin receptor substrate (IRS), 137, 138, 256, 326
Iodine, 241–243, 278, 279, 349, 363
  deficits, 363
  iodine uptake, 363
Ion channels, 44, 174, 247, 254, 343, 344
  potassium channel, 88, 147, 313
    osmoregulation, 88
  sodium channels, 177, 228, 334, 336, 338, 369
Ion transport peptide (ITP), 136, 137
IRS. *See* Insulin receptor substrate
Islet cell tumor, 364
Isocaproic acid, 203
Isopentenyl-isomerase, 193
Isoprenylcysteine carboxyl methyltransferase (ICMT), 211
Isomerase, 144
Isotoxin, 85, 88
ITP. *See* Ion transport peptide

J

Jejunum, 103–105
Jensen, Elwood, 248
Juvenile hormone (JH), 192–200, 327–332
  binding protein, 199
  biosynthesis, 181, 194, 196
  synthesis, blocked by PTSH, 179
  epoxidase, 194
  esterase, 329, 330
  hydrolysis, 199
  metabolism, 178–181
  methyl transferase, 194
  release stimulated by PSP, 163
  synthesis, 192–194
  downregulation to pupate, 327

K

KAL1 gene, 372
Kallman fold, 206
Kallmann syndrome, 36, 359, 370
Kidney, 35, 45, 48, 53, 88, 108, 114, 118, 119, 126–131, 208, 209, 223, 227, 228, 232, 235–237, 280–282, 287–288, 304, 305, 334–337, 362, 365, 368, 381, 382
  glomeruli, 287, 338
  water retention, 88

Kinase, 51, 66, 89, 91, 92, 94, 116, 117, 120, 121, 137, 139, 148, 150, 193, 199, 205, 213, 216, 245, 247, 255–258, 311, 313, 327, 340, 341, 369
Kinins, 148–158
Kisspeptins, 50–51, 71, 174, 178, 306
  neurons, 50, 88
Klinefelter syndrome, 369

L
Lactation, 81, 100, 275
  calcitonin, 100
Lactotropic cells, 31, 79–81, 83, 118, 239, 274, 275, 360
Lampetra fluvialis, 37, 111
Lanosterol, 194, 195, 201, 202, 223, 230, 233, 351, 352
Lanosterol-14α-demethylase (CYP51), 194, 202, 206, 352, 394–395. *See also* CYP, CYP51
Lateral hypothalamic area (LHA), 54, 55, 271, 272, 318
Lateral nucleus of the amygdala, 116
Lepidopterans, 150, 161, 179, 196, 197, 38
Leptin (LEP), 2, 16, 57, 79, 82, 95, 96, 130–132, 257, 258, 272, 315, 317–319, 358, 371
  satiety, 315
Leptinotarsa decemlineata, 186
Leucokinins, 149, 153–155
Leucophaea maderae, 154, 156, 157, 350, 384
LH. *See* Luteinizing hormone
LHX3 protein, 359, 360, 372
Light adaption, 106, 181–183, 278
Light-dark cycle, 200, 245, 277, 339, 342, 345
  melatonin, 245
  rhythm, 342
Limbic system, 31, 33, 302, 338
  NPY, stress, 302
  SP, 302
Limulus polyphemus, 188
Lipid storage droplet protein-1 (LSDP-1), 140
Lipocortin (LPH), 26, 59, 65
Litoria caerulea, 104
Liver, 45, 76, 89, 91–94, 118, 122, 202, 208, 209, 215, 226, 227, 232, 235, 240, 248, 249, 266, 311–313, 364, 379, 380, 382
Locus coeruleus (LC), 302, 334, 335, 337
Locusta migratoria, 197, 384
Loop of Henle, 287, 338
Lumbricus terrestris, 384
Luteinizing hormone (LH), 15, 23, 24, 30, 35–37, 44, 51, 67, 69–73, 75, 119, 121, 251, 264–266, 274–275, 307–310, 360, 369–371, 373, 379
  defect mutation, 370, 373
  follistatin, regulation, 119, 308
  inhibition, 71
  surge, 71, 308
Lymnokinins, 149, 153–155
Lymnaea stagnalis, 293
Lys-Conopressin, 85
Lys-Vasopressin, 85
Lysosome, 243, 279, 289, 387

M
Macrobrachium rosenbergii, 183
Macrocallista nimbosa, 174, 384
Macula densa, 287, 288, 334
Magnocellular neurons
  hypothamic, 87
  neurosecretory, 336
  supraoptic, 85–87, 335
Male development, 212
  gonads, 7, 288–292
  PSP expression
    ductus ejaculatorius, 162
    heart, 162
Malpighian tubules, 147, 148, 153, 155
Mammary gland
  maturation with galanin, 52, 53
  milk ejection, 47
  oxytocin, 87
Manduca sexta, 140, 142, 147, 161, 170, 178, 179, 181, 196, 328, 329, 331, 384
Marsupenaeus japonicus, 167
Median eminence, 14, 15, 30, 31, 33, 35, 40–42, 44, 50, 80, 239, 271–273, 276, 302, 305, 307, 309, 316, 318, 344
Medullary cortex
  kidney, 131
  oxygen sensing, 131
Meiosis, 195, 308, 369, 388
Melanin-concentrating hormone (MCH), 54–56, 318, 319
  feeding control, 55
  food intake regulator, 55
  knockout mice, 55
Melanocortins (MSH), 26, 46, 49, 59, 60–62, 65–67, 81, 110, 250, 276,
  α−MSH, 34, 49, 61, 62, 66, 316, 317, 319
  β−endorphin, 26, 29, 34, 43, 49, 59, 60–61, 64–65, 112, 302, 305–306
  inhibiting central food regulation, 319
  link between neuroendocrine and immune system, 66
  receptors block by AgRP, 46

stimulating pigmentation, 67
Melanocytes, 45, 61, 62, 67, 250,
prohormone convertase (PC2), 65–67
Melatonin, 243–245, 251, 277–278, 339, 345–346
integration, 334
serotonin, 237, 278
Menarche, 309, 374
Menopause, 310
Menstrual cycle, 73, 121, 266, 306–310, 344, 360, 361, 369, 373
Menstruation, 309, 310
Mesotocin, 85, 88–89
Messenger RNA (mRNA), 22, 32, 55, 73, 99, 159, 386
Metamorphosis, 166–169, 179, 199, 295, 299, 325, 328–330, 332
Methoprene-tolerant, 200, 332
JH receptor, 196, 200
Methylfarnesoate, 17, 136, 192, 194, 198–199, 294, 297
Mevalonate kinase, 193
Major histocompability complex (MHC), 55, 213, 218, 364
Mineralocorticoids, 17, 64, 123, 213, 226, 228, 249, 282, 303, 352, 366, 368
stimulation by ACTH, 64
Mitochondria, 17, 22, 202, 203, 213–216, 226, 228, 232, 235, 274, 275, 279, 280, 282, 284, 286, 312, 386
Molting, 17, 136, 142, 163, 164, 166–169, 178, 196, 199, 200, 230, 294, 295, 320, 324–329, 331, 332, 345
Molt-inhibiting hormone (MIH), 135, 136, 166–168, 171, 189
Monooxygenases, 36, 139, 185, 201–203, 206, 263, 296, 350
Mood changes, 309
Multiple endocrine neoplasia (MEN), 368, 369
MEN1 gene, 368, 369
Muscle contractions, 114, 143–146, 155, 158, 174, 177, 187, 188
inhibition, 144, 146
Myocyte enhancer factor 2 (MEF2), 159
Myoinhibitory peptide 1 (MIP1), 161
Mytilus edulis, 146, 384
Mytilus inhibitory peptide (MIP), 135, 145–147, 189

N

NADPH-cytochrome P450 reductase, 234
Na^+/K^+-ATPase, 287, 338
19-Nandrolone, 378
Nematodes, 101, 105, 137, 151–153, 155, 158, 174, 177, 181, 182, 196, 326, 350
Neobellieria bullata, 156, 166, 384
Neoendorphins, 112
Nerve growth factors (NGF), 68, 70
Neurohemal organs, 14, 16, 30, 135, 141, 171, 240, 272, 276, 292, 318, 332
Neuromedin C, 115, 116
Neurons
α_1—adrenergic, 32, 34, 41, 45, 237–239, 245, 302, 344
AgRP, 97, 318, 319
CAP, 196
cholinergic, 382,
corazonin, 194
dopaminergic, 230, 238, 239, 307
dorsomedial
corazonin, 168
ventral cord, 168, 319, 331
EH, 196, 197
GnRH, 359, 372
defective migration, 359
NPY, 45, 66, 67, 110, 319
orcokinin, 174
pacemaker, 182
parvocellular, 356
POMC/CART, 318
PTTH, 182
serotonergic, 43
tuberoinfundibulary dopimergic (TIDA), 306, 307
SST, 50
Neuropeptide-F (NPF), 184–187, 319, 320, 326
Neuropeptide Y (NPY), 41, 45, 66, 67, 95, 97, 108–112, 184–187, 250, 302, 305–307, 314–319
antagonist of NPY and hunger, 95
homologue, 184
leptin, 112, 318, 319
neurotransmitter, 110, 302, 305, 316
PNP, 108, 110–111
receptors, 108, 111, 184, 185, 317
structure, 45, 108
Neurophysins, 24, 86, 88
mutation, 86
Neurosecretion, 5, 14, 51, 134, 173, 269, 292–297
Neurotransmitters, 13, 14, 16, 31, 32, 34, 36, 42, 44, 51, 55, 56, 88, 101, 105, 106, 110, 115, 116, 144, 151, 153, 158, 177, 184–186, 239, 240, 263, 264, 278, 295, 302, 305–307, 316, 317, 337, 344, 353, 390

Nicotine adenine dinucleotide (NAD^+), 207
3β-HSD, 204–206, 213, 214, 221
Nicotine adenine dinucleotide phosphate ($NADP^+$), 241
Nicotine adenine dinucleotide phosphate hydrogenated (NADPH), 203, 207, 214, 216
Nidation, 12, 72, 221, 308, 309, 369
Noradrenaline (NA), 5, 13, 16, 17, 34, 41, 42, 53, 110, 237–240, 250, 282, 300–302, 306, 307, 316, 337, 353, 365, 367, 377, 379, 381
AVP release, 337
19-Norandrostendione, 37
Norepinephrine. *See* Noradrenaline
N-terminus, 33, 60, 74, 75, 83, 85, 107, 162, 188, 191, 202, 219
Nucleus
arcuate (NArc), 40, 41, 45–47, 50, 51, 53–55, 61, 66, 70, 97, 110, 257, 271, 272, 276, 314–319
bed nucleus of the stria terminalis (NIST), 314, 316, 329, 331
central nucleus of the amygdala, 314
paraventricular (NPV), 31, 33, 41, 44, 45, 47, 50, 51, 58, 85, 86, 88, 272, 273, 276, 302, 314–316, 318, 319, 334–337, 343
perifornical, 57
raphe, 45, 57, 335, 336
reticular nucleus of the thalamus, 31
sexually dimorphic, 31, 223
solitary tract, 61, 314, 319, 334, 335, 337, 338
supraoptic (NSO), 31, 85–88, 271–272, 276, 334–337, 341, 342
circadian zeitgeber, 41, 327
suprachiasmatic (SCN), 31, 41, 244, 245, 271, 272, 278, 342, 343, 346,
ventromedial (NVM), 40, 41, 47, 54, 171, 271–273, 276

O
Octopressin, 85
Octreotide, 48, 361
Olfactory bulb, 31
Oncopeltus fasciatus, 328
Onthophagus taurus, 328
Oocyte, 219, 290–292, 308, 369
maturation, 230, 360
Opiates, endogenous, 64, 306
Optic nerve, 245, 270, 272, 278, 361, 373
Orcokinin, 135, 149, 151–153, 189
Orconectes immunis, 141, 143, 183, 384
Orconectes limosus, 384
Orexins
CNS, 58
ENS, 56
Organum vasculosum laminae terminalis, 272, 334–338
Osmolarity
Na_x, 336
sensor, 288
sodium channel, 336
Osmotic stimuli
blood concentration, 87
regulation by AVP, 86
sodium, 334
Osteoblasts, 322, 323
Osteoporosis, 100, 308
treatment by calcitonin, 100
Ovarian ducts, 121, 224, 308
Ovarian granulosa cells, 120
Ovaries, 4, 45, 74, 121, 207, 224, 266, 290–292, 307, 361, 369
Ovulation, 71, 81, 166, 221, 230, 290, 292, 308–310, 370, 374
Oxidoreductase. *See* P450-Oxidoreductase
Oxytocin (OXT), 4, 5, 24, 53, 84–89, 135, 142, 148, 189, 215, 239, 250, 251, 272, 276, 334–337, 359
antagonist, inhibing brood care, 87
facilitates pairing, 87
knockout mice, 86
prepartal burst, 87
Oxytocinase, 87

P
Pagurus bernhardus, 136
PACAP. *See* Pituitary adenylate cyclase activating peptide
Pancreas
cells, 101, 311
dysfunction, 364–365
Langerhans islets, 90, 285–286
Pancreatic polypeptide (PNP), 49, 108–112, 115, 364
NPY, 108
PYY, 108
Pandalus borealis, 183
Pandalus jordani, 182
Paracrine, 12, 47, 48, 51, 94, 95, 116, 118–120, 124, 126, 189, 194, 264, 283, 294, 315, 321, 322, 324, 344
Parathormone (PTH), 101, 127, 128, 232, 279, 280, 304, 305, 321, 322, 349, 367
evolution, 128
synthesis, 305

Parathormone-related peptide (PTHrP), 128, 321, 322, 324
  from bone, 321
Parathyroid glands, 101, 127, 128, 279–280, 304–305, 349
  calcium metabolism, 304–305
  parathormone, 101, 127, 280, 304, 349
Pars intercerebralis, 138, 142, 148, 332
Pars intermedia
  MSH, 276
  POMC, 59
Pars tuberalis, 244, 273, 345, 346
PCT. *See* Proctolin
PDH. *See* Pigment dispersing hormone (PDH)
Penis retraction muscle, 146
PEPCK. *See* Phosphoenolpyruvate carboxy kinase
Peptidases
  insulin, 90
  protein maturation, 22
Peptide motives
  allatostatin B
    A:**Y/F-x-FG-L/I**, 135, 180
    B:**Wx$_6$W**, 135, 160, 180
  allatotropin
    TARGF, 135, 179
  β—PDH
    NSELINSxLxxSxxxxxA, 182, 183
  di-basic
    KK, 26–27, 61, 65, 93, 102, 152, 175
    KR, 26–28, 31, 52, 61, 65, 92, 93, 100, 104, 109, 128, 136, 137, 152, 159, 175, 179, 180
    RK, 26–27, 65, 93, 94, 175
    RR, 26–27, 65, 83, 84, 93, 94, 105
  enterin
    PxxxHxxFV, 135, 144, 145
  ETH
    I/LKxxKxI/VPRx, 170
  FMRF/RF-amide
    FMRF, 135, 174–177, 359
    RF, 135, 174, 175, 177,
  furin, 27, 28, 47, 107, 112, 115, 125, 127, 129, 157, 159, 180
    Rx$_m$R, Rx$_n$K, 27, 175, 177, 178, 185, 188
  leucokinin
    FxxWG, 135, 149, 153–155
  lymnokinin
    FxxWx, 154, 155
  MIP
    PxFF/I/V, 135, 145, 146
  orcokinin
    NxDEI, 135, 149, 152
  pyrokinin
    FxPRL, 135, 149, 150
  racemase
    LxF, 144
  sNPF
    xxxxRLRF, 135
  sulfakinin
    YGHMRF, 135, 157, 350
  tachykinin
    FxGLM, 113, 135,
  tachykinin-like peptide
    FxGxR, 156
Peptidylglycine alpha-amidating monooxygenase (PAM), 28, 36, 40, 66, 100, 103, 106
Peptidylglycine alpha-hydroxylating monooxigenase (PHM), 94, 139, 175, 176, 185, 296
Peptide tyrosyl tyrosyl (PYY), 110
Pericardial organ, 136, 142, 294, 295
Period (PER), 340, 341, 346
  PER/cryptochrome dimer, 340, 341
  PER/ICER seasonally controlled, 346
Periplaneta americana, 146, 183
PETH. *See* Pre-ecdysis triggering hormone
Petramyzon marinus, 111
Phagosome, 279, 388
Phenylethanolamine *N*-methyltransferase (PNMT), 238
Pheochromocytoma, 367, 368
Pheromone, 12–14, 98, 148, 149, 151, 161–162, 200, 230
Pheromone-biosynthesis-activating neuropeptide (PBAN), 98, 149–151, 162, 163, 189
Pheromonostatic peptide (PSP), 161–163
  receptor, 163
PHM. *See* Peptidylglycine alpha-hydroxylating monooxigenase
Phornia regina, 327
Phosphoenolpyruvate carboxykinase (PEPCK), 312
Phospholipases, 13, 254, 304
Phosphomevalonate kinase, 193
Phosphorylase, 94
Phosphorylase kinase, 94
Phyllomedusa sauvagii, 104
Pigment-dispersing factor (PDF), 181–183
Pigment-dispersing hormone (PDH), 181–184, 189
Pineal gland, 31, 32, 243, 244, 270, 277–278, 336, 346
  acervuli, 277–278
Pinocytosis, 243, 388

Pituitary, 4–6, 12, 15, 26, 29, 30, 32, 34, 39–41, 44, 46, 47, 53, 59–88, 96–98, 116–120, 131, 225, 226, 241, 243, 264, 266, 270, 273–276, 280, 289, 292, 300–302, 307–309, 339, 344, 345, 363, 366, 369, 370, 372
anterior (adenohypophysis), 59–80, 266, 274–275, 322,
defects, 359
combined pituitary hormone deficiency, 360
pars intermedia, 81–83, 276
posterior (neurohypophysis), 14, 84–88, 215, 334, 336–337
stalk (infundibulum), 14, 272–239, 245, 273
tumors, 360–361
Pituitary adenylate cyclase activating peptide (PACAP), 41, 42, 94, 105, 106
Placenta, 12, 33–35, 50–52, 59, 62, 115, 116, 121, 205, 208, 209, 219, 220, 308, 309, 365, 370, 373
Plant steroids, 233
Platelet-derived growth factor (PDGF), 70, 123, 256
Plexus choroid, 335
PNMT. *See* Phenylethanolamine *N*-methyltransferase
Polyprolin (PP)-fold, 108, 110
Polyuria, 4, 362
POMC. *See* Proopiomelanocortin
Ponasterone A, 230, 294
Portal system, 3, 30, 44, 64, 69, 84, 239, 271–273, 276, 302, 308, 360
Posttranslational modifications
hormone maturation, 23–29
protein, 16
Potassium, 53, 88, 126, 147, 155, 156, 188, 216, 228, 254, 287, 313, 337, 338, 344, 368
P450 Oxidoreductase, 205, 225
PP cells. *See* Cells, pancreatic islet, PP
PP-fold. *See* Polyprolin-fold
Preantral follicles, 120, 121
antimüllerian hormone (AMH), 120
Pre-ecdysis behavior, 331, 332
Pre-ecdysis-triggering hormone (PETH), 169, 170
Pregnancy, 69, 71, 72, 80, 82, 87, 100, 219, 221, 222, 226, 275, 309, 310, 364, 370, 379, 380
Pregnenolone, 17, 201–203, 205, 213, 221, 223, 229, 230, 353
Preoptic area, 31, 32, 306, 359
Pressure control, blood volume regulation, 334
Pressure determination, 334
Primary aldosteronism, 368
Primary polydipsia. *See* Diabetes insipidus
Primordial follicle, 290, 291, 310
oocyte, 290
PRL. *See* Prolactin
PRLRP. *See* Prolactin-releasing peptide
Procambarus clarkii, 152
ProCT
infection parameter, 100
procaltitonin, 100
Proctodeal nerve, 171
Proctolin (PCT)
neurons, 188
receptor, 187
GPcR, 44
Product feedback, 40
Progesterone, 15, 71, 72, 80, 201, 204, 205, 211–213, 221–223, 226, 229, 230, 249, 266, 292, 307–309, 352–353, 360, 366, 370
bifunctional, 205–206
isoforms, 222
natural, 233
synthesis, 72, 308, 309, 370
Prohormone convertases (PC), 16, 25–27, 31, 35, 40, 43, 47, 50, 52, 57, 61, 63, 65, 84, 87, 93, 107, 109, 112–114, 125, 127, 136, 137, 144, 149, 152, 154, 175–177, 179, 180, 182, 185, 350
PC1, 26, 27, 31, 35, 42, 43, 47, 52, 57–59, 61, 62, 64–66, 83, 92–94, 100, 106, 107, 110, 114, 128, 129, 137, 138, 140, 148, 152, 159–161, 175, 178, 180, 184, 185, 370
corticotropic cells, 26
PC2, 26, 27, 47, 57–59, 61, 65–67, 83, 90, 92, 100, 102, 103, 105, 114, 115, 125, 129, 136, 139, 148, 152, 175, 185
Prolactin (PRL), 5, 15, 31, 46, 49, 53, 78–83, 239, 244, 257, 274, 275, 345, 346, 360–361, 393
stimulation, 53
mediated by galanin, 159
TRH, 80, 84
Prolactinoma, 53, 81, 360, 368
Prolactin-releasing peptide (PRLRP), 83–84, 174, 178
Proopiomelanocortin (POMC), 26, 34, 49, 54, 55, 59–67, 81, 110, 112
Prostaglandins, 13, 72, 306
Protein kinase A (PKA), 51, 66, 94, 139, 140, 245, 313

Prothoracicostatic hormone (PTSH), MIP 1, 160–161
Prothoracicotropic hormone (PTTH), 135, 158–161, 170, 189, 327–330
  release, 327–330
    circadian control, 327
Psacothea hilaris, 196, 385
Pseudohermaphroditism, 207, 212, 365, 371, 374
PSP. *See* Pheromonostatic peptide
PTSH. *See* Prothoracicostatic hormone
PTTH. *See* Prothoracicotropic hormone
Puberty, 121, 264, 309–310, 321, 360, 361, 369, 371, 373–375
Pulse amplitude, 339, 342, 343, 380
Pulse frequency, 323, 339, 342–343
Pupation, 150, 196, 200, 325, 326
Pylorus, 151, 153
Pyroglutamate, 16, 29, 57, 102, 136–139, 145, 146, 149, 150, 162
  hormone maturation, 29
  N-terminal glutamine cyclization, 16, 57, 136, 139, 149, 150, 160, 181
Pyrokinins, 135, 150, 151, 320
PYY. *See* Peptide tyrosyl tyrosyl

R
Rathke's pouch, 359
Receptor
  adrenergic, 240, 379
    α—antagonist, 41, 245, 381
    β_2—agonist, 41, 379
  bursicon, 172, 173
  calcium sensing (CASR), 128, 130
  dimerization, 79, 224, 247, 249
  erythropoietin, 131, 132, 258
  estrogens, 224, 225, 249, 259, 307, 324, 352, 370, 375, 376
  fast reaction, 259
  FMRF, 174–177, 350
  FSH, 72, 118, 252, 374
  GABA, 230
  galanin, 52, 53, 251
    galanin receptor family, 181
  glucocorticoids, 249, 303, 304
  glutamate, 230
  gonadotropin-releasing hormone (GnRH), 35, 44, 250, 251, 370, 373–374
  G-protein coupled membrane, 35, 42, 44, 54, 56, 57, 59, 63, 64, 66, 67, 69, 70, 72, 73, 83, 85, 86, 93, 94, 96, 99, 100, 102–105, 107, 108, 115, 122, 127, 128, 139, 141–143, 147, 148, 151–153, 155–157, 162, 163, 168–170, 172–174, 177, 178, 180–182, 184, 185, 187, 188, 216, 247, 248, 251, 254, 258, 259, 350, 373
  GPR54, 50
  GH, 40, 76, 79, 83, 131, 322, 323, 358, 361
  growth hormone secretagogue, 48, 96–98, 316, 317
  insulin, 89, 92, 93, 137, 138, 255–257, 312, 313, 326, 364
  IGF, 322, 323
  kisspeptin, 50
  leptin, 2, 95, 257, 258, 315, 317, 358, 371
  LH, 69, 73, 373, 379
  lymnokinin, 153
  mechanoreceptors, 126, 314, 317–318, 328
  melanocortins, 62, 250, 316
    MCR1, 45, 62
    MCR4, 45, 46, 62, 319
  membrane, 7, 31, 33, 40, 51, 52, 247, 253, 255, 257–259, 304, 316, 323, 350, 351, 373, 374
  PTHrP, 128, 321, 322, 324, 349
  PTTH (torso), 159
  retinoic acid (ROR), 340
  SH3 binding, 222
  SST, 41, 47, 48, 251
    differentially expressed, 48, 254–255
  stretch receptors, 228, 330, 334
  tyhyroxine/triiodothyronine (ErbA, TR), 74, 241, 243, 249, 323
Receptor activity modulating protein (RAMP-1), 100
Receptor protein tyrosine kinase (RET), 369
Red-pigment-concentrating hormone (RPCH), 139–141
5α-Reductase (SRD5A), 202, 211–212, 221, 229, 351, 352, 375
  SRD5A1, 211, 212
  SRD5A2, 211, 212
Regulator circuits, 36, 301–305
Renal glomeruli, 338
Renin, 5, 122–123, 282, 287, 288, 334, 336, 362, 367, 368
  endopeptidase for angiotensinogen, 228, 336
RET. *See* Receptor protein tyrosine kinase
Rhodnius prolixus, 327, 328, 385
Ribonucleic acid (RNA), 2, 16, 21, 22, 40, 42, 44, 48, 51, 55, 59, 83, 136, 156, 157, 186, 316, 340, 385, 386, 388, 390
  cap, 22, 386
  interference, 44, 156, 186
  tRNA, 22, 386
RNA. *See* Ribonucleic acid

S
Salbutamol, 379
Salt export, 155
  leucokinin, 155
Salt-wasting disorders, 214, 228
Satiety, 66, 311, 314–318
Sclerotization, 332
SDR. *See* Small dehydrogenase/reductase family
Secretin, 4, 42, 56, 58, 94, 101, 104–107, 147, 148, 182, 251, 285
Secretory granules/vesicles, 24, 69, 132, 263, 264, 266, 274, 275, 279, 280, 282–286, 308, 313, 337, 343, 344, 386, 387
  pH, 86
Selye, Hans, 5, 301, 302
sER, *See* Endoplasmatic reticulum, smooth ER
Serotonin, 17, 34, 186, 244, 250, 278, 284, 300, 317
Sex-determining region Y (SRY), 223, 224, 288, 375
Sex hormone binding protein (SHBG), 265, 266
Sex peptide, 14, 161–163, 320
SF-1. *See* Steroidogenic factor 1
SHBG. *See* Sex hormone binding protein
Shock, 23, 49, 240, 362
Short neuropeptide-F (sNPF), 27, 135, 184–187, 326
Sieve plates, 271, 282, 286, 337
Signal peptidase, 23, 35, 40, 110, 115, 125, 129, 139, 165, 171, 185, 187, 188, 390
Signal peptide, 16, 23, 25, 39, 40, 43, 46, 52, 54–56, 58, 67, 68, 84, 96, 97, 107, 109, 112, 120, 127, 131, 134, 136–141, 145–147, 149, 150, 152, 154, 156, 157, 159, 160, 163, 167, 169, 171–173, 175, 176, 179, 184, 185, 187, 390
Signal transducers and activators of transcription (STAT), 76, 257
Silk moth. *See Bombyx mori*
Sinuses of the carotid, 334
Sinus gland, 135, 136, 140, 163, 166, 167, 182, 294
Sleep rhythm, 40, 41
SMAD, 117, 122
Small dehydrogense/reductase family (SDR), 211, 257
  17β-HSD, 206
Smith-Lemli-Opitz syndrome, 233
SNARE. *See* Soluble *N*-ethylmaleimide sensitive factor attachment receptor
sNPF. *See* Short neuropeptide-F
Sodium resorption, 177, 228, 334, 338, 368
Soluble *N*-ethylmaleimide sensitive factor attachment receptor (SNARE), 264, 387
Somatolactin, 79, 81–83
Somatostatin (SST), 5, 15, 32, 40, 41, 46–49, 101, 105, 127, 180, 181, 250, 251, 254–255, 272, 275, 279, 284–286, 300, 322, 323, 360, 361, 364
  agonist, 48
  long lasting agonist:octreotide, 361
  SST-14, 47, 48
  SST-28, 47, 48
SP. *See* Substance P
Spermatogenesis, 219, 360, 373, 374
Sphingomyelinase, 254
Splanchnic nerve, 315
Splicing, 21, 26, 31, 35, 42, 57, 76, 92, 97, 99, 115, 119, 136, 179, 219, 258, 316
Spodoptera frugiperda, 181
Squalene, 17, 192, 194–195, 201, 223, 230, 351, 352
  synthase, 194
Src homology domain 2 (SH2), 256, 257
SRD5A. *See* 5α-Reductase
SST. *See* Somatostatin
Stanniocalcin (STC), 128–130
STAR. *See* Steroidogenic acute regulatory protein
STAT. *See* Signal transducers and activators of transcription
Steroids, 7, 17, 22, 40, 64, 191–236, 258–259, 263–267, 302, 309, 351–353, 357, 364, 366
  anabolic, 378
Steroid hormones, evolution, 351–353
Steroidogenesis, 63, 118, 202, 214, 220, 226, 229, 230, 365, 369, 371, 374–376
Steroidogenic acute regulatory protein (STAR), 202, 203, 216, 230, 374
Steroidogenic enzymes, 202–221, 225, 347, 358
Steroidogenic factor 1 (SF-1), 359, 371, 373
Stomach, 96–98, 101–105, 113, 283, 284, 317, 318
  antrum, 101
Stratum granulosum (SG), 291, 292
Stress, 5, 31–34, 41, 49, 57, 64, 66, 100, 101, 110, 226, 227, 240, 243, 301–304, 306, 307, 346, 381
Strongylocentrotus purpuratus, 61, 385
Subfornical organ, 334, 335, 337
Substance P (SP), 49, 53, 113–115, 156, 272, 302, 306, 307
Substantia nigra, 31, 239
Sulfakinin, 135, 149, 156–158, 189

Sulfatase, 229
Sulfotransferase, 158, 229
Sympathetic nervous system, 240, 302, 353, 381
Synapses, 50, 55, 174, 239, 240, 264, 271, 272, 282, 294, 314–316, 335, 337, 390

T

Tachykinin, 113–115, 135, 156, 157, 186, 189
Tachykinin-related peptides (TRP), 155–157, 189
Tenebrio molitor, 197, 385
Terpenes, 17, 327
Testosterone, 5, 15, 17, 71, 121, 192, 201, 202, 206, 207, 211, 212, 216–221, 223–225, 229, 249, 258, 264, 288, 290, 292, 300, 305, 308–310, 322, 352, 366, 370, 373–375, 377–379
  analogs, 378, 379
3α,5α-Tetrahydroprogesterone, 229, 230
TGF-α. *See* Transforming growth factor-α
TGF-β. *See* Transforming growth factor-β
TH. *See* Tyrosine hydroxylase
Thalamus, 31, 40, 57, 277, 315, 316
Thermoregulation, TRH, 32
Third ventricle, 270, 336
Thirst, 88, 123, 335, 337–338, 362
Thymine, 21, 385, 386
Thyroglobulin, 240–243, 278, 279, 349, 388
Thyroid gland
  diseases, 4, 363
  follicular colloid, 241, 243, 278
  phagosome, 279
  thyroid peroxidas, 241, 242, 279, 349, 363
    evolution, 349
Thyroid-stimulating hormone (TSH), 5, 15, 23, 24, 26, 30–32, 67–70, 73–75, 135, 241, 243, 274, 300, 301, 361, 363
  acidophilic dye, 275
  dopamine, 32, 300
  prolactin, 274
  secretory vesicles, 274
  T4 inhibition, 300
  TRH, 31, 32, 67, 69, 74, 300, 363
Thyrotoxicosis, 4, 363
Thyrotroph adenoma, 361, 363
Thyrotropic cells, 67, 81, 274, 275, 359, 361
Thyrotropin, 12, 15, 30, 31, 363
Thyrotropin-releasing hormone (TRH), 5, 7, 15, 26–28, 30–32, 41, 67, 69, 74, 84, 170, 239, 251, 272, 275, 300, 301, 318, 350, 361, 363
  noradrenaline stimulation, 300
  serotonin stimulation, 300
  T4 inhibition, 300
Thyroxine, 5, 7, 15, 17, 74, 75, 240–243, 248, 278, 279, 300, 301, 321–323, 349, 361, 363, 386, 388
  release, 75, 243, 349
TIDA. *See* Neurons, tuberoinfundibular dopaminergic
TMOF. *See* Trypsin-modulating oostatic factor
TNF-α. *See* Transforming growth factor-α
Tobacco hornworm. *See Manduca sexta*
TPST. *See* Tyrosylprotein sulfotransferase
Tracheae, 169, 170
Transcription factors
  ATF-1, 216
  CREB, 69, 216
  Egr1, 359
  GATA2, 359
  HESX1, 359, 360, 372
  LHX3, 359, 360, 372
  LHX4, 359
  NGF1B, 216
  NROB1 (DAX1), 371
  NURR1, 216
  Pit1, 69, 81, 96, 359
  Pitx1, 359, 360
  Pitx2, 359
  P-LIM, 69
  PROP1, 359, 360, 370, 372, 373
  SF1, 359, 371, 373
  SOX9, 371
  SRY, 223, 224, 288, 375
Transforming growth factor-β (TGF-β), 70, 117, 119–121, 257, 322–324
Transsphenoidal surgery, 360, 361
TRH. *See* Thyrotropin-releasing hormone
Tribolium castaneum, 173, 385
Trichoplax adhaerens, 385
Triglycerol lipase, 140
Triiodothyronine, 15, 17, 74, 240–243, 248, 279, 300, 321–323, 386
  biosynthesis, 240
  thyroxine, 17, 241, 243, 279, 323, 386
Trophoblastic cells
  CG synthesis, 309, 370
  nidation, 12, 72, 309
TRP. *See* Tachykinin-related peptides
Trypsin-modulating oostatic factor (TMOF), 164–166, 189
  gut enzymes, 164, 165
  receptor, 165
Tryptophan, 17, 143, 144, 244, 263, 392
TSH. *See* Thyroid-stimulating hormone
Tumor, 3, 4, 7, 12, 14, 50, 51, 53, 59, 79, 81, 91, 103, 116, 128, 130, 220, 224, 266, 279,

290, 323, 324, 344, 358–362, 364–368, 380, 390
Tumor necrosis factor α (TNF-α), 220
Turner syndrome, 369
Tyrosine, 17, 32, 89, 92, 102, 104, 135, 137, 157, 158, 184, 237, 240–242, 244, 255–257, 263, 279, 349, 353, 369, 392
Tyrosine hydroxylase (TH), 32, 90, 91, 142, 238
  α−methyl-*p*-tyrosine inhibotor, 32
Tyrosine kinase, 89, 92, 137, 158, 255–257, 369
  membrane, 255
    insulin receptor, 255, 256
Tyrosylprotein sulfotransferase (TPST), 157

U
Ubiquitin, 130, 338, 340
Uca puligator, 183
Ultradian rhythms, 342–345
Urocortin, 34, 251
Urotensin, 34, 49
UV light, 67, 233, 234, 348, 350

V
Vagus nerve, 111, 314, 315, 317–319
Vasoactive intestinal peptide (VIP), 94, 105–106, 239, 251, 272, 364
  neurotransmitter, 105, 106
Vasopressin (AVP), 33, 34, 84, 86–88, 302, 303, 334–338, 342, 343
  neurotransmitter, 13, 88
  and oxytocin, 5, 84–89, 135, 142, 148, 189, 215, 276, 359
Ventral gland, 230
Ventral nerve cord, 138, 186, 330, 332
  pars intercerebralis, 142
Vesicles. *See* Secretory granules
Vesalius, 4, 6
Vesicular monoamine transporters, 238
VIH. *See* Vitellogenesis-inhibiting peptide
Vinclozolin, 224
VIP. *See* Vasoactive intestinal peptide
Vitamin D_2, 235
Vitamin D_3, 17, 194, 203, 232–235, 258, 266, 279, 350, 353
Vitamin D_3-binding protein, 235
Vitellogenesis, 164, 166, 196, 230, 294, 320
  inhibition by colloostatin, 166
Vitellogenesis-inhibiting peptide (VIH), 163–164
Vitellogenin incorporation, 200

W
Water resorption, 334, 337
Wolfram syndrome, 362

X
X organ, 135, 136, 140, 163, 166, 189, 294

Y
Y organ, 230, 294, 326

Z
Zeitgeber, 41, 245, 302, 327, 339, 341, 342
Zona fasciculata
  ACTH, 214, 215, 282
  cortisol, 214, 221, 282
Zona glomerulosa
  aldosterone, 213, 216, 221, 228, 282
  mineralcorticoids, 226, 282
Zona pellucida (ZP), 290–292
Zona reticularis, 213, 221, 225, 281, 282

MIX
Papier aus verantwortungsvollen Quellen
Paper from responsible sources
FSC® C105338

If you have any concerns about our products,
you can contact us on
ProductSafety@springernature.com

In case Publisher is established outside the EU,
the EU authorized representative is:
Springer Nature Customer Service Center GmbH
Europaplatz 3, 69115 Heidelberg, Germany

Printed by Libri Plureos GmbH
in Hamburg, Germany